AF679607

# Integral Equation Methods for Electromagnetics

# Integral Equation Methods for Electromagnetics

John L. Volakis and Kubilay Sertel
The Ohio State University

SciTech Publishing, Inc.
Raleigh, NC
scitechpub.com

SciTech Publishing, Inc.

Published by SciTech Publishing, Inc.
911 Paverstone Drive, Suite B
Raleigh, NC 27615
(919) 847-2434, fax (919) 847-2568
scitechpublishing.com

Copyright © 2012 by SciTech Publishing, Raleigh, NC. All rights reserved.

No part of this publication may be reproduced, stored in a retrieval system or transmitted in any form or by any means, electronic, mechanical, photocopying, recording, scanning or otherwise, except as permitted under Sections 107 or 108 of the 1976 United Stated Copyright Act, without either the prior written permission of the Publisher, or authorization through payment of the appropriate per-copy fee to the Copyright Clearance Center, 222 Rosewood Drive, Danvers, MA 01923, (978) 750-8400, fax (978) 646-8600, or on the web at copyright.com. Requests to the Publisher for permission should be addressed to the Publisher, SciTech Publishing, Inc., 911 Paverstone Drive, Suite B, Raleigh, NC 27615, (919) 847-2434, fax (919) 847-2568, or email editor@scitechpub.com.

The publisher and the author make no representations or warranties with respect to the accuracy or completeness of the contents of this work and specifically disclaim all warranties, including without limitation warranties of fitness for a particular purpose.

Editor: Dudley R. Kay
Production Manager: Robert Lawless
Typesetting: MPS Limited, a Macmillan Company
Cover Design: Brent Beckley
Printer: Sheridan Books, Inc.

This book is available at special quantity discounts to use as premiums and sales promotions, or for use in corporate training programs. For more information and quotes, please contact the publisher.

Printed in the United States of America
10 9 8 7 6 5 4 3 2 1

ISBN: 978-1-891121-93-7

**Library of Congress Cataloging-in-Publication Data**

Volakis, John Leonidas, 1956-
Integral equation methods for electromagnetics / John L. Volakis and Kubilay Sertel.
p. cm.
Includes bibliographical references and index.
ISBN 978-1-891121-93-7 (hardcover : alk. paper) 1. Electromagnetic fields–Mathematical models. 2. Integral equations. I. Sertel, Kubilay. II. Title.
TK7867.2.V653 2011
530.14'1–dc23

2011041801

# Contents

# Preface

Integral equation methods have been the workhorse of electromagnetic analysis in core areas such as antennas, scattering, and microwave circuits. These days, integral equation toolsets are found in most commercial electromagnetic software packages, and their value is unquestioned. However, until now, no comprehensive and detailed text exists for graduate students or as a reference for developers and practitioners. This book has been in various stages of development and refinement for over ten years and is now ready to fill that need.

## Approach and Organization

A valuable aspect of the book is its detailed treatment on formulating integral equations for electromagnetic applications. For example, Chapters 2 and 3 are unique in providing extensive coverage of the various integral expressions and representations and their derivation directly from Maxwell's equations. Both two-dimensional and three-dimensional integral *representations*—not *solutions*—are provided in the spatial and spectral domain. In this manner, readers become comfortable with the development of integral equations that will be eventually solved. Both scalar and dyadic representations are given, along with spectral forms that will serve the reader well in future sections when dealing with layered media and the classic Sommerfeld problems. Chapter 3 is virtually encyclopedic in providing a thorough presentation of the Kirchhoff and Stratton Chu integral equations. Both open and closed surface integral equations are covered. This is followed by the Electric Field (EFIE), Magnetic Field (MFIE) and Combined Field (CFIE) Integral Equations. A variety of metallic, dielectric and composite structures are covered. These chapters also include many homework problems that often expand upon the coverage within the body of the chapters.

Having covered the derivation of integral equations for various classic and practical problems, Chapter 4, 5 and 6 provide the solution process in a step-by-step manner. Specifically, Chapter 4 covers wire antennas and scatterers for linear and non-linear wires, using a variety of basis functions. It starts with the piecewise linear basis for didactic purposes and proceeds to introduce other basis functions. Using this approach, the readers are introduced concurrently to various numerical challenges of solving the final linear matrix system. Numerous figures are included in this chapter, giving a variety of results that relate input impedance, radiation, and scattering calculations. Among them also are results related to the wire grid modeling of complex three-dimensional structures.

Chapter 5 is devoted to two-dimensional scattering and is almost encyclopedic in providing many integral equation formulation and solution approaches. It is intended to serve as a didactic approach to teaching the many aspects of integral equation solutions for metallic, impedance, dielectric or composite scatterers using pulse, linear, and higher order basis functions. It includes 40 figures and covers all forms of integral equations for most problems encountered in electromagnetics. Many scattering and current density curves are provided, along with analytical solutions for the narrow strip problem using Hilbert transforms.

Chapter 6 covers three-dimensional integral equation solutions. It starts with the traditional Rao-Wilton-Glisson (RWG) surface basis functions and elements, and progresses to the fully volumetric curvilinear basis functions for dielectric and composite structures (surface and volumetric). It is over 90 pages in length and includes over 70 figures of results: current density data, scattering and antenna problem solutions, solution convergence plots and color surface maps. Both surface and volume integral equation solutions are covered in detail. Thus readers can follow in a step-by-step manner all aspects of their numerical implementation. Naturally, the chapter contains extensive coverage of the moment method matrix element evaluations. The convergence of iterative methods is discussed as well, and a step-by-step example is given to help in programming three-dimensional moment method solutions.

An important companion of Chapter 6 is the Appendix that describes the development of integral equations for microstrip antennas. A detailed derivation of the associated Green's function is given along with matrix element evaluations. Explicit far zone field representations are also given by invoking the reciprocity theorem.

Chapter 7 is devoted to fast solvers and the fast multipole method. It is fairly didactic in its coverage and, together with Chapter 5, provides readers with a complete coverage of modern approaches to fast integral methods in electromagnetics. Both the standard FMM and the multilevel FMM (MLFMM) are presented, and a discussion of other fast methods is included. In summary, Chapters 5 to 7 provide a detailed coverage of moment method solutions for a rich set of basis functions, numerical integration methods, and applications dealing with metallic and dielectric structures. In the authors' experience, when using the book as a text, Chapter 1–4 can serve as a first course, whereas Chapter 5 to 7 might serve as a continuation course for further study or as a special topics course.

## Distinctive Features

Overall, the authors believe that this book provides the student, casual, or advanced reader with a fairly complete and up-to-date coverage of integral methods for composite scatterers and antenna structures. It is unique in providing a detailed process in formulating radiation and scattered field expressions for a large set of 2D and 3D problems that can accommodate both beginners and advanced readers. Chapter 2 and 3 are unique in providing the student with a didactic walk through the maze of integral equations formulations and solutions for all sorts of situations. Throughout, we provide historical perspectives on the development of the formulations and solutions approaches. Coupled with the integral equation mechanics and numerical applications given in Chapters 4 to 6 for wire, 2D and 3D metallic and dielectric structures, this book will remain useful for many years to come.

## The Book's Development

It is important to note that this book has been a 'labor of love' and has been evolving since the late 1980s. Its writing began when first author John Volakis was teaching at the University of Michigan and initiated a new class called *EECS 633: Analytical and Numerical Techniques for Electromagnetics*. This class was intended to provide electromagnetics students with an understanding of various standard analytical and numerical methods commonly used in electromagnetics. In 1993 a coursepack was developed for the class that covered applications in radiation, scattering, and microwave circuits. That original coursepack constitutes a major portion of this book. The goal was for students to develop an expertise on the fundamentals and applications of numerical methods to analyze arbitrary shape composite structures. An equally important aspect of the course was to provide an understanding on the limitations of the underlying methods, now extensively used in commercial numerical packages. Initially, the class included both integral and partial differential methods, but as the author's interest shifted to finite element methods[1,2], it became essential to separate the integral and differential methods. By the late 1990s as iterative solution methods, fast integral methods, and matrix compression techniques became an essential part of integral formulations. Therefore, it became clear that any book on the subject needed a more careful exposition of these techniques. Chapters 6 and 7, written by the second author, Kubilay Sertel, provide a careful account of these methods and their applications to electromagnetics. The melding of chapters into a coherent whole was assisted by expert reviews obtained by our publisher SciTech, and we are grateful for their many helpful suggestions.

[1]J.L.Volakis, A. Chatterjee and L. Kempel, *Finite Element Methods for Electromagnetics*, IEEE Press, New York and Wiley-InterScience (ISBN 0-7803-3425-6).

[2]J.L. Volakis, K. Sertel and B. Usner, *Frequency Domain Hybrid Finite Element Methods in Electromagnetics*, Morgan & Claypool Publishing (ISBN: 1598290800), 2006; ISBN-10: 1598290800, ISBN-13: 978-1598290806)

# Acknowledgements

The authors are indebted to many of their students, teachers, and collaborators who played an important role in the writing and refinement of the book. The first two chapters and Chapter Five are certainly influenced by the first author's teachings at Ohio State during his graduate years, particularly while taking courses from Prof. Kouyoumjian and Prof. Richmond.

Many thanks goes to Richard Carnes and Jim Ryan, who wrote the first Latex version of chapters 1–5 of the book. Mr. Carnes' expertise in Latex helped make the writing enjoyable, and Mr. Ryan's figure rendering was superb. Also, the authors are indebted to the many students who generated the data for the figures in Chapters Four and Five. Dr. John Natzke, now an Associate Professor at George Fox University (Oregon, USA), generated the data for several graphs in Chapter Four. The late Dr. Kasra Barkeshli generated the data for some of the graphs in Chapters Four and Five, and Dr. Leo Kempel, now Professor and Associate Dean for Research at Michigan State University generated some of the data curves in Chapter Five for strip scattering. We are indebted to these individuals and the many students who participated in the development of this book and its many coursepack versions.

In closing this preface, we hope that this book will serve as a text and/or a reference guide for many years to come.

**John L. Volakis**
Kubilay Sertel
Columbus, Ohio, USA
October, 2011

# Publisher Acknowledgements

A book such as this requires special care. Its intended use for advanced study of electromagnetics at the graduate level means the content must be clear, the mathematical notation accurate, and the page format open and inviting. Practitioners will use the book for self-study and as an ongoing reference for important work in their fields. The expectation is that this text-reference will enjoy a long life before any revisions are needed, so it should be done right the first time. To ensure this quality and accuracy, SciTech Publishing is blessed by talented reviewers who give their time and labor in support of good books that serve the electromagnetics community. We therefore acknowledge and thank the outstanding contributions of these reviewers:

Andrew Peterson – Georgia Institute of Technology

Cynthia Furse – University of Utah

Natalia Nikolova – McMaster University

Karl Warnick – Brigham Young University

Branislav Notaros – Colorado State University

It is said that the perfect book has yet to be rewritten. Despite the best efforts of the authors and reviewers, errors and ambiguities will be found. SciTech takes great pride in its speed of corrections and minor improvements on each reprinting of the book. We therefore welcome and encourage reporting of suspected errors, as well as suggestions for improvements.

Dudley R. Kay – President
editor@scitechpub.com

CHAPTER 1

# Fundamental Concepts and Theorems

**Chapter Outline**



It is assumed that the reader is already familiar with basic electromagnetic theory usually covered in courses on general electromagnetics (Harrington, 1961; Balanis, 1989; Van Bladel, 2007; Stratton, 2007) and antennas beyond the required undergraduate material. In this chapter, we simply provide a review of Maxwell's equations and related theorems that are of importance in this book. This chapter also provides the required notational definitions for the various field quantities.

## 1.1 | MAXWELL'S EQUATION IN DIFFERENTIAL TIME DOMAIN FORM

In 1861 James Clerk Maxwell presented the electromagnetic field equations in the form that is known today. Maxwell formulated these equations on the basis of the electromagnetic laws already developed by great scientists such as Johann Carl Friedrich Gauss, André-Marie Ampere, Michael Faraday, and Joseph Henry. Other than the compact form of the field relations that he provided, Maxwell's only new contribution in this respect was the introduction of the displacement current to supplement Ampere's law. This was essentially postulated to make the equations compatible with Gauss's law and the equation of continuity.

Maxwell's equations in differential time domain form are

$$\nabla \times \mathcal{H} = \mathcal{J} + \frac{\partial \mathcal{D}}{\partial t} \quad \text{(Ampere-Maxwell law)} \tag{1.1}$$

$$\nabla \times \mathcal{E} = -\mathcal{M} - \frac{\partial \mathcal{B}}{\partial t} \quad \text{(Faraday's law)} \tag{1.2}$$

$$\nabla \cdot \mathcal{B} = \varrho_m \quad \text{(Gauss's law for magnetic fields)} \tag{1.3}$$

$$\nabla \cdot \mathcal{D} = \varrho \quad \text{(Gauss's law)} \tag{1.4}$$

where

$t$ = time
$\mathcal{E}$ = electric field intensity in volts/meter (V/m)
$\mathcal{D}$ = electric flux density in coulombs/meter$^2$ (C/m$^2$)
$\mathcal{H}$ = magnetic field intensity in amperes/meter (A/m)
$\mathcal{B}$ = magnetic field density in webers/meter$^2$ (Wb/m$^2$)
$\mathcal{J}$ = electric current density in amperes/meter$^2$ (A/m$^2$)
$\mathcal{M}$ = magnetic current density in volts/meter$^2$ (V/m$^2$)
$\varrho$ = electric charge density in coulombs/meter$^3$ (C/m$^3$)
$\varrho_m$ = magnetic charge density in webers/m$^3$ (Wb/m$^3$)

Each of the field, current or charge quantities is, of course, assumed to be a function of position and time, and will be measured in the rationalized meter, kilogram and second (MKS) system of units. The introduction of the magnetic currents and charges in (1.1) through (1.4) is a matter of mathematical utility since to date there is no evidence of their existence. However, they have been found very useful for constructing mathematical models of electromagnetic problems, where they are often employed as equivalent fictitious sources, and this is the primary reason for including them in the definition of Maxwell's equations.

By taking the divergence of (1.1) and making use of (1.4) we obtain

$$\nabla \cdot \mathcal{J} + \frac{\partial \varrho}{\partial t} = 0 \tag{1.5}$$

where we have also invoked the identity $\nabla \cdot (\nabla \times \mathbf{A}) = 0$, which holds for any vector $\mathbf{A}$. Expression (1.5) is known as the continuity equation and implies conservation of charge. Similarly, from (1.2) and (1.3) we obtain the continuity equation for magnetic changes given by

$$\nabla \cdot \mathcal{M} + \frac{\partial \varrho_m}{\partial t} = 0. \tag{1.6}$$

Conversely, one can derive (1.4) by taking the divergence of (1.1) and making use of (1.5). Equation (1.4) then follows by setting the integration constant with respect to time equal to zero. Likewise, (1.3) can be obtained from (1.2) and (1.6). Consequently, we may take (1.1) through (1.4) as the fundamental equations of electromagnetism or, alternatively, (1.1), (1.2), (1.5), and (1.6) could be used to form an independent set of equations for the solution of the fields. We will choose here (1.1) through (1.4) as the fundamental set of equations.

The four independent equations (1.1) through (1.4) actually consist of eight scalar ones since (1.1) and (1.2) are vector equations. Noting then that each vector field has three components, we have 20 scalar unknown functions appearing in (1.1) through (1.4). Thus, additional field relations are required for a unique solution of the field quantities. These are provided from the constitutive relations given by

$$\mathcal{D} = \epsilon \mathcal{E} = \epsilon_o \epsilon_r \mathcal{E} \tag{1.7}$$

$$\mathcal{B} = \mu \mathcal{H} = \mu_o \mu_r \mathcal{H} \tag{1.8}$$

$$\mathcal{J} = \sigma \mathcal{E} \tag{1.9}$$

$$\mathcal{M} = \sigma_m \mathcal{H} \tag{1.10}$$

In (1.7) through (1.10),

$\epsilon_o$ = free space permittivity = $8.854 \times 10^{-12}$ farads/meter (F/m)
$\mu_o$ =free space permeability = $4\pi \times 10^{-7}$ henrys/meter (H/m)
$\epsilon_r$ = medium's relative permittivity constant
$\mu_r$ = medium's relative permeability constant
$\sigma$ = electric current conductivity in siemens/m (S/m)
$\sigma_m$ = magnetic current conductivity in ohms/m ($\Omega$/m)

which characterize the electrical properties of the medium. They are referred to as the constitutive parameters of that medium, and based on their spatial or frequency dependence, the medium can be characterized as linear or nonlinear, homogeneous or inhomogeneous, dispersive or nondispersive, and isotropic or anisotropic.

- If the constitutive parameters are independent of the field intensities, the material is said to be linear.
- If the constitutive parameters are constant or uniform throughout a material section, that section of the medium is said to be homogeneous.
- If the constitutive parameters are dependent on the oscillating frequency of the fields, that medium is said to be dispersive.
- If the constitutive parameters are independent of the field polarity, that medium is referred to as isotropic.

In the case of anisotropic media, the constitutive relations (1.7) and (1.8) take the form

$$\mathcal{D} = \overline{\overline{\epsilon}} \cdot \mathcal{E} \tag{1.11}$$

$$\mathcal{B} = \overline{\overline{\mu}} \cdot \mathcal{H} \tag{1.12}$$

where $\overline{\overline{\epsilon}}$ and $\overline{\overline{\mu}}$ are $3 \times 3$ matrices or tensors and can be more explicitly written as

$$\overline{\overline{\epsilon}} = \begin{pmatrix} \epsilon_{xx} & \epsilon_{xy} & \epsilon_{xz} \\ \epsilon_{yx} & \epsilon_{yy} & \epsilon_{yz} \\ \epsilon_{zx} & \epsilon_{zy} & \epsilon_{zz} \end{pmatrix} \tag{1.13}$$

$$\overline{\overline{\mu}} = \begin{pmatrix} \mu_{xx} & \mu_{xy} & \mu_{xz} \\ \mu_{yx} & \mu_{yy} & \mu_{yz} \\ \mu_{zx} & \mu_{zy} & \mu_{zz} \end{pmatrix}. \tag{1.14}$$

We remark that for physically realizable materials

$$\epsilon_{ij} = \epsilon_{ji}^*, \qquad \mu_{ij} = \mu_{ij}^* \qquad \text{for} \qquad (i \neq j) \tag{1.15}$$

implying that the permittivity and permeability matrices are Hermitian. The "* " in (1.15) denotes complex conjugation, and in the case of real $\overline{\overline{\epsilon}}$ or $\overline{\overline{\mu}}$ the corresponding Hermitian matrices are also symmetric. Depending on the specific value or relation of the permittivity or permeability matrix components, classifications for the material include uniaxial, biaxial, and gyrotropic. In particular, for biaxial materials $\overline{\overline{\epsilon}}$ and $\overline{\overline{\mu}}$ take the form

$$\overline{\overline{\epsilon}} = \begin{pmatrix} \epsilon_x & 0 & 0 \\ 0 & \epsilon_y & 0 \\ 0 & 0 & \epsilon_z \end{pmatrix} \qquad \overline{\overline{\mu}} = \mu\overline{\mathbf{I}} = \begin{pmatrix} \mu & 0 & 0 \\ 0 & \mu & 0 \\ 0 & 0 & \mu \end{pmatrix} \tag{1.16}$$

where $\overline{\mathbf{I}}$ denotes the unit dyad and can be alternatively written as $\overline{\mathbf{I}} = \hat{x}\hat{x} + \hat{y}\hat{y} + \hat{z}\hat{z}$. When $\epsilon_x = \epsilon_y = \epsilon$ in (1.16), the material is referred to as uniaxial, with the $z$ axis being the optical axis. Gyrotropic media are encountered in various plasma studies, and in this case

$$\overline{\overline{\epsilon}} = \begin{pmatrix} \epsilon & -j\epsilon_g & 0 \\ j\epsilon_g & \epsilon & 0 \\ 0 & 0 & \epsilon_z \end{pmatrix} \qquad \overline{\overline{\mu}} = \mu\overline{\mathbf{I}}. \tag{1.17}$$

Generally, biological media such as human tissues are anisotropic and dispersive. For example, retina fibers are thought to be uniaxial, whereas the cornea appears to be similar to biaxial crystals.

Finally, when the constitutive relations are of the form (Kong, 1975)

$$\mathcal{D} = \overline{\overline{\epsilon}} \cdot \mathcal{E} + \overline{\overline{\zeta}} \cdot \mathcal{B} \tag{1.18}$$

$$\mathcal{B} = \overline{\overline{\mu}} \cdot \mathcal{H} + \overline{\overline{\xi}} \cdot \mathcal{E} \tag{1.19}$$

where $\overline{\overline{\xi}}$ and $\overline{\overline{\zeta}}$ denote tensors, the corresponding medium is referred to as bianisotropic. The representations (1.18) and (1.19) are encountered in the theory of relativity (moving media). However, special forms of these have also been found to hold for stationary magnetoelectric materials (Astrov, 1960). For example, Dzyalashinskie (1960) proposed that $\overline{\overline{\epsilon}} = \hat{x}\hat{x}\epsilon + \hat{y}\hat{y}\epsilon + \hat{z}\hat{z}\epsilon_z$, $\overline{\overline{\zeta}} = \overline{\overline{\xi}} = \hat{x}\hat{x}\xi + \hat{y}\hat{y}\xi + \hat{z}\hat{z}\xi_z$ and $\overline{\overline{\mu}} = \hat{x}\hat{x}\mu + \hat{y}\hat{y}\mu + \hat{z}\hat{z}\mu_z$ for chromium oxide. When $\overline{\overline{\epsilon}} = \epsilon\overline{\mathbf{I}}$, $\overline{\overline{\mu}} = \mu\overline{\mathbf{I}}$ and $\overline{\overline{\zeta}} = -\overline{\overline{\xi}}/\mu = -j\chi\overline{\mathbf{I}}$ the material is referred to as chiral isotropic (Jaggard, Mickelson, and Pappas, 1978; Lakhtakia, Varadan, and Varadan, 1988; Post, 1962) with $\chi$ being the chirality admittance parameter. Chiral material (material that exhibits right- or left-handed waves) have the characteristic property of not supporting linearly polarized waves and are also associated with different propagation constants for the right- and left-hand waves. This has been experimentally demonstrated at optical frequencies, but to date we have not found natural materials with this behavior at microwave frequencies.

## 1.2 | MAXWELL'S EQUATIONS IN INTEGRAL FORM

The integral form of Maxwell's equations can be directly obtained from their differential counterparts. By integrating (1.1) and (1.2) over a surface $S$, we obtain

$$\iint_S (\nabla \times \mathcal{H}) \cdot d\mathbf{s} = \iint_S \mathcal{J} \cdot d\mathbf{s} + \iint_S \frac{\partial \mathcal{D}}{\partial t} \cdot d\mathbf{s} \tag{1.20}$$

$$\iint_S (\nabla \times \mathcal{E}) \cdot d\mathbf{s} = -\iint_S \mathcal{M} \cdot d\mathbf{s} - \iint_S \frac{\partial \mathcal{B}}{\partial t} \cdot d\mathbf{s} \tag{1.21}$$

where $d\mathbf{s} = \hat{n} ds$, in which $\hat{n}$ is the outward unit normal, and $ds$ denotes the differential surface element. Similarly, by integrating (1.3) and (1.4) over a volume $V$ we have

$$\iiint_V (\nabla \cdot \mathcal{D})\, dv = \iiint_V \varrho\, dv \tag{1.22}$$

$$\iiint_V (\nabla \cdot \mathcal{B})\, dv = \iiint_V \varrho_m\, dv. \tag{1.23}$$

Employing now Stokes's theorem in (1.20) and (1.21) and the divergence theorem in (1.22) and (1.23) yields the customary integral form of Maxwell's equations given by

$$\oint_C \mathcal{H} \cdot d\bar{\ell} = \iint_S \left(\mathcal{J} + \frac{\partial \mathcal{D}}{\partial t}\right) \cdot d\mathbf{s} \tag{1.24}$$

$$\oint_C \mathcal{E} \cdot d\bar{\ell} = -\iint_S \left(\mathcal{M} + \frac{\partial \mathcal{B}}{\partial t}\right) \cdot d\mathbf{s} \tag{1.25}$$

$$\oiint_{S_c} \mathcal{D} \cdot d\mathbf{s} = \iiint_V \varrho\, dv \tag{1.26}$$

$$\oiint_{S_c} \mathcal{B} \cdot d\mathbf{s} = \iiint_V \varrho_m\, dv \tag{1.27}$$

where $C$ is the contour bounding the open surface $S$ illustrated in Figure 1-1. As usual, the circle through the single integral implies integration over a closed contour, whereas the same symbol through the surface integral implies integration over the

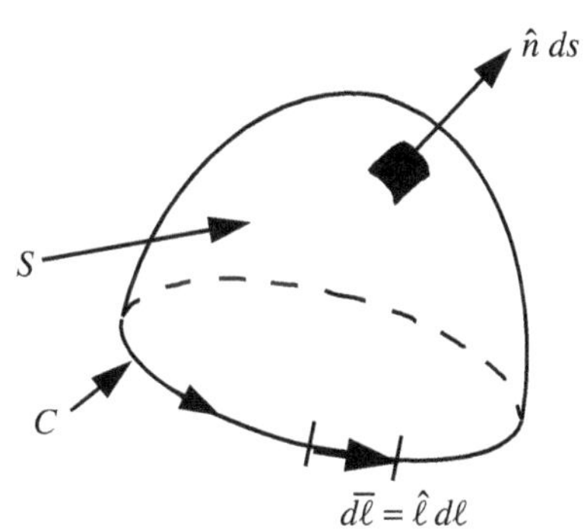


**FIGURE 1-1** Illustration of the differential element ds and the contour $C$. (After Volakis, Chatterjee, and Kempel, 1998. With permission.)

surface $S_c$ enclosing the corresponding volume $V$. We remark that the surface $S$ associated with the integral (1.24) and (1.25) is completely unrelated to that enclosing the volume $V$.

## 1.3 | MAXWELL'S EQUATIONS IN PHASOR FORM

Of primary interest in this book is the study of harmonically varying fields with an oscillating angular frequency $\omega = 2\pi f$ rad/sec. The electric field then takes the form ($j = \sqrt{-1}$)

$$\begin{aligned}\mathcal{E}(x, y, z; t) &= Re\left[\mathbf{E}(x, y, z)e^{jwt}\right] \\ &= \hat{x}E_{xo}\cos(\omega t + \phi_x) + \hat{y}E_{yo}\cos(\omega t + \phi_y) + \hat{z}E_{zo}\cos(\omega t + \phi_z)\end{aligned} \tag{1.28}$$

where the complex vector

$$\mathbf{E}(x, y, z) = \hat{x}E_{xo}e^{j\phi_x} + \hat{y}E_{yo}e^{j\phi_y} + \hat{z}E_{zo}e^{j\phi_z} \tag{1.29}$$

is referred to as the field phasor, and similar representations can be employed for the other field quantities $\mathcal{D}$, $\mathcal{H}$, $\mathcal{B}$, $\mathcal{J}$, $\mathcal{M}$, $\varrho$, and $\varrho_m$. Introducing these into (1.1) through (1.4) we obtain the simplified set of Maxwell's equations

$$\nabla \times \mathbf{H} = \mathbf{J} + j\omega\epsilon\mathbf{E} \tag{1.30}$$

$$\nabla \times \mathbf{E} = -\mathbf{M} - j\omega\mu\mathbf{H} \tag{1.31}$$

$$\nabla \cdot (\mu\mathbf{H}) = \rho_m \tag{1.32}$$

$$\nabla \cdot (\epsilon\mathbf{E}) = \rho \tag{1.33}$$

where **E**, **H**, **J**, **M**, $\rho$, and $\rho_m$ are the corresponding field phasors. In terms of these, the constitutive relations become

$$\mathbf{D} = \epsilon\mathbf{E} \tag{1.34}$$

$$\mathbf{B} = \mu\mathbf{H} \tag{1.35}$$

$$\mathbf{J} = \sigma\mathbf{E} \tag{1.36}$$

$$\mathbf{M} = \sigma_m\mathbf{H} \tag{1.37}$$

where the first two were combined with (1.3) and (1.4) to yield (1.32) and (1.33). Similarly, the phasor forms of the continuity equations are

$$\nabla \cdot \mathbf{J} + j\omega\rho = 0 \tag{1.38}$$

$$\nabla \cdot \mathbf{M} + j\omega\rho_m = 0. \tag{1.39}$$

Once (1.30) and (1.31) are solved in conjunction with (1.36) and (1.37) for the phasor quantities, the associated instantaneous fields can be obtained from (1.28).

In general, the current densities **J** and **M** appearing in (1.30) and (1.31) can be written as a sum of impressed (or excitation) and induced (or conduction) currents. That is, we may write

$$\mathbf{J} = \mathbf{J}_i + \mathbf{J}_c = \mathbf{J}_i + \sigma \mathbf{E} \tag{1.40}$$

$$\mathbf{M} = \mathbf{M}_i + \mathbf{M}_c = \mathbf{M}_i + \sigma_m \mathbf{H} \tag{1.41}$$

where the subscript $i$ denotes impressed currents, and the subscript $c$ implies conduction currents. When these are substituted into (1.30) and (1.31), we obtain

$$\nabla \times \mathbf{H} = \mathbf{J}_i + j\omega\epsilon_o\dot{\epsilon}_r\mathbf{E} \tag{1.42}$$

$$\nabla \times \mathbf{E} = -\mathbf{M}_i - j\omega\mu_o\dot{\mu}_r\mathbf{H} \tag{1.43}$$

where

$$\dot{\epsilon}_r = \epsilon_r - j\frac{\sigma}{\omega\epsilon_o} = \epsilon' - j\epsilon'' = \epsilon'(1 - j\tan\delta) \tag{1.44}$$

and

$$\dot{\mu}_r = \mu_r - j\frac{\sigma_m}{\omega\mu_o} = \mu' - j\mu'' = \mu'(1 - j\tan\delta_m) \tag{1.45}$$

represent equivalent relative complex permittivity and permeability constants. Anyone of the representations given in (1.44) and (1.45) is likely to be found in the literature with the quantities

$$\tan\delta = \frac{\epsilon''}{\epsilon'} \tag{1.46}$$

$$\tan\delta_m = \frac{\mu''}{\mu'} \tag{1.47}$$

referred to as the electric and magnetic loss tangents, respectively, of the material.

To summarize, in phasor form, Maxwell's equations for isotropic media are

$$\nabla \times \mathbf{H} = \mathbf{J}_i + j\omega\dot{\epsilon}\mathbf{E} \tag{1.48}$$

$$\nabla \times \mathbf{E} = -\mathbf{M}_i - j\omega\dot{\mu}\mathbf{H} \tag{1.49}$$

$$\nabla \cdot (\dot{\mu}\mathbf{H}) = -(\nabla \cdot \mathbf{M}_i)/j\omega \tag{1.50}$$

$$\nabla \cdot (\dot{\epsilon}\mathbf{E}) = -(\nabla \cdot \mathbf{J}_i)/j\omega \tag{1.51}$$

where we employed the phasor form of (1.38) and (1.39) to rewrite (1.32) and (1.33). The corresponding integral representations of (1.48) through (1.51) are (see (1.20) through (1.23))

$$\oint_C \mathbf{H} \cdot d\bar{\ell} = \iint_S (\mathbf{J}_i + j\omega\dot{\epsilon}\mathbf{E}) \cdot d\mathbf{s} \tag{1.52}$$

$$\oint_C \mathbf{E} \cdot d\bar{\ell} = -\iint_S (\mathbf{M}_i + j\omega\dot{\mu}\mathbf{H}) \cdot d\mathbf{s} \tag{1.53}$$

$$\oiint_{S_c} \dot{\mu}\mathbf{H} \cdot d\mathbf{s} = -\iiint_V \frac{\nabla \cdot \mathbf{M}_i}{j\omega}\, dv = \iiint_V \rho_m\, dv \tag{1.54}$$

$$\oiint_{S_c} \dot{\epsilon}\mathbf{E} \cdot d\mathbf{s} = -\iiint_V \frac{\nabla \cdot \mathbf{J}_i}{j\omega}\, dv = \iiint_V \rho\, dv \tag{1.55}$$

The parameters $\dot{\epsilon}$ and $\dot{\mu}$ in (1.48) through (1.55) are assumed to be given in terms of the equivalent complex constants (1.44) and (1.45); that is, $\dot{\epsilon} = \epsilon_o \dot{\epsilon}_r$, and $\dot{\mu} = \mu_o \dot{\mu}_r$. For notational convenience, from here on we will drop the dot over the relative constitutive parameters with the understanding that these will still represent all possible material losses.

We observe that given $\mathbf{J}_i$ and $\mathbf{M}_i$, (1.48) and (1.49) imply six scalar equations for the solution of the six components associated with $\mathbf{E}$ and $\mathbf{H}$. Thus, for time harmonic fields, (1.48) and (1.49) or (1.52) and (1.53) are sufficient for a solution of the electric and magnetic fields, and (1.50) and (1.51), or their integral counterparts, are superfluous. In fact, the last two equations follow directly from the first two upon taking their divergence and noting again that $\nabla \cdot (\nabla \times \mathbf{A}) = 0$ for any vector $\mathbf{A}$.

Equations (1.48) through (1.55) can be easily modified to apply for anisotropic materials as well. This requires that $\epsilon\mathbf{E}$ and $\mu\mathbf{H}$ be replaced by $\overline{\overline{\epsilon}} \cdot \mathbf{E}$ and $\overline{\overline{\mu}} \cdot \mathbf{H}$, respectively. In the case of bianisotropic media, the same quantities should be replaced by $\overline{\overline{\epsilon}} \cdot \mathbf{E} + \overline{\overline{\zeta}} \cdot \mathbf{B}$ and $\overline{\overline{\mu}} \cdot \mathbf{H} + \overline{\overline{\xi}} \cdot \mathbf{E}$, respectively, as dictated by the constitutive relations (1.18) and (1.19).

## 1.4 | NATURAL BOUNDARY CONDITIONS

Maxwell's differential equations cannot be solved without the specification of the required boundary conditions at material interfaces. The pertinent boundary conditions can be derived directly from the integral form of Maxwell's equations. Specifically, (1.52) is applied to the contour illustrated in Figure 1-2a with $S$ being the area enclosed by $C$. Assuming $\Delta\ell$ is small, $\Delta h \to 0$ and $\Delta\ell \gg \Delta h$, (1.52) gives

$$(\mathbf{H}_1 - \mathbf{H}_2) \cdot \hat{t} = [\mathbf{J}_i \cdot (\hat{n}_1 \times \hat{t})]\Delta h \tag{1.56}$$

and in deriving this we set

$$\lim_{\Delta h \to 0} \frac{1}{2}\Delta h\, [\epsilon_1 \mathbf{E}_1 + \epsilon_2 \mathbf{E}_2] \cdot \hat{n}_1 = 0$$

which is valid provided $\epsilon\mathbf{E}$ is finite at the interface. When (1.53) is applied to the same contour in Figure 1-2a we find that

$$(\mathbf{E}_1 - \mathbf{E}_2) \cdot \hat{t} = -[\mathbf{M}_i \cdot (\hat{n}_1 \times \hat{t})]\Delta h \tag{1.57}$$

where we have again set

$$\lim_{\Delta h \to 0} \frac{1}{2}\Delta h\, [\mu_1 \mathbf{H}_1 + \mu_2 \mathbf{H}_2] = 0$$

implying that $\mu\mathbf{H}$ is finite at the dielectric interface.

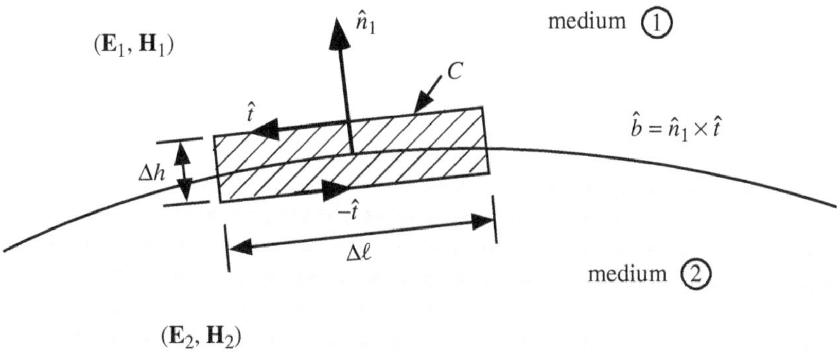


(a)

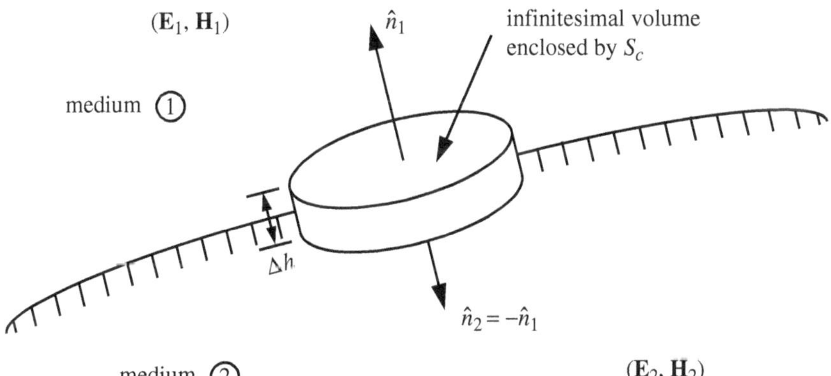


(b)

**FIGURE 1-2** Geometries for deriving the boundary conditions (a) for tangential components and (b) for normal components. (Volakis, Chatterjee, and Kempel, 1998)

The condition (1.56) and (1.57) can be rewritten in vector form and more compactly by introducing the definitions

$$\mathbf{J}_{is} = \mathbf{J}_i \Delta h \tag{1.58}$$

$$\mathbf{M}_{is} = \mathbf{M}_i \Delta h. \tag{1.59}$$

The quantities $\mathbf{J}_{is}$ and $\mathbf{M}_{is}$ are referred to as the impressed electric and (equivalent) magnetic current densities in A/m and Wb/m, respectively, at the interface. Using (1.58) and (1.59) the scalar conditions can be rewritten in vector form as

$$\hat{n}_1 \times (\mathbf{H}_1 - \mathbf{H}_2) = \mathbf{J}_{is} \tag{1.60}$$

$$\hat{n}_1 \times (\mathbf{E}_1 - \mathbf{E}_2) = -\mathbf{M}_{is}. \tag{1.61}$$

To generate the boundary conditions corresponding to (1.54) and (1.55), we select $S_c$ to be the surface of a small pillbox, shown in Figure 1-2b, enclosing the volume $V$. The pillbox is positioned at the dielectric interface so that half of its volume is in medium 1 and the other half in medium 2. It is again assumed that $\Delta h \to 0$ so that only its flat surfaces need be considered in performing the integrations. Through direct integration of (1.54) and (1.55) we readily obtain

$$\hat{n}_1 \cdot (\mu_1 \mathbf{H}_1 - \mu_2 \mathbf{H}_2) = \rho_{ms} \tag{1.62}$$

$$\hat{n}_1 \cdot (\epsilon_1 \mathbf{E}_1 - \epsilon_2 \mathbf{E}_2) = \rho_s \tag{1.63}$$

where $\rho_s$ denotes the electric surface charge density in C/m$^2$ at the interface, and $\rho_{ms}$ is the corresponding fictitious surface magnetic charge density in Wb/m$^2$.

The boundary conditions (1.60) through (1.63), although derived for time harmonic fields, are applicable for instantaneous fields as well. As noted earlier, in the time harmonic case, only (1.60) and (1.61) are required in conjunction with (1.48) and (1.49) for a unique solution of the fields.

If we ignore the fictitious magnetic currents and charges appearing in (1.60) through (1.63) we have

$$\hat{n}_1 \times (\mathbf{H}_1 - \mathbf{H}_2) = \mathbf{J}_{is} \tag{1.64}$$

$$\hat{n}_1 \times (\mathbf{E}_1 - \mathbf{E}_2) = 0 \tag{1.65}$$

$$\hat{n}_1 \cdot (\mu_1 \mathbf{H}_1 - \mu_2 \mathbf{H}_2) = 0 \tag{1.66}$$

$$\hat{n}_1 \cdot (\epsilon_1 \mathbf{E}_1 - \epsilon_2 \mathbf{E}_2) = \rho_s. \tag{1.67}$$

The first two of these state that the tangential electric fields are continuous across the interface, whereas the tangential magnetic fields are discontinuous at the same location by an amount equal to the impressed electric current. Unless a source (i.e., free charge) is actually placed at the interface, $\mathbf{J}_{is}$ is also zero and, in that case, the tangential magnetic fields will be continuous across the media as well.

When medium 2 is a perfect electric conductor then $\mathbf{E}_2 = \mathbf{H}_2 = 0$. In addition, $M_{is}$ and $\rho_{ms}$ vanish, and (1.60) through (1.63) reduce to

$$\hat{n}_1 \times \mathbf{H}_1 = \mathbf{J}_{is} \tag{1.68}$$

$$\hat{n}_1 \times \mathbf{E}_1 = 0 \tag{1.69}$$

$$\hat{n}_1 \cdot (\mu_1 \mathbf{H}_1) = 0 \tag{1.70}$$

$$\hat{n}_1 \cdot (\epsilon_1 \mathbf{E}_1) = \rho_s. \tag{1.71}$$

The first two of these now imply that the tangential electric field vanishes on the surface of the perfect electric conductor, whereas the tangential magnetic field is equal to the impressed electric surface current on the conductor.

## 1.5 POYNTING'S THEOREM

We observe that the quantity

$$\mathcal{S} = \mathcal{E} \times \mathcal{H} \tag{1.72}$$

has units of power density (i.e., watts/m$^2$) and can therefore be associated with the energy carried by the instantaneous wave per square unit. This is in the direction coincident with that of the vector $\mathcal{S}$. As stated already, $\mathcal{S}$ denotes instantaneous power density, but generally of interest is the time-averaged power. We therefore consider the quantity

$$\langle \mathcal{S} \rangle = \frac{1}{T} \int_0^T (\mathcal{E} \times \mathcal{H})\, dt \tag{1.73}$$

where $T$ is the period of the oscillation, that is, $\omega = 2\pi/T$. Assuming time harmonic fields, we set

$$\mathcal{E} = Re(\mathbf{E}e^{jwt}) = \hat{e}|E| \cos(\omega t + \alpha) \tag{1.74}$$

$$\mathcal{H} = Re(\mathbf{H}e^{jwt}) = \hat{h}|H| \cos(\omega t + \beta) \tag{1.75}$$

with $\mathbf{E} = \hat{e}|E| = \hat{e}Ee^{j\alpha}$ and $\mathbf{H} = \hat{h}H = \hat{h}|H|e^{j\beta}$. We then have

$$\mathcal{E} \times \mathcal{H} = \frac{1}{2}\hat{e} \times \hat{h}|E||H| \left[\cos(\alpha - \beta) + \cos(2\omega t + \alpha + \beta)\right] \tag{1.76}$$

and by substituting this into (1.73) and integrating, we obtain

$$\langle \mathcal{S} \rangle = \frac{1}{2}\hat{e} \times \hat{h}|E||H| \cos(\alpha - \beta) = \frac{1}{2}Re(\mathbf{E} \times \mathbf{H}^*) \tag{1.77}$$

where the superscript star implies the complex conjugate of the vector.

The quantity

$$\mathbf{S} = \frac{1}{2}\, \mathbf{E} \times \mathbf{H}^* \tag{1.78}$$

is known as the complex Poynting vector. Since it is equal to the complex power density of the wave, it is important that we qualify the source and nature of this power. To do this we refer to (1.48) and (1.49), where by dotting each equation with $\mathbf{E}$ or $\mathbf{H}^*$ we have

$$\mathbf{E} \cdot \nabla \times \mathbf{H}^* = \mathbf{J}_i^* \cdot \mathbf{E} - j\omega\epsilon^*\mathbf{E}^* \cdot \mathbf{E} = \mathbf{J}_i^* \cdot \mathbf{E} - j\omega\epsilon^*|\mathbf{E}|^2 \tag{1.79}$$

$$\mathbf{H}^* \cdot \nabla \times \mathbf{E} = -\mathbf{M}_i \cdot \mathbf{H}^* - j\omega\mu\mathbf{H} \cdot \mathbf{H}^* = -\mathbf{M}^i \cdot \mathbf{H}^* - j\omega\mu|\mathbf{H}|^2. \tag{1.80}$$

From the vector identity

$$\nabla \cdot (\mathbf{E} \times \mathbf{H}^*) = \mathbf{H}^* \cdot \nabla \times \mathbf{E} - \mathbf{E} \cdot \nabla \times \mathbf{H}^* \tag{1.81}$$

we then obtain

$$\nabla \cdot (\mathbf{E} \times \mathbf{H}^*) = j\omega\epsilon^*|\mathbf{E}|^2 - j\omega\mu|\mathbf{H}|^2 - \mathbf{J}_i^* \cdot \mathbf{E} - \mathbf{M}_i \cdot \mathbf{H}^* \tag{1.82}$$

which is an identity valid everywhere in space. Integrating both sides of this over a volume $V$ containing all sources and invoking the divergence theorem yields

$$\frac{1}{2}\oiint_{S_c}(\mathbf{E}\times\mathbf{H}^*)\cdot d\mathbf{s}=\frac{1}{2}\iiint_V\left[j\omega\epsilon^*|\mathbf{E}|^2+j\omega\mu|\mathbf{H}|^2-\mathbf{J}_i^*\cdot\mathbf{E}-\mathbf{M}_i\cdot\mathbf{H}^*\right]dv \tag{1.83}$$

which is commonly referred to as Poynting's theorem. Since $S_c$ is closed, based on energy conservation, one deduces that the right-hand side of (1.83) must represent the sum of the power stored or radiated (i.e., escaping) out of the volume $V$. Each term of the volume integral of (1.83) is associated with a specific type of power, but before proceeding with their identification it is instructive that $\epsilon^*$ be first replaced by $\epsilon_o\epsilon_r+j\frac{\sigma}{w}$. Equation (1.83) can then be rewritten as

$$\frac{1}{2}Re\oiint_{S_c}(\mathbf{E}\times\mathbf{H}^*)\cdot d\mathbf{s}=P_{ei}+P_{mi}-P_d \tag{1.84}$$

$$\frac{1}{2}Im\oiint_{S_c}(\mathbf{E}\times\mathbf{H}^*)\cdot d\mathbf{s}=2\omega[W_e-W_m]-\frac{1}{2}Im\iiint\left[\mathbf{J}_i^*\cdot\mathbf{E}+\mathbf{M}_i\cdot\mathbf{H}^*\right]dv \tag{1.85}$$

where

$$P_{ei}=-\frac{1}{2}\iiint_V Re(\mathbf{J}_i^*\cdot\mathbf{E})\,dv=\text{average outgoing power due to the source }\mathbf{J}_i \tag{1.86}$$

$$P_{mi}=-\frac{1}{2}\iiint_V Re(\mathbf{M}_i\cdot\mathbf{H}^*)\,dv=\text{average outgoing power due to the source }\mathbf{M}_i \tag{1.87}$$

$$P_d=\frac{1}{2}\iiint_V\sigma|\mathbf{E}|^2\,dv=\text{average power dissipated in }V \tag{1.88}$$

$$W_e=\frac{1}{4}\iiint_V\epsilon_o\epsilon_r|\mathbf{E}|^2\,dv=\text{average electric energy in }V \tag{1.89}$$

$$W_m=\frac{1}{4}\iiint_V\mu_o\mu_r|\mathbf{H}|^2\,dv=\text{average magnetic energy in }V. \tag{1.90}$$

The time-averaged power delivered to the electromagnetic field outside $V$ is clearly the sum of $P_{ei}$ and $P_{mi}$, whereas $P_d$ is that dissipated in $V$ due to conductor losses. Thus, we may consider

$$P_{av}=\frac{1}{2}Re\oiint_{S_c}(\mathbf{E}\times\mathbf{H}^*)\cdot d\mathbf{s} \tag{1.91}$$

to be the average radiated power outside $V$ if $\sigma$ is zero in $V$. Expression (1.85) gives the reactive power (i.e., stored within $V$ and not allowed to escape outside the boundary of $S_c$).

## 1.6 UNIQUENESS THEOREM

Whenever one pursues a solution to a set of equations it is important to know a priori whether this solution is unique and, if not, the required conditions for a unique solution. This is important because, depending on the application, different analytical or numerical methods will likely be used for the solution of Maxwell's equations. Given that Maxwell's equations (subject to the appropriate boundary conditions) yield a unique solution, we know that any convenient method of analysis will yield the correct solution to the problem.

The most common form of the uniqueness theorem is as follows:

> *In a region $V$ completely occupied with dissipative media, a harmonic field ($\mathbf{E}$, $\mathbf{H}$) is uniquely determined by the impressed currents in that region plus the tangential components of the electric or magnetic fields on the closed surface $S_c$ bounding $V$.*

This theorem may be proved by assuming for the moment that two solutions exist, denoted by $(\mathbf{E}_1, \mathbf{H}_1)$ and $(\mathbf{E}_2, \mathbf{H}_2)$. Both fields must, of course, satisfy Maxwell's equations (1.48) and (1.49) with the same impressed currents $(\mathbf{J}_i, \mathbf{M}_i)$. We have

$$\nabla \times \mathbf{H}_1 = \mathbf{J}_i + j\omega\epsilon\mathbf{E}_1, \quad \nabla \times \mathbf{H}_2 = \mathbf{J}_i + j\omega\epsilon\mathbf{E}_2$$

$$\nabla \times \mathbf{E}_1 = -\mathbf{M}_i - j\omega\mu\mathbf{H}_1, \quad \nabla \times \mathbf{E}_2 = -\mathbf{M}_i - j\omega\mu\mathbf{H}_2$$

and when these are subtracted we obtain

$$\nabla \times \mathbf{H}' = j\omega\epsilon\mathbf{E}' \tag{1.92}$$

$$\nabla \times \mathbf{E}' = -j\omega\mu\mathbf{H}' \tag{1.93}$$

where $\mathbf{E}' = \mathbf{E}_1 - \mathbf{E}_2$, and $\mathbf{H}' = \mathbf{H}_1 - \mathbf{H}_2$. To prove the theorem it is then necessary to show that $(\mathbf{E}', \mathbf{H}')$ are zero or, equivalently, if no sources are enclosed by a volume $V$, the fields in that volume are zero for a given set of tangential electric and magnetic fields on $S_c$. Recalling Poynting's theorem for a source-free region, from (1.84) through (1.90) (with $\mathbf{J}_i = \mathbf{M}_i = 0$) we have

$$Re \oint\!\!\!\oint_{S_c} (\mathbf{E}' \times \mathbf{H}'^*) \cdot d\mathbf{s} = -\iiint_V \sigma|\mathbf{E}'|^2\, dv \tag{1.94}$$

$$Im \oint\!\!\!\oint_{S_c} (\mathbf{E}' \times \mathbf{H}'^*) \cdot d\mathbf{s} = \iiint_V \omega(\epsilon_o\epsilon_r|\mathbf{E}'|^2 - \mu_o\mu_r|\mathbf{H}'|^2)\, dv. \tag{1.95}$$

Since we assumed that the tangential $\mathbf{E}$ and $\mathbf{H}$ are also given on $S_c$, it follows that $(\mathbf{E}' \times \mathbf{H}'^*) \cdot d\mathbf{s}$ vanishes everywhere on $S_c$. Thus, the right-hand side of (1.94) must be zero, and since $|\mathbf{E}'|^2$ is always greater than zero we deduce that $\mathbf{E}'$ must vanish everywhere in $V$. Consequently, $\mathbf{H}'$ must also vanish in $V$, proving the previously stated uniqueness theorem.

As a corollary to the uniqueness theorem, it can be shown from (1.94):

> *If a harmonic field has zero tangential $\mathbf{E}$ or zero tangential $\mathbf{H}$ field on a surface enclosing a source-free region $V$ occupied by dissipative media, the field vanishes everywhere within $V$.*

In the case of lossless media, (1.94) can no longer be used to show that **E** vanishes in $V$ if no sources are enclosed. We must then resort to (1.95), which states that no time-averaged power enters or leaves the region. It is consequently possible that the fields be nonzero in the volume $V$ even when the tangential **E** and **H** are zero on different parts of the surface $S_c$. This implies the existance of resonant-mode fields within $V$ for which the time-averaged energy stored in the electric field is equal to the time-averaged energy stored in the magnetic field within that region. By definition, the volume integral vanishes in (1.95) for such a field. This nonuniqueness for lossless media will be discussed in later chapters when we consider numerical solutions for the scattered field from specific structures. In general, it will be seen that the numerical solution fails when resonant modes are excited unless some remedies are taken to suppress these nonphysical modes.

## 1.7 SUPERPOSITION THEOREM

The superposition theorem states that, for a linear medium, the total field intensity due to two or more sources is equal to the sum of the field intensities attributed to each individual source radiating independently of the others. In particular, let us consider two electric sources $\mathbf{J}_1$ and $\mathbf{J}_2$. On the basis of the superposition theorem, to find the total field caused by the simultaneous presence of both sources we can consider the field due to each individual source in isolation. The fields $(\mathbf{E}_1, \mathbf{H}_1)$ due to $\mathbf{J}_1$ satisfy the equations

$$\nabla \times \mathbf{H}_1 = \mathbf{J}_1 + j\omega\epsilon\mathbf{E}_1 \tag{1.96}$$

$$\nabla \times \mathbf{E}_1 = -j\omega\mu\mathbf{H}_1 \tag{1.97}$$

and the fields corresponding to $\mathbf{J}_2$ satisfy

$$\nabla \times \mathbf{H}_2 = \mathbf{J}_2 + j\omega\epsilon\mathbf{E}_2 \tag{1.98}$$

$$\nabla \times \mathbf{E}_2 = -j\omega\mu\mathbf{H}_2. \tag{1.99}$$

By adding these two sets of equations, it is clear that the total field due to both sources combined is given by

$$\mathbf{E} = \mathbf{E}_1 + \mathbf{E}_2 \qquad \mathbf{H} = \mathbf{H}_1 + \mathbf{H}_2 \tag{1.100}$$

where $(\mathbf{E}_1, \mathbf{H}_1)$ and $(\mathbf{E}_2, \mathbf{H}_2)$ are obtained by solving separately (1.96) through (1.97) and (1.98) through (1.99), respectively.

## 1.8 DUALITY THEOREM

The duality theorem relates to the interchangability of the electric and magnetic fields, currents, charges, or material properties. We observe from (1.30) and (1.31) that the first can be obtained from the second via the interchanges

$$\mathbf{M} \to -\mathbf{J}$$
$$\mathbf{E} \to \mathbf{H}$$

$$\begin{aligned} \mathbf{H} &\to -\mathbf{E} \\ \mu &\to \epsilon. \end{aligned} \tag{1.101a}$$

Similarly, (1.31) can be obtained from (1.30) via the interchanges

$$\begin{aligned} \mathbf{J} &\to \mathbf{M} \\ \mathbf{E} &\to \mathbf{H} \\ \mathbf{H} &\to -\mathbf{E} \\ \epsilon &\to \mu. \end{aligned} \tag{1.101b}$$

Based on these observations, we can state that if $(\mathbf{E}_1, \mathbf{H}_1)$ represent the fields due to the harmonic sources $(\mathbf{J}_1, \mathbf{M}_1)$ radiating in a linear medium $(\mu_1, \epsilon_1)$, then the sources $(\mathbf{J}_2 = -\mathbf{M}_1,\ \mathbf{M}_2 = \mathbf{J}_1)$ will generate the fields $(\mathbf{E}_2 = -\mathbf{H}_1,\ \mathbf{H}_2 = \mathbf{E}_1)$ if radiating in a linear medium having $(\mu_2 = \epsilon_1,\ \epsilon_2 = \mu_1)$. This is a statement of the duality theorem. If we assume $\mathbf{M}_1 = 0$, it simply states that the fields due to a magnetic source of equal strength, that is, $\mathbf{M}_2 = \mathbf{J}_1$, radiating in the dual medium are $(\mathbf{E}_2 = -\mathbf{H}_1,\ \mathbf{H}_2 = \mathbf{E}_1)$. The material parameters dual to $(\epsilon_1, \mu_1)$ are $(\epsilon_2 = \mu_1,\ \mu_2 = \epsilon_1)$, and the fields dual to $(\mathbf{E}_1, \mathbf{H}_1)$ are $(\mathbf{E}_2 = -\mathbf{H}_1, \mathbf{H}_2 = \mathbf{E}_2)$. Alternatively, if $\mathbf{J}_1 = 0$, the dual fields to $(\mathbf{E}_1, \mathbf{H}_1)$ are generated by the electric source $\mathbf{J}_2 = -\mathbf{M}_1$ radiating in the dual medium.

The duality theorem can reduce computational time and effort when one is able to invoke it for a particular application.

## 1.9 | VOLUME EQUIVALENCE THEOREM

The volume equivalence theorem shows how one can replace field disturbances caused by the presence of a material body with some prescribed equivalent currents.

Let us consider the harmonic sources $(\mathbf{J}_i, \mathbf{M}_i)$ radiating in the presence of a material structure as depicted in Figure 1-3. When the source fields $(\mathbf{E}^i, \mathbf{H}^i)$ reach the inhomogeneous (linear) body, additional fields, referred to as $(\mathbf{E}^s, \mathbf{H}^s)$, will be generated as dictated by the continuity boundary conditions. That is, because the fields penetrating the material will not be equal to the external excitation fields $(\mathbf{E}^i, \mathbf{H}^i)$, it will be necessary that new external fields be generated to ensure satisfaction of the tangential field continuity conditions. These new fields $(\mathbf{E}^s, \mathbf{H}^s)$ are commonly referred to as the *scattered fields*, and the total field internal and external to the material body can be written as

$$\mathbf{E} = \mathbf{E}^i + \mathbf{E}^s \qquad \mathbf{H} = \mathbf{H}^i + \mathbf{H}^s. \tag{1.102}$$

The incident field is, of course, expected to satisfy the equations

$$\nabla \times \mathbf{H}^i = \mathbf{J}_i + j\omega\epsilon_o\mathbf{E}^i, \qquad \nabla \times \mathbf{E}^i = -j\omega\mu_o\mathbf{H}^i \tag{1.103}$$

whereas the total field satisfies

$$\nabla \times \mathbf{H} = \mathbf{J}_i + j\omega\epsilon\mathbf{E}, \qquad \nabla \times \mathbf{E} = -j\omega\mu\mathbf{H} \tag{1.104}$$

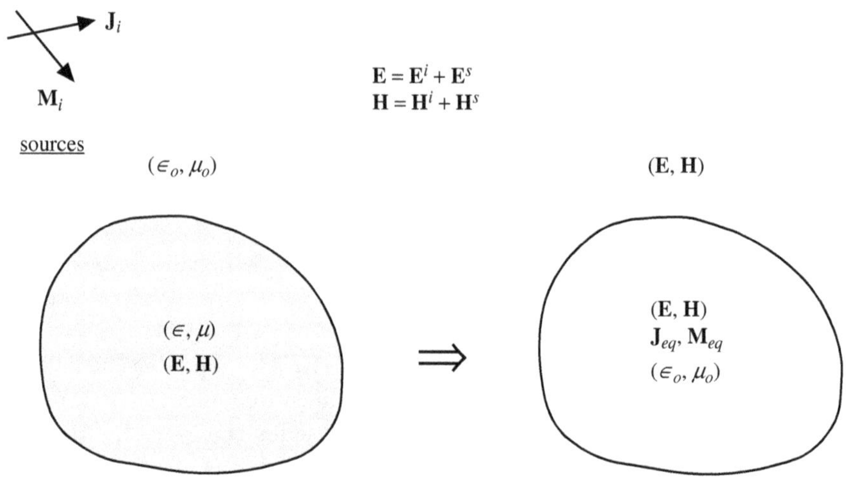


FIGURE 1-3 Volume equivalence theorem.

when measured within the material body having the material parameters $(\epsilon, \mu)$. Subtracting (1.103) from (1.104) and making use of (1.102) we obtain

$$\nabla \times \mathbf{H}^s = j\omega(\epsilon - \epsilon_o)\mathbf{E} + j\omega\epsilon_o\mathbf{E}^s \tag{1.105}$$

$$\nabla \times \mathbf{E}^s = -j\omega(\mu - \mu_o)\mathbf{H} - j\omega\mu_o\mathbf{H}^s. \tag{1.106}$$

When these are compared with (1.30) and (1.31), we conclude that the scattered fields $(\mathbf{E}^s, \mathbf{H}^s)$ can be thought as generated by the equivalent harmonic sources

$$\mathbf{J}_{eq} = j\omega(\epsilon - \epsilon_o)\mathbf{E} = j\omega\epsilon_o(\epsilon_r - 1)\mathbf{E} \tag{1.107}$$

$$\mathbf{M}_{eq} = j\omega(\mu - \mu_o)\mathbf{H} = j\omega\mu_o(\mu_r - 1)\mathbf{H} \tag{1.108}$$

which occupy the original volume of the material body and radiate in free space. That is, the material body can be removed and instead be replaced by the equivalent volume currents defined in (1.107) and (1.108).

It would appear from (1.107) and (1.108) that six independent current components are required to model a ferrite material scatterer. However, a closer examination reveals otherwise, and this is important in numerical simulations of material scatterers. By taking the curl of equations (1.104) and replacing $(\mathbf{J}_i, \mathbf{M}_i)$ with $(\mathbf{J}_{eq}, \mathbf{M}_{eq})$, we have

$$\nabla \times (\nabla \times \mathbf{E}) - k_o^2\mathbf{E} = -j\omega\mu_o\mathbf{J}_{eq} - \nabla \times \mathbf{M}_{eq} \tag{1.109}$$

$$\nabla \times (\nabla \times \mathbf{H}) - k_o^2\mathbf{H} = -j\omega\epsilon_o\mathbf{M}_{eq} + \nabla \times \mathbf{J}_{eq} \tag{1.110}$$

where $k_o = \omega\sqrt{\mu_o\epsilon_o} = 2\pi/\lambda_o$ is the free-space propagation constant, and $\lambda_o$ is the wavelength. Clearly, (1.109) can be solved independent of (1.110) and implies that the curl of the magnetic current appears as another electric current given by

$$\widetilde{\mathbf{J}}_{eq} = \nabla \times \mathbf{M}_{eq}/j\omega\mu_o. \tag{1.111}$$

Thus, instead of the equivalent currents (1.107) and (1.108), one could replace the presence of a material body by the radiation of the equivalent electric current density

$$\begin{aligned}\mathbf{J}_{eq} &= j\omega\epsilon_o(\epsilon_r - 1)\mathbf{E} + \nabla \times [(\mu_r - 1)\mathbf{H}] \\ &= \frac{(\epsilon_r - 1)}{\epsilon_r}\nabla \times \mathbf{H} + \nabla \times [(\mu_r - 1)\mathbf{H}]\end{aligned} \tag{1.112}$$

again occupying the original volume of the body. Similarly, we could replace the presence of the material body by the radiation of the equivalent magnetic current density

$$\begin{aligned}\mathbf{M}_{eq} &= j\omega\mu_o(\mu_r - 1)\mathbf{H} - \nabla \times [(\epsilon_r - 1)\mathbf{E}] \\ &= -\frac{(\mu_r - 1)}{\mu_r}\nabla \times \mathbf{E} - \nabla \times [(\epsilon_r - 1)\mathbf{E}]\end{aligned} \tag{1.113}$$

which is obviously the dual of (1.112). Thus, we observe that a material body can be modeled with only electric or magnetic currents, implying that three independent current components are required in either case. The aforementioned equivalence between electric and magnetic currents was originally presented by Mayes (1958), although in a different context. It should be noted, however, that in (1.109) through (1.113), $(\mathbf{J}_{eq}, \mathbf{M}_{eq})$ are defined over a finite domain, and this needs to be considered when taking the indicated curls. In other words, the fact that $\mu_r$ and $\epsilon_r$ are discontinuities at the outer boundary of the scatterer or perhaps at other boundaries internal to the scatterer must be accounted for when carrying out the curl operations. We will discuss this in more detail when we later make use of these equivalences.

## 1.10 | SURFACE EQUIVALENCE THEOREM

The surface equivalence theorem follows directly from the uniqueness theorem stating that a harmonic field $(\mathbf{E}, \mathbf{H})$ in a source-free region $V$ can be uniquely determined by a knowledge of the tangential fields on an imaginary surface $S_c$ bounding $V$. This implies that it is not necessary to explicitly know the sources of $(\mathbf{E}, \mathbf{H})$ so that this field be determined in regions outside those sources. Instead, one could specify equivalent currents on a surface $S_c$ bounding the source-free region $V$ by invoking Ampere's law (see (1.60) and (1.61)). This, of course, provides an alternative formulation that is often useful since it could lead to simplifications in the analysis.

To illustrate the surface equivalence theorem, let us consider a material region enclosed by the surface as shown in Figure 1-4. We are interested in computing the field outside the material region, and thus $V$ is the volume between $S_c$ and a surface at infinity. For the region interior to $S_c$ we may arbitrarily assume that the fields are zero as illustrated in Figure 1-4b. In accordance with the boundary conditions given by (1.60) and (1.61), the fields $(\mathbf{E}, \mathbf{H})$ outside $S_c$ can then be thought as generated by the equivalent surface currents

$$\mathbf{J}_s = \hat{n} \times \mathbf{H} \tag{1.114}$$

$$\mathbf{M}_s = \mathbf{E} \times \hat{n} \tag{1.115}$$

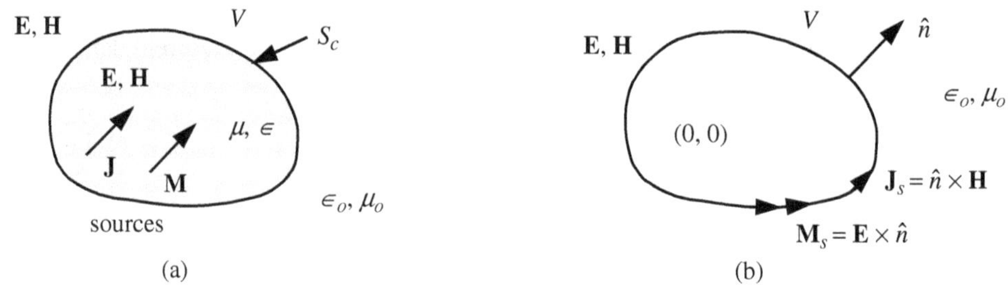


**FIGURE 1-4** ■ Love's equivalence principle: (a) original problem and (b) equivalent problem.

where $\hat{n}$ is the unit normal to $S_c$ pointing toward $V$. This form of the surface equivalence theorem is known as Love's equivalence principle (Love, 1903). It assumes zero field within $S_c$, a condition that can be satisfied when this region is occupied by a perfect electric or magnetic conductor. For a planar electric conductor, image theory dictates that the net electric current on $S_c$ will be zero, whereas $M_s$ is doubled (see Figure 1-5). Thus, the field outside $S_c$ can be also represented by the radiation of the equivalent surface magnetic current

$$\mathbf{M}_s = 2\mathbf{E} \times \hat{n} \tag{1.116}$$

where $\mathbf{E}$ is the value of the electric field on the planar surface $S_c$. Alternatively, the null field condition can be satisfied when $S_c$ is the surface of a planar perfect magnetic conductor, and in this case $\mathbf{M}_s$ vanishes whereas $\mathbf{J}_s = 2\hat{n} \times \mathbf{H}$.

When setting up the equivalent surface currents one can also select the field within $S_c$ to be other than zero. For example, referring to Figure 1-6, if these are set to $(\mathbf{E}^c, \mathbf{H}^c)$, the corresponding surface equivalent currents are

$$\mathbf{J}_s = \hat{n} \times (\mathbf{H} - \mathbf{H}^c) \tag{1.117a}$$

$$\mathbf{M}_s = (\mathbf{E} - \mathbf{E}^c) \times \hat{n}. \tag{1.117b}$$

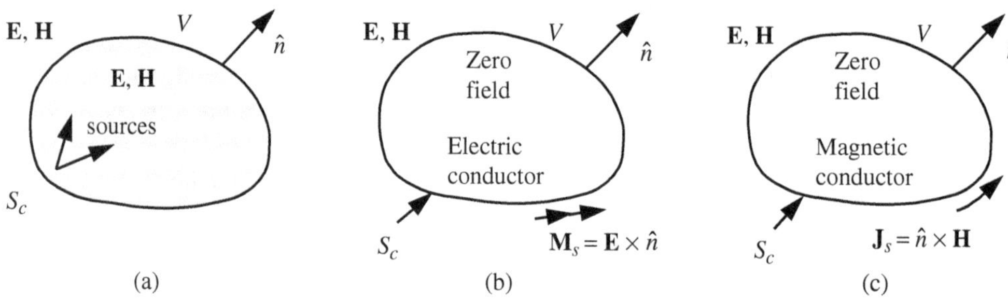


**FIGURE 1-5** ■ Other forms of Love's equivalence principle: (a) original problem, (b) equivalent problem with a magnetic current on an electric conductor, and (c) equivalent problem with an electric current on a magnetic conductor.

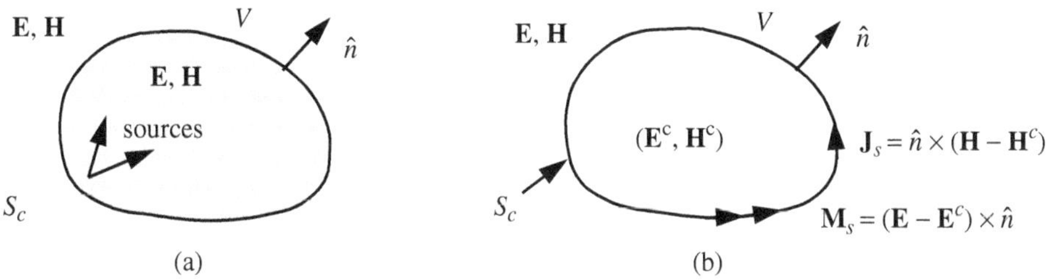


**FIGURE 1-6** General form of the equivalence theorem: (a) original problem and (b) equivalence problem.

These will now generate the field ($\mathbf{E}$, $\mathbf{H}$) exterior to $S_c$ and the field ($\mathbf{E}^c$, $\mathbf{H}^c$) interior to $S_c$.

# 1.11 | RECIPROCITY AND REACTION THEOREMS

The reciprocity theorem for electromagnetics parallels the familiar theorem in circuit theory. It simply states that the fields and sources can be interchanged in a given problem or set-up without affecting the system's response. This implies that the transmitting and receiving antenna patterns are the same, even though in the first case the source was at the feed whereas for the receiving antenna the source is at infinity. Another example refers to the case of plane wave scattering illustrated in Figure 1-7. Let us assume that the far-zone scattered field $\mathbf{E}^s$ is measured along $\hat{r}$ and is caused by a plane wave excitation $\mathbf{E}^i$ incident along $\hat{r}^i$. Based on the reciprocity theorem, one can then state that the scattered field is unchanged when we let $\hat{r}^i \to -\hat{r}$ and $\hat{r} \to -\hat{r}^i$.

To derive a mathematical statement of the reciprocity theorem, we assume the existence of two sets of fields caused by two different sets of sources radiating in the same environment. In particular, suppose that the field ($\mathbf{E}_1$, $\mathbf{H}_1$) are associated with the sources ($\mathbf{J}_1$, $\mathbf{M}_1$), whereas the fields ($\mathbf{E}_2$, $\mathbf{H}_2$) are due to the sources ($\mathbf{J}_2$, $\mathbf{M}_2$). Each set of these fields and sources will then satisfy the equations

$$\nabla \times \mathbf{E}_{\substack{1\\2}} = -j\omega\mu\mathbf{H}_{\substack{1\\2}} - \mathbf{M}_{\substack{1\\2}} \tag{1.118a}$$

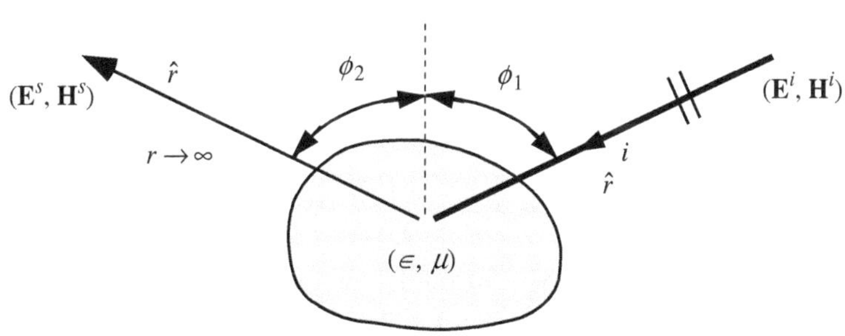


**FIGURE 1-7** Illustration of reciprocity for plane wave incidence and far-zone observation.

and

$$\nabla \times \mathbf{H}_{\frac{1}{2}} = j\omega\epsilon \mathbf{E}_{\frac{1}{2}} - \mathbf{J}_{\frac{1}{2}}. \tag{1.118b}$$

By invoking the identity

$$\nabla \cdot (\mathbf{A} \times \mathbf{B}) = \mathbf{B} \cdot \nabla \times \mathbf{A} - \mathbf{A} \cdot \nabla \times \mathbf{B} \tag{1.119}$$

we then have

$$\nabla \cdot (\mathbf{E}_1 \times \mathbf{H}_2 - \mathbf{E}_2 \times \mathbf{H}_1) = \mathbf{H}_2 \cdot \nabla \times \mathbf{E}_1 - \mathbf{E}_1 \cdot \nabla \times \mathbf{H}_2 - \mathbf{H}_1 \cdot \nabla \times \mathbf{E}_2 + \mathbf{E}_2 \cdot \nabla \times \mathbf{H}_1. \tag{1.120}$$

This can be simplified by introducing (1.118a), giving, for example,

$$\mathbf{H}_2 \cdot \nabla \times \mathbf{E}_1 = \mathbf{H}_2 \cdot (-j\omega\mu \mathbf{H}_1 - \mathbf{M}_1) = -j\omega\mu \mathbf{H}_1 \cdot \mathbf{H}_2 - \mathbf{M}_1 \cdot \mathbf{H}_2$$

$$-\mathbf{H}_1 \cdot \nabla \times \mathbf{E}_2 = -\mathbf{H}_1 \cdot (-j\omega\mu \mathbf{H}_2 - \mathbf{M}_2) = j\omega\mu \mathbf{H}_1 \cdot \mathbf{H}_2 + \mathbf{M}_2 \cdot \mathbf{H}_1$$

which upon substitution into (1.120) yield

$$-\nabla \cdot (\mathbf{E}_1 \times \mathbf{H}_2 - \mathbf{E}_2 \times \mathbf{H}_1) = \mathbf{E}_1 \cdot \mathbf{J}_2 + \mathbf{H}_2 \cdot \mathbf{M}_1 - \mathbf{E}_2 \cdot \mathbf{J}_1 - \mathbf{H}_1 \cdot \mathbf{M}_2. \tag{1.121}$$

Integrating both sides of this equation over a volume $V$ enclosed by the surface $S_c$, and applying the divergence theorem, it is further deduced that

$$-\oiint_{S_c} (\mathbf{E}_1 \times \mathbf{H}_2 - \mathbf{E}_2 \times \mathbf{H}_1) \cdot \hat{n}\, ds = \iiint_V (\mathbf{E}_1 \cdot \mathbf{J}_2 + \mathbf{H}_2 \cdot \mathbf{M}_1 - \mathbf{E}_2 \cdot \mathbf{J}_1 - \mathbf{H}_1 \cdot \mathbf{M}_2)\, dv. \tag{1.122}$$

It will be shown in the next chapter that in the far field (let $S_c$ become an infinite circle),

$$\mathbf{E} = -Z_o \hat{r} \times \mathbf{H} \quad \text{and} \quad \mathbf{H} = \frac{1}{Z_o} \hat{r} \times \mathbf{E}$$

where $Z_o = \sqrt{\mu_o/\epsilon_o}$ is the free-space intrinsic impedance, and thus

$$\mathbf{E}_1 \times \mathbf{H}_2 = \mathbf{E}_1 \times \frac{(\hat{r} \times \mathbf{E}_2)}{Z_o} = \frac{1}{Z_o}[(\mathbf{E}_1 \cdot \mathbf{E}_2)\hat{r} - (\mathbf{E}_1 \cdot \hat{r})\mathbf{E}_2] = \frac{1}{Z_o}(\mathbf{E}_1 \cdot \mathbf{E}_2)\hat{r}$$

$$\mathbf{E}_2 \times \mathbf{H}_1 = \frac{\mathbf{E}_2 \times (\hat{r} \times \mathbf{E}_1)}{Z_o} = \frac{1}{Z_o}[(\mathbf{E}_2 \cdot \mathbf{E}_1)\hat{r} - (\mathbf{E}_1 \cdot \hat{r})\mathbf{E}_1] = \frac{1}{Z_o}(\mathbf{E}_2 \cdot \mathbf{E}_1)\hat{r}$$

implying that the surface integral in (1.122) vanishes when $S_c$ is a sphere of infinite radius. Consequently, we conclude that

$$\iiint_V (\mathbf{E}_1 \cdot \mathbf{J}_2 - \mathbf{H}_1 \cdot \mathbf{M}_2)\, dv = \iiint_V (\mathbf{E}_2 \cdot \mathbf{J}_1 - \mathbf{H}_2 \cdot \mathbf{M}_1)\, dv \tag{1.123}$$

which is a mathematical statement of the reciprocity theorem (special case of the Lorentz reciprocity theorem given by (1.122)). It states that the fields and sources

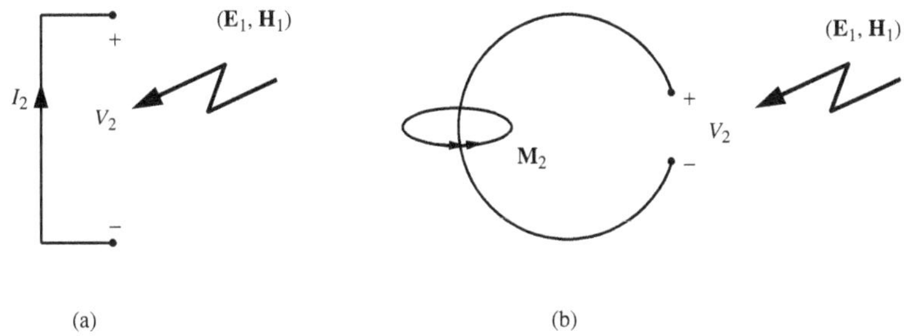


**FIGURE 1-8** Illustration of circuit source: (a) current source and (b) voltage source.

can be interchanged without altering the outcome of (1.123). Integrals of the type in (1.123) are also referred to as reactions of one set of sources with the fields caused by another set of sources. Based on this reasoning, (1.123) is often written as

$$\langle 1, 2 \rangle = \langle 2, 1 \rangle \tag{1.124}$$

where

$$\langle 1, 2 \rangle = \iiint_V (\mathbf{E}_1 \cdot \mathbf{J}_2 - \mathbf{H}_1 \cdot \mathbf{M}_2)\, dv. \tag{1.125}$$

The symbolism $\langle 1, 2 \rangle$ denotes the reaction of fields $(\mathbf{E}_1, \mathbf{H}_1)$ with the sources $(\mathbf{J}_2, \mathbf{M}_2)$ and (1.125) is often referred to as the reaction theorem. If $\mathbf{J}_2$ represents a linear source of strength $I_2$ (i.e., $\mathbf{J}_2\, dv = I_2 \hat{\ell}\, d\ell$) and $\mathbf{M}_2 = 0$, (1.125) reduces to

$$\langle 1, 2 \rangle = \iiint_V \mathbf{E}_1 \cdot \mathbf{J}_2\, dv = I_2 \int \mathbf{E}_1 \cdot \hat{\ell}\, d\ell = -I_2 V_2^{(1)} \tag{1.126}$$

where $V_2^{(1)}$ is the voltage across the terminals of source 2 due to some unspecified source 1. Similarly, across the terminals of a magnetic source $\mathbf{M} = K\hat{\ell}$ (current loop), shown in Figure 1-8, $V = -K$, and if we set $\mathbf{M}_2\, dv = K_2 \hat{\ell}\, d\ell$ and $\mathbf{J}_2 = 0$, (1.125) gives

$$\langle 1, 2 \rangle = -\iiint \mathbf{H}_1 \cdot \mathbf{M}_2\, dv = -K_2 \int \mathbf{H}_1 \cdot \hat{\ell}\, d\ell = +V_2 I_2^{(1)} \tag{1.127}$$

where $I_2^{(1)}$ is now the current flowing to the terminal of source 2 due to the field excitation $\mathbf{H}_1$ from some unspecified source 1.

To illustrate the application of the reciprocity theorem in electromagnetics, we consider the radiation of two antenna elements in free space as illustrated in Figure 1-9. Each of these radiates the fields $(\mathbf{E}_1, \mathbf{H}_1)$ and $(\mathbf{E}_2, \mathbf{H}_2)$, respectively, and their equivalent circuit parameters can be characterized by the usual system

$$\begin{bmatrix} V_1 \\ V_2 \end{bmatrix} = \begin{bmatrix} Z_{11} & Z_{12} \\ Z_{21} & Z_{22} \end{bmatrix} \begin{bmatrix} I_1 \\ I_2 \end{bmatrix} \tag{1.128}$$

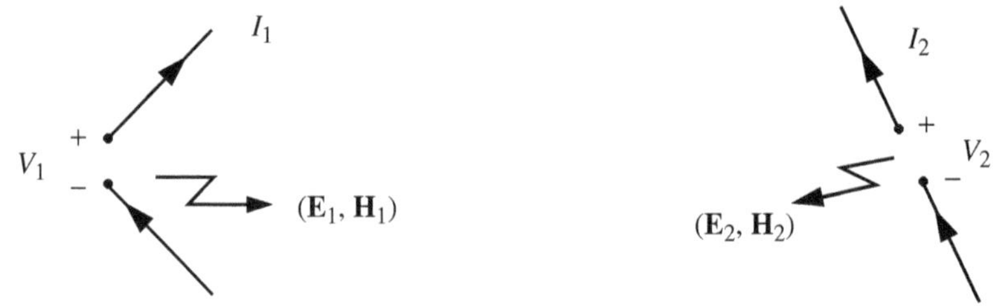


**FIGURE 1-9** Reaction between two antennas.

which is identical to that for a two-port network in circuit theory. Reciprocity and the reaction theorem will now prove useful in determining the elements $Z_{ij}$ of the impedance matrix. These elements can be easily determined by shorting or open-circuiting the antennas one at a time. Setting $I_2 = 0$, gives

$$Z_{21} = \frac{V_2^{(1)}}{I_1}$$

and by referring to (1.126) we may express $Z_{21}$ as

$$Z_{21} = -\frac{\langle 1, 2 \rangle}{I_1 I_2}. \tag{1.129}$$

By invoking the reciprocity theorem (1.123), we also have $Z_{12} = Z_{21}$ and in general

$$Z_{ij} = -\frac{\langle j, i \rangle}{I_i I_j}. \tag{1.130}$$

This expression is valid for computing the self-impedance elements $Z_{ii}$ as well and is useful in numerical simulations of antenna and scattering problems.

## 1.12 | APPROXIMATE BOUNDARY CONDITIONS

In Section 1.4, we discussed the boundary conditions that must be imposed on material interfaces. These are the usual natural or exact boundary conditions. However, in many cases, it is possible to employ approximate boundary conditions that effectively account for the presence of some inhomogeneous interface, a material coating on a conductor, or a dielectric layer without actually having to include their geometry explicitly in the analysis.

### 1.12.1 Impedance Boundary Conditions

The most common approximate boundary condition (ABC) is the impedance boundary condition attributed to Leontovich (1948), which often carries his name in the literature. It can be derived by considering the simple problem of a plane wave incidence on a material half space. Choosing the interface to be the plane $y = 0$ with the $y$ axis directed out of the half space, the Leontovich impedance boundary condition takes the form

$$E_z = -\eta Z_o H_x, \qquad E_x = \eta Z_o H_z \tag{1.131}$$

where $Z_o = \sqrt{\frac{\mu_o}{\epsilon_o}}$, and $\eta$ is a function of the material properties of the half space. These conditions are applied at $y = 0^+$ (just above the interface) and can be combined to yield the vector form

$$\hat{n} \times (\hat{n} \times \mathbf{E}) = -\eta Z_o \hat{n} \times \mathbf{H} \tag{1.132}$$

where $\hat{n}$ is the unit vector normal in the outward direction (see Figure 1-10). As can be seen, the form of the impedance boundary condition is independent of the geometry of the interface or the boundary where it is enforced and is thus applicable to planar as well as curved surfaces. Further, it can be generalized to the case of anisotropic material surfaces by writing it as

$$\hat{n} \times (\hat{n} \times \mathbf{E}) = -Z_o \overline{\overline{\eta}} \cdot \hat{n} \times \mathbf{H} \tag{1.133}$$

where $\overline{\overline{\eta}}$ is a tensor (a $2 \times 2$ matrix).

One way to derive the appropriate normalized impedance parameter $\eta$ is to demand that the equivalent impedance surface satisfying the condition (1.132) reproduces the same reflected field. In doing so, for the planar dielectric interface we readily find that

$$\eta = \sqrt{\frac{\mu_r}{\epsilon_r}} \tag{1.134}$$

and for this choice of $\eta$ the condition (1.132) becomes an approximation for simulating curved dielectric boundaries (see Figure 1-10b) provided (Senior, 1960)

$$\left|Im(\sqrt{\epsilon_r \mu_r})\right| k_o \rho_i \gg 1 \tag{1.135}$$

where $\rho_i$ are the principal radii of curvature associated with the surface. This ensures that the material is sufficiently lossy so that the fields penetrating the surface do not reemerge at some other point.

For the coated conductor in Figure 1-10c, the value of $\eta$ is generally chosen to be the actual impedance of the corresponding planar structure illuminated by a plane wave, typically at normal incidence. Accordingly, for a homogeneous coating of thickness $\tau$ (Harrington and Mautz, 1975)

$$\eta = j\sqrt{\frac{\mu_r}{\epsilon_r}} \tan(k_o \sqrt{\epsilon_r \mu_r} \tau) \tag{1.136}$$

and we can readily compute the corresponding impedance for multilayer coatings. However, as can be expected, the accuracy of the proposed impedance boundary condition deteriorates for oblique angles of incidences, requiring that $\tau$ be kept small with respect to the wavelength to achieve resonable accuracies.

Provided the material parameters change slowly from one point of the simulated surface to another, the impedance boundary condition (1.132) is still applicable. In this case, the normalized surface impedance for the coating is computed from (1.136) with the material parameters now being functions of the location on the surface. For

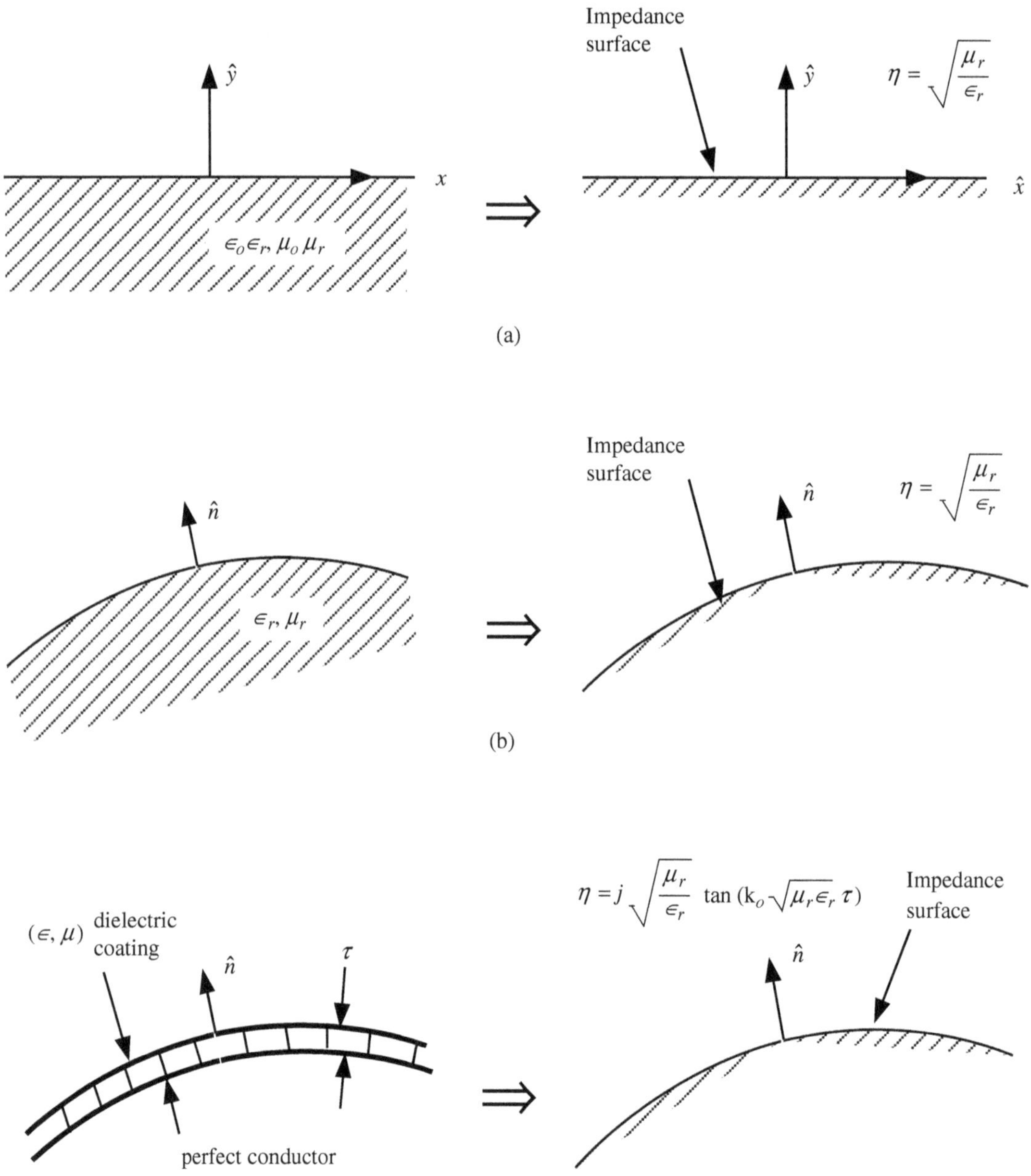

(c)

**FIGURE 1-10** ■ Simulation of dielectric boundaries and coatings with equivalent impenetrable impedance surfaces. (Volakis, Chatterjee, and Kempel, 1998)

a planar interface, if $\epsilon_r$ and $\mu_r$ vary with respect to $y$, Rytov (1940) has shown that

$$\eta = \sqrt{\frac{\mu_r}{\epsilon_r}}\left\{1 + \frac{1}{2jk_oN}\frac{\partial}{\partial y}\ln(Z_oN) + O(N^{-2})\right\} \tag{1.137}$$

where $N = \sqrt{\mu_r\epsilon_r}$ is the refractive index, and the derivative is evaluated at the surface.

## 1.12.2 Resistive and Conductive Sheet Transition Conditions

For certain applications, it is desirable to replace a thin dielectric layer with an equivalent model in an effort to simplify the analysis. To illustrate this idea, let us consider a thin dielectric slab of thickness $\tau$ as shown in Figure 1-11. The slab has a conductivity $\sigma$, and it will thus support a current density given by (see (1.36))

$$\mathbf{J} = \sigma \mathbf{E} \tag{1.138}$$

where $\mathbf{E}$ denotes the field within the slab. However, since $\tau \ll \lambda$, we may replace $\mathbf{J}$ by an equivalent sheet current (having units in $A/m$)

$$\mathbf{J}_s = \tau \mathbf{J} \tag{1.139}$$

and thus from (1.138)

$$\mathbf{E} = \mathbf{J}_s/\sigma\tau = Z_o R_e \mathbf{J}_s. \tag{1.140}$$

This condition is a mathematical definition for a *resistive* sheet supporting a sheet current $\mathbf{J}_s$. The parameter $Z_o R_e$ is referred to as the *resistivity* of the sheet and is measured in $\Omega$/square (e.g., $\Omega/\text{cm}^2$ or $\Omega/\text{m}^2$) (Senior and Volakis, 1995).

In deriving (1.140) it has been assumed that $\mathbf{E}$ is tangential to the layer or sheet, and therefore a more precise definition of the condition is

$$\hat{n} \times (\hat{n} \times \mathbf{E}) = -Z_o R_e \mathbf{J}_s \tag{1.141}$$

where $\hat{n}$ denotes the upper unit normal to the sheet. Further, it is desirable to work with field quantities that are measured outside the layer or sheet, and since $\hat{n} \times \mathbf{E}$ is continuous across the layer we may rewrite (1.141) as

$$\hat{n} \times [\hat{n} \times (\mathbf{E}^+ + \mathbf{E}^-)] = -2Z_o R_e \mathbf{J}_s \tag{1.142a}$$

$$\hat{n} \times (\mathbf{E}^+ - \mathbf{E}^-) = 0 \tag{1.142b}$$

The superscripts $\pm$ denote the fields above and below the sheet or layer, and it was necessary to introduce (1.142b) to maintain the equivalence of (1.142) with (1.141). Alternatively, by employing the natural boundary condition (1.60), (1.142) can be rewritten as

$$\hat{n} \times [\hat{n} \times (\mathbf{E}^+ + \mathbf{E}^-)] = -2Z_o R_e\, \hat{n} \times (\mathbf{H}^+ - \mathbf{H}^-) \tag{1.143a}$$

$$\hat{n} \times (\mathbf{E}^+ - \mathbf{E}^-) = 0 \tag{1.143b}$$

By allowing $\hat{n}$ to be other than constant, these can be employed for the simulation of curved layers, provided again there is sufficient loss in the layer to suppress field repenetration from one layer location to another.

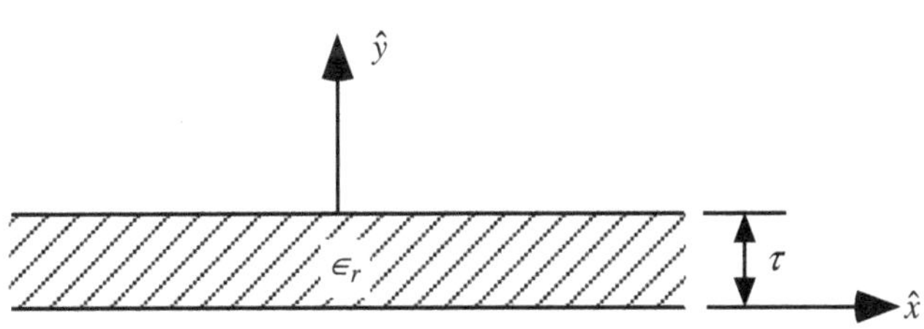


**FIGURE 1-11** A thin sheet of dielectric material.

The dual to (1.143) are

$$\begin{aligned} \hat{n} \times [\hat{n} \times (\mathbf{H}^{+} + \mathbf{H}^{-})] &= 2Y_o R_m \, \hat{n} \times (\mathbf{E}^{+} - \mathbf{E}^{-}) \\ \hat{n} \times (\mathbf{H}^{+} - \mathbf{H}^{-}) &= 0 \end{aligned} \tag{1.144}$$

and these define a *conductive* sheet capable of supporting a magnetic current $\mathbf{M}_s = -\hat{n} \times (\mathbf{E}^{+} - \mathbf{E}^{-})$. The parameter, $Y_o R_m$, is now referred to as the *conductivity* (Senior and Volakis, 1995) of the magnetic sheet measured in Siemens/square. The utility of this sheet is not yet apparent but it will be shown to be essential for a sheet simulation of dielectric layers with nontrivial permeability. Also, it has been shown (Senior, 1985) that a special combination of coincident electric and magnetic current sheets is equivalent to an impenetrable impedance sheet. This equivalence holds when we set

$$R_e = \frac{\eta}{2}, \qquad R_m = \frac{1}{2\eta} \tag{1.145}$$

implying $4R_e R_m = 1$, where $\eta$ is the normalized impedance of the sheet. Because co-planar electric and magnetic currents are independent of each other, (1.145) is important in simplifying the analysis with flat impedance surfaces.

Let us now consider a dielectric layer having a relative permittivity $\epsilon_r$ and thickness $\tau$ such that $k_o\tau \ll 1$ as illustrated in Figure 1-11. Based on the volume equivalence theorem, this layer can be replaced by the equivalent polarization currents

$$\begin{aligned} J_x &= jk_o Y_o(\epsilon_r - 1)E_x \\ J_y &= jk_o Y_o(\epsilon_r - 1)E_y \\ J_z &= jk_o Y_o(\epsilon_r - 1)E_z. \end{aligned} \tag{1.146}$$

On the assumption of $k_o\tau \ll 1$, the $J_y$ component may be neglected, and the current densities $J_{x,z}$ can then be replaced by the equivalent sheet currents

$$J_{sx} = \tau J_x, \qquad J_{sz} = \tau J_z. \tag{1.147}$$

From (1.140), it now follows that

$$E_x = Z_o R_e J_{sx}, \qquad E_z = Z_o R_e J_{sz} \tag{1.148}$$

with

$$R_e = \frac{-j}{k_o\tau(\epsilon_r - 1)}. \tag{1.149}$$

Equation (1.148) are clearly identical to (1.140) except that $R_e$ is now complex. Coordinate independent transition conditions for the dielectric layer are thus given by (1.141), (1.142), or (1.143) with the new definition for $R_e$.

When the dielectric slab is associated with nonunity $\mu_r$, (1.143) must be complemented with a conductive sheet defined by (1.144) and in accordance with the volume equivalence theorem the normalized conductivity $R_m$ is given by (Senior and Volakis, 1995)

$$R_m = \frac{-j}{k_o\tau(\mu_r - 1)}. \tag{1.150}$$

Thus, possible sheet transition conditions for a thin ferrite layer are

$$\begin{aligned} \hat{n} \times [\hat{n} \times (\mathbf{E}^+ + \mathbf{E}^-)] &= -2Z_o R_e \hat{n} \times (\mathbf{H}^+ - \mathbf{H}^-) \\ \hat{n} \times [\hat{n} \times (\mathbf{H}^+ + \mathbf{H}^-)] &= +2Y_o R_m \hat{n} \times (\mathbf{E}^+ - \mathbf{E}^-) \end{aligned} \tag{1.151}$$

where $R_e$ and $R_m$ are defined in (1.149) and (1.150), and $\hat{n}$ denotes the upward unit normal to the layer. More accurate, higher-order transition and impedance conditions can also be derived. These involve higher-order derivatives of the field above and below the sheet or impedance surface and can permit modeling of thicker or higher contrast material layers and coatings (see, e.g., Senior and Volakis, 1987, 1991).

## PROBLEMS

**1.** A chiral medium has the constitutive relations

$$\mathbf{D} = \epsilon \mathbf{E} - j\chi \mathbf{B}, \quad \mathbf{H} = -j\chi \mathbf{E} + \frac{\mathbf{B}}{\mu}$$

where $\chi$ is the chirality parameter.

**(a)** Show that the vector wave equation for this medium takes the form (see (1.109) and (1.110))

$$\nabla \times \nabla \times \mathbf{F} - k^2 \mathbf{F} + \mathbf{V} = 0$$

where $\mathbf{F} = \mathbf{E}, \mathbf{H}, \mathbf{D}$, or $\mathbf{B}$, and $\mathbf{V}$ is a vector to be found from your solution.

**(b)** Assume now a circularly polarized plane wave (RCP or LCP) propagating along the $z$-direction. Find the propagation constants $k_{\text{RCP}}$ and $k_{\text{LCP}}$ so that the wave equation found in (a) is satisfied.

**(c)** If a linearly polarized plane wave is incident upon a chiral interface as shown in Figure 1.P1, find the reflected and transmitted fields (by enforcing tangential field continuity at the interface).

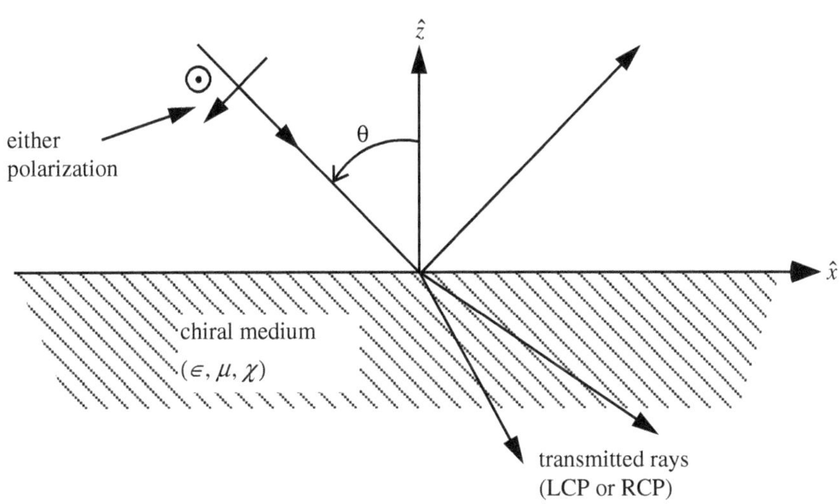


**FIGURE 1.P1** Plane wave incident on a chiral interface.

**2.** For the geometry illustrated in Figure 1.P2, use the divergence theorem

$$\iiint_V \nabla \cdot \mathbf{A}\, dv = \oiint_{S_c} \mathbf{A} \cdot d\mathbf{s}$$

**(a)** to prove that

$$\iiint_V \nabla \times \mathbf{B}\, dv = \oiint_{S_c} (\hat{n} \times \mathbf{B}) \cdot d\mathbf{s} \qquad \text{(vector Stokes theorem)}$$

where $d\mathbf{s} = \hat{n}\, ds$. Hint: Set $\mathbf{A} = \hat{u} \times \mathbf{B}$, and make use of the identity $\nabla \cdot (\mathbf{a} \times \mathbf{b}) = -\mathbf{a} \cdot \nabla \times \mathbf{b}$, where $\hat{u}$ is an arbitrary unit vector.
and

**(b)** that

$$\iiint_V \nabla f\, dv = \oiint_{S_c} f \hat{n}\, ds$$

Hint: Set $\mathbf{A} = \hat{u} f$, and note that $\nabla \cdot (\hat{u} f) = \hat{u} \cdot \nabla f$.

**FIGURE 1.P2** Geometry for Problem 2.

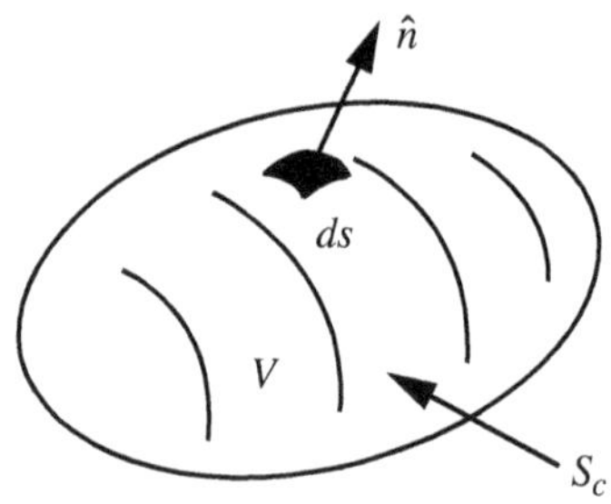


**3.** Consider a surface whose unit normal is $\hat{n} = \hat{u}_3$ and $\hat{u}_{1,2}$ denote the principal directions of that surface, as illustrated in Figure 1.P3. The directions $\hat{u}_{1,2,3}$ form an orthonormal set and are associated with the metric coefficients $h_1, h_2, h_3$. We are interested in obtaining the surface gradient, divergence, and curl, and these are defined by taking the pertinent vector or scalar function to be independent of $v_3$ (i.e., invarient along $\hat{u}_3 = \hat{n}$) and setting $h_3 = 1$.

**FIGURE 1.P3** Parameters and geometry for Problem 3.

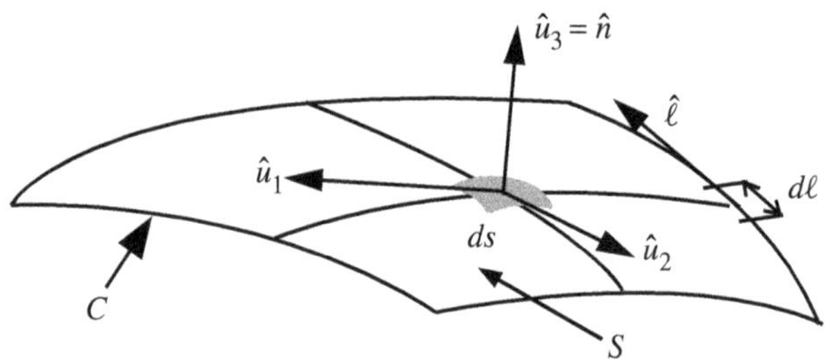


For example, in the case of the gradient we have

$$\nabla_s f = \frac{1}{h_1}\frac{\partial f}{\partial v_1}\hat{u}_1 + \frac{1}{h_2}\frac{\partial f}{\partial v_2}\hat{u}_2 + \frac{1}{h_3}\frac{\partial f}{\partial v_3}\hat{u}_3 = \frac{1}{h_1}\frac{\partial f}{\partial v_1}\hat{u}_1 + \frac{1}{h_2}\frac{\partial f}{\partial v_2}\hat{u}_2$$

since $f = f(v_1, v_2)$ on $S$. Similarly, for the divergence we have $(h_3 = 1)$

$$\nabla \cdot \mathbf{A} = \frac{1}{h_1 h_2}\left[\frac{\partial}{\partial v_1}(h_2 A_1) + \frac{\partial}{\partial v_2}(h_1 A_2)\right] + \frac{1}{h_1 h_2}\frac{\partial (h_1 h_2 A_3)}{\partial v_3}.$$

But

$$\frac{1}{h_1 h_2}\frac{\partial (h_1 h_2 A_3)}{\partial v_3} = \frac{1}{h_1 h_2}\left[\frac{\partial (h_1 h_2)}{\partial v_3} A_3 + h_1 h_2 \frac{\partial A_3}{\partial v_3}\right]$$

$$= \left[\frac{1}{h_1}\frac{\partial h_1}{\partial v_3} A_3 + \frac{1}{h_2}\frac{\partial h_2}{\partial v_3} A_3 + \frac{\partial A_3}{\partial v_3}\right]$$

where $A_i = \mathbf{A} \cdot \hat{u}_i$, and if we set $\frac{\partial A_3}{\partial v_3} = 0$ as noted already, we obtain the surface divergence of $\mathbf{A}$ as

$$\nabla_s \cdot \mathbf{A} = \frac{1}{h_1 h_2}\left[\frac{\partial}{\partial v_1}(h_2 A_1) + \frac{\partial}{\partial v_2}(h_1 A_2)\right] - J A_3 = \nabla_s \cdot \mathbf{A}_t - J A_n$$

where $A_3 = \hat{n} \cdot \mathbf{A} = A_n$, $\mathbf{A}_t = \mathbf{A} - \hat{n} A_n$ and

$$J = \left(\frac{1}{R_1} + \frac{1}{R_2}\right) = -\left[\frac{1}{h_1}\frac{\partial h_1}{\partial v_3} + \frac{1}{h_2}\frac{\partial h_2}{\partial v_3}\right]$$

is the curvature of the surface with $R_1$ and $R_2$ being the principal radii of $S$ (max and min radii at each point on $S$). More specifically, $R_1$ is the radius of curvature of the curve on $S$ associated with $\hat{u}_1$, and $R_2$ is the same for the curve whose tangent is $\hat{u}_2$.

**(a)** Using a similar procedure, derive the surface curl $\nabla_s \times \mathbf{A}$ from $\nabla \times \mathbf{A}$.

**(b)** A generalization of Gauss's theorem for the surface divergence is $(\hat{b} = \hat{\ell} \times \hat{n})$

$$\iint_S \nabla_s \cdot \mathbf{A}_t ds = \oint_C \hat{b} \cdot \mathbf{A}_t \, d\ell$$

or

$$\iint_S \nabla_s \cdot \mathbf{A} ds = \oint_C \hat{b} \cdot \mathbf{A} d\ell - \iint_S J(\mathbf{A} \cdot \hat{n})\, ds.$$

Starting with this result, show that

$$\iint_S \nabla_s f \, ds = \oint_C \hat{b} f d\ell - \iint_S J f \hat{n} \, ds$$

and that

$$\iint_S \nabla_s \times \mathbf{A} \, ds = \oint_C \hat{b} \times \mathbf{A} \, d\ell - \iint_S J(\hat{n} \times \mathbf{A})\, ds.$$

(c) If the surface $S$ in (b) coincides with a curved metal plate and $\mathbf{A}$ represents the current on the plate, what can you say about the value of the surface integral

$$\iint_S \nabla_s \cdot \mathbf{A} \, ds = ?$$

**4.** Consider a finite length electric current sheet (as shown in Figure 1.P4) of the form

$$\mathbf{J}(x) = \hat{y} J_o P_{2\Delta}(x - x_o)$$

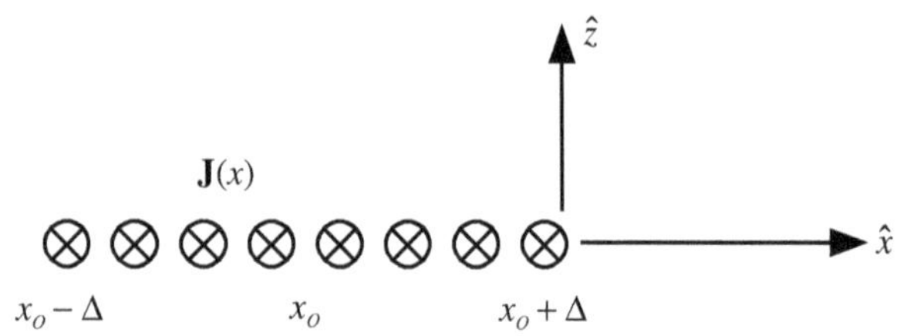


**FIGURE 1.P4** Figure for Problem 4. Finite width current sheet.

where

$$P_{2\Delta}(x) = \begin{cases} 1 & -\Delta < x < \Delta \\ 0 & \text{elsewhere} \end{cases}$$

**(a)** Find an equivalent magnetic current distribution whose radiation will be the same as that of the given electric current sheet.

**(b)** Repeat part (a) for $\mathbf{J}(x) = \hat{y} J_o f(x) P(x - x_o)$.

**5.** Give expressions for the equivalent surface currents on $S_{diel.}$ and $S_{cond.}$ for the structure shown in Figure 1.P5 under the following assumptions:

**(a)** $(\mathbf{E}_i, \mathbf{H}_i) \neq (0, 0)$

**(b)** $\mathbf{E}_i = 0$, but $\mathbf{H}_i \neq 0$

**(c)** $\mathbf{E}_i \neq 0$, but $\mathbf{H}_i = 0$

Consider the exterior and interior regions separately.

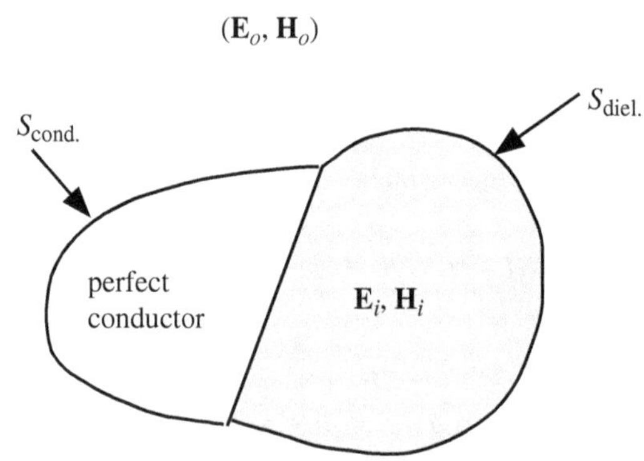


**FIGURE 1.P5** Geometry for Problem 5.

**6.** The electric field generated by an infinitesimal dipole $\mathbf{J}_a = \hat{x} I \ell$ $(\ell \to 0)$ in the presence of a dielectric scatterer is found to be $\mathbf{E}_a(x, y, z)$. Give an expression of the $x$-component of the corresponding field generated by the source $\mathbf{J}_b(x, y, z)$.

**7.** Consider an antenna in the presence of an ideal dipole whose current density is given by $\mathbf{J}_2 = \hat{\ell}_2 I_2 \delta(\mathbf{r} - \mathbf{r}_2)$, where $\delta(\mathbf{r})$ denotes the Dirac delta function (as shown in Figure 1.P7).

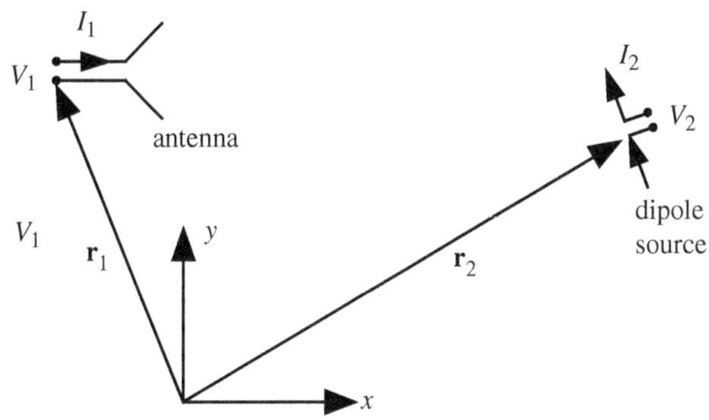


**FIGURE 1.P7** Geometry for Problem 7.

Use the reciprocity theorem to derive the expression

$$V_1^{oc} = -\frac{1}{I_1}\hat{\ell}_2 \cdot \mathbf{E}_1(\mathbf{r}_2)$$

where $\mathbf{E}_1(\mathbf{r}_2)$ is the field radiated by the antenna alone at $\mathbf{r}_2$, and the superscript $oc$ implies that the voltage is measured with $I_2 = 0$. When $|\mathbf{r}_1 - \mathbf{r}_2| \to \infty$, this expression can be simplified to read

$$V_1^{oc} = \mathbf{h}_1 \cdot \mathbf{E}_2(\mathbf{r}_1)$$

where $h_1$ is referred to as the antenna height.

**8.** Assume the plane wave

$$\mathbf{E}^i = \hat{\theta}e^{j\mathbf{k}_i\cdot\mathbf{r}} = \hat{\theta}e^{jk_o(x\sin\theta\cos\phi + y\sin\theta\sin\phi + z\cos\theta)}$$

is incident upon the planar resistive sheet of (normalized) resistivity $R_e$ as shown in Figure 1.P8.

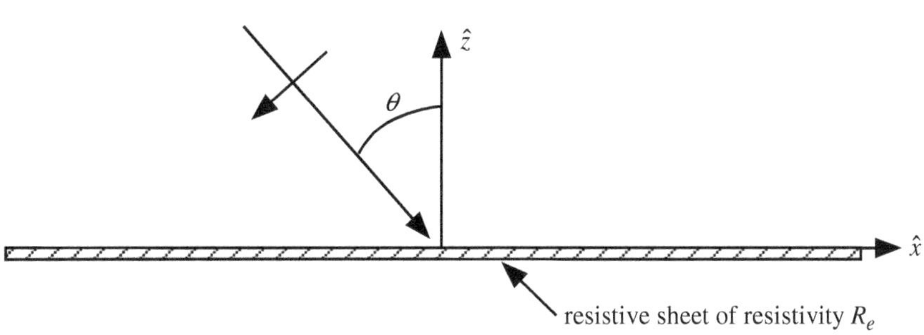


**FIGURE 1.P8** Geometry for Problem 8.

**(a)** Determine the reflected and transmitted fields.

**(b)** Repeat (a) for a conductive sheet of (normalized) conductivity $R_m$.

**(c)** Combine the results of (a) and (b) to find the reflected field from a planar surface having a (normalized) impedance $\eta$.

**9.** Consider the material slab shown in Figure 1.P9 having relative permittivity $\epsilon_r$ and relative permeability $\mu_r$.

FIGURE 1.P9 ■ Geometry for Problem 9.

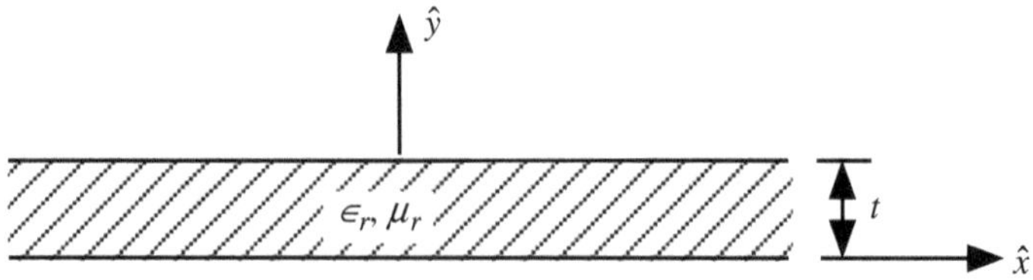


On the assumption that $kt \ll 1$, where $k = \omega\sqrt{\mu_o\epsilon_o}\sqrt{\mu_r\epsilon_r} = k_o\sqrt{\mu_r\epsilon_r}$, introduce the approximations

$$\frac{\partial E_x}{\partial y} \approx \frac{\Delta E_x}{\Delta y} \approx \frac{E_x(t^-) - E_x(0^+)}{t}$$

$$\frac{\partial E_y}{\partial y} = \frac{\Delta E_y}{\Delta y} \approx \frac{E_y(t^-) - E_y(0^+)}{t}$$

$$H_{x,z} \approx [H_{x,z}(t^-) + H_{x,z}(0^+)]/2$$

and

$$E_y \approx [E_y(t^-) + E_y(0^+)]/2$$

where $F(t^-)$ denotes the field component $F$ at $y = t - 0$ (i.e., just inside the layer), and, similarly $F(0^+)$ denotes the field component at $y = 0^+$, just inside the layer.

**(a)** Using the boundary conditions

$$\epsilon_r E_y(t^-) = E_y(t^+)$$

$$\epsilon_r E_y(0^+) = E_y(0^-)$$

along with Maxwell's equations show that

$$E_x(t^+) - E_x(0^-) = \frac{jk_o t\mu_r Z_o}{2}[H_z(t^+) + H_z(0^-)] + \frac{t}{2\epsilon_r}\frac{\partial}{\partial x}[E_y(t^+) + E_y(0^-)]$$

$$E_z(t^+) - E_z(0^-) = -\frac{jk_o t\mu_r Z_o}{2}[H_x(t^+) + H_x(0^-)] + \frac{t}{2\epsilon_r}\frac{\partial}{\partial z}[E_y(t^+) + E_y(0^-)]$$

**(b)** The derived conditions in (a) are in terms of the fields at the top and bottom of the layer. This layer has been equivalently/mathematically replaced by these conditions. To derive sheet conditions similar to those for the resistive and conductive sheet, the fields at $y = t^+$ and $y = 0^-$ can be transferred to $y = t/2$. To do so employ the two-term Taylor series expansion

$$F(t) = F\left(\frac{t^+}{2}\right) + \frac{t}{2}\frac{\partial}{\partial y}F\left(\frac{t^+}{2}\right)$$

$$F\left(0^-\right) = F\left(\frac{t^-}{2}\right) - \frac{t}{2}\frac{\partial}{\partial y}F\left(\frac{t^-}{2}\right)$$

for all field components (since the layer is no longer present) along with Maxwell's equations to derive the sheet conditions

$$E_x\left(\frac{t^+}{2}\right) - E_x\left(\frac{t^-}{2}\right) = Z_o\left(\frac{1}{2R_m} + \frac{1}{2R_e'}\right)\left[H_z\left(\frac{t^+}{2}\right) + H_z\left(\frac{t^-}{2}\right)\right] - \frac{1}{2jk_oR_e'}\frac{\partial}{\partial y}\left[E_x\left(\frac{t^+}{2}\right) + E_x\left(\frac{t^-}{2}\right)\right] + O(t^2)$$

$$E_z\left(\frac{t^+}{2}\right) - E_z\left(\frac{t^-}{2}\right) = -Z_o\left(\frac{1}{2R_m} + \frac{1}{2R_e'}\right)\left[H_x\left(\frac{t^+}{2}\right) + H_x\left(\frac{t^-}{2}\right)\right] - \frac{1}{2jk_oR_e'}\frac{\partial}{\partial y}\left[E_z\left(\frac{t^+}{2}\right) + E_z\left(\frac{t^-}{2}\right)\right] + O(t^2)$$

in which

$$R_m = \frac{-j}{k_o t(\mu_r - 1)} \quad \text{and} \quad R_e' = \frac{-j\epsilon_r}{k_o t(\epsilon_r - 1)}.$$

For a complete modeling of the layer, these conditions must be supplemented by their dual. Note that the aforementioned sheet conditions can be written in a compact vector form as

$$\hat{n} \times \hat{n} \times \left(\frac{1}{2R_m} + \frac{1}{2R_e'}\right)[\mathbf{H}^+ + \mathbf{H}^-] + \frac{j}{2k_oR_e'}\hat{n} \times \frac{\partial}{\partial n}\left(\mathbf{E}^+ + \mathbf{E}^-\right) = \hat{n} \times (\mathbf{E}^+ - \mathbf{E}^-)$$

where $\hat{n}$ is the normal to the dielectric layer (possibly curved).

**(c)** Compare the conditions derived in (b) with those for the standard resistive and conductive sheets. Explain which of the two sets is more accurate and why.

# BIBLIOGRAPHY


Astrov, D. N. (1960). The magnetoelectric effect of antiferromagnetics. *Zh. Eksp. Teor. Fiz.*, *38*, 984–985.

Balanis, C. (1989). *Advanced Engineering Electromagnetics*. New York: Wiley.

Dzyalashinskie, I. E. (1960). On the magnetoelectrical effect of antiferromagnets. *Soviet Phys. JETP*, *10*, 628–669.

Harrington, R. F. (1961). *Time-Harmonic Electromagnetic Fields*. New York: McGraw-Hill.

Harrington, R. F. and Mautz, J. R. (1975). An impedance sheet approximation for thin dielectric shells. *IEEE Trans. Antennas Propagat.*, *20*(10), 531–534.

Jaggard, D. L., Mickelson, A. R., and Pappas, C. H. (1978). On electromagnetic waves in chiral media. *Appl. Phys.*, *18*, 211–216.

Kong, J. A. (1975). *Theory of Electromagnetic Waves*. New York: Wiley InterScience.

Lakhtakia, A., Varadan, V. V., and Varadan, V. K. (1988). Field equations, Huygens' principle, integral equations, and theorems for radiation and scattering of electromagnetic waves in isotropic chiral media. *J. Opt. Soc. Am. -A*, *5*, 175–183.

Leontovich, M. (1948). *"Approximate boundary conditions for the electromagnetic field on the surface of a good conductor"* (in *Investigations on Radio Wave Propagat. Part II*), Moscow: Printing House of the USSR Academy of Sciences. 5–12.

Love, A. E. H. (1903). The integration of the equation of propagation of electric waves. *Proc. London Math. Soc. Ser. 2*, *1*.

Mayes, P. E. (1958). The equivalences of electric and magnetic sources. *IEEE Trans. Antennas Propagat.*, *AP-6*, 295–296.

Post, E. J. (1962). *Formal Structure of Electromagnetics*. Amsterdam: North-Holland.

Rytov, S. M. T. (1940). Computation of the skin effect by the pertubation method. *J. Exp. Theor. Phys.*, *10*, 180.

Senior, T. B. A. (1960). Impedance boundary conditions for imperfectly conducting surfaces. *Appl. Sci. Res.*, *8-B*, 418–436.

Senior, T. B. A. (1985). Combined resistive and conductive sheets. *IEEE Trans. Antennas Propagat.*, *AP-33*(5), 577–579.

Senior, T. B. A. and Volakis, J. L. (1987). Sheet simulation of a thin dielectric layer. *Radio Sci.*, *22*, 1261–1272.

Senior, T. B. A. and Volakis, J. L. (1991). Generalized boundary conditions in scattering. *Proc. IEEE*, *79*(10), 1413–1420.

Senior, T. B. A. and Volakis, J. L. (1995). *Approximate Boundary Conditions in Electromagnetics*. IET Press.

Stratton, J. A. (2007). *Electromagnetic Theory*. New York: Wiley-Interscience, Classic Reissue.

Van Bladel, J. (2007). *Electromagnetic Fields*. Wiley-Interscience, 2nd ed.

Volakis, J. L., Chatterjee, A., and Kempel, L. (1998). *Finite Element Methods for Electromagnetics*. New York: IEEE Press.

CHAPTER 2

# Field Solutions and Representations

**Chapter Outline**



In this chapter we present formal solutions to Maxwell's equations. These can be cast into a variety of integral representations for computing the radiated field by an antenna or the scattered field from a composite structure. Some of these representations are given here, and we will refer to them in the other chapters.

## 2.1 | FIELD SOLUTIONS IN TERMS OF VECTOR AND HERTZ POTENTIALS

In the absence of magnetic currents and charges $\nabla \cdot (\mu \mathbf{H}) = 0$, implying that in a homogeneous medium the magnetic field due to electric sources can be written as

$$\mathbf{H} = \mathbf{H}_e = \nabla \times \mathbf{A} \tag{2.1}$$

where $\mathbf{A}$ is an arbitrary unknown vector and is referred to as the *magnetic vector potential*. Its complete specification would, of course, require all point derivatives of $\mathbf{A}$, and we note that (2.1) involves only a subset of those derivatives. Thus, one could also independently define the $\nabla \cdot \mathbf{A}$ without interfering with the definition (2.1), and this will be exploited later in the solution for $\mathbf{A}$.

Substituting (2.1) in (1.48) yields

$$\nabla \times \nabla \times \mathbf{A} = \mathbf{J}_i + j\omega\epsilon\mathbf{E}_e \tag{2.2}$$

where $\mathbf{E}_e$ denotes the electric field due to the electric source $\mathbf{J}_i$. Also, from (1.49) we have

$$\nabla \times (\mathbf{E}_e + j\omega\mu\mathbf{A}) = 0 \tag{2.3}$$

implying

$$\mathbf{E}_e + j\omega\mu\mathbf{A} = -\nabla\Phi_e \tag{2.4}$$

with $\Phi_e$ being again an arbitrary scalar function commonly referred to as the electric scalar potential. Combining now (2.2) and (2.4) we obtain

$$\nabla \times \nabla \times \mathbf{A} - k^2\mathbf{A} = \mathbf{J}_i - j\omega\epsilon\nabla\Phi_e \tag{2.5}$$

where $k = \omega\sqrt{\mu\epsilon} = 2\pi/\lambda$ is the medium propagation constant associated with the wavelength $\lambda$. This can be further simplified by employing the identity

$$\nabla \times \nabla \times \mathbf{A} = \nabla\nabla \cdot \mathbf{A} - \nabla^2\mathbf{A} \tag{2.6}$$

which when introduced in (2.5) gives

$$\nabla^2\mathbf{A} + k^2\mathbf{A} = -\mathbf{J}_i + j\omega\epsilon\nabla\Phi_e + \nabla\nabla \cdot \mathbf{A} \tag{2.7}$$

and in rectangular coordinates

$$\nabla^2\mathbf{A} = \hat{x}\nabla^2 A_x + \hat{y}\nabla^2 A_y + \hat{z}\nabla^2 A_z. \tag{2.8}$$

As stated already, since the derivatives involved in the definition of $\nabla \times \mathbf{A}$ are different from those associated with $\nabla \cdot \mathbf{A}$, we may arbitrarily set

$$\nabla \cdot \mathbf{A} = -j\omega\epsilon\Phi_e \tag{2.9}$$

which is a relation that is often referred to as the Lorentz gauge condition. Substituting (2.9) into (2.7) we obtain

$$\nabla^2\mathbf{A} + k^2\mathbf{A} = -\mathbf{J}_i \tag{2.10}$$

or

$$(\nabla^2 + k^2)\begin{Bmatrix} A_x \\ A_y \\ A_z \end{Bmatrix} = -\begin{Bmatrix} J_{ix} \\ J_{iy} \\ J_{iz} \end{Bmatrix}$$

in conjunction with the definitions (2.1) and (2.8). Also from (2.4)

$$\mathbf{E}_e = -j\omega\mu\mathbf{A} + \frac{1}{j\omega\epsilon}\nabla\nabla \cdot \mathbf{A} = -jkZ\left(\mathbf{A} + \frac{1}{k^2}\nabla\nabla \cdot \mathbf{A}\right) \tag{2.11}$$

where $Z = \sqrt{\mu/\epsilon}$ denotes the impedance of the medium, and upon substituting this into (1.51) or (1.33), it can be readily shown from (2.9) that $\Phi_e$ satisfies

$$\nabla^2\Phi_e + k^2\Phi_e = -\frac{\rho}{\epsilon}. \tag{2.12}$$

Given the electric potential $\mathbf{A}$, (2.11) provides the solution for the electric field due to the electric sources $\mathbf{J}_i$. The corresponding magnetic field is obtained from (2.1), and to complete the solution of Maxwell's equations we must also find similar expressions in the presence of magnetic sources. If for the moment we assume that only magnetic sources are present, we could then invoke duality to find the corresponding field expressions. In particular, in comparison with (2.1) we now set

$$\mathbf{E}_m = -\nabla \times \mathbf{F} \tag{2.13}$$

where $\mathbf{F}$ is referred to as the *electric vector potential*, and $\mathbf{E}_m$ is the field due to $\mathbf{M}_i$. Following, a procedure similar to that in connection with the electric sources or simply by invoking duality we obtain

$$\mathbf{H}_m = -j\omega\epsilon\mathbf{F} + \frac{1}{j\omega\mu}\nabla\nabla\cdot\mathbf{F} = -jkY\left(\mathbf{F} + \frac{1}{k^2}\nabla\nabla\cdot\mathbf{F}\right) \tag{2.14}$$

where $Y = 1/Z = \sqrt{\epsilon/\mu}$ denotes the medium admittance, and $\mathbf{F}$ satisfies the partial differential equation

$$\nabla^2\mathbf{F} + k^2\mathbf{F} = -\mathbf{M}_i$$

or

$$(\nabla^2 + k^2)\begin{Bmatrix} F_x \\ F_y \\ F_z \end{Bmatrix} = -\begin{Bmatrix} M_{ix} \\ M_{iy} \\ M_{iz} \end{Bmatrix} \tag{2.15}$$

In the process of deriving (2.15) we introduced the magnetic scalar potential $\Phi_m$ such that

$$\nabla\cdot\mathbf{F} = -j\omega\mu\Phi_m. \tag{2.16}$$

By employing this into (2.13) and making use of (1.32) we find that $\Phi_m$ satisfies the differential equation

$$\nabla^2\Phi_m + k^2\Phi_m = -\frac{\rho_m}{\mu} \tag{2.17}$$

which is the dual of (2.12).

The fields due to the presence of both electric and magnetic sources $(\mathbf{J}_i, \mathbf{M}_i)$ can now be found by invoking the superposition principle. That is

$$\mathbf{E} = \mathbf{E}_e + \mathbf{E}_m, \qquad \mathbf{H} = \mathbf{H}_e + \mathbf{H}_m \tag{2.18}$$

and from (2.1), (2.11), (2.13), and (2.14) we obtain

$$\mathbf{E} = -\nabla\times\mathbf{F} - j\omega\mu\mathbf{A} + \frac{1}{j\omega\epsilon}\nabla\nabla\cdot\mathbf{A} \tag{2.19a}$$

$$\mathbf{H} = \nabla\times\mathbf{A} - j\omega\epsilon\mathbf{F} + \frac{1}{j\omega\mu}\nabla\nabla\cdot\mathbf{F}. \tag{2.19b}$$

The vectors

$$\mathbf{\Pi}_e = \frac{1}{j\omega\epsilon}\mathbf{A}, \quad \mathbf{\Pi}_m = \frac{1}{j\omega\mu}\mathbf{F} \tag{2.20}$$

and usually referred to as the *Hertz potentials* and in terms of these

$$\begin{aligned}\mathbf{E} &= -jkZ\nabla\times\mathbf{\Pi}_m + k^2\mathbf{\Pi}_e + \nabla\nabla\cdot\mathbf{\Pi}_e \\ &= \nabla\times\nabla\times\mathbf{\Pi}_e - jkZ\nabla\times\mathbf{\Pi}_m \end{aligned} \tag{2.21a}$$

$$\mathbf{H} = jkY\nabla\times\mathbf{\Pi}_e + k^2\mathbf{\Pi}_m + \nabla\nabla\cdot\mathbf{\Pi}_m \tag{2.21b}$$

## 2.2 | SOLUTION FOR THE VECTOR AND SCALAR POTENTIALS

Let us first solve (2.10) for the case where $\mathbf{J}_i$ is a small infinitesimal source occupying the volume $dv'$ located at the origin (as illustrated in Figure 2-1). Given that $\mathbf{J}_i = \hat{a} J_a$, (2.10) becomes

$$\nabla^2 A_a + k^2 A_a = -J_a \tag{2.22}$$

where $a$ denotes one of the rectangular coordinates $x$, $y$, or $z$ and away from the origin $A_a$ satisfies

$$\nabla^2 A_a + k^2 A_a = 0. \tag{2.23}$$

Further since $J_a$ is infinitesimal, $A_a$ is expected to be independent of the spherical angles $\phi$ and $\theta$, implying that (2.23) can be replaced by

$$\frac{1}{r^2}\frac{\partial}{\partial r}\left(r^2 \frac{\partial A_a}{\partial r}\right) + k^2 A_a = 0 \tag{2.24}$$

upon substituting for the spherical form of the Laplace operator, $\nabla^2$. Setting

$$A_a = \frac{f(r)}{r} \tag{2.25}$$

(2.24) becomes

$$\frac{\partial^2 f}{\partial r^2} + k^2 f = 0; \quad r \neq 0 \tag{2.26}$$

and by inspection a solution for $f(r)$ is

$$f(r) = C_1 e^{-jkr}. \tag{2.27}$$

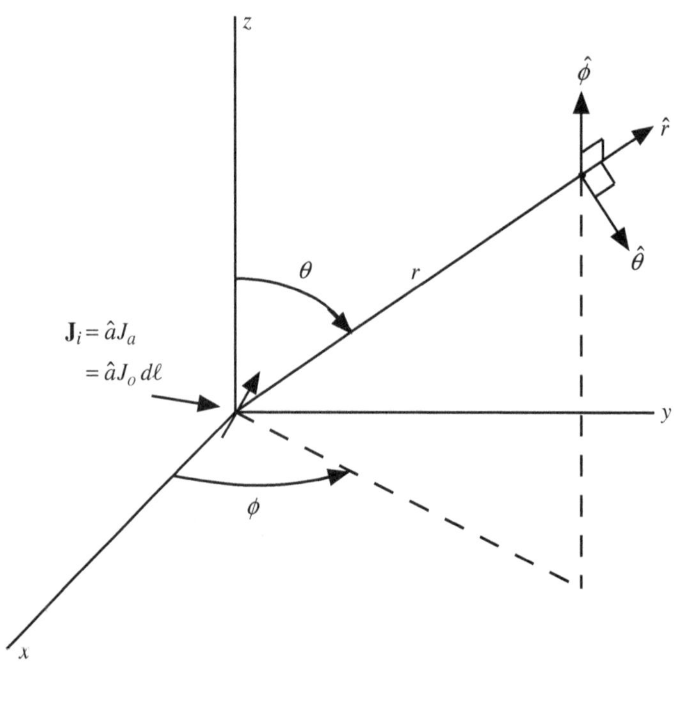


**FIGURE 2-1** Illustration of an infinitesimal source (dipole) at the origin.

Upon restoring the suppressed time dependence $e^{j\omega t}$ it follows that

$$f(r,t) = Re\left\{C_1 e^{-jkr} e^{j\omega t}\right\}$$
$$= C_1 \cos(\omega t - kr) = C_1 \cos\left[\omega\left(t - \frac{r}{c}\right)\right] \tag{2.28}$$

describing an outward propagating wave where $r/c$ denotes the time required for the wave to travel a distance $r$ and consequently $c$ is the speed of the wave. A second solution of (2.25) is associated with the inward propagating wave and is given by

$$f(r) = C_2 e^{+jkr} \tag{2.29}$$

implying that the complete solution for (2.24) is

$$A_a(r) = C_1 \frac{e^{-jkr}}{r} + C_2 \frac{e^{+jkr}}{r}. \tag{2.30}$$

For the particular case considered here, the source is expected to generate outward propagating waves, and we will thus set $C_2 = 0$. To find the value of the constant $C_1$ we must return to (2.22). One approach is to note that for $k = 0$ the resulting field obtained from (2.11) should reduce to the known static solution implying that

$$C_1 = \frac{J_a\, dv'}{4\pi} \tag{2.31}$$

with $J_a\, dv'$ being the strength of the equivalent point source. Alternatively, we could integrate both sides of (2.22) over a small spherical volume of radius $r_o \to 0$, which encloses the source $J_a$. In doing so we obtain

$$\iiint_{V_o} \nabla^2 A_a\, dv + k^2 \iiint_{V_o} A_a\, dv = -J_a\, dv' \tag{2.32}$$

Setting $dv = r^2 \sin\theta\, d\theta\, d\phi\, dr$ and substituting for $A_a = C_1 \frac{e^{-jkr}}{r}$, the second integral vanishes since $r_o \to 0$. The first integral can be rewritten as

$$\iiint_V \nabla^2 A_a\, dv = \int_o^{r_o} \int_0^{2\pi} \int_0^{\pi} \nabla \cdot (\nabla A_a)\, dv$$
$$= \int_o^{2\pi} \int_0^{\pi} \nabla A_a \cdot \hat{r}_o r_o^2 \sin\theta\, d\theta\, d\phi.$$

Evaluating the gradient of $A_a$ gives

$$\nabla A_a \cdot \hat{r} = \frac{\partial A(r_o)}{\partial r_o} = -(1 + jkr_o) C_1 \frac{e^{-jkr_o}}{r_o^2}$$

and when this is introduced into the previous integral we obtain

$$\iiint_{V_o} \nabla^2 A_a\, dv = -\int_o^{2\pi} \int_o^{\pi} C_1 e^{-jkr_o} \sin\theta\, d\theta\, d\phi + O\left(r_o^2\right)$$
$$= -4\pi C_1$$

for $r_o \to 0$. Substituting this result into (2.32) yields (2.31).

Based on the previous, the electric vector potential is given by

$$\mathbf{A} = \mathbf{J}_i \frac{e^{-jkr}}{4\pi r} dv' \tag{2.33}$$

where $\mathbf{J}_i$ is the electric current density of the source occupying the infinitesimal volume $dv'$ located at the origin. If the infinitesimal volume is moved to $\mathbf{r}'$, then, via a coordinate transformation, the corresponding vector potential is given by

$$\mathbf{A} = \mathbf{J}_i \frac{e^{-jk|\mathbf{r}-\mathbf{r}'|}}{4\pi |\mathbf{r}-\mathbf{r}'|} dv' = \mathbf{J}_i \frac{e^{-jkR}}{4\pi R} dv' \tag{2.34}$$

where $R = |\mathbf{r} - \mathbf{r}'|$ is the distance from the source to the observer (see Figure 2-2). Finally, if the source occupies an arbitrary volume $V$, then through the superposition principle the vector potential can be written as an integral over the source (see Figure 2-3), i.e.

$$\mathbf{A} = \iiint_V \mathbf{J}_i(\mathbf{r}') \frac{e^{-jk|\mathbf{r}-\mathbf{r}'|}}{4\pi |\mathbf{r}-\mathbf{r}'|} dv' = \iiint_V \mathbf{J}_i(\mathbf{r}') \frac{e^{-jkR}}{4\pi R} dv'. \tag{2.35a}$$

In the case of a surface current density $\mathbf{J}_s$, the electric vector potential is given by

$$\mathbf{A} = \iint_S \mathbf{J}_s(\mathbf{r}') \frac{e^{-jkR}}{4\pi R} ds' \tag{2.35b}$$

and similarly for a linear source $I(\ell)$ we have

$$\mathbf{A} = \int_C \hat{\ell}' \, I(\ell') \frac{e^{-jkR}}{4\pi R} d\ell' \tag{2.35c}$$

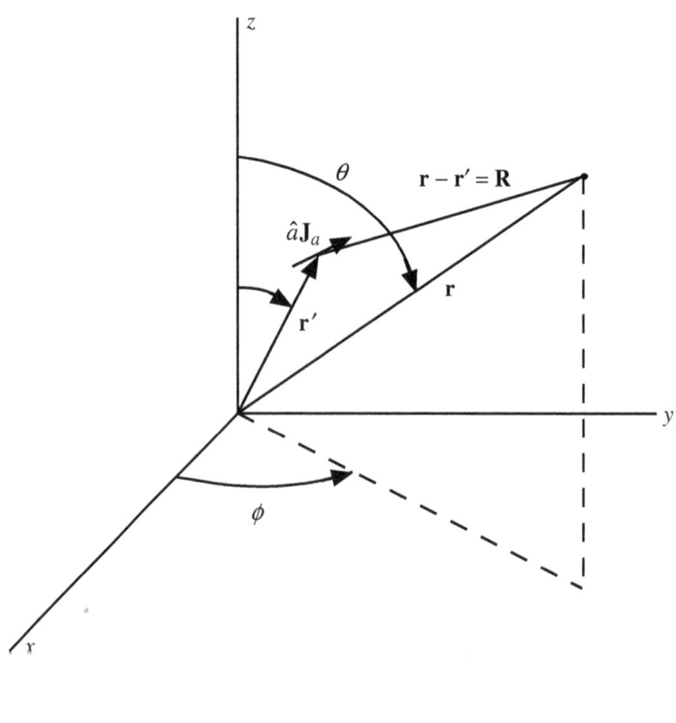


**FIGURE 2-2** Illustration of an infinitesimal source (dipole) away from the origin at $\mathbf{r}'$. (Volakis, Chatterjee, and Kempel, 1998)

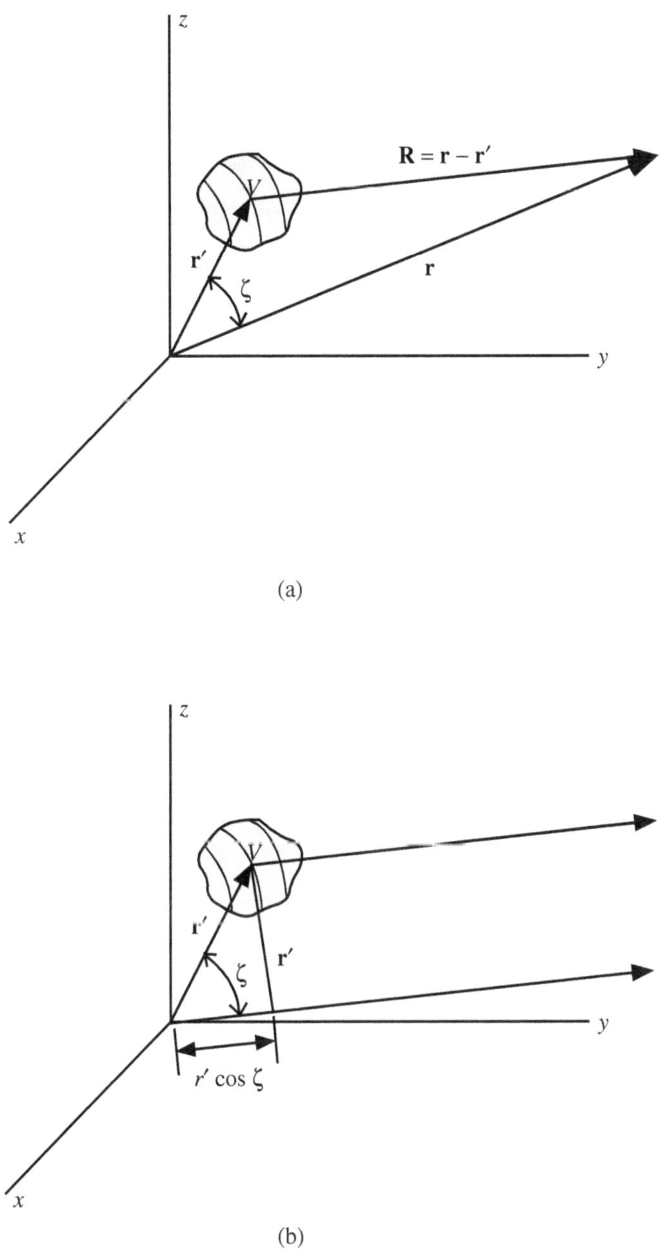


**FIGURE 2-3** Illustration of the geometrical parameters associated with field representations.

in which the contour $C$ coincides with the wire supporting the line current $I(\ell)$, and $\hat{\ell}$ is the unit vector tangent to the wire. Expression (2.35b) can be reduced directly from (2.35a) by setting $\mathbf{J}_i = \mathbf{J}_s\delta(\xi - \xi')$, where $\delta(\xi)$ denotes the Dirac delta function, and (2.35c) can be obtained in a similar manner. The electric field due to a volume, a surface, or a linear electric source is now readily obtained from (2.11) upon substituting for $\mathbf{A}$ as given in (2.35). In case of a combination of volume, surface, or linear electric sources, the magnetic vector potential is given by the sum of the corresponding integrals.

To solve for the magnetic vector potential we can follow a procedure parallel to that employed for the solution of **A**. For a volume magnetic source we obtain

$$\mathbf{F} = \iiint_V \mathbf{M}_i(\mathbf{r}') \frac{e^{-jkR}}{4\pi R}\, dv'. \tag{2.36a}$$

Similarly, the appropriate solutions for surface and linear magnetic sources are given by

$$\mathbf{F} = \iint_S \mathbf{M}_s(\mathbf{r}') \frac{e^{-jkR}}{4\pi R}\, ds' \tag{2.36b}$$

and

$$\mathbf{F} = \int_C \hat{\ell}' I_m(\ell') \frac{e^{-jkR}}{4\pi R}\, dl' \tag{2.36c}$$

in which $I_m(\ell)$ denotes the linear magnetic current source.

We remark that the common kernel

$$G(\mathbf{r}, \mathbf{r}') = \frac{e^{-jk|\mathbf{r}-\mathbf{r}'|}}{4\pi |\mathbf{r}-\mathbf{r}'|} = G(R) = \frac{e^{-jkR}}{4\pi R} \tag{2.37}$$

appearing in the integrals (2.35) and (2.36) is referred to as the *Green's function*. It satisfies the differential equation

$$\nabla^2 G(\mathbf{r}, \mathbf{r}') + k^2 G(\mathbf{r}, \mathbf{r}') = -\delta(\mathbf{r}-\mathbf{r}') \tag{2.38}$$

and the radiation boundary condition

$$\lim_{r\to\infty} r \left( jkG + \frac{\partial G}{\partial r} \right) = 0. \tag{2.39}$$

In (2.38) $\delta(\mathbf{r}-\mathbf{r}')$ denotes the Dirac delta function satisfying the identity

$$\iiint_V f(\mathbf{r}')\delta(\mathbf{r}-\mathbf{r}')\, dv' = \begin{cases} f(\mathbf{r}) & \mathbf{r} \text{ in } V \\ 0 & \mathbf{r} \text{ not in } V. \end{cases} \tag{2.40}$$

When the ambient medium is free space, then $k = k_o = 2\pi/\lambda_o$, and $G(\mathbf{r}, \mathbf{r}')$ is generally referred to as the free-space Green's function; otherwise it is simply the Green's function of the unbounded homogeneous medium.

Were we to begin with the solution of (2.38), it is a straightforward task to derive the solution for the vector potentials as given in (2.35) and (2.36). To do so for **A**, we first multiply (2.22) by $G(\mathbf{r}, \mathbf{r}')$ and (2.38) by $A_a(r)$, and upon subtracting the resulting expressions we obtain

$$A_a(\mathbf{r})\nabla^2 G(\mathbf{r}, \mathbf{r}') - G(\mathbf{r}, \mathbf{r}')\nabla^2 A_a(\mathbf{r}) = -A_a(\mathbf{r})\delta(\mathbf{r}-\mathbf{r}') + G(\mathbf{r}, \mathbf{r}')J_a(\mathbf{r}). \tag{2.41}$$

Integrating the left side of this equation over the volume $V$ enclosing the source and invoking Green's second identity it follows that

$$\iiint_V \left[A_a(\mathbf{r})\nabla^2 G(\mathbf{r}, \mathbf{r}') - G(\mathbf{r}, \mathbf{r}')\nabla^2 A_a(\mathbf{r})\right] dv$$
$$= \oiint_{S_c} \left[ A_a(\mathbf{r}) \frac{\partial G(\mathbf{r}, \mathbf{r}')}{\partial n} - G(\mathbf{r}, \mathbf{r}') \frac{\partial A_a(\mathbf{r})}{\partial n} \right] ds \tag{2.42}$$

where $\frac{\partial}{\partial n}$ denotes differentiation in the direction of the outward surface normal. If $S_c$ (which encloses $V$) is now allowed to be a sphere of infinite radius, $G(\mathbf{r}, \mathbf{r}')$ satisfies the radiation condition (2.39) at each point $S_c$ and if we assume that $A_a(r)$ satisfies the same condition as well, we observe that the right-hand side of (2.42) vanishes. Thus, from (2.41) we have

$$\begin{aligned} 0 &= \iiint_V \left[A_a(\mathbf{r})\nabla^2 G(\mathbf{r}, \mathbf{r}') - G(\mathbf{r}, \mathbf{r}')\nabla^2 A_a(\mathbf{r})\right] dv' \\ &= -\iiint_V A_a(r)\delta(\mathbf{r} - \mathbf{r}')\, dv + \iiint_V G(\mathbf{r}, \mathbf{r}') J_a(r)\, dv \end{aligned} \tag{2.43}$$

and upon applying (2.40) we obtain that

$$A_a(\mathbf{r}') = \iiint_V J_a(\mathbf{r}) G(\mathbf{r}, \mathbf{r}')\, dv. \tag{2.44}$$

Since $G(\mathbf{r}, \mathbf{r}') = G(\mathbf{r}', \mathbf{r})$, which is a consequence of reciprocity, we may rewrite (2.44) as

$$A_a(\mathbf{r}) = \iiint_V J_a(\mathbf{r}') G(\mathbf{r}, \mathbf{r}')\, dv' \tag{2.45}$$

which can be readily generalized to (2.35).

Before closing this section we should note that the solution for the electric scalar potential, $\Phi_e$, can be readily derived by solving (2.12) through the same procedure followed for the solution of (2.22). This gives

$$\Phi_e(\mathbf{r}) = \frac{1}{\epsilon} \iiint_V \rho(\mathbf{r}') G(\mathbf{r}, \mathbf{r}')\, dv' \tag{2.46}$$

and by invoking duality we find that the appropriate expression for the magnetic scalar potential is

$$\Phi_m(\mathbf{r}) = \frac{1}{\mu} \iiint_V \rho_m(\mathbf{r}') G(\mathbf{r}, \mathbf{r}')\, dv'. \tag{2.47}$$

Corresponding expressions due to the surface charge densities are obtained directly from (2.46) and (2.47) upon replacing the volume integrals with surface integrals over the domain of the surface charges.

## 2.3 | NEAR- AND FAR-ZONE FIELD EXPRESSIONS

### 2.3.1 Near-Zone Fields

Using the vector potential expressions (2.35) and (2.36), from (2.19) the electric field can be more explicitly written as

$$\begin{aligned} \mathbf{E} = \iiint_V \Big\{ &-\nabla \times \left[\mathbf{M}(\mathbf{r}') G(\mathbf{r}, \mathbf{r}')\right] - jkZ\mathbf{J}(\mathbf{r}') G(\mathbf{r}, \mathbf{r}') \\ &- \frac{jZ}{k} \nabla \left[\nabla \cdot \{\mathbf{J}(\mathbf{r}') G(\mathbf{r}, \mathbf{r}')\}\right] \Big\}\, dv' \end{aligned} \tag{2.48}$$

where as usual $\mathbf{r}$ and $\mathbf{r}'$ denote the observation and source points, respectively. For simplicity, in this we have dropped the subscript $i$ from the symbols denoting the electric and magnetic currents.

To simplify (2.48), it is necessary to carry out the indicated differentiation, and as a first step toward this we note that

$$\begin{aligned}\nabla \times \left[\mathbf{M}(\mathbf{r}')G(\mathbf{r}, \mathbf{r}')\right] &= G(\mathbf{r}, \mathbf{r}')\nabla \times \mathbf{M}(\mathbf{r}') - \mathbf{M}(\mathbf{r}') \times \nabla G(\mathbf{r}, \mathbf{r}') \\ &= -\mathbf{M}(\mathbf{r}') \times \nabla G(\mathbf{r}, \mathbf{r}') \end{aligned} \tag{2.49}$$

since the $\nabla$ operator denotes differentiation only with respect to $\mathbf{r}$ and not the primed coordinates. For the same reason we also have

$$\begin{aligned}\nabla \cdot \left[\mathbf{J}(\mathbf{r}')G(\mathbf{r}, \mathbf{r}')\right] &= G(\mathbf{r}, \mathbf{r}')\nabla \cdot \mathbf{J}(\mathbf{r}') + \mathbf{J}(\mathbf{r}') \cdot \nabla G(\mathbf{r}, \mathbf{r}') \\ &= \mathbf{J}(\mathbf{r}') \cdot \nabla G(\mathbf{r}, \mathbf{r}') \end{aligned} \tag{2.50}$$

and thus

$$\begin{aligned}\nabla \left[\nabla \cdot \mathbf{J}(\mathbf{r}')G(\mathbf{r}, \mathbf{r}')\right] &= \nabla \mathbf{J}(\mathbf{r}') \cdot \nabla G(\mathbf{r}, \mathbf{r}') + \mathbf{J}(\mathbf{r}') \cdot \nabla\nabla G(\mathbf{r}, \mathbf{r}') \\ &= \mathbf{J}(\mathbf{r}') \cdot \nabla\nabla G(\mathbf{r}, \mathbf{r}'). \end{aligned} \tag{2.51}$$

When the identities (2.49) through (2.51) are introduced into (2.48) we obtain

$$\begin{aligned}\mathbf{E}(\mathbf{r}) = \iiint_V \Bigg[&\mathbf{M}(\mathbf{r}') \times \nabla G(\mathbf{r}, \mathbf{r}') - jkZ\mathbf{J}(\mathbf{r}')G(\mathbf{r}, \mathbf{r}') \\ &- \frac{jZ}{k}\mathbf{J}(\mathbf{r}') \cdot \nabla\nabla G(\mathbf{r}, \mathbf{r}')\Bigg] \, dv' \end{aligned} \tag{2.52a}$$

and by invoking duality we also have that

$$\begin{aligned}\mathbf{H}(\mathbf{r}) = \iiint_V \Bigg[&-\mathbf{J}(\mathbf{r}') \times \nabla G(\mathbf{r}, \mathbf{r}') - jkY\mathbf{M}(\mathbf{r}')G(\mathbf{r}, \mathbf{r}') \\ &- \frac{jY}{k}\mathbf{M}(\mathbf{r}') \cdot \nabla\nabla G(\mathbf{r}, \mathbf{r}')\Bigg] \, dv'. \end{aligned} \tag{2.52b}$$

To proceed further we must carry out the differentiation on the Green's function. By applying the chain rule of differentiation and noting that $R = |\mathbf{r} - \mathbf{r}'|$, we have

$$\nabla G(\mathbf{r}, \mathbf{r}') = \frac{d}{dR}G(R)\nabla R = -\left(jk + \frac{1}{R}\right) G(R)\nabla R. \tag{2.53}$$

In addition,

$$\nabla R = \hat{x}\frac{\partial R}{\partial x} + \hat{y}\frac{\partial R}{\partial y} + \hat{z}\frac{\partial R}{\partial z} = \frac{\hat{x}(x - x') + \hat{y}(y - y') + \hat{z}(z - z')}{R} = \frac{\mathbf{R}}{R} = \hat{R} \tag{2.54}$$

where $\hat{R}$ denotes the unit vector along the direction joining the integration and observation points. Using this result in (2.53) we have

$$\nabla G(\mathbf{r},\mathbf{r}') = -\left(jk + \frac{1}{R}\right) G(\mathbf{r},\mathbf{r}')\hat{R}. \tag{2.55}$$

Also,

$$\begin{aligned}\nabla\nabla G(\mathbf{r},\mathbf{r}') &= -\nabla\left[\left(jk + \frac{1}{R}\right) G(\mathbf{r},\mathbf{r}')\hat{R}\right]\\ &= -\nabla\left[\left(jk + \frac{1}{R}\right) G(\mathbf{r},\mathbf{r}')\right]\hat{R} - \left(jk + \frac{1}{R}\right) G(\mathbf{r},\mathbf{r}')\nabla\hat{R}\end{aligned} \tag{2.56}$$

where the gradient of the unit vector $\hat{R}$ is interpreted to imply the operation

$$\nabla\mathbf{V} = \hat{x}\nabla V_x + \hat{y}\nabla V_y + \hat{z}\nabla V_z. \tag{2.57}$$

Clearly, $\nabla\mathbf{V}$ is a product of two vectors, and in that respect it is a *dyadic* whose actual meaning can be realized only when dotted with a vector as indicated in the last term of (2.52). To carry out the $\nabla$ operation on $\hat{R}$ we first employ chain rule differentiation to rewrite it as

$$\begin{aligned}\nabla\hat{R} &= \nabla\left(\frac{\mathbf{R}}{R}\right) = \nabla\left(\frac{\mathbf{r}-\mathbf{r}'}{R}\right) = \frac{\nabla(\mathbf{r}-\mathbf{r}')}{R} + (\mathbf{r}-\mathbf{r}')\nabla\left(\frac{1}{R}\right)\\ &= \frac{\nabla\mathbf{r}}{R} - \frac{\mathbf{R}\hat{R}}{R^2}.\end{aligned} \tag{2.58}$$

In accordance with (2.57) we can now express $\nabla\mathbf{r}$ as

$$\begin{aligned}\nabla\mathbf{r} &- \hat{x}\nabla(x) + \hat{y}\nabla(y) + \hat{z}\nabla(z)\\ &= \hat{x}\hat{x} + \hat{y}\hat{y} + \hat{z}\hat{z} = \bar{\mathbf{I}}\end{aligned} \tag{2.59}$$

which is referred to as the unit dyad satisfying the identity

$$\mathbf{V}\cdot\bar{\mathbf{I}} = \mathbf{V}. \tag{2.60}$$

Substituting (2.59) into (2.58) and then into (2.56) we have

$$\begin{aligned}\nabla\nabla G(\mathbf{r},\mathbf{r}') = \hat{R}\hat{R}&\left[\frac{1}{R^2} + \left(jk + \frac{1}{R}\right)^2\right] G(\mathbf{r},\mathbf{r}')\\ &- (\bar{\mathbf{I}} - \hat{R}\hat{R})\left(jk + \frac{1}{R}\right)\frac{G(\mathbf{r},\mathbf{r}')}{R}\end{aligned} \tag{2.61}$$

and thus

$$\begin{aligned}-jkZ\mathbf{J}(\mathbf{r})G(\mathbf{r},\mathbf{r}') - \frac{jZ}{k}\mathbf{J}(\mathbf{r}')\cdot\nabla\nabla G(\mathbf{r},\mathbf{r}') &= -jkZ\left\{\left[1 - \frac{j}{kR} - \frac{1}{(kR)^2}\right]\mathbf{J}(\mathbf{r}')\right.\\ &\left. - \left[1 - \frac{3j}{kR} - \frac{3}{(kR)^2}\right](\mathbf{J}(\mathbf{r}')\cdot\hat{R})\hat{R}\right\}G(\mathbf{r},\mathbf{r}').\end{aligned} \tag{2.62}$$

When this result along with (2.55) is introduced into (2.52) we obtain

$$\begin{aligned}\mathbf{E} = &-jk \iiint_V \left[\mathbf{M}(\mathbf{r}') \times \hat{R}\right] \left(1 + \frac{1}{jkR}\right) G(\mathbf{r}, \mathbf{r}')\, dv' \\ &- jkZ \iiint_V \left\{ \left[1 - \frac{j}{kR} - \frac{1}{(kR)^2}\right] \mathbf{J}(\mathbf{r}') \right. \\ &\left. - \left[1 - \frac{3j}{kR} - \frac{3}{(kR)^2}\right] (\mathbf{J}(\mathbf{r}') \cdot \hat{R})\hat{R} \right\} G(\mathbf{r}, \mathbf{r}')\, dv' \end{aligned} \tag{2.63a}$$

and similarly

$$\begin{aligned}\mathbf{H} = &+jk \iiint_V \left[\mathbf{J}(\mathbf{r}') \times \hat{R}\right] \left(1 + \frac{1}{jkR}\right) G(\mathbf{r}, \mathbf{r}')\, dv' \\ &- jkY \iiint_V \left\{ \left[1 - \frac{j}{kR} - \frac{1}{(kR)^2}\right] \mathbf{M}(\mathbf{r}') \right. \\ &\left. - \left[1 - \frac{3j}{kR} - \frac{3}{(kR)^2}\right] (\mathbf{M}(\mathbf{r}') \cdot \hat{R})\hat{R} \right\} G(\mathbf{r}, \mathbf{r}')\, dv'. \end{aligned} \tag{2.63b}$$

Before closing this section we remark that if $J(\mathbf{r})$ is replaced by an infinitesimal dipole of length $\Delta\ell \to 0$ and carrying a current $I$, that is, $\mathbf{J}(\mathbf{r}) = I\hat{\ell}$, from (2.52) the associated near-zone fields are given by

$$\begin{aligned}\mathbf{E}_e = &-jkZ(I\Delta\ell)\left[1 - \frac{j}{kR} - \frac{1}{(kR)^2}\right]\frac{e^{-jkR}}{4\pi R}\hat{\ell} \\ &+ jkZ(I\Delta\ell)\left[1 - \frac{3j}{kR} - \frac{3}{(kR)^2}\right]\frac{e^{-jkR}}{4\pi R}(\hat{\ell}\cdot\hat{R})\hat{R}\end{aligned} \tag{2.64a}$$

and

$$\mathbf{H}_e = +jk(I\Delta\ell)\left[1 + \frac{1}{jkR}\right]\frac{e^{-jkR}}{4\pi R}(\hat{\ell}\times\hat{R}). \tag{2.64b}$$

If the dipole is $z$-directed ($\hat{\ell} = \hat{z}$) and is located at the origin ($\mathbf{R} = \mathbf{r}$), then upon setting $\hat{z} = \hat{r}\cos\theta - \hat{\theta}\sin\theta$, (2.64) reduces to the usual dipole radiated fields. Similarly, if $\mathbf{M}(\mathbf{r})$ is set equal to an infinitesimal $(I_m)\hat{\ell}$, by referring back to (2.63) or by invoking duality we find that the corresponding radiated fields are given by

$$\mathbf{E}_m = -jk(I_m\Delta\ell)\left[1 + \frac{1}{jkR}\right]\frac{e^{-jkR}}{4\pi R}(\hat{\ell}\times\hat{R}) \tag{2.65a}$$

$$\begin{aligned}\mathbf{H}_m = &-jkY(I_m\Delta\ell)\left[1 - \frac{j}{kR} - \frac{1}{(kR)^2}\right]\frac{e^{-jkR}}{4\pi R}\hat{\ell} \\ &+ jkY(I_m\Delta\ell)\left[1 - \frac{3j}{kR} - \frac{3}{(kR)^2}\right]\frac{e^{-jkR}}{4\pi R}(\hat{\ell}\cdot\hat{R})\hat{R}.\end{aligned} \tag{2.65b}$$

Again, upon setting $\hat{\ell} = \hat{z}$ and $\mathbf{r}' = 0$ (implying $\mathbf{R} = \mathbf{r}$), these reduce to the usual magnetic dipole fields.

## 2.3.2 Field Evaluation in the Source Region

The field expressions (2.63) can be readily evaluated for $r \neq r'$. However, care is required in evaluating the integrals when $\mathbf{r} \to \mathbf{r}'$. In that case $R \to 0$ and the integrand has a nonintegrable $1/R^3$ singularity. To circumvent this difficulty we can rewrite the portion of the integral due to the electric current as

$$\mathbf{E}_e = -jkZ \iiint_{V-V_o} \left[ \mathbf{J}(\mathbf{r}')G(\mathbf{r}_o, \mathbf{r}') + \frac{1}{k^2}\mathbf{J}(\mathbf{r}') \cdot \nabla\nabla G(\mathbf{r}_o, \mathbf{r}') \right] dv'$$
$$- jkZ \iiint_{V_o} \left[ \mathbf{J}(\mathbf{r}')G(\mathbf{r}_o, \mathbf{r}') + \frac{1}{k^2} J(\mathbf{r}') \cdot \nabla\nabla G(\mathbf{r}_o, \mathbf{r}') \right] dv' \quad (2.66)$$

where $V_o$ is an infinitesimally small spherical volume centered at the observation point $\mathbf{r}_o$ (see Figure 2-4) and $\mathbf{r}'$ is in $V_o$. It can then be shown that the first term of the second integrand over $V_o$ vanishes and to deal with the second term of the same integral we invoke the identity (Van Bladel, 1985; Collin, 1986a,b).

$$\iiint_{V_o} \nabla f \, dv = \oiint_{S_o} \hat{n}_o f \, ds. \quad (2.67)$$

This is based on a generalization of Gauss's theorem, and accordingly $\hat{n}_o$ denotes the outward normal unit vector to the surface $S_o$ enclosing $V_o$. Using (2.67) and

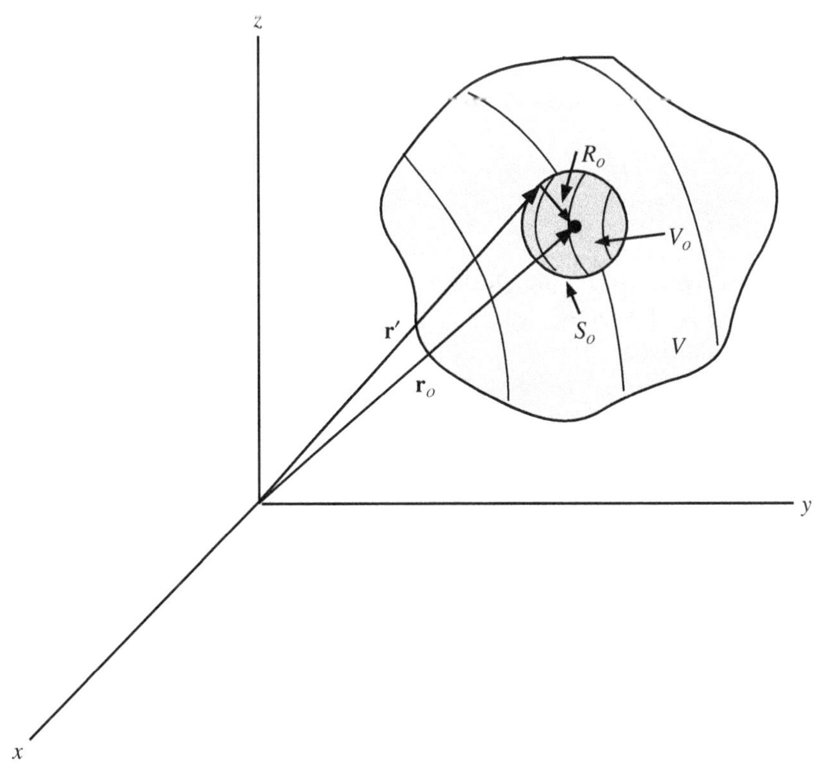


**FIGURE 2-4** Geometry for evaluating the field in the source region.

recalling that $V_o$ is a small spherical volume we have

$$\iiint_{V_o\to 0} \frac{1}{k^2}\mathbf{J}(\mathbf{r}')\cdot\nabla\nabla G(\mathbf{r}_o,\mathbf{r}')\,dv' \simeq -\frac{1}{k^2}\mathbf{J}(\mathbf{r}_o)\cdot\nabla\iiint_{V_o\to 0}\nabla' G(\mathbf{r}_o,\mathbf{r}')\,dv'$$

$$= +\frac{1}{k^2}\oiint_{S_o}(\hat{R}_o\cdot\mathbf{J}(\mathbf{r}_o))\nabla G(\mathbf{r}_o,\mathbf{r}')\,ds'$$

where $\hat{R}_o = (\mathbf{r}_o - \mathbf{r}')/|\mathbf{r}_o - \mathbf{r}'|$. For $R_o = |\mathbf{r}_o - \mathbf{r}'| \to 0$, $\nabla G \approx -\frac{\hat{R}_o}{4\pi R_o^2}$ and upon transferring to spherical coordinates we obtain

$$-\lim_{R_o\to 0}\oiint_{S_o}(\hat{R}_o\cdot\mathbf{J}(r_o))\nabla G(\mathbf{r}_o,\mathbf{r}')\,ds' = \frac{1}{4\pi}\int_0^\pi\int_0^{2\pi}(\hat{R}_o\cdot\mathbf{J}(r_o))\,\hat{R}_o\,\sin\theta_o\,d\theta_o\,d\phi_o.$$

Setting now $\hat{R}_o = -[\hat{x}\cos\phi_o\,\sin\theta_o + \hat{y}\sin\phi_o\,\sin\theta_o + \hat{z}\cos\theta_o]$ and dropping the integrals that vanish, we have

$$-\lim_{R_o\to 0}\oiint_{S_o}(\hat{R}_o\cdot\mathbf{J}(r_o))\nabla G(\mathbf{r}_o,\mathbf{r}')\,ds' = \hat{x}(\hat{x}\cdot\mathbf{J}(\mathbf{r}_o))\int_o^\pi\int_0^{2\pi}\cos^2\phi_o\,\sin^3\theta_o\,d\theta_o\,d\phi_o$$

$$+\hat{y}(\hat{y}\cdot\mathbf{J}(\mathbf{r}_o))\int_o^\pi\int_0^{2\pi}\sin^2\phi_o\,\sin^3\theta_o\,d\theta_o\,d\phi_o$$

$$+\hat{z}(\hat{z}\cdot\mathbf{J}(\mathbf{r}_o))\int_o^\pi\int_0^{2\pi}\cos^2\theta_o\,\sin\theta_o\,d\theta_o\,d\phi_o.$$

Upon evaluating these integrals we obtain

$$\lim_{R_o\to 0}\oiint_{S_o}(\hat{R}_o\cdot\mathbf{J}(r_o))\nabla G(\mathbf{r}_o,\mathbf{r}')\,ds' = -\frac{1}{3}\mathbf{J}(\mathbf{r}_o). \tag{2.68}$$

Consequently, we may rewrite (2.66) as

$$\mathbf{E}_e(\mathbf{r}) = -jkZ\,\fint\!\!\!\iiint_V\left[\mathbf{J}(\mathbf{r}')G(\mathbf{r},\mathbf{r}') + \frac{1}{k^2}\mathbf{J}(\mathbf{r}')\cdot\nabla\nabla G(\mathbf{r},\mathbf{r}')\right]dv' + \frac{jZ}{3k}\mathbf{J}(\mathbf{r}) \tag{2.69}$$

where the horizontal bar through the integral denotes the principal value of that integral. That is, if $\mathbf{r} = \mathbf{r}_o$, where $\mathbf{r}_o$ is located in $V$, the principal value of the integral is evaluated as

$$\fint\!\!\!\iiint_V f(\mathbf{r}')\,dv' = \lim_{V\to 0}\iiint_{V-V_o} f(\mathbf{r}')\,dv'. \tag{2.70}$$

Otherwise, if $\mathbf{r}$ is not in the source region $V$, the principal value integral is evaluated as an ordinary integral. This simply implies that for $\mathbf{r}$ not in $V$, no special care is required for the evaluation of the integrals in (2.63).

### 2.3.3 Fresnel and Far-Zone Fields

When $R = |\mathbf{r} - \mathbf{r}'|$ is very large, the field expressions (2.63) can be simplified substantially by neglecting those terms whose amplitude is $O(1/R^2)$ or less. In addition, we may approximate $\hat{R}$ by $\hat{r}$ (see Figure 2-3b), permitting us to simplify (2.63) to

$$\mathbf{E} = -jk \iiint_V \left[\mathbf{M}(\mathbf{r}') \times \hat{r}\right] \frac{e^{-jk|\mathbf{r}-\mathbf{r}'|}}{4\pi r}\, dv' + jkZ \iiint_V \hat{r} \times \left[\hat{r} \times \mathbf{J}(\mathbf{r}')\right] \frac{e^{-jk|\mathbf{r}-\mathbf{r}'|}}{4\pi r}\, dv' \quad (2.71a)$$

$$\mathbf{H} = jk \iiint_V \left[\mathbf{J}(\mathbf{r}') \times \hat{r}\right] \frac{e^{-jk|\mathbf{r}-\mathbf{r}'|}}{4\pi r}\, dv' + jkY \iiint_V \hat{r} \times \left[\hat{r} \times \mathbf{M}(\mathbf{r}')\right] \frac{e^{-jk|\mathbf{r}-\mathbf{r}'|}}{4\pi r}\, dv' \quad (2.71b)$$

upon also invoking the vector identity

$$\mathbf{a} \times (\mathbf{b} \times \mathbf{c}) = (\mathbf{a} \cdot \mathbf{c})\mathbf{b} - (\mathbf{a} \cdot \mathbf{b})\mathbf{c}. \quad (2.72)$$

To proceed further, it is necessary to obtain a more explicit expression for $|\mathbf{r} - \mathbf{r}'|$, and from Figure 2-3, we obtain that

$$R = |\mathbf{r} - \mathbf{r}'| = \sqrt{r^2 + r'^2 - 2rr' \cos\zeta} \quad (2.73)$$

where $r' \cos\zeta = \mathbf{r}' \cdot \hat{r}$ as illustrated in Figure 2-3b. By employing a binomial expansion for the square root, $R$ can be written as

$$R \approx r + \frac{1}{2}\frac{1}{r}(-2rr' \cos\zeta + r'^2) + \frac{\frac{1}{2}\left(\frac{-1}{2}\right)}{2}(r^2)^{\frac{-3}{2}}(-2rr' \cos\zeta + r'^2) + \cdots$$

and upon some rearrangement we have

$$R \approx r - r' \cos\zeta + \frac{r'^2 \sin^2\zeta}{2r} + \frac{r'^3 \cos\zeta}{2r^2} - \frac{(r')^4}{8r^3} - \frac{r'^3 \cos^3\zeta}{2r^2} + \cdots \quad (2.74)$$

When the first three terms of this expansion are kept, the resulting field expressions become

$$\mathbf{E} \simeq jk\frac{e^{-jkr}}{4\pi r} \iiint_V \left[\hat{r} \times \mathbf{M}(\mathbf{r}') + Z\hat{r} \times \hat{r} \times \mathbf{J}(\mathbf{r}')\right] e^{jk\left[(\mathbf{r}'\cdot\hat{r}) + r'(\mathbf{r}'\cdot\hat{r}) - (r')^2\right]}\, dv' \quad (2.75a)$$

$$\mathbf{H} \simeq jk\frac{e^{-jkr}}{4\pi r} \iiint_V \left[-\hat{r} \times \mathbf{J}(\mathbf{r}') + Y\hat{r} \times \hat{r} \times \mathbf{M}(\mathbf{r}')\right] e^{jk\left[(\mathbf{r}'\cdot\hat{r}) + r'(\mathbf{r}'\cdot\hat{r}) - (r')^2\right]}\, dv'. \quad (2.75b)$$

These expressions give the Fresnel zone fields, and if we restrict the maximum phase error due to the three-term approximation for $R$ to be less than $\frac{\pi}{8k}$, this demands that the next higher-order term of the expansion must contribute a phase smaller than $\pi/8$. From (2.74), this implies that the sum of the fourth and sixth terms must always be less than $\frac{\pi}{8k}$, that is,

$$\max\left\{\frac{k\left(\frac{D}{2}\right)^3}{2r^2} \cos\zeta\,(1 - \cos^2\zeta)\right\} \leq \frac{\pi}{8}$$

where $D$ denotes the maximum linear dimension of the source. Differentiating the left side of this inequality with respect to $\zeta$ and setting the result to zero gives $\zeta_1 = 0$ and $\zeta_2 = \cos^{-1}(1/\sqrt{3})$. Of these solutions, the second is associated with a maximum, and we thus have

$$\frac{k\left(\frac{D}{2}\right)^3}{2r^2}0.385 \leq \frac{\pi}{8}$$

implying that

$$r \geq 0.62\sqrt{\frac{D^3}{\lambda}} \tag{2.76}$$

if the phase error associated with the approximate expressions (2.75) is to be kept to less than $\pi/8$.

If we retain only the first two terms of the expansion (2.73) the resulting field expressions are

$$\mathbf{E} = \mathbf{E}_{ff} = jk\frac{e^{-jkr}}{4\pi r}\iiint_V \left[\hat{r} \times \mathbf{M}(\mathbf{r}') + Z\hat{r} \times \hat{r} \times \mathbf{J}(\mathbf{r}')\right] e^{jk\mathbf{r}'\cdot\hat{r}}\, dv' \tag{2.77a}$$

$$\mathbf{H} = \mathbf{H}_{ff} = -jkY\frac{e^{-jkr}}{4\pi r}\iiint_V \left[Z\hat{r} \times \mathbf{J}(\mathbf{r}') - \hat{r} \times \hat{r} \times \mathbf{M}(\mathbf{r}')\right] e^{jk\mathbf{r}'\cdot\hat{r}}\, dv'. \tag{2.77b}$$

These are referred to as the *far-zone fields*, and, as is the case with (2.75), they are also valid for surface and linear sources upon replacing the volume integrals with surface or line integrals over the domain of the source. Again, by demanding that the maximum phase error due to the two-term approximation for $kR$ is less than $\pi/8$, we can find the corresponding minimum value for $r$. In this case, the third term of the expansion (2.74) must satisfy the inequality

$$\frac{kD^2}{8r} \leq \frac{\pi}{8}$$

implying that

$$r \geq \frac{2D^2}{\lambda} \tag{2.78}$$

and this relation is referred to as the *far-zone criterion.*

By comparing the $\mathbf{E}$ and $\mathbf{H}$ expressions given in (2.77) and making use of the identity (2.72), it follows that

$$\mathbf{H}_{ff} = Y\hat{r} \times \mathbf{E}_{ff}, \quad \mathbf{E}_{ff} = -Z\hat{r} \times \mathbf{H}_{ff}. \tag{2.79}$$

These are the usual relations for the far-zone fields and imply that these fields do not have a component along $\hat{r}$, the direction of propagation. Through comparison of (2.77) with (2.35) and (2.36), it also follows that

$$\mathbf{E}_{ff} = jk\hat{r} \times \mathbf{F} + jkZ\hat{r} \times (\hat{r} \times \mathbf{A}) = -Z\hat{r} \times \mathbf{H}_{ff} \tag{2.80a}$$

$$\mathbf{H}_{ff} = -jk\hat{r} \times \mathbf{A} + jkY\hat{r} \times (\hat{r} \times \mathbf{F}) = Y\hat{r} \times \mathbf{E}_{ff}. \tag{2.80b}$$

**TABLE 2-1** Dual quantities in electromagnetics.

| | | |
|---|---|---|
| $\mathbf{E}$ | $\rightarrow$ | $\mathbf{H}$ |
| $\mathbf{H}$ | $\rightarrow$ | $-\mathbf{E}$ |
| $\mathbf{J}$ | $\rightarrow$ | $\mathbf{M}$ |
| $\mathbf{M}$ | $\rightarrow$ | $-\mathbf{J}$ |
| $\mathbf{A}$ | $\rightarrow$ | $\mathbf{F}$ |
| $\mathbf{F}$ | $\rightarrow$ | $-\mathbf{A}$ |
| $\epsilon$ | $\rightarrow$ | $\mu$ |
| $\mu$ | $\rightarrow$ | $\epsilon$ |
| $\eta$ | $\rightarrow$ | $1/\eta$ |

We can also express the far-zone fields in terms of the spherical components of the vector potential as

$$\mathbf{E}_{ff} = -\hat{\theta} jk(F_\phi + ZA_\theta) + \hat{\phi} jk(F_\theta - ZA_\phi) \tag{2.81a}$$

$$\mathbf{H}_{ff} = \hat{\theta} jk(A_\phi - YF_\theta) - \hat{\phi} jk(A_\theta + YF_\phi). \tag{2.81b}$$

By comparing (2.80a) and (2.80b) we observe that the dual quantity of $\mathbf{A}$ is $\mathbf{F}$ and the dual of $\mathbf{F}$ is $-\mathbf{A}$. A more complete summary of the dual field quantities is given in Table 2-1.

## 2.4 | DIRECT SOLUTION OF THE VECTOR WAVE EQUATION

### 2.4.1 Vector Wave Equations

In many cases it is advantageous and, thus, desirable to work directly with the electric and magnetic fields rather than the vector potentials. To obtain a solution for the $\mathbf{E}$ and $\mathbf{H}$ fields we must then pursue a direct solution of Maxwell's equations (1.48) through (1.51) without invoking the vector or scalar potentials. Because of duality let us concentrate on the electric field solution.

Dividing (1.49) by $\mu_r$, taking its curl, and making use of (1.48) we obtain

$$\nabla \times \left( \frac{1}{\mu_r} \nabla \times \mathbf{E} \right) - k_o^2 \epsilon_r \mathbf{E} = -j\omega\mu_o \mathbf{J} - \nabla \times \left( \frac{1}{\mu_r} \mathbf{M} \right) \tag{2.82a}$$

where we have again dropped the subscripts $i$ from $\mathbf{J}$ and $\mathbf{M}$. This can be alternatively written as

$$\nabla \times \nabla \times \mathbf{E} - k^2 \mathbf{E} + \mu_r \nabla\left( \frac{1}{\mu_r} \right) \times \nabla \times \mathbf{E} = -j\omega\mu \mathbf{J} - \nabla \times \mathbf{M} - \mu_r \nabla\left( \frac{1}{\mu_r} \right) \times \mathbf{M} \tag{2.82b}$$

where we have made use of the identity

$$\nabla \times (\phi \mathbf{A}) = \nabla\phi \times \mathbf{A} + \phi \nabla \times \mathbf{A}. \tag{2.83}$$

Equations (2.82) are the most general forms of the electric field *wave equation* for isotropic media. The corresponding magnetic field wave equations are

$$\nabla \times \left(\frac{1}{\epsilon_r}\nabla \times \mathbf{H}\right) - k_o^2\mu_r\mathbf{H} = -j\omega\epsilon_o\mathbf{M} + \nabla \times \left(\frac{1}{\epsilon_r}\mathbf{J}\right) \tag{2.84a}$$

or

$$\nabla \times \nabla \times \mathbf{H} - k^2\mathbf{H} + \epsilon_r\nabla\left(\frac{1}{\epsilon_r}\right) \times \nabla \times \mathbf{H} = -j\omega\epsilon\mathbf{M} + \nabla \times \mathbf{J} + \epsilon_r\nabla\left(\frac{1}{\epsilon_r}\right) \times \mathbf{J}. \tag{2.84b}$$

In the case of homogeneous media, $\nabla\left(\frac{1}{\epsilon_r}\right) = \nabla\left(\frac{1}{\mu_r}\right) = 0$, and the wave equation then reduces to

$$\begin{aligned} \nabla \times \nabla \times \mathbf{E} - k^2\mathbf{E} &= -j\omega\mu\mathbf{J} - \nabla \times \mathbf{M} \\ \nabla \times \nabla \times \mathbf{H} - k^2\mathbf{H} &= -j\omega\epsilon\mathbf{M} + \nabla \times \mathbf{J}. \end{aligned} \tag{2.85}$$

By employing the identity (2.6) and the relation (1.51), it follows that

$$\begin{aligned} \nabla^2\mathbf{E} + k^2\mathbf{E} &= j\omega\mu\mathbf{J} - \frac{\nabla\nabla \cdot \mathbf{J}}{j\omega\epsilon} + \nabla \times \mathbf{M} \\ \nabla^2\mathbf{H} + k^2\mathbf{H} &= j\omega\epsilon\mathbf{M} - \frac{\nabla\nabla \cdot \mathbf{M}}{j\omega\mu} - \nabla \times \mathbf{J}. \end{aligned} \tag{2.86a}$$

In the source-free region, these reduce to

$$\begin{aligned} \nabla^2\mathbf{E} + k^2\mathbf{E} &= 0 \\ \nabla^2\mathbf{H} + k^2\mathbf{H} &= 0 \end{aligned} \tag{2.86b}$$

which are commonly referred to as the wave equations. These imply that each field component satisfies the scalar Helmholtz equation

$$\nabla^2\psi + k^2\psi = 0 \tag{2.87}$$

where $\psi$ denotes $E_x$, $E_y$, $E_z$, $H_x$, $H_y$, or $H_z$.

### 2.4.2 Dyadic Representation

To solve (2.85) we may instead consider the solution of the dyadic equation

$$\nabla \times \nabla \times \overline{\mathbf{\Gamma}}(\mathbf{r}, \mathbf{r}') - k^2\overline{\mathbf{\Gamma}}(\mathbf{r}, \mathbf{r}') = -\overline{\mathbf{I}}\delta(\mathbf{r} - \mathbf{r}') \tag{2.88}$$

where $\overline{\mathbf{\Gamma}}(\mathbf{r}, \mathbf{r}')$ is referred to as the free-space dyadic Green's function and satisfies the radiation condition

$$\lim_{r\to\infty} r\left[jk\overline{\mathbf{\Gamma}} + \hat{r} \times (\nabla \times \overline{\mathbf{\Gamma}})\right] = 0. \tag{2.89}$$

By setting $\overline{\mathbf{\Gamma}} = \mathbf{T}\hat{a}$, where $\hat{a}$ is some arbitrary unit vector, this can be alternatively written as

$$\lim_{r\to\infty} r\left[jk\mathbf{T} + \hat{r} \times (\nabla \times \mathbf{T})\right]\hat{a} = 0. \tag{2.90}$$

By comparing (2.85) and (2.88) it is readily seen that **T** represents a vector field, and for (2.89) to be a valid radiation condition (2.90) must be satisfied when **T** is replaced by the **E** or **H** far-zone fields. This is easily verified by reverting to the far-zone relation (2.74) and in conjunction with the source free Maxwell's equation, we find that

$$\nabla \times \mathbf{T} = -jk\hat{r} \times \mathbf{T} \tag{2.91}$$

where **T** represents the **E** or **H** fields. From (2.91) it is seen that (2.90) is indeed satisfied by the far-zone **E** and **H** fields.

To solve for $\overline{\boldsymbol{\Gamma}}$ we first take the divergence of both sides of (2.88), giving

$$-\nabla \cdot \left[k^2 \overline{\boldsymbol{\Gamma}}(\mathbf{r}, \mathbf{r}')\right] = -\nabla \cdot \left[\overline{\mathbf{I}}\delta(\mathbf{r} - \mathbf{r}')\right]. \tag{2.92}$$

Also, by invoking the identity (2.6), equation (2.88) can be written as

$$\nabla^2 \overline{\boldsymbol{\Gamma}}(\mathbf{r}, \mathbf{r}') + k^2 \overline{\boldsymbol{\Gamma}}(\mathbf{r}, \mathbf{r}') = \overline{\mathbf{I}}\delta(\mathbf{r} - \mathbf{r}') + \nabla \left[\nabla \cdot \overline{\boldsymbol{\Gamma}}(\mathbf{r}, \mathbf{r}')\right] \tag{2.93}$$

which can be combined with (2.92) to obtain

$$(\nabla^2 + k^2)\overline{\boldsymbol{\Gamma}}(\mathbf{r}, \mathbf{r}') = \left[\overline{\mathbf{I}} + \frac{\nabla\nabla}{k^2}\right] \delta(\mathbf{r} - \mathbf{r}'). \tag{2.94}$$

Setting

$$\overline{\boldsymbol{\Gamma}}(\mathbf{r}, \mathbf{r}') = -\left[\overline{\mathbf{I}} + \frac{\nabla\nabla}{k^2}\right] G(\mathbf{r}, \mathbf{r}') \tag{2.95}$$

it follows from (2.94) that $G(\mathbf{r}, \mathbf{r}')$ must satisfy the differential equation

$$(\nabla^2 + k^2)G(\mathbf{r}, \mathbf{r}') = -\delta(\mathbf{r} - \mathbf{r}').$$

This is identical to (2.38), and, since $G(\mathbf{r}, \mathbf{r}')$ must also satisfy the scalar radiation condition (2.39), it is then given by (2.37).

By referring to Section 2.3.1, it is readily seen that $\overline{\boldsymbol{\Gamma}}$ can be expressed as

$$\begin{aligned}\overline{\boldsymbol{\Gamma}}(\mathbf{r}, \mathbf{r}') = &\left(\frac{j}{kR} + \frac{1}{(kR)^2} - 1\right) G(R)\overline{\mathbf{I}} \\ &+ \left(1 - \frac{3}{(kR)^2} - \frac{3j}{kR}\right) G(R)\hat{R}\hat{R}.\end{aligned} \tag{2.96}$$

Also,

$$\begin{aligned}\nabla \times \overline{\boldsymbol{\Gamma}}(\mathbf{r}, \mathbf{r}') &= -\nabla \times \left(\overline{\mathbf{I}} + \frac{\nabla\nabla}{k^2}\right) G(R) \\ &= -\nabla \times \overline{\mathbf{I}}G(R) + \nabla \times \frac{\nabla\nabla}{k^2} G(R) \\ &= -\nabla \times \overline{\mathbf{I}}G(R) = -\nabla G(R) \times \overline{\mathbf{I}}\end{aligned} \tag{2.97a}$$

$$\nabla \times \overline{\boldsymbol{\Gamma}}(\mathbf{r}, \mathbf{r}') = \left(jk + \frac{1}{R}\right) G(R)\hat{R} \times \overline{\mathbf{I}} \tag{2.97b}$$

since $\nabla R = \hat{R}$. However, before proceeding with field representations in terms of the dyadic Green's function, it is instructive to look at more explicit expressions of $\overline{\mathbf{\Gamma}}$ aimed at clarifying its use in subsequent integral equations.

Any dyadic can be written in terms of its components, and for the case of $\overline{\mathbf{\Gamma}}$ we have

$$\begin{aligned}\overline{\mathbf{\Gamma}} = &\ \hat{x}\hat{x}\Gamma_{xx} + \hat{x}\hat{y}\Gamma_{xy} + \hat{x}\hat{z}\Gamma_{xz} \\ &+ \hat{y}\hat{x}\Gamma_{yx} + \hat{y}\hat{y}\Gamma_{yy} + \hat{y}\hat{z}\Gamma_{yz} \\ &+ \hat{z}\hat{x}\Gamma_{zx} + \hat{z}\hat{y}\Gamma_{zy} + \hat{z}\hat{z}\Gamma_{zz}\end{aligned} \tag{2.98a}$$

conveniently written in matrix form as

$$\overline{\mathbf{\Gamma}} = \begin{pmatrix} \Gamma_{xx} & \Gamma_{xy} & \Gamma_{xz} \\ \Gamma_{yx} & \Gamma_{yy} & \Gamma_{yz} \\ \Gamma_{zx} & \Gamma_{zy} & \Gamma_{zz} \end{pmatrix}. \tag{2.98b}$$

Again, it should be noted that the meaning of the dyadic can be realized only when it is dotted with another vector. Using (2.98b), this operation becomes the matrix-vector product

$$\overline{\mathbf{\Gamma}} \cdot \mathbf{V} = \begin{pmatrix} \Gamma_{xx} & \Gamma_{xy} & \Gamma_{xz} \\ \Gamma_{yx} & \Gamma_{yy} & \Gamma_{yz} \\ \Gamma_{zx} & \Gamma_{zy} & \Gamma_{zz} \end{pmatrix} \begin{pmatrix} V_x \\ V_y \\ V_z \end{pmatrix}$$

which is convenient for numerical implementations.

To find the individual matrix components of $\overline{\mathbf{\Gamma}}$ we may return to (2.95), and upon replacing the $\nabla$ operators with their Cartesian forms we find that

$$\overline{\mathbf{\Gamma}}(\mathbf{r}, \mathbf{r}') = -\frac{1}{4\pi k^2} \begin{pmatrix} k^2 + \dfrac{\partial^2}{\partial x^2} & \dfrac{\partial^2}{\partial x \partial y} & \dfrac{\partial^2}{\partial x \partial z} \\ \dfrac{\partial^2}{\partial y \partial x} & k^2 + \dfrac{\partial^2}{\partial y^2} & \dfrac{\partial^2}{\partial y \partial z} \\ \dfrac{\partial^2}{\partial z \partial x} & \dfrac{\partial^2}{\partial z \partial y} & k^2 + \dfrac{\partial^2}{\partial z^2} \end{pmatrix} \frac{e^{-jkR}}{R} \tag{2.99}$$

and

$$\nabla \times \overline{\mathbf{\Gamma}}(\mathbf{r}, \mathbf{r}') = -\frac{1}{4\pi} \begin{pmatrix} 0 & -\dfrac{\partial}{\partial z} & \dfrac{\partial}{\partial y} \\ \dfrac{\partial}{\partial z} & 0 & -\dfrac{\partial}{\partial x} \\ -\dfrac{\partial}{\partial y} & \dfrac{\partial}{\partial x} & 0 \end{pmatrix} \frac{e^{-jkR}}{R}. \tag{2.100}$$

One could now carry out the differentiations, or more conveniently (2.96) and (2.97) could be employed along with the Cartesian expressions of $\hat{R}$ and $\overline{\mathbf{I}}$ given by (2.59) and (2.54), respectively. The Cartesian components of $\overline{\mathbf{\Gamma}}$ are then readily found by collecting the like terms. The corresponding matrix components of $\overline{\Gamma}$ in the cylindrical coordinate systems are found by replacing the $\nabla$ operators in (2.95) with their corresponding cylindrical forms. Upon collecting like terms, this gives

$$\overline{\Gamma}(\mathbf{r},\mathbf{r}') = -\frac{1}{4\pi k^2}\begin{pmatrix} k^2+\dfrac{\partial^2}{\partial\rho^2} & \dfrac{1}{\rho}\dfrac{\partial}{\partial\rho\partial\phi} & \dfrac{\partial^2}{\partial\rho\partial z} \\ \dfrac{1}{\rho}\dfrac{\partial^2}{\partial\phi\partial\rho} & k^2+\dfrac{1}{\rho^2}\dfrac{\partial^2}{\partial\phi^2} & \dfrac{1}{\rho}\dfrac{\partial^2}{\partial\phi\partial z} \\ \dfrac{\partial^2}{\partial z\partial\rho} & \dfrac{1}{\rho}\dfrac{\partial^2}{\partial z\partial\phi} & k^2+\dfrac{\partial^2}{\partial z^2} \end{pmatrix}\frac{e^{-jkR}}{R} \tag{2.101}$$

where $(\rho,\phi,z)$ denote the usual cylindrical coordinates. In carrying out the differ entiations, it should be noted that

$$R = \sqrt{\rho^2+\rho'^2-2\rho\rho'\cos(\phi-\phi')+(z-z')^2}$$

in terms of the cylindrical coordinates.

To express the **E** and **H** fields in terms of the free-space dyadic Green's function, we may examine (2.52) and identify the presence of $\overline{\Gamma}(\mathbf{r},\mathbf{r}')$ and $\nabla\times\overline{\Gamma}(\mathbf{r},\mathbf{r}')$ which are given in (2.95) and (2.97a), respectively. Then, upon making use of the identity (2.60) and that $-\nabla G\times\mathbf{M} = -\nabla G\times\overline{\mathbf{I}}\cdot\mathbf{M}$, we can rewrite (2.52) as

$$\mathbf{E}(\mathbf{r}) = \iiint\left\{\left[\nabla\times\overline{\Gamma}(\mathbf{r},\mathbf{r}')\right]\cdot\mathbf{M}(\mathbf{r}') + jkZ\overline{\Gamma}(\mathbf{r},\mathbf{r}')\cdot\mathbf{J}(\mathbf{r}')\right\}dv' \tag{2.102a}$$

$$\mathbf{H}(\mathbf{r}) = \iiint\left\{jkY\overline{\Gamma}(\mathbf{r},\mathbf{r}')\cdot\mathbf{M}(\mathbf{r}') - \left[\nabla\times\overline{\Gamma}(\mathbf{r},\mathbf{r}')\right]\cdot\mathbf{J}(\mathbf{r}')\right\}dv'. \tag{2.102b}$$

A more formal derivation of (2.102) directly from the vector wave equations (2.85) and the dyadic equations (2.88) requires use of the vector-dyadic second Green's identity (Tai, 1992). This is given by

$$\iiint_V\left\{\left[\nabla\times\nabla\times\mathbf{E}(\mathbf{r})\right]\cdot\overline{\Gamma}(\mathbf{r},\mathbf{r}') - \mathbf{E}(\mathbf{r})\cdot\nabla\times\nabla\times\overline{\Gamma}(\mathbf{r},\mathbf{r}')\right\}dv =$$
$$\oiint_{S_c}\left\{\left[\hat{n}\times\mathbf{E}(\mathbf{r})\right]\cdot\nabla\times\overline{\Gamma}(\mathbf{r},\mathbf{r}') + \left[\hat{n}\times\nabla\times\mathbf{E}(\mathbf{r})\right]\cdot\overline{\Gamma}(\mathbf{r},\mathbf{r}')\right\}ds \tag{2.103}$$

where $\hat{n}$ is the outward unit normal to the surface $S_c$ enclosing $V$. Also, **E** and $\overline{\Gamma}$ can be any vector or dyadic, but in this case they are chosen to represent the electric field and the free-space dyadic Green's function. We note that the identity (2.103) can be derived directly from the scalar second Green's identity (2.42).

Upon making use of (2.85) and (2.88) in (2.103) along with the property of the delta function, the left-hand side of (2.103) becomes

$$\mathbf{E}(\mathbf{r}') - \iiint_V\left\{j\omega\mu\mathbf{J}(\mathbf{r})\cdot\overline{\Gamma}(\mathbf{r},\mathbf{r}') + \left[\nabla\times\mathbf{M}(\mathbf{r})\right]\cdot\overline{\Gamma}(\mathbf{r},\mathbf{r}')\right\}dv.$$

To evaluate the right-hand side surface integral we may assume that $S_c$ is a spherical surface at infinity since $V$ encompasses all space. Then, upon invoking the radiation condition (2.89) and (2.91), it can be shown that the right-hand side of (2.103) vanishes. Thus, we have

$$\mathbf{E}(\mathbf{r}') = \iiint_V \{j\omega\mu\mathbf{J}(\mathbf{r}) \cdot \overline{\mathbf{\Gamma}}(\mathbf{r}, \mathbf{r}') + [\nabla \times \mathbf{M}(\mathbf{r})] \cdot \overline{\mathbf{\Gamma}}(\mathbf{r}, \mathbf{r}')\} \, dv \quad (2.104)$$

and by noting that

$$\iiint_V \nabla \cdot \{\mathbf{M}(\mathbf{r}) \times \overline{\mathbf{\Gamma}}(\mathbf{r}, \mathbf{r}')\} \, dv = \oiint_{S_c} \{\mathbf{M}(\mathbf{r}) \times \overline{\mathbf{\Gamma}}(\mathbf{r}, \mathbf{r}')\} \cdot \hat{n} \, ds = 0 \quad (2.105)$$

(since $\mathbf{M}(\mathbf{r})$ is zero on $S_c$) and the identity

$$\nabla \cdot \{\mathbf{M}(\mathbf{r}) \times \overline{\mathbf{\Gamma}}(\mathbf{r}, \mathbf{r}')\} = \nabla \times \mathbf{M}(\mathbf{r}) \cdot \overline{\mathbf{\Gamma}}(\mathbf{r}, \mathbf{r}') - \mathbf{M}(\mathbf{r}) \cdot \nabla \times \overline{\mathbf{\Gamma}}(\mathbf{r}, \mathbf{r}') \quad (2.106)$$

it follows that

$$\mathbf{E}(\mathbf{r}') = \iiint_V \{j\omega\mu\mathbf{J}(\mathbf{r}) \cdot \overline{\mathbf{\Gamma}}(\mathbf{r}, \mathbf{r}') + \mathbf{M}(\mathbf{r}) \cdot [\nabla \times \overline{\mathbf{\Gamma}}(\mathbf{r}, \mathbf{r}')]\} \, dv. \quad (2.107)$$

To cast this into the integral form given in (2.102), it is necessary to invoke some properties of the dyadic Green's function. We note that

$$\tilde{\overline{\mathbf{\Gamma}}}(\mathbf{r}, \mathbf{r}') = \overline{\mathbf{\Gamma}}(\mathbf{r}, \mathbf{r}') \quad (2.108a)$$

$$\widetilde{\nabla \times \overline{\mathbf{\Gamma}}}(\mathbf{r}, \mathbf{r}') = -\nabla \times \overline{\mathbf{\Gamma}}(\mathbf{r}, \mathbf{r}') = \nabla' \times \overline{\mathbf{\Gamma}}(\mathbf{r}, \mathbf{r}') \quad (2.108b)$$

where the tilde over the dyadic quantity denotes the transpose of their associated matrix. Thus

$$\mathbf{M}(\mathbf{r}) \cdot \nabla \times \overline{\mathbf{\Gamma}}(\mathbf{r}, \mathbf{r}') = -\nabla \times \overline{\mathbf{\Gamma}}(\mathbf{r}, \mathbf{r}') \cdot \mathbf{M}(\mathbf{r}) = \nabla' \times \overline{\mathbf{\Gamma}}(\mathbf{r}, \mathbf{r}') \cdot \mathbf{M}(\mathbf{r})$$

and

$$\mathbf{J}(\mathbf{r}) \cdot \overline{\mathbf{\Gamma}}(\mathbf{r}, \mathbf{r}') = \tilde{\overline{\mathbf{\Gamma}}}(\mathbf{r}, \mathbf{r}') \cdot \mathbf{J}(\mathbf{r}) = \overline{\mathbf{\Gamma}}(\mathbf{r}, \mathbf{r}') \cdot \mathbf{J}(\mathbf{r}).$$

Further, since $\overline{\mathbf{\Gamma}}(\mathbf{r}, \mathbf{r}') = \overline{\mathbf{\Gamma}}(\mathbf{r}', \mathbf{r})$ and $\nabla' \times \overline{\mathbf{\Gamma}}(\mathbf{r}, \mathbf{r}') = \nabla \times \overline{\mathbf{\Gamma}}(\mathbf{r}', \mathbf{r})$, it follows that (2.107) is identical to (2.102).

When making practical use of the integral expressions (2.102), it is necessary to work with the individual field components. The Cartesian field components can be readily found by introducing the matrices (2.99) and (2.100) for $\overline{\mathbf{\Gamma}}$ and $\nabla \times \overline{\mathbf{\Gamma}}$, respectively. We have

$$E_x(\mathbf{r}) = \frac{-jZ}{k} \iiint_V \left[ J_x(\mathbf{r}') \left( k^2 + \frac{\partial^2}{\partial x^2} \right) + J_y(\mathbf{r}') \frac{\partial^2}{\partial x \partial y} + J_z(\mathbf{r}') \frac{\partial^2}{\partial x \partial z} \right] \cdot G(R) \, dv' - \iiint_V \left[ M_z(\mathbf{r}') \frac{\partial}{\partial y} - M_y(\mathbf{r}') \frac{\partial}{\partial z} \right] G(R) \, dv' \quad (2.109a)$$

$$E_y(\mathbf{r}) = \frac{-jZ}{k} \iiint_V \left[ J_x(\mathbf{r}') \frac{\partial^2}{\partial y \partial x} + J_y(\mathbf{r}') \left( k^2 + \frac{\partial^2}{\partial y^2} \right) + J_z(\mathbf{r}') \frac{\partial^2}{\partial y \partial z} \right]$$

$$\cdot G(R)\, dv' - \iiint_V \left[ M_x(\mathbf{r}') \frac{\partial}{\partial z} - M_z(\mathbf{r}') \frac{\partial}{\partial x} \right] G(R)\, dv' \qquad (2.109b)$$

$$E_z(\mathbf{r}) = \frac{-jZ}{k} \iiint_V \left[ J_x(\mathbf{r}') \frac{\partial^2}{\partial z \partial x} + J_y(\mathbf{r}') \frac{\partial^2}{\partial z \partial y} + J_z(\mathbf{r}') \left( k^2 + \frac{\partial^2}{\partial z^2} \right) \right]$$

$$\cdot G(R)\, dv' - \iiint_V \left[ -M_x(\mathbf{r}') \frac{\partial}{\partial y} + M_y(\mathbf{r}') \frac{\partial}{\partial x} \right] G(R)\, dv' \qquad (2.109c)$$

and the corresponding Cartesian components of **H** follow by duality. To obtain the ideal dipole fields from (2.102) or (2.109) and their dual, we simply set

$$\mathbf{J}(\mathbf{r}) = \hat{\ell}(I \Delta \ell) \delta(\mathbf{r} - \mathbf{r}') = d\mathbf{p}_e \delta(\mathbf{r} - \mathbf{r}')$$
$$\mathbf{M}(\mathbf{r}) = \hat{\ell}(I_m \Delta \ell) \delta(\mathbf{r} - \mathbf{r}') = d\mathbf{p}_m \delta(\mathbf{r} - \mathbf{r}') \qquad (2.110)$$

where $d\mathbf{p}_{e,m}$ are referred to as the electric and magnetic dipole moments. From (2.102) we then have

$$\mathbf{E}(\mathbf{r}) = jkZ\overline{\mathbf{\Gamma}}(\mathbf{r}, \mathbf{r}') \cdot d\mathbf{p}_e + \nabla \times \overline{\mathbf{\Gamma}}(\mathbf{r}, \mathbf{r}') \cdot d\mathbf{p}_m$$
$$\mathbf{H}(\mathbf{r}) = jkY\overline{\mathbf{\Gamma}}(\mathbf{r}, \mathbf{r}') \cdot d\mathbf{p}_m - \nabla \times \overline{\mathbf{\Gamma}}(\mathbf{r}, \mathbf{r}') \cdot d\mathbf{p}_e \qquad (2.111)$$

which can be shown to be identical to the expressions (2.64) and (2.65) upon making use of (2.96) and (2.97).

## 2.5 | TWO-DIMENSIONAL FIELDS

Two-dimensional fields generally refer to fields that do not exhibit a dependence on one coordinate variable. For example, the problem of propagation in a parallel plate waveguide is two-dimensional in nature and so is that of plane wave scattering by a cylinder (see Figure 2-5). Of course, neither the parallel plate waveguide nor the infinitely long cylinder are physically realizable structures, but nevertheless their study can provide important results that are applicable to the three-dimensional structures they approximate. For example, results based on the parallel plate waveguide are useful in the analysis of striplines and the general theory of transmission lines. In the case of scattering, elongated bodies can often be treated as infinite in one dimension to simplify the analysis. Results based on the two-dimensional model can then be applied to the corresponding three-dimensional problem through scaling.

### 2.5.1 Two-Dimensional Sources

The most elemental two-dimensional source is an electric or a magnetic line source. Referring to Figure 2-6, these are $z$-directed current filaments carrying a constant

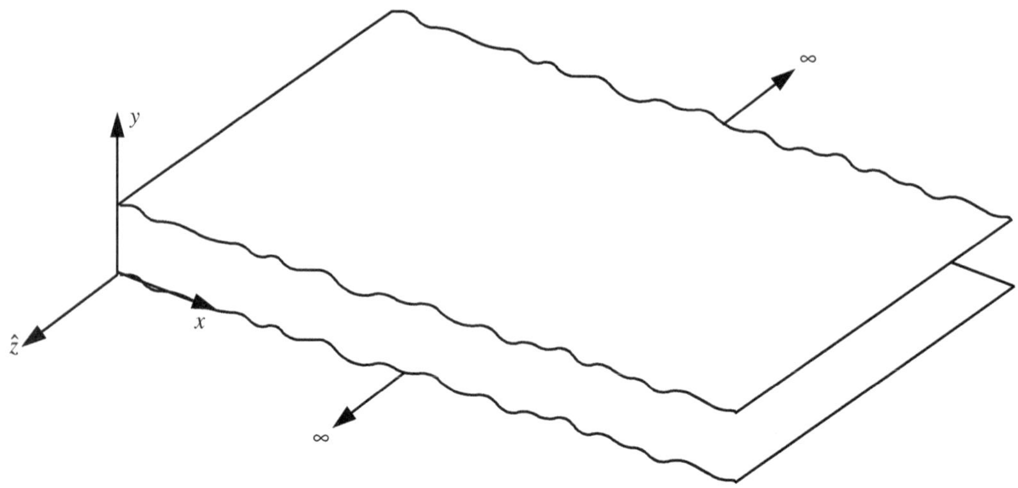


(a)

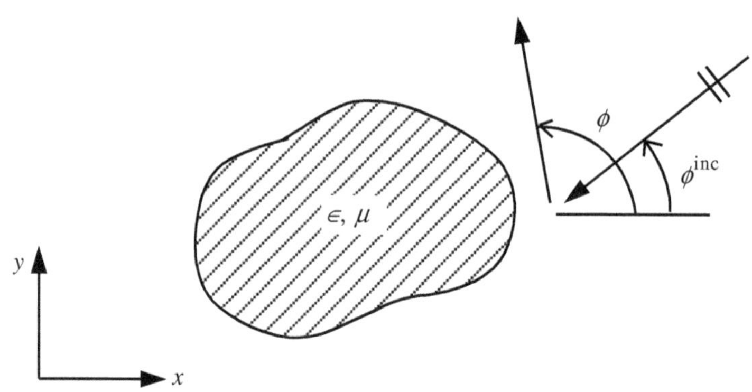


(b)

**FIGURE 2-5** Examples of two-dimensional problems: (a) propagation in a parallel plate waveguide whose plates are infinite in the *z*-direction and (b) scattering by an infinitely long cylinder infinite in the *z*-direction.

current. They may, thus, be represented as

$$\mathbf{J}(\mathbf{r}) = \hat{z} I_e \delta(\overline{\rho} - \overline{\rho}_o) = \hat{z} I_e \frac{\delta(\rho - \rho_o)\delta(\phi - \phi_o)}{\rho} \tag{2.112a}$$

$$\mathbf{M}(\mathbf{r}) = \hat{z} I_m \delta(\overline{\rho} - \overline{\rho}_o) = \hat{z} I_m \frac{\delta(\rho - \rho_o)\delta(\rho - \phi_o)}{\rho} \tag{2.112b}$$

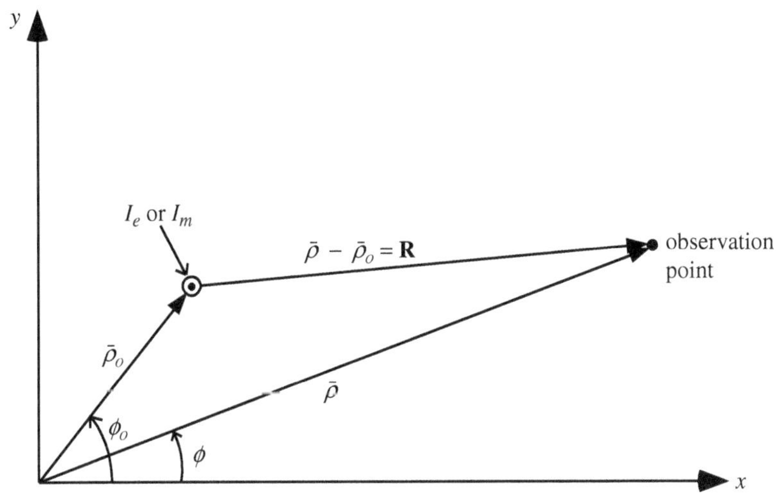


**FIGURE 2-6** Illustration of geometrical parameters associated with a line source.

where $\overline{\rho} = x\hat{x} + y\hat{y} = \rho(\hat{x}\cos\phi + \hat{y}\sin\phi) = \rho\hat{\rho}$ denotes the vector to the observation point and $\overline{\rho}_o = x_o\hat{x} + y_o\hat{y} = \rho_o(\hat{x}\cos\phi_o + \hat{y}\sin\phi_o)$ is the vector to the source point. To find the field radiated by these sources we refer to expression (2.52) or (2.102). Upon setting $dv = \rho\, d\rho\, d\phi\, dz$ and noting that

$$\int_{-\infty}^{\infty} \frac{e^{-jkR}}{R} dz' = \int_{-\infty}^{\infty} \frac{e^{-jk\sqrt{|\overline{\rho}-\overline{\rho}_o|^2+(z')^2}}}{\sqrt{|\overline{\rho}-\overline{\rho}_o|^2+(z')^2}} dz' = -j\pi H_o^{(2)}\left(k|\overline{\rho}-\overline{\rho}_o|\right) \quad (2.113)$$

the field due to the electric line source is found to be

$$\mathbf{E}_e = -\hat{z}\frac{jkZI_e}{4\pi}\left[-j\pi H_o^{(2)}(k|\overline{\rho}-\overline{\rho}_o|)\right] - \frac{jZI_e}{4\pi k}\hat{z}\cdot\nabla\nabla\{-j\pi H_o^{(2)}(k|\overline{\rho}-\overline{\rho}_o|)\}$$
$$= -\hat{z}I_e\frac{kZ}{4}H_o^{(2)}\left(k|\overline{\rho}-\overline{\rho}_o|\right) = -\hat{z}jkZI_eG_{2d}(R) \quad (2.114)$$

since the dot product in the second term is zero. In this $H_o^{(2)}(\cdot)$ denotes the zeroth-order Hankel function of the second kind. The quantity $R = |\overline{\rho} - \overline{\rho}_o| = \sqrt{\rho^2 + \rho_o^2 - 2\rho\rho_o\cos(\phi - \phi_o)} = \sqrt{(x - x_o)^2 + (y - y_o)^2}$ is now the distance between the source and the observer and

$$G_{2d}(R) = G_{2d}(\overline{\rho}, \overline{\rho}_o) = -\frac{j}{4}H_o^{(2)}\left(k|\overline{\rho}-\overline{\rho}_o|\right) = -\frac{j}{4}H_o^{(2)}(kR) \quad (2.115)$$

is referred to as the two-dimensional Green's function for the unbounded homogeneous media. It satisfies the differential equation

$$\nabla^2 G_{2d}(\overline{\rho}, \overline{\rho}') + k^2 G_{2d}(\overline{\rho}, \overline{\rho}') = -\delta(\overline{\rho} - \overline{\rho}') \quad (2.116)$$

and the radiation condition

$$\lim_{\rho\to\infty} \sqrt{\rho}\left(jkG_{2d} + \frac{\partial G_{2d}}{\partial\rho}\right) = 0. \quad (2.117)$$

As can be expected (2.115) can be obtained by integrating (2.38) with respect to $z$ and setting $z' = 0$. Also, the validity of the radiation condition (2.117) can be readily established by noting that as $\rho \to \infty$

$$H_o^{(2)}(k|\overline{\rho} - \overline{\rho}'|)_{\rho \to \infty} \sqrt{\frac{2j}{\pi}} e^{jk\hat{\rho}\cdot\overline{\rho}'} \frac{e^{-jk\rho}}{\sqrt{\rho}} \tag{2.118}$$

and when this is used in (2.115) it follows that (2.117) is satisfied.

To find the electric field radiated by the magnetic line source, we substitute (2.112b) into (2.52), yielding

$$\mathbf{E}_m = \frac{-j}{4} I_m \hat{z} \times \nabla H_o^{(2)}(k|\overline{\rho} - \overline{\rho}_o|). \tag{2.119}$$

Since (see Section 2.3.1)

$$\nabla H_o^{(2)}(k|\overline{\rho} - \overline{\rho}_o|) = \nabla H_o^{(2)}(kR) = k\frac{d}{d(kR)} H_o^{(2)}(kR)\nabla R = -kH_1^{(2)}(kR)\hat{R} \tag{2.120}$$

where $H_1^{(2)}(\cdot)$ is the first-order Hankel function of the second kind, (2.119) simplifies to

$$\mathbf{E}_m = I_m \frac{jk}{4} H_1^{(2)}(kR)\hat{z} \times \hat{R} \tag{2.121}$$

where $\hat{R} = (\overline{\rho} - \overline{\rho}_o)/|\overline{\rho} - \overline{\rho}_o|$. The corresponding magnetic field due to the magnetic line source is given by

$$\mathbf{H}_m = -\hat{z} I_m \frac{kY}{4} H_o^{(2)}(k|\overline{\rho} - \overline{\rho}_o|) \tag{2.122}$$

which is the dual of (2.114). Also by invoking duality, the magnetic field due to an electric current is given by

$$\begin{aligned} \mathbf{H}_e &= \frac{j}{4} I_e \hat{z} \times \nabla H_o^{(2)}(k|\overline{\rho} - \overline{\rho}_o|) \\ &= -I_e \frac{jk}{4} H_1^{(2)}(kR)\hat{z} \times \hat{R}. \end{aligned} \tag{2.123}$$

Upon comparing this with (2.114), it is observed that for two-dimensional $z$-directed electric sources

$$\mathbf{H}_e = -\frac{jY}{k} \hat{z} \times \nabla(\mathbf{E}_e \cdot \hat{z}) \tag{2.124a}$$

and for $z$-directed magnetic sources

$$\mathbf{E}_m = \frac{jZ}{k} \hat{z} \times \nabla(\mathbf{H}_m \cdot \hat{z}). \tag{2.124b}$$

Previously, we discussed the radiated fields by two-dimensional $z$-directed line sources. However, it should be noted that in subsequent studies we will encounter two-dimensional sources that may be invariant in $z$ but are not necessarily $z$-directed. That is,

$$\mathbf{J}(\mathbf{r}) = \hat{a} I_e \delta(\overline{\rho} - \overline{\rho}_o) \tag{2.125}$$

and this represents a current filament as illustrated in Figure 2-7. Such sources, although not physical, are useful as equivalent sources in formulating a given radiation or scattering problem (as is the case with magnetic sources). Their

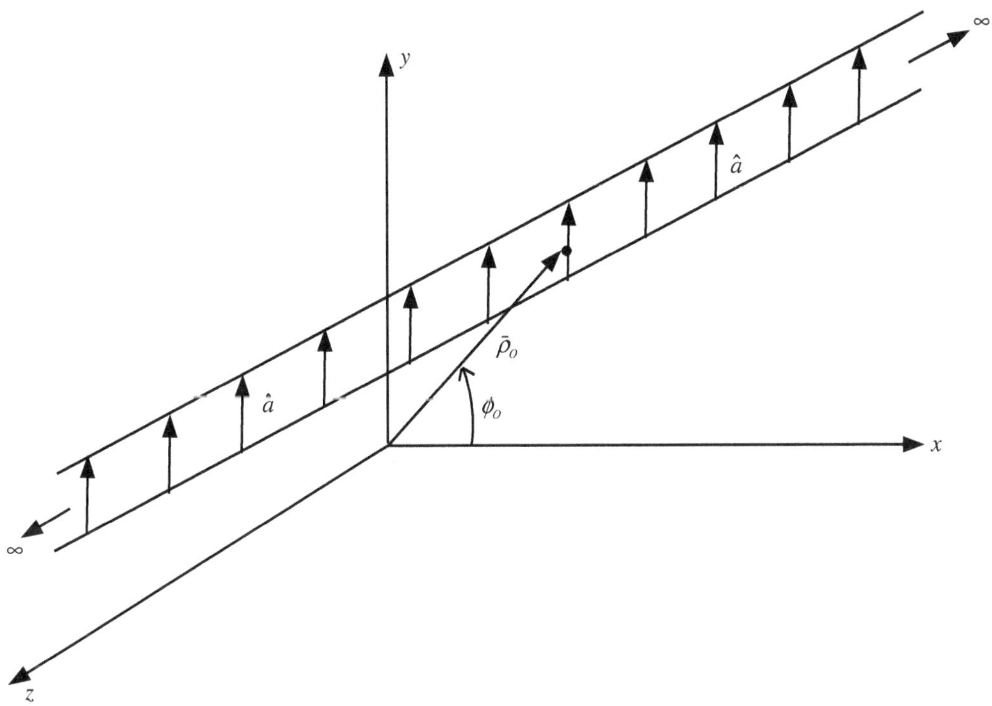


**FIGURE 2-7** Illustration of two-dimensional current filament.

associated fields can be found by following a procedure similar to that outlined for the $z$-directed two-dimensional sources. Alternatively, they can be obtained from the more general expressions given in the next section that refer to completely arbitrary two-dimensional source distributions.

## 2.5.2 Exact Integral Expressions

Let us assume the presence of the general two-dimensional sources $\{\mathbf{J}(\overline{\rho}), \mathbf{M}(\overline{\rho})\}$ occupying the cross sectional areas $A$, as illustrated in Figure 2-8.

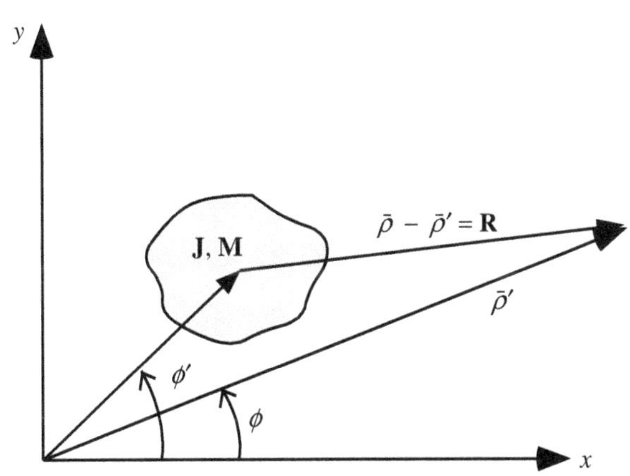


**FIGURE 2-8** Radiation by two-dimensional sources.

As discussed in the previous section, to find the field radiated by these sources, we refer to the integral expressions (2.52) and set

$$\begin{aligned}\mathbf{J}(\mathbf{r}) &\to \mathbf{J}(\overline{\rho})\\ \mathbf{M}(\mathbf{r}) &\to \mathbf{M}(\overline{\rho}).\end{aligned} \tag{2.126}$$

By making use of the identity (2.113) we then readily find

$$\mathbf{E}(\overline{\rho}) = \iint_A \Big[\mathbf{M}(\overline{\rho}') \times \nabla G_{2d}(\overline{\rho}, \overline{\rho}') - jkZ\mathbf{J}(\overline{\rho}')G_{2d}(\overline{\rho}, \overline{\rho}') - \frac{jZ}{k}\mathbf{J}(\overline{\rho}') \cdot \nabla\nabla G_{2d}(\overline{\rho}, \overline{\rho}')\Big]\, ds' \tag{2.127a}$$

$$\mathbf{H}(\overline{\rho}) = \iint_A \Big[-\mathbf{J}(\overline{\rho}') \times \nabla G_{2d}(\overline{\rho}, \overline{\rho}') - jkY\mathbf{M}(\overline{\rho}')G_{2d}(\overline{\rho}, \overline{\rho}') - \frac{jY}{k}\mathbf{M}(\overline{\rho}') \cdot \nabla\nabla G_{2d}(\overline{\rho}, \overline{\rho}')\Big]\, ds' \tag{2.127b}$$

where $G_{2d}(\overline{\rho}, \overline{\rho}')$ is given by (2.115) upon replacing $\overline{\rho}_o$ by $\overline{\rho}' = \rho'(\hat{x}\cos\phi' + \hat{y}\sin\phi')$.

Clearly, (2.127) are identical to (2.52) other than the replacement of the three-dimensional Green's function with the two-dimensional one. Also, the volume integral has been replaced by a double integral over the domain of the sources in the $x$-$y$ plane. For computational purposes it is necessary to carry out the gradient operations, which necessitates that we write out the dyadic $\nabla\nabla G_{2d}$ in its explicit form. This can be readily done by following the procedure employed for the three-dimensional case along with the identities (2.120) and

$$\begin{aligned}H_\nu'^{(2)}(kR) &= H_{\nu-1}^{(2)}(kR) - \frac{\nu}{kR}H_\nu^{(2)}(kR)\\ H_\nu'^{(2)}(kR) &= -H_{\nu+1}^{(2)}(kR) + \frac{\nu}{kR}H_\nu^{(2)}(kR)\end{aligned} \tag{2.128}$$

where $\nu$ denotes the order of the Hankel function. We obtain

$$\nabla\nabla G_{2d}(\overline{\rho}, \overline{\rho}') = \frac{jk}{4}\left\{\left[kH_o^{(2)}(kR) - \frac{2}{R}H_1^{(2)}(kR)\right]\hat{R}\hat{R} + \frac{H_1^{(2)}(kR)}{R}\overline{\mathbf{I}}_{2d}\right\} \tag{2.129}$$

and in this case $\overline{\mathbf{I}}_{2d} = \hat{x}\hat{x} + \hat{y}\hat{y}$, $R = |\overline{\rho} - \overline{\rho}'| = \sqrt{(x - x')^2 + (y - y')^2}$ and $\hat{R} = (\overline{\rho} - \overline{\rho}')/|\overline{\rho} - \overline{\rho}'|$. Substituting (2.129) into (2.127) yields

$$\begin{aligned}\mathbf{E}(\overline{\rho}) = \frac{jk}{4}\iint_A \mathbf{M}(\overline{\rho}') \times \hat{R}\; H_1^{(2)}(kR)\, ds' &- \frac{kZ}{4}\iint_A \mathbf{J}(\overline{\rho}')H_0^{(2)}(kR)\, ds' + \frac{kZ}{4}\iint_A \mathbf{J}(\overline{\rho}') \cdot \overline{\mathbf{I}}_{2d}\frac{H_1^{(2)}(kR)}{kR}\, ds'\\ &+ \frac{kZ}{4}\iint_A \hat{R}(\mathbf{J}(\overline{\rho}') \cdot \hat{R})\left[H_o^{(2)}(kR) - \frac{2}{kR}H_1^{(2)}(kR)\right] ds'\end{aligned} \tag{2.130a}$$

$$\mathbf{H}(\overline{\rho}) = -\frac{jk}{4}\iint_A \mathbf{J}(\overline{\rho}') \times \hat{R}\; H_1^{(2)}(kR)\, ds'$$
$$-\frac{kY}{4}\iint_A \mathbf{M}(\overline{\rho}') H_0^{(2)}(kR)\, ds' + \frac{kY}{4}\iint_A \mathbf{M}(\overline{\rho}')\cdot \bar{\mathbf{I}}_{2d}\frac{H_1^{(2)}(kR)}{kR}\, ds'$$
$$+\frac{kY}{4}\iint_A \hat{R}(\mathbf{M}(\overline{\rho}')\cdot\hat{R})\left[H_0^{(2)}(kR) - \frac{2}{kR}H_1^{(2)}(kR)\right] ds'. \qquad (2.130b)$$

Alternatively, we may expand the $\nabla$ operators in their Cartesian form, and in doing so we can rewrite (2.127) as

$$E_x = \frac{-jZ}{k}\iint_A \left[J_x(\overline{\rho}')\left(k^2 + \frac{\partial^2}{\partial x^2}\right) + J_y(\overline{\rho}')\frac{\partial}{\partial x \partial y}\right] G_{2d}(\overline{\rho},\overline{\rho}')\, ds'$$
$$-\iint_A M_z(\overline{\rho}')\frac{\partial}{\partial y}G_{2d}(\overline{\rho},\overline{\rho}')\, ds'$$

$$E_y = \frac{-jZ}{k}\iint_A \left[J_x(\overline{\rho}')\frac{\partial^2}{\partial y \partial x} + J_y(\overline{\rho}')\left(k^2 + \frac{\partial^2}{\partial y^2}\right)\right] G_{2d}(\overline{\rho},\overline{\rho}')\, ds'$$
$$+\iint_A M_z(\overline{\rho}')\frac{\partial}{\partial x}G_{2d}(\overline{\rho},\overline{\rho}')\, ds'$$

$$E_z = -jkZ\iint_A J_z(\overline{\rho}')G_{2d}(\overline{\rho},\overline{\rho}')\, ds'$$
$$-\iint_A \left[M_y(\overline{\rho}')\frac{\partial}{\partial x} - M_x(\overline{\rho}')\frac{\partial}{\partial y}\right] G_{2d}(\overline{\rho},\overline{\rho}')\, ds'$$

$$H_x = -\frac{jY}{k}\iint_A \left[M_x(\overline{\rho}')\left(k^2 + \frac{\partial^2}{\partial x^2}\right) + M_y\frac{\partial^2}{\partial x \partial y}\right] G_{2d}(\overline{\rho},\overline{\rho}')\, ds'$$
$$+\iint_A J_z(\overline{\rho}')\frac{\partial}{\partial y}G_{2d}(\overline{\rho},\overline{\rho}')\, ds'$$

$$H_y = -\frac{jY}{k}\iint_A \left[M_x(\overline{\rho}')\frac{\partial^2}{\partial y \partial x} + M_y(\overline{\rho}')\left(k^2 + \frac{\partial^2}{\partial y^2}\right)\right] G_{2d}(\overline{\rho},\overline{\rho}')\, ds'$$
$$-\iint_A J_z(\overline{\rho}')\frac{\partial}{\partial x}G_{2d}(\overline{\rho},\overline{\rho}')\, ds'$$

$$H_z = -jkY\iint_A M_z(\overline{\rho}')G_{2d}(\overline{\rho},\overline{\rho}')\, ds'$$
$$+\iint_A \left[J_y(\overline{\rho}')\frac{\partial}{\partial x} - J_x(\overline{\rho}')\frac{\partial}{\partial y}\right] G_{2d}(\overline{\rho},\overline{\rho}')\, ds'. \qquad (2.131)$$

Obviously, (2.130) are more useful for computing the radiation of given sources, but the representations (2.131) are more appropriate in numerical solutions to be discussed later. We note that (2.131) can be deduced directly from the corresponding three-dimensional expressions (2.109) upon setting to zero all derivatives with respect to $z$ and making use of the identity (2.113).

### 2.5.3 Far-Zone Fields

In the far zone, (2.130) can be simplified by introducing the asymptotic approximation

$$H_\nu^{(2)}(kR) \underset{k\rho\to\infty}{\approx} \sqrt{\frac{2j}{\pi}}\, j^\nu \frac{e^{-jk|\overline{\rho}-\overline{\rho}'|}}{\sqrt{k|\overline{\rho}-\overline{\rho}'|}} \approx \sqrt{\frac{2j}{\pi k}}\, j^\nu \frac{e^{-jk\rho}}{\sqrt{\rho}} e^{jk\hat{\rho}\cdot\overline{\rho}'}. \tag{2.132}$$

Then, on letting $R \to \rho$, $\hat{R} \to \hat{\rho}$ and keeping only terms of $O(1/\sqrt{\rho})$ we have

$$\mathbf{E}_{ff} \approx +\sqrt{\frac{k}{8\pi}}\frac{e^{-j(k\rho-\frac{\pi}{4})}}{\sqrt{\rho}} \iint_A \left[\hat{\rho}\times\mathbf{M}(\overline{\rho}') + Z\hat{\rho}\times\hat{\rho}\times\mathbf{J}(\overline{\rho}')\right] e^{jk\hat{\rho}\cdot\overline{\rho}'}\, ds' \tag{2.133a}$$

and

$$\mathbf{H}_{ff} \approx \sqrt{\frac{k}{8\pi}}\frac{e^{-j(k\rho-\frac{\pi}{4})}}{\sqrt{\rho}} \iint_A \left[\mathbf{J}(\overline{\rho}')\times\hat{\rho} + Y\hat{\rho}\times\hat{\rho}\times\mathbf{M}(\overline{\rho}')\right] e^{jk\hat{\rho}\cdot\overline{\rho}'}\, ds' \tag{2.133b}$$

and from Figure 2-9, it is seen that $\hat{\rho}\cdot\overline{\rho}' = x'\cos\phi + y'\sin\phi = \rho\cos(\phi-\phi')$.

By comparing (2.133a) with (2.133b), and making use of the identity (2.72), it is then readily seen that

$$\mathbf{H}_{ff} = Y\hat{\rho}\times\mathbf{E}_{ff}, \quad \mathbf{E}_{ff} = -Z\hat{\rho}\times\mathbf{H}_{ff} \tag{2.134}$$

which are identical to (2.79) except for the replacement of $\hat{r}$ with $\hat{\rho}$. We can rewrite (2.133) in a more compact form by introducing the two-dimensional vector potential definitions

$$\mathbf{A} = \frac{-j}{4}\iint_A \mathbf{J}(\overline{\rho}')H_0^{(2)}(k|\overline{\rho}-\overline{\rho}'|)\, ds' \tag{2.135}$$

$$\mathbf{F} = \frac{-j}{4}\iint_A \mathbf{M}(\overline{\rho}')H_0^{(2)}(k|\overline{\rho}-\overline{\rho}'|)\, ds'. \tag{2.136}$$

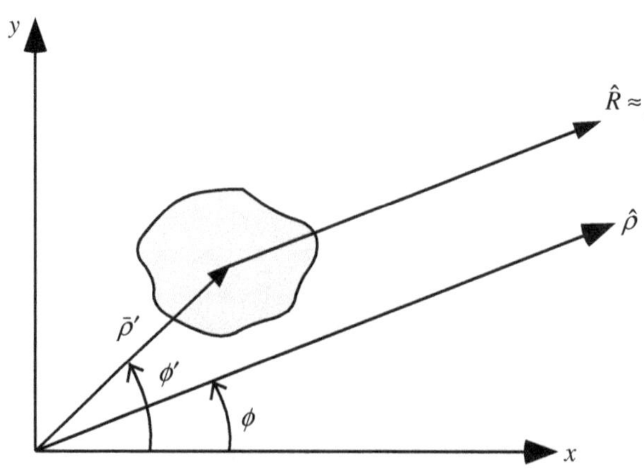


**FIGURE 2-9** Far-zone radiation by two-dimensional sources.

On making use of (2.132) we obtain the far-zone expressions

$$\mathbf{A} \approx \frac{-j}{\sqrt{8\pi k}} \frac{e^{-j(k\rho-\pi/4)}}{\sqrt{\rho}} \iint_A \mathbf{J}(\overline{\rho}')e^{jk\hat{\rho}\cdot\overline{\rho}'}\, ds' \tag{2.137}$$

$$\mathbf{F} \approx \frac{-j}{\sqrt{8\pi k}} \frac{e^{-j(k\rho-\pi/4)}}{\sqrt{\rho}} \iint_A \mathbf{M}(\overline{\rho}')e^{jk\hat{\rho}\cdot\overline{\rho}'}\, ds'. \tag{2.138}$$

From (2.133) we then deduce that

$$\mathbf{E}_{ff} = +jk\hat{\rho} \times \mathbf{F} + jkZ\hat{\rho} \times \hat{\rho} \times \mathbf{A} = -Z\hat{\rho} \times \mathbf{H}_{ff} \tag{2.139a}$$

$$\mathbf{H}_{ff} = -jk\hat{\rho} \times \mathbf{A} + jkY\hat{\rho} \times \hat{\rho} \times \mathbf{F} - Y\hat{\rho} \times \mathbf{E}_{ff}. \tag{2.139b}$$

These are identical to the three-dimensional far-zone expressions given in (2.80). They imply that the two-dimensional far-zone fields have only $\phi$ and $z$ components since the field propagation is along the $\hat{\rho}$ direction. More explicitly, upon carrying out the cross products in (2.139) we find that

$$\mathbf{E}_{ff} = \hat{z}jk(F_\phi - ZA_z) - \hat{\phi}jk(F_z + ZA_\phi) \tag{2.140a}$$

$$\mathbf{H}_{ff} = -\hat{z}jk(A_\phi + YF_z) + \hat{\phi}jk(A_z - YF_\phi). \tag{2.140b}$$

### 2.5.4 Field Evaluation in the Source Region

When the observation point is in the source region (i.e., $\rho \to \rho'$), the integrands in (2.127) and (2.130) contain a nonintegrable $\frac{1}{R^2}$ singularity because

$$H_1^{(2)}(kR) \underset{R \to 0}{\simeq} \frac{j}{\pi}\frac{2}{kR} \tag{2.141}$$

for $kR \to 0$. To treat this situation we must proceed in a manner analogous to that described in Section 2.3.2. That is, we first rewrite (2.127a) as

$$\begin{aligned}\mathbf{E}_e = &-jkZ \iint_{A-A_o} \left[\mathbf{J}(\overline{\rho}')G_{2d}(\overline{\rho}_o, \overline{\rho}') + \frac{1}{k^2}\mathbf{J}(\overline{\rho}') \cdot \nabla\nabla G_{2d}(\overline{\rho}_o, \overline{\rho}')\right] ds' \\ &-jkZ \iint_{A_o} \left[\mathbf{J}(\overline{\rho}')G_{2d}(\overline{\rho}_o, \overline{\rho}') + \frac{1}{k^2}\mathbf{J}(\overline{\rho}') \cdot \nabla\nabla G_{2d}(\overline{\rho}_o, \overline{\rho}')\right] ds'\end{aligned} \tag{2.142}$$

where we kept only those terms associated with electric currents. Further, $A_o$ is arbitrarily chosen as the cross section of a circular cylinder of radius $a \to 0$ and is centered at the observation point $\overline{\rho} = \overline{\rho}_o$. Setting $ds' = \rho' d\rho' d\phi'$ and noting that

$$H_o^{(2)}(kR) \underset{R \to 0}{\simeq} \frac{2}{\pi} \ln(1.123kR) \tag{2.143}$$

it is readily seen that the first term of the integral over $A_o$ vanishes as $A_o \to 0$. To deal with the second term of the same integral, we invoke the identity

$$\iint_{A_o} \nabla f\, ds = \oint_{C_o} \hat{n}_o f\, d\ell \tag{2.144}$$

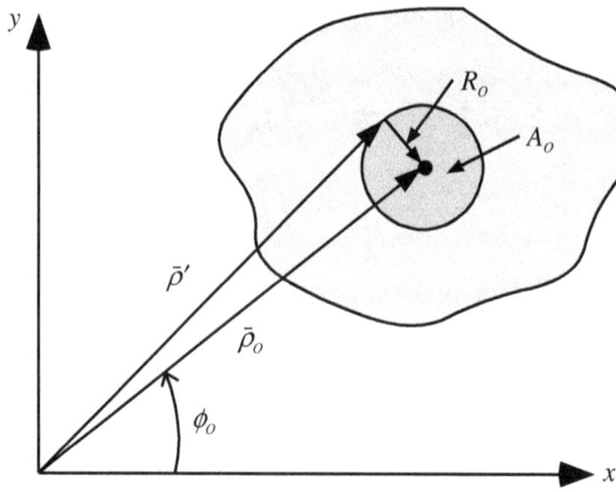


**FIGURE 2-10** Geometry for evaluating the field in the source region.

which is deduced directly from (2.67), and thus $\hat{n}_o$ is the outward unit normal to the contour, $C_o$, enclosing $A_o$ (see Figure 2-10). Thus,

$$\frac{1}{k^2}\iint_{A_o\to 0}\mathbf{J}(\bar{\rho}')\cdot\nabla\nabla G_{2d}(\bar{\rho}_o,\bar{\rho}')\,ds' \approx -\frac{1}{k^2}\mathbf{J}(\bar{\rho}_o)\cdot\nabla\iint_{A_o\to 0}\nabla' G_{2d}(\bar{\rho}_o,\bar{\rho}')\,ds'$$
$$\approx +\frac{1}{k^2}\oint_{C_o}\left[\hat{R}_o\cdot\mathbf{J}(\bar{\rho}_o)\right]\nabla G_{2d}(\bar{\rho}_o,\bar{\rho}')\,d\ell' \tag{2.145}$$

where $\hat{R}_o = (\bar{\rho}_o - \bar{\rho}')/|\bar{\rho}_o - \bar{\rho}'|$, and $\bar{\rho}'$ is on $C_o$. For $|\bar{\rho}_o - \bar{\rho}'| = a \to 0$, $\nabla G_{2d} \approx -\left(\frac{-j}{4}\right)\left(\frac{j2}{\pi ka}\right)\nabla(kR_o) = -\frac{\hat{R}_o}{2\pi a}$ and upon setting $d\ell' = ad\phi_o$, we get

$$\frac{1}{k^2}\iint_{A_o\to 0}\mathbf{J}(\bar{\rho}')\cdot\nabla\nabla G_{2d}(\bar{\rho}_o,\bar{\rho}')\,ds' = -\frac{1}{2\pi k^2}\int_0^{2\pi}\left(\hat{R}_o\cdot\mathbf{J}(\bar{\rho}_o)\right)\hat{R}_o d\phi_o$$
$$= -\frac{1}{2\pi k^2}\left[\hat{x}J_x(\bar{\rho}_o)\int_0^{2\pi}\cos^2\phi_o d\phi_o + \hat{y}J_y(\bar{\rho}_o)\int_0^{2\pi}\sin^2\phi_o d\phi_o\right].$$

We thus conclude that

$$\frac{1}{k^2}\iint_{A_o\to 0}\mathbf{J}(\bar{\rho}')\cdot\nabla\nabla G_{2d}(\bar{\rho}_o,\bar{\rho}')\,ds' = -\frac{\mathbf{J}(\bar{\rho}_o)\cdot\bar{\mathbf{I}}_{2d}}{2k^2} \tag{2.146}$$

and, consequently, (2.142) can be rewritten as

$$\mathbf{E}_e = -jkZ\,⨍⨍_A\left[\mathbf{J}(\bar{\rho}')G_{2d}(\bar{\rho},\bar{\rho}') + \frac{1}{k^2}\mathbf{J}(\bar{\rho}')\cdot\nabla\nabla G_{2d}(\bar{\rho},\bar{\rho}')\right]ds' + \frac{jZ}{2k}\mathbf{J}\cdot\bar{\mathbf{I}}_{2d} \tag{2.147}$$

As usual, the horizontal bar through the integral denotes its principal value.

## 2.6 SPECTRAL FIELD REPRESENTATIONS

One readily observes that the field representations given in this chapter are convolution integrals involving the Green's function and the source current density. Consequently, by taking the Fourier transform of the integrals in (2.109) or (2.131), it follows that the Fourier transform of the fields is a simple algebraic function. From the convolution theorem, this function is the product of the transforms associated with the Green's function and the current densities. To find explicit expressions for the field transform or spectrum we must first introduce the appropriate Fourier transform pair and evaluate the transform of the Green's function. For simplicity, let us first consider the two-dimensional case.

### 2.6.1 Two-Dimensions

In two dimensions there is no dependence on $z$, and the Fourier transform pair is thus defined as

$$\tilde{f}(k_x, k_y) = \int_{-\infty}^{\infty}\int_{-\infty}^{\infty} f(x, y)e^{-j(k_x x + k_y y)}\, dx\, dy \tag{2.148a}$$

$$f(x, y) = \frac{1}{(2\pi)^2}\int_{-\infty}^{\infty}\int_{-\infty}^{\infty} \tilde{f}(k_x, k_y)e^{j(k_x x + k_y y)}\, dk_x\, dk_y. \tag{2.148b}$$

In these, $k_x$ and $k_y$ are the Fourier or spectral variables, and in the future we shall use the notation

$$\tilde{f}(k_x, k_y) = \mathcal{F}\{f(x, y)\} \tag{2.149a}$$

$$f(x, y) = \mathcal{F}^{-1}\{\tilde{f}(k_x, k_y)\} \tag{2.149b}$$

to imply the integral expressions (2.148). By differentiating both sides of (2.148b) with respect to $x$ or $y$, it readily follows that, for example,

$$\mathcal{F}\left\{\frac{\partial}{\partial x} f(x, y)\right\} = jk_x \tilde{f}(k_x, k_y) \tag{2.150}$$

$$\mathcal{F}\left\{\frac{\partial}{\partial y} f(x, y)\right\} = jk_y \tilde{f}(k_x, k_y) \tag{2.151}$$

$$\mathcal{F}\left\{\frac{\partial}{\partial x \partial y} f(x, y)\right\} = -k_x k_y \tilde{f}(k_x, k_y) \tag{2.152}$$

$$\mathcal{F}\left\{\frac{\partial^2}{\partial x^2} f(x, y)\right\} = -k_x^2 \tilde{f}(k_x, k_y). \tag{2.153}$$

Also, from the convolution theorem we have

$$\begin{aligned} f(x, y) * g(x, y) &= \int_{-\infty}^{\infty}\int_{-\infty}^{\infty} f(x', y')g(x - x', y - y')\, dx'\, dy' \\ &= \tilde{f}(k_x, k_y)\tilde{g}(k_x, k_y) \end{aligned} \tag{2.154}$$

in which $\tilde{g}(k_x, k_y) = \mathcal{F}\{g(x, y)\}$. When this identity, in conjunction with (2.150) through (2.153), is now applied to the integral expressions (2.131) we obtain

$$\begin{pmatrix} \widetilde{E}_x \\ \widetilde{E}_y \\ \widetilde{E}_z \end{pmatrix} = -\frac{jZ}{k}\widetilde{G}_{2d}(k_x, k_y) \begin{pmatrix} k^2 - k_x^2 & -k_x k_y & 0 \\ -k_y k_x & k^2 - k_y^2 & 0 \\ 0 & 0 & k^2 \end{pmatrix} \begin{pmatrix} \widetilde{J}_x \\ \widetilde{J}_y \\ \widetilde{J}_z \end{pmatrix}$$

$$+ \widetilde{G}_{2d}(k_x, k_y) \begin{pmatrix} 0 & 0 & -jk_y \\ 0 & 0 & +jk_x \\ jk_y & -jk_x & 0 \end{pmatrix} \begin{pmatrix} \widetilde{M}_x \\ \widetilde{M}_y \\ \widetilde{M}_z \end{pmatrix} \tag{2.155a}$$

$$\begin{pmatrix} \widetilde{H}_x \\ \widetilde{H}_y \\ \widetilde{H}_z \end{pmatrix} = -\frac{jY}{k}\widetilde{G}_{2d}(k_x, k_y) \begin{pmatrix} k^2 - k_x^2 & -k_x k_y & 0 \\ -k_y k_x & k^2 - k_y^2 & 0 \\ 0 & 0 & k^2 \end{pmatrix} \begin{pmatrix} \widetilde{M}_x \\ \widetilde{M}_y \\ \widetilde{M}_z \end{pmatrix}$$

$$+ \widetilde{G}_{2d}(k_x, k_y) \begin{pmatrix} 0 & 0 & jk_y \\ 0 & 0 & -jk_x \\ -jk_y & +jk_y & 0 \end{pmatrix} \begin{pmatrix} \widetilde{J}_x \\ \widetilde{J}_y \\ \widetilde{J}_z \end{pmatrix}. \tag{2.155b}$$

In these, $\widetilde{G}_{2d}(k_x, k_y)$ is the Fourier transform of the two-dimensional Green's function with $\overline{\rho}' = 0$, that is,

$$\widetilde{G}_{2d}(k_x, k_y) = \mathcal{F}\left\{\frac{-j}{4} H_o^{(2)}\left(k\sqrt{x^2 + y^2}\right)\right\} \tag{2.156}$$

and likewise $\widetilde{J}_{x,y,z}$ and $\widetilde{M}_{x,y,z}$ denote the Fourier transforms of the electric and magnetic current densities.

One can rewrite (2.155) more compactly by referring to (2.127) and noting that the transform of the gradient function $\nabla f(x, y)$ is

$$\mathcal{F}\{\nabla f(x, y)\} = \widetilde{\nabla} \tilde{f}(k_x, k_y) = j(\hat{x}k_x + \hat{y}k_y)\tilde{f}(k_x, k_y) = j\mathbf{k}_{2d}\tilde{f}(k_x, k_y)$$

where

$$\mathbf{k}_{2d} = \hat{x}k_x + \hat{y}k_y$$

and we may also write that $\widetilde{\nabla} = j\mathbf{k}_{2d}$. Thus, upon application of the convolution theorem we obtain

$$\begin{aligned} \widetilde{\mathbf{E}}(k_x, k_y) &= \Big[ -jkZ\widetilde{\mathbf{J}}(k_x, k_y) + \frac{jZ}{k}\widetilde{\mathbf{J}}(k_x, k_y) \cdot \mathbf{k}_{2d}\mathbf{k}_{2d} \\ &\quad + j\widetilde{\mathbf{M}}(k_x, k_y) \times \mathbf{k}_{2d} \Big] \widetilde{G}_{2d}(k_x, k_y) \\ \widetilde{\mathbf{H}}(k_x, k_y) &= \Big[ -jkY\widetilde{\mathbf{M}}(k_x, k_y) + \frac{jY}{k}\widetilde{\mathbf{M}}(k_x, k_y) \cdot \mathbf{k}_{2d}\mathbf{k}_{2d} \\ &\quad - j\widetilde{\mathbf{J}}(k_x, k_y) \times \mathbf{k}_{2d} \Big] \widetilde{G}_{2d}(k_x, k_y) \end{aligned} \tag{2.157}$$

where $\widetilde{\mathbf{J}} = \hat{x}\widetilde{\mathbf{J}}_x + \hat{y}\widetilde{\mathbf{J}}_y + \hat{z}\widetilde{\mathbf{J}}_z$ and $\widetilde{\mathbf{M}} = \hat{x}\widetilde{\mathbf{M}}_x + \hat{y}\widetilde{\mathbf{M}}_y + \hat{z}\widetilde{\mathbf{M}}_z$. It remains to compute the transform of the Green's function, and to do so we recall the differential equation

$$\left(\frac{\partial^2}{\partial x^2} + \frac{\partial^2}{\partial y^2} + k^2\right) G_{2d}(\overline{\rho}, 0) = -\delta(x)\delta(y) \tag{2.158}$$

obtained from (2.116) by setting $\overline{\rho}' = 0$. Also, from (2.156)

$$\begin{aligned} G_{2d}(\overline{\rho}, 0) &= \frac{-j}{4} H_o^{(2)}\left(k\sqrt{x^2+y^2}\right) \\ &= \frac{1}{(2\pi)^2}\int_{\infty}^{\infty}\int_{\infty}^{\infty} \widetilde{G}_{2d}(k_x, k_y) e^{j(k_x x + k_y y)}\, dk_x\, dk_y \end{aligned} \tag{2.159}$$

and in accordance with distribution theory (Papoulis, 1962) we may express the Dirac delta function as

$$\delta(x)\delta(y) = \frac{1}{(2\pi)^2}\int_{-\infty}^{\infty}\int_{-\infty}^{\infty} e^{j(k_x x + k_y y)}\, dk_x\, dk_y \tag{2.160}$$

Substituting (2.159) and (2.160) into (2.158) and taking the derivatives inside the integral, we obtain

$$\frac{1}{(2\pi)^2}\int_{-\infty}^{\infty}\int_{-\infty}^{\infty} (-k_x^2 - k_y^2 + k^2)\widetilde{G}_{2d}(k_x, k_y) e^{j(k_x x + k_y y)}\, dk_x\, dk_y$$
$$= -\frac{1}{(2\pi)^2}\int_{-\infty}^{\infty}\int_{-\infty}^{\infty} e^{j(k_x x + k_y y)}\, dk_x\, dk_y$$

from which it follows that

$$\widetilde{G}_{2d}(k_x, k_y) = \frac{1}{k_y^2 + k_x^2 - k^2}. \tag{2.161}$$

Consequently we may write

$$H_o^{(2)}\left(k\sqrt{x^2+y^2}\right) = \frac{j}{\pi^2}\int_{-\infty}^{\infty}\int_{-\infty}^{\infty} \frac{1}{k_y^2 + k_x^2 - k^2} e^{j(k_x x + k_y y)}\, dk_x\, dk_y \tag{2.162}$$

which can be referred to as the plane wave spectral representation of the Hankel function. This is because the Fourier integral can be thought as a sum of inhomogeneous plane waves, that is, plane waves whose amplitude is a function of the spectral variables $k_x$ and $k_y$, which can be associated with the propagation constants along the $x$ and $y$ directions, respectively.

For the existence of the Fourier transform pair in the classical sense, it is necessary that the function and its transform be integrable. However, as seen from (2.162) the transform of $H_o^{(2)}(k_o\sqrt{x^2+y^2})$ has poles in the $k_y$ plane located at (see Figure 2-11)

$$k_y = \pm\sqrt{k^2 - k_x^2}. \tag{2.163}$$

For real $k$, these poles are on the $Re(k_y)$ axis, and in this case the integral in (2.162) is non-convergent. This difficulty, though, can be avoided by introducing a small

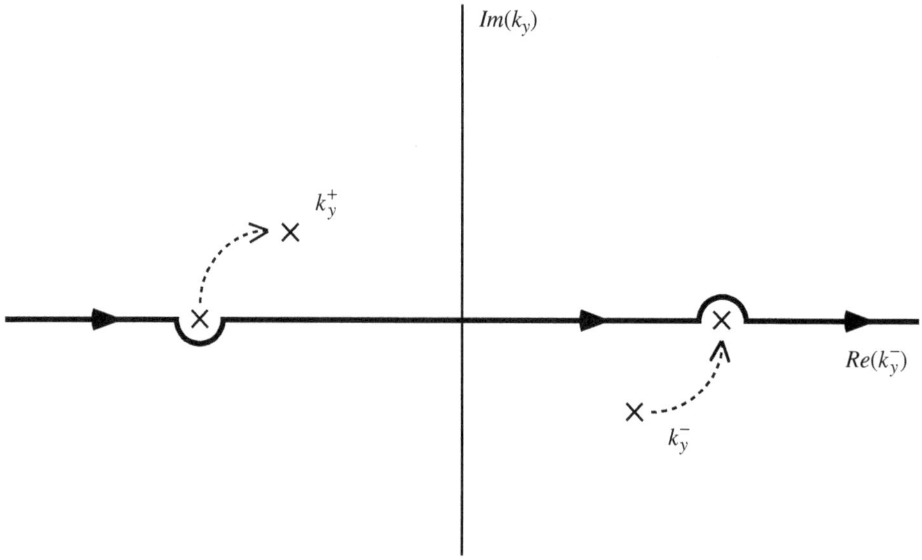


**FIGURE 2-11** Integration path for Fourier transform integral.

loss in the medium. That is, we assume that $k = k' - jk''$, where $k'' \to 0^+$, and the poles (2.163) are then located off the real axis.

By closing the integral over $k_y$ with a semi-infinite contour in the upper half of the $k_y$ plane, Cauchy's theorem gives

$$H_o^{(2)}\left(k_o\sqrt{x^2+y^2}\right) = -\frac{1}{\pi}\int_{-\infty}^{\infty}\frac{e^{jk_y^+ y}}{k_y^+}e^{jk_x x}\,dk_x;\quad y>0 \tag{2.164}$$

in which

$$k_y^+ = k_y,\quad Im(k_y) > 0. \tag{2.165}$$

This is valid only for $y > 0$ to ensure a vanishing semi-infinite contour in the application of Cauchy's theorem. For $y < 0$, the path of integration is closed in the lower half of the $k_y$ plane. This gives

$$H_o^{(2)}\left(k_o\sqrt{x^2+y^2}\right) = \frac{1}{\pi}\int_{-\infty}^{\infty}\frac{e^{jk_y^- y}}{k_y^-}e^{jk_x x}\,dk_x;\quad y<0 \tag{2.166}$$

in which

$$k_y^- = k_y,\quad Im(k_y) < 0 \tag{2.167}$$

and as such $k_y^- = -k_y^+$. Combining (2.164) through (2.167) we may write

$$H_o^{(2)}\left(k\sqrt{x^2+y^2}\right) = \frac{1}{\pi}\int_{-\infty}^{\infty}\frac{e^{-jk_y^-|y|}}{k_y^-}e^{jk_x x}\,dk_x \tag{2.168}$$

and since $Im(k) < 0$, then $Im\left(\sqrt{k^2-k_x^2}\right) < 0$, implying that $k_y^- = \sqrt{k^2-k_x^2}$. In the case of a lossless medium and for $k > k_x$, the factor $k_y^-$ becomes real, and from Figure 2-11, it is now clear how the path of integration is chosen in (2.162) and (2.168). In particular, it must be indented around the poles or branch points to ensure the convergence of the integrals and to avoid the branch cuts of $k_y^-$.

When $|y| \to 0$, from (2.168) we obtain

$$H_o^{(2)}(k|x|) = \frac{1}{\pi}\int_{-\infty}^{\infty} \frac{e^{jk_x x}}{k_y^-}\, dk_x \tag{2.169}$$

and in accordance with (2.167)

$$k_y^- = \begin{cases} \sqrt{k^2 - k_x^2} & k > k_x \\ -j\sqrt{k_x^2 - k^2} & k < k_x. \end{cases} \tag{2.170}$$

Introducing the one-dimensional Fourier transform pair

$$\tilde{f}(k_x) = \int_{-\infty}^{\infty} f(x) e^{-jk_x x}\, dx \tag{2.171a}$$

$$f(x) = \frac{1}{2\pi}\int_{-\infty}^{\infty} \tilde{f}(k_x) e^{+jk_x x}\, dk_x. \tag{2.171b}$$

the result in (2.169) simply implies that the transform of $H_o^{(2)}(k|x|)$ is given by

$$\widetilde{H}_o^{(2)}(k_x) = \mathcal{F}\{H_o^{(2)}(k|x|)\} = \begin{cases} \frac{2}{\sqrt{k^2-k_x^2}} & k > k_x \\ \frac{2j}{\sqrt{k_x^2-k^2}} & k < k_x. \end{cases} \tag{2.172}$$

We note that the integral relations (2.168) and (2.169) are often the starting point for deriving a variety of integral identities relating to the Hankel and Bessel functions. For example, by taking the real part of (2.169) we obtain

$$J_o(k|x|) = Re\{H_o^{(2)}(k|x|)\} = \frac{1}{\pi}\int_{-k}^{k} \frac{e^{jk_x x}}{\sqrt{k^2 - k_x^2}}\, dk_x \tag{2.173}$$

where $J_o(k)$ denotes the zeroth-order Bessel function. Introducing now the variable transformation $k_x = k\cos\alpha$ permits us to rewrite (2.173) as

$$\begin{aligned} J_o(k|x|) &= \frac{1}{\pi}\int_{-\pi/2}^{\pi/2} e^{jkx\sin\alpha}\, d\alpha = \frac{2}{\pi}\int_0^{\pi/2} \cos(kx\sin\alpha)\, d\alpha \\ &= \frac{1}{2\pi}\int_0^{2\pi} e^{jkx\cos\alpha}\, d\alpha = \frac{1}{2\pi}\int_0^{2\pi} e^{jk\rho\cos(\alpha-\phi)}\, d\alpha \end{aligned} \tag{2.174}$$

which is one of the usual integral expressions for the Bessel function. Other formulas can be derived in a similar manner.

Now that we have evaluated the transform of the two-dimensional Green's function and its implications, we are in a position to examine the behavior of the field spectra given in (2.157). Substituting (2.161) into the first of (2.157) yields

$$\widetilde{\mathbf{E}}(k_x, k_y) = -\frac{jZ}{k}\widetilde{\mathbf{J}} \cdot \frac{(k^2\bar{\mathbf{I}}_{2d} - \mathbf{k}_{2d}\mathbf{k}_{2d})}{k_x^2 + k_y^2 - k^2} + j\frac{\widetilde{\mathbf{M}}(k_x, k_y) \times \mathbf{k}_{2d}}{k_x^2 + k_y^2 - k^2}. \tag{2.175}$$

Taking the limit of this expression as $k_x \to \infty$ with $k_y$ finite or alternatively as $k_y \to \infty$ with $k_x$ finite yields

$$
\begin{aligned}
\widetilde{\mathbf{E}}(k_x \to \infty, k_y) &\to \frac{jZ}{k}\widetilde{J}_x \\
\widetilde{\mathbf{E}}(k_x, k_y \to \infty) &\to \frac{jZ}{k}\widetilde{J}_y.
\end{aligned}
\tag{2.176}
$$

These results imply that the transformation of the electric field is bandlimited only if the transform of the current is also bandlimited. Thus, strictly speaking, the inverse Fourier integral is not convergent for impulsive currents, which is a consequence of the singular kernel associated with the field integrals given by (2.127). The evaluation of these integrals in the source region required special treatment as discussed in Section 2.5.4, and a similar procedure can be used for the evaluation of the inverse Fourier integral. Specifically, we may express $\mathbf{E}(\overline{\rho})$ as

$$
\begin{aligned}
\mathbf{E}(\overline{\rho}) = \mathbf{E}(x, y) = &\frac{-jZ}{(2\pi)^2 k}\int_{-\infty}^{\infty}\int_{-\infty}^{\infty}\widetilde{\mathbf{J}}(k_x, k_y) \\
&\cdot\left[\left(\frac{k^2\overline{\mathbf{I}}_{2d} - \mathbf{k}_{2d}\mathbf{k}_{2d}}{k_x^2 + k_y^2 - k^2}\right) + \overline{\mathbf{I}}_{2d}\right] e^{j(k_x x + k_y y)}\, dk_x\, dk_y \\
&+ \frac{jZ}{(2\pi)^2 k}\int_{-\infty}^{\infty}\int_{-\infty}^{\infty}(\widetilde{\mathbf{J}}(k_x, k_y)\cdot\overline{\mathbf{I}}_{2d})e^{j(k_x x + k_y y)}\, dk_x\, dk_y \\
&+ \frac{j}{(2\pi)^2}\int_{-\infty}^{\infty}\int_{-\infty}^{\infty}\frac{\widetilde{\mathbf{M}}(k_x, k_y)\times\mathbf{k}_{2d}}{k_x^2 + k_y^2 - k^2}e^{j(k_x x + k_y y)}\, dk_x\, dk_y
\end{aligned}
\tag{2.177}
$$

where we have added and subtracted the term

$$
\begin{aligned}
&\frac{jZ}{(2\pi)^2 k}\int_{-\infty}^{\infty}\int_{-\infty}^{\infty}(\widetilde{\mathbf{J}}(k_x, k_y)\cdot\overline{\mathbf{I}}_{2d})e^{j(k_x x + k_y y)}\, dk_x\, dk_y \\
&\qquad = \frac{jZ}{(2\pi)^2 k}\int_{-\infty}^{\infty}\int_{-\infty}^{\infty}\widetilde{\mathbf{J}}(k_x, k_y)e^{j(k_x x + k_y y)}\, dk_x\, dk_y
\end{aligned}
\tag{2.178}
$$

so that the first integral of (2.177) is now convergent, and the same is true for the third integral. In accordance with the definitions (2.148), the second integral is simply equal to $\frac{jZ}{k}\mathbf{J}$, and thus we can rewrite (2.177) as

$$
\begin{aligned}
\mathbf{E}(\overline{\rho}) = &\frac{jZ}{k}\mathbf{J}(\overline{\rho}) - \frac{jZ}{(2\pi)^2 k}\int_{-\infty}^{\infty}\int_{-\infty}^{\infty}\widetilde{\mathbf{J}}(k_x, k_y) \\
&\cdot\left[\frac{k^2\overline{\mathbf{I}}_{2d} - \mathbf{k}_{2d}\mathbf{k}_{2d}}{k_x^2 + k_y^2 - k^2} + \overline{\mathbf{I}}_{2d}\right] e^{j(k_x x + k_y y)}\, dk_x\, dk_y \\
&+ \frac{j}{(2\pi)^2}\int_{-\infty}^{\infty}\int_{-\infty}^{\infty}\frac{\widetilde{\mathbf{M}}(k_x, k_y)\times\mathbf{k}_{2d}}{k_x^2 + k_y^2 - k^2}e^{j(k_x x + k_y y)}\, dk_x\, dk_y
\end{aligned}
\tag{2.179}
$$

which is compatible and completely equivalent to the representation given by (2.147), provided we set $\mathbf{M} = 0$.

## 2.6.2 Three Dimensions

The Fourier transform pair in three dimensions is defined by

$$\tilde{f}(k_x, k_y, k_z) = \int_{-\infty}^{\infty}\int_{-\infty}^{\infty}\int_{-\infty}^{\infty} f(x, y, z)e^{-j(k_x x + k_y y + k_z z)}\, dx\, dy\, dz$$

$$f(\mathbf{r}) = f(x, y, z) = \frac{1}{(2\pi)^3}\int_{-\infty}^{\infty}\int_{-\infty}^{\infty}\int_{-\infty}^{\infty} \tilde{f}(k_x, k_y, k_z)e^{j(k_x x + k_y y + k_z z)}\, dk_x\, dk_y\, dk_z \tag{2.180}$$

with $k_x$, $k_y$, and $k_z$ being the Fourier or spectral variables. In subsequent references we shall more compactly write this transform pair using the notation

$$\begin{aligned} \tilde{f}(k_x, k_y, k_z) &= \mathcal{F}\{f(x, y, z)\} \\ f(x, y, z) &= \mathcal{F}^{-1}\{\tilde{f}(k_x, k_y, k_z)\}. \end{aligned} \tag{2.181}$$

It follows directly from (2.180) that

$$\begin{aligned} \mathcal{F}\left\{\frac{\partial^2}{\partial\alpha\,\partial\beta} f(x, y, z)\right\} &= (+jk_\alpha)(+jk_\beta)\tilde{f}(k_x, k_y, k_z) \\ &= -k_\alpha k_\beta \tilde{f}(k_x, k_y, k_z) \end{aligned} \tag{2.182}$$

where $\alpha$ and $\beta$ represent any of the Cartesian coordinates $x$, $y$, or $z$. Also, from the convolution theorem we have

$$\begin{aligned} f(x, y, z) * g(x, y, z) &= \int_{-\infty}^{\infty}\int_{-\infty}^{\infty}\int_{-\infty}^{\infty} f(x', y', z') \\ &\quad \cdot g(x - x', y - y', z - z')\, dx'\, dy'\, dz' \\ &= \mathcal{F}^{-1}\left\{\tilde{f}(k_x, k_y, k_z)\tilde{g}(k_x, k_y, k_z)\right\}. \end{aligned} \tag{2.183}$$

When the identities (2.181) through (2.183) are applied to the integral representations (2.109) we obtain

$$\begin{aligned} \widetilde{\mathbf{E}}(k_x, k_y, k_z) = {} & \frac{-jZ}{k}\widetilde{G}(k_x, k_y, k_z)\begin{pmatrix} (k^2 - k_x^2) & -k_x k_y & -k_x k_z \\ -k_y k_x & (k^2 - k_y^2) & -k_y k_z \\ -k_z k_x & -k_z k_y & (k^2 - k_z^2) \end{pmatrix}\begin{pmatrix} \widetilde{J}_x \\ \widetilde{J}_y \\ \widetilde{J}_z \end{pmatrix} \\ & - j\widetilde{G}(k_x, k_y, k_z)\begin{pmatrix} 0 & -k_z & k_y \\ k_z & 0 & -k_x \\ -k_y & k_x & 0 \end{pmatrix}\begin{pmatrix} \widetilde{M}_x \\ \widetilde{M}_y \\ \widetilde{M}_z \end{pmatrix} \end{aligned} \tag{2.184}$$

as the transform of the electric field. In this,

$$\widetilde{G}(k_x, k_y, k_z) = \mathcal{F}\left\{\frac{e^{-jk\sqrt{x^2+y^2+z^2}}}{4\pi\sqrt{x^2 + y^2 + z^2}}\right\} \tag{2.185}$$

denotes the transform of the unbounded-space Green's function,

$$\widetilde{\mathbf{J}}(k_x, k_y, k_z) = \hat{x}\widetilde{J}_x + \hat{y}\widetilde{J}_y + \hat{z}\widetilde{J}_z \tag{2.186}$$

is the transform of the electric current, and likewise

$$\widetilde{\mathbf{M}}(k_x, k_y, k_z) = \hat{x}\widetilde{M}_x + \hat{y}\widetilde{M}_y + \hat{z}\widetilde{M}_z \tag{2.187}$$

is the transform of the magnetic current. By duality the transform of the magnetic field is explicitly given by

$$\begin{aligned}\widetilde{\mathbf{H}}(k_x, k_y, k_z) = \frac{-jY}{k}\widetilde{G}(k_x, k_y, k_z) &\begin{pmatrix} k^2 - k_x^2 & -k_x k_y & -k_x k_z \\ -k_y k_x & k^2 - k_y^2 & -k_y k_z \\ -k_z k_x & -k_z k_y & k^2 - k_z^2 \end{pmatrix} \begin{pmatrix} \widetilde{M}_x \\ \widetilde{M}_y \\ \widetilde{M}_z \end{pmatrix} \\ + j\widetilde{G}(k_x, k_y, k_z) &\begin{pmatrix} 0 & -k_z & k_y \\ k_z & 0 & -k_x \\ -k_y & k_x & 0 \end{pmatrix} \begin{pmatrix} \widetilde{J}_x \\ \widetilde{J}_y \\ \widetilde{J}_z \end{pmatrix}.\end{aligned} \tag{2.188}$$

As was the case in two dimensions, these matrices may be written more compactly by introducing the transform pair

$$\begin{aligned}\mathcal{F}\{\nabla f(x, y, z)\} &= \widetilde{\nabla}\tilde{f}(k_x, k_y, k_z) \\ &= j\mathbf{k}\tilde{f}(k_x, k_y, k_z)\end{aligned} \tag{2.189}$$

in which

$$\mathbf{k} = \hat{x}k_x + \hat{y}k_y + \hat{z}k_z. \tag{2.190}$$

From (2.52) we then have

$$\begin{aligned}\widetilde{\mathbf{E}}(k_x, k_y, k_z) &= -jkZ\widetilde{\mathbf{J}}(k_x, k_y, k_z) \cdot \left(\bar{\mathbf{I}} - \frac{\mathbf{kk}}{k^2}\right)\widetilde{G}(k_x, k_y, k_z) \\ &\quad + j\widetilde{\mathbf{M}}(k_x, k_y, k_z) \times \mathbf{k}\widetilde{G}(k_x, k_y, k_z) \\ \widetilde{\mathbf{H}}(k_x, k_y, k_z) &= -jkY\widetilde{\mathbf{M}}(k_x, k_y, k_z) \cdot \left(\bar{\mathbf{I}} - \frac{\mathbf{kk}}{k^2}\right)\widetilde{G}(k_x, k_y, k_z) \\ &\quad - j\widetilde{\mathbf{J}}(k_x, k_y, k_z) \times \mathbf{k}\widetilde{G}(k_x, k_y, k_z).\end{aligned} \tag{2.191}$$

Also, from (2.95) we find that the transform of the dyadic Green's function is given by

$$\widetilde{\bar{\mathbf{\Gamma}}}(k_x, k_y, k_z) = \mathcal{F}\{\bar{\mathbf{\Gamma}}(\mathbf{r}, \mathbf{r}')\} = -\left(\bar{\mathbf{I}} - \frac{\mathbf{kk}}{k^2}\right)\widetilde{G}(k_x, k_y, k_z). \tag{2.192}$$

Needless to mention, (2.191) are identical to the matrix representations (2.184) and (2.188).

It remains to explicitly determine $\widetilde{G}(k_x, k_y, k_z)$ defined in (2.185). However, instead of integrating the unbounded Green's function directly we recall the identity (2.38)

$$(\nabla^2 + k^2)G(\mathbf{r}, 0) = -\delta(\mathbf{r}) \tag{2.193}$$

where $\delta(\mathbf{r}) = \delta(x)\delta(y)\delta(z)$. Taking the Fourier transform of this equation yields

$$(-\mathbf{k}\cdot\mathbf{k} + k^2)\widetilde{G}(k_x, k_y, k_z) = -1 \tag{2.194}$$

which gives

$$\widetilde{G}(k_x, k_y, k_z) = \frac{1}{k_x^2 + k_y^2 + k_z^2 - k^2} \tag{2.195}$$

and when this result is compared with (2.185) we deduce the identity

$$\frac{e^{-jkr}}{4\pi r} = \frac{e^{-jk\sqrt{x^2+y^2+z^2}}}{4\pi\sqrt{x^2+y^2+z^2}} = \frac{1}{(2\pi)^3}\int_{-\infty}^{\infty}\int_{-\infty}^{\infty}\int_{-\infty}^{\infty}\frac{e^{j(k_x x+k_y y+k_z z)}}{k_x^2+k_y^2+k_z^2-k^2}\,dk_x\,dk_y\,dk_z. \tag{2.196}$$

As was the case for two dimensions, the Fourier integral in (2.196) does not exist in the classical sense because of the poles at

$$k_z = \pm\sqrt{k^2 - k_x^2 - k_y^2}. \tag{2.197}$$

This difficulty, though, can be alleviated by assuming a small loss in the medium, which amounts to setting $k = k' - jk''$ with $k'' \neq 0$. The poles (2.197) are then shifted away from the real axis. Consequently, by closing the integral over $k_z$ with a semi-infinite contour in the upper half of the $k_z$ plane, we capture one of the poles. Upon invoking Cauchy's theorem we obtain

$$\frac{e^{-jkr}}{4\pi r} = \frac{j}{2(2\pi)^2}\int_{-\infty}^{\infty}\int_{-\infty}^{\infty}\frac{e^{jk_z^+ z}}{k_z^+}e^{j(k_x x+k_y y)}\,dk_x\,dk_y; \quad z > 0 \tag{2.198}$$

in which

$$k_z^+ = k_z, \quad Im(k_z) > 0. \tag{2.199}$$

This result is valid only for $z > 0$ to ensure a vanishing semi-infinite contour in the application of Cauchy's theorem. For $z < 0$, the path of integration is closed by a semi-infinite contour in the lower half of the $k_z$ plane giving

$$\frac{e^{-jkr}}{4\pi r} = \frac{-j}{2(2\pi)^2}\int_{-\infty}^{\infty}\int_{-\infty}^{\infty}\frac{e^{jk_z^- z}}{k_z^-}e^{j(k_x x+k_y y)}\,dk_x\,dk_y; \quad z < 0 \tag{2.200}$$

in which

$$k_z^- = k_z, \quad Im(k_z) < 0 \tag{2.201}$$

and as such $k_z^- = -k_z^+$. Combining (2.198) through (2.201) we deduce the well-known Weyl identity

$$\frac{e^{-jkr}}{4\pi r} = \frac{-j}{8\pi^2} \int_{-\infty}^{\infty} \int_{-\infty}^{\infty} \frac{e^{-jk_z^-|z|}}{k_z^-} e^{j(k_x x + k_y y)}\, dk_x\, dk_y. \tag{2.202}$$

In the case of a lossless medium and for $k^2 > k_x^2 + k_y^2$, the factor $k_z^-$ is real, and the path of integration must then be identical to that shown in Figure 2-11, thus avoiding the branch cuts and ensuring the convergence of the integrals.

We remark that Weyl's identity is the starting point for deriving a variety of other identities involving Bessel and Hankel functions whose arguments are functions of both $x$ and $y$ or the standard cylindrical variables $(\rho, \phi)$. For example, by setting

$$\mathbf{k}_{2d} = \hat{x} k_x + \hat{y} k_y = k_\rho (\hat{x} \cos\alpha + \hat{y} \sin\alpha) \tag{2.203}$$

with $k_\rho = \sqrt{k_x^2 + k_y^2}$, it can be shown from (2.202) that

$$\frac{e^{-jkr}}{4\pi r} = \frac{-j}{8\pi^2} \int_0^{\infty} \int_0^{2\pi} \frac{k_\rho e^{-j[k_\rho \rho \cos(\alpha - \phi) + k_z^-|z|]}}{k_z^-}\, d\alpha\, dk_\rho. \tag{2.204}$$

Further, by invoking the identity (2.174) it follows that

$$\frac{e^{-jkr}}{4\pi r} = \frac{-j}{4\pi} \int_0^{\infty} \frac{k_\rho}{k_z^-} J_o(k_\rho \rho) e^{-jk_z^-|z|}\, dk_\rho \tag{2.205}$$

and in accordance with (2.201),

$$k_z^- = \begin{cases} \sqrt{k^2 - k_\rho^2} = \sqrt{k^2 - k_x^2 - k_y^2} & k^2 > k_\rho^2 \\ -j\sqrt{k_\rho^2 - k^2} = -j\sqrt{k_x^2 + k_y^2 - k^2} & k^2 < k_\rho^2. \end{cases} \tag{2.206}$$

Expression (2.205) is known as Sommerfeld's identity, and several variations of this can be derived by employing various relations among the Bessel and Hankel functions. For example, we may introduce in (2.205) the relation

$$J_o(k_\rho \rho) = \frac{H_o^{(1)}(k_\rho \rho) + H_o^{(2)}(k_\rho \rho)}{2} \tag{2.207}$$

where $H_o^{(1)}$ is the zeroth-order Hankel function of the first kind, and since

$$H_o^{(1)}(-k_\rho \rho) = H_o^{(1)}(e^{j\pi} k_\rho \rho) = -H_o^{(2)}(k_\rho \rho) \tag{2.208}$$

it follows that

$$\frac{e^{-jkr}}{4\pi r} = \frac{-j}{8\pi} \int_{-\infty}^{\infty} \frac{k_\rho}{k_z^-} H_o^{(2)}(k_\rho \rho) e^{-jk_z^-|z|}\, dk_\rho. \tag{2.209}$$

From the wave equation or through a change of variables in (2.209), it can also be shown that

$$\frac{e^{-jkr}}{4\pi r} = \frac{-j}{8\pi} \int_{-\infty}^{\infty} H_0^{(2)}\left(\sqrt{k^2 - k_z^2}\, \rho\right) e^{-jk_z|z|}\, dk_z \tag{2.210}$$

with the integration path as defined in Figure 2-11.

In closing this section we also note that when $|z| \to 0$, from (2.202) we deduce the identity

$$\frac{e^{-jk\sqrt{x^2+y^2}}}{4\pi\sqrt{x^2+y^2}} = \frac{-j}{8\pi^2}\int_{-\infty}^{\infty}\int_{-\infty}^{\infty}\frac{e^{+j(k_x x+k_y y)}}{k_z^-}\,dk_x\,dk_y \qquad (2.211)$$

and upon recalling the definitions for the two-dimensional transform pair, it is seen that this identity implies

$$\mathcal{F}\left\{\frac{e^{-jk\sqrt{x^2+y^2}}}{4\pi\sqrt{x^2+y^2}}\right\} = \begin{cases} \frac{1}{2j\sqrt{k^2-k_x^2-k_y^2}} & k^2 > k_x^2+k_y^2 \\ \frac{1}{2\sqrt{k_x^2+k_y^2-k^2}} & k^2 < k_x^2+k_y^2. \end{cases} \qquad (2.212)$$

This transform will be found useful in the study and modeling of thin planar layers of material and flat metallic plates.

## 2.7 | RADIATION OVER A DIELECTRIC HALF SPACE

A particular utility of the identity (2.202) is in solving boundary value problems associated with boundary conditions that are invariant with respect to $x$ and $y$. As an example, let us consider the radiation of an electric source $\mathbf{J} = \hat{x}J_x(x, y)$ located at a dielectric interface as shown in Figure 2-12. Transforming the fields with respect to $x$ and $y$ in accordance with (2.148), from (2.202) we can readily write down the transform of the field generated by $\hat{x}J_x(x, y)$. For example, in the case of the $x$-component we have

$$\widetilde{E}_x^g(k_x, k_y, z) = \frac{-jZ_1}{k_1}\widetilde{J}_x(k_x, k_y)\left(k_1^2 - k_x^2\right)\left(\frac{-j}{2}\right)\frac{e^{-jk_{1z}^-|z|}}{k_{1z}^-} \qquad (2.213)$$

in which $k_1 = \omega\sqrt{\mu_1\epsilon_1}$ is the wave number in region 1, and $Z_1 = \sqrt{\mu_1/\epsilon_1}$ is the intrinsic impedance in the same region. Clearly, this represents a plane wave traveling along the $z$-direction with a propagation constant of $k_{1z}^- = \sqrt{k_1^2 - k_\rho^2}$. To consider the reflection of this wave from the dielectric interface, it is instructive to decompose it into transverse-electric (TE) and transverse-magnetic (TM) components. For the

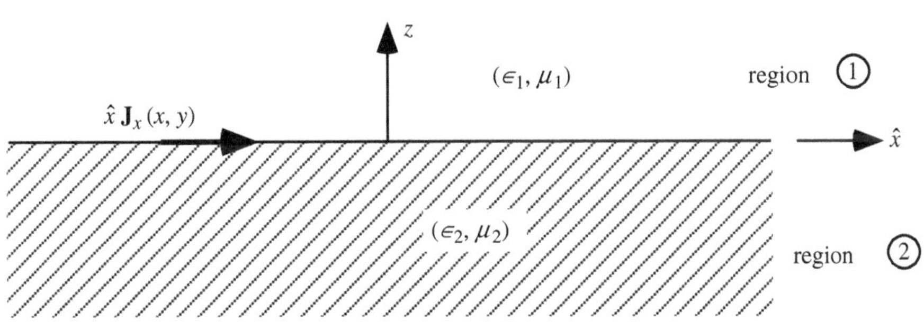


**FIGURE 2-12** Geometry of a source on a dielectric interface.

TE field, $E_z = 0$, and for the TM field $H_z = 0$, and thus such a decomposition can be accomplished once the fields are expressed in terms of their $E_z$ and $H_z$ components. Since $\frac{\partial}{\partial z} \to \pm jk_{1z}^{-}$, from Maxwell's equations it is not difficult to show that[1]

$$\widetilde{\mathbf{E}}_t(k_x, k_y, z) = \frac{1}{k_1^2 - (k_{1z}^{-})^2}\left[\widetilde{\nabla}_t\left(\frac{\partial \widetilde{E}_z}{\partial z}\right) + jk_1 Z_1 \hat{z} \times \nabla_t \widetilde{H}_z\right] \tag{2.214}$$

where $\widetilde{\mathbf{E}}_t = \widetilde{\mathbf{E}} - \hat{z}\widetilde{E}_z$, and $\nabla_t = \nabla - \hat{z}\frac{\partial}{\partial z}$. From (2.184) and (2.202), the transforms of the $z$-components generated by the source alone are found to be

$$\begin{aligned}\widetilde{E}_z^g(k_x, k_y, z) &= \widetilde{E}_z^g(k_x, k_y, z = 0^{\pm})e^{\mp jk_{1z}^{-} z}\\ &= \mp \frac{jZ_1}{k_1}\widetilde{J}_x(k_x, k_y)k_x k_{1z}^{-}\left(\frac{-j}{2}\right)\frac{e^{\mp jk_{1z}^{-} z}}{k_{1z}^{-}}\end{aligned} \tag{2.215}$$

$$\widetilde{H}_z^g(k_x, k_y, z) = -j\widetilde{J}_x(k_x, k_y)k_y\left(\frac{-j}{2}\right)\frac{e^{-jk_{1z}^{-}|z|}}{k_{1z}^{-}}. \tag{2.216}$$

Each of these wave components in (2.215) and (2.216) independently gives rise to a reflected field, which is simply given by

$$E_z^r(k_x, k_y, z > 0) = R^{TM}\widetilde{E}_z^g(k_x, k_y, z = 0^-)e^{-jk_{1z}^{-} z} \tag{2.217}$$

$$H_z^r(k_x, k_y, z > 0) = R^{TE}\widetilde{H}_z^g(k_x, k_y, z = 0^-)e^{-jk_{1z}^{-} z} \tag{2.218}$$

where

$$R^{TM} = -\frac{\epsilon_1 k_{2z}^{-} - \epsilon_2 k_{1z}^{-}}{\epsilon_1 k_{2z}^{-} + \epsilon_2 k_{1z}^{-}} \tag{2.219}$$

$$R^{TE} = \frac{\mu_2 k_{1z}^{-} - \mu_1 k_{2z}^{-}}{\mu_2 k_{1z}^{-} + \mu_1 k_{2z}^{-}} \tag{2.220}$$

are the TM and TE plane wave reflection coefficients, respectively, in which $k_{2z}^{-} = \sqrt{(k_2)^2 - k_\rho^2}$ with $k_2 = \omega\sqrt{\mu_2\epsilon_2}$. They are readily derived by employing transmission line theory concepts in conjunction with the TE and TM wave impedances $\left(Z^{TM} = \frac{k_z^{-}}{\omega\epsilon} \text{ and } Z^{TE} = \frac{\omega\mu}{k_z^{-}}\right)$. By combining (2.115) through (2.218), the $z$-components of the total TE and TM fields above the interface can now be expressed as

$$\widetilde{E}_z(k_x, k_y, z > 0) = \widetilde{E}_z^g + \widetilde{E}_z^r = -\frac{jZ_1}{k_1}\widetilde{J}_x(k_x, k_y)k_x k_{1z}^{-}(1 - R^{TM})\left(\frac{-j}{2}\right)\frac{e^{-jk_{1z}^{-} z}}{k_{1z}^{-}} \tag{2.221}$$

$$\widetilde{H}_z(k_x, k_y, z > 0) = \widetilde{H}_z^g + \widetilde{H}_z^r = -j\widetilde{J}_x(k_x, k_y)k_y(1 + R^{TE})\left(\frac{-j}{2}\right)\frac{e^{-jk_{1z}^{-} z}}{k_{1z}^{-}}. \tag{2.222}$$

---

[1] Separate in Maxwell's equation the $z$- and transverse components. Then by crossing one of them with $\hat{z}$, eliminate from the other the transverse electric/magnetic field.

The spectra of the other field components can be obtained by substituting (2.221) and (2.222) into (2.214). Note that when medium 2 is a conductor, then $R^{TM} \rightarrow +1$ and $R^{TE} \rightarrow -1$.

To obtain the fields in terms of the spatial variables $x$, $y$, and $z$, each spectral field component must be inverse transformed as dictated by (2.148b). This results in an integral of the form

$$I(\Omega) = \int_{-\infty}^{\infty} \int_{-\infty}^{\infty} f(x, y) e^{j\Omega g(x,y)}\, dx\, dy \tag{2.223}$$

(where $g(x, y)$ is real) whose evaluation must take into account any possible integrand poles and branch cuts when the observation point is near or on the surface of the dielectric interface (Collin, 1991; Felsen & Marcuvitz, 1973 [for a discussion on branch cuts and branch point singularities, see pp. 459–476 and for asymptotic integral evaluations see Chapter 4]). The residues of the integrand poles provide the contribution of surface/guided and leaky wave modes, which will be present in the case of a layered reflecting medium (i.e., a grounded slab) or simply any reflecting surface that can support guided waves below the dielectric interface. For the particular problem illustrated in Figure 2-12, the reflecting surface is a homogeneous half space and cannot support any surface waves other than those below the Brewster angle, which can be considered as part of the reflected field. The latter, along with direct source contribution are the only fields present in the far zone and can be explicitly determined as $r \rightarrow \infty$ by considering an asymptotic evaluation of the integral (2.223). This type of evaluation relies on the principle that, since $\Omega g(x, y) = r g(x, y)$, the exponential part of the integrand is rapidly oscillating, and thus any contribution must come from the portion of integration near the region where

$$\frac{\partial g(x, y)}{\partial x} = 0, \quad \frac{\partial g(x, y)}{\partial y} = 0. \tag{2.224}$$

These relations define one or more points $(x_s, y_s)$ referred to as the stationary points. Based on 2D stationary phase method analysis (Felsen and Marcuvitz, 1973; Senior and Volakis, 1995), it follows that for large $\Omega$

$$I(\Omega) \sim f(x_s, y_s) \left(\frac{2\pi}{\Omega}\right) e^{j\Omega g(x_s, y_s)} \frac{e^{j(\pi/4)\sigma}}{\left| det\left(\dfrac{\partial^2 g}{\partial x, \partial y}\right)\right|^{1/2}_{x=x_s, y=y_s}} \tag{2.225}$$

where

$$det\left(\frac{\partial^2 g}{\partial x, \partial y}\right) = \begin{vmatrix} \dfrac{\partial^2 g}{\partial x^2} & \dfrac{\partial^2 g}{\partial x \partial y} \\ \dfrac{\partial^2 g}{\partial y \partial x} & \dfrac{\partial^2 g}{\partial y^2} \end{vmatrix}$$

and

$$\sigma = sgn(d_1) + sgn(d_2)$$

in which $(d_1, d_2)$ are the eigenvalues of the matrix comprising the elements of $det\left(\frac{\partial^2 g}{\partial x, \partial y}\right)$ evaluated at $x = x_s$, $y = y_s$. Also, the function $sgn(x) = sign(x) = \pm 1$.

From (2.214) and (2.148b), the pertinent inverse Fourier integral to be evaluated for the TM field is

$$\mathbf{E}^{TM}(x, y, z) = -\frac{Z_1}{2k_1}\left(\frac{1}{2\pi}\right)^2 \int_{-\infty}^{\infty}\int_{-\infty}^{\infty}\left(\left[\frac{\hat{x}k_x k_{1z}^- + \hat{y}k_y k_{1z}^-}{k_1^2 - (k_{1z}^-)^2}\right] + \hat{z}\right) k_x$$

$$\cdot (1 - R^{TM})\widetilde{J}_x(k_x, k_y)e^{-jk_{1z}^- z}e^{j(k_x x + k_y y)}\, dk_x\, dk_y. \qquad (2.226)$$

Upon setting $x = r\sin\theta\cos\phi$, $y = r\sin\theta\sin\phi$, and $z = r\cos\theta$, $g(k_x, k_y)$ is identified to be (with $\Omega = r$)

$$g(k_x, k_y) = k_x \sin\theta\cos\phi + k_y\sin\theta\sin\phi - \cos\theta\sqrt{k_1^2 - k_x^2 - k_y^2} \qquad (2.227)$$

from which we find that the stationary point is located at

$$k_x = k_{xs} = -k_1\sin\theta\cos\phi \quad k_y = k_{ys} = -k_1\sin\theta\sin\phi. \qquad (2.228)$$

Consequently

$$\left| det\left(\frac{\partial^2 q}{\partial k_x, \partial k_y}\right)\right|^{1/2}_{k_x=k_{xs}, k_y=y_{ys}} = \frac{1}{k_1\cos\theta} \qquad (2.229)$$

and $\sigma = 2$. From the asymptotic formula (2.225), it then follows that the TM far-zone field is given by

$$E_\theta(r, \phi, \theta) = \hat{\theta}\cdot\mathbf{E}^{TM}(x, y, z) \sim -\frac{jk_o Z_o}{4\pi}\frac{e^{-jkr}}{r}\cos\theta\cos\phi(1 - R^{TM})\widetilde{J}_x(k_{xs}, k_{ys}) \qquad (2.230)$$

where we have set $k_1 = k_o = 2\pi/\lambda_o$ (i.e. $\epsilon_1 = \epsilon_o, \mu_1 = \mu_o$) and $Z_1 = Z_o = \sqrt{\mu_o/\epsilon_o}$. Also, upon setting $\epsilon_r = \epsilon_2/\epsilon_o$ and $\mu_r = \mu_2/\mu_o$, $R_{TM}$ can be reduced to

$$R^{TM} = -\frac{\sqrt{\epsilon_r\mu_r - \sin^2\theta} - \epsilon_r\cos\theta}{\sqrt{\epsilon_r\mu_r - \sin^2\theta} + \epsilon_r\cos\theta} \qquad (2.231)$$

which is the plane wave reflection coefficient associated with the planar dielectric interface. Clearly, this result is the sum of the far-zone field generated directly from the source and that reflected from the interface.

By following a similar analysis, we find that in general the far-zone fields due to the current distribution $\mathbf{J}(x, y, z) = \left[\hat{x}J_x(x, y) + \hat{y}J_y(x, y)\right]\delta(z)$, located on a dielectric interface, can be expressed as

$$E_\theta(r, \phi, \theta) = -\frac{jk_o Z_o}{4\pi}\frac{e^{-jk_o r}}{r}(1 - R^{TM})\Big[\cos\theta\cos\phi\widetilde{J}_x(k_{xs}, k_{ys})$$

$$+ \cos\theta\sin\phi\widetilde{J}_y(k_{xs}, k_{ys})\Big] \qquad (2.232)$$

$$E_\phi(r, \phi, \theta) = -\frac{jk_o Z_o}{4\pi}\frac{e^{-jk_o r}}{r}(1 + R^{TE})\Big[-\sin\phi\widetilde{J}_x(k_{xs}, k_{ys})$$

$$+ \cos\phi\widetilde{J}_y(k_{xs}, k_{ys})\Big], \qquad (2.233)$$

The first of these is the TM (to $z$) field, and the second represents the TE (to $z$) component. The expressions for $k_{xs}$ and $k_{ys}$ are again given by (2.228) whereas $R^{TM}$ is as given by (2.231) and

$$R^{TE} = \frac{\mu_r \cos\theta - \sqrt{\mu_r \epsilon_r - \sin^2\theta}}{\mu_r \cos\theta + \sqrt{\mu_r \epsilon_r - \sin^2\theta}} \tag{2.234}$$

is the TE plane wave reflection coefficient.

The previous elementary exercise demonstrates, rather well, the usefulness of the spectral representations presented in this chapter. Of importance is that the outlined method can be easily generalized to other source distributions and layered dielectric planar interfaces. Further, for near-zone observations, the spectral integrand characteristics yield the physical wave mechanisms, which take place near and below the dielectric interface. However, a more elementary and customary approach for this analysis is to consider the radiation of an infinitesimal dipole source above the dielectric interface, and in this case we may write the result in terms of a single integral by invoking one of Sommerfeld's identities. This problem was first solved by Sommerfeld and is briefly discussed next.

Consider again the problem illustrated in Figure 2-12 with the source now being replaced by the horizontal infinitesimal dipole

$$\mathbf{J}(x, y) = \hat{x} I \Delta\ell\, \delta(x)\, \delta(y) \tag{2.235}$$

and as such $\widetilde{\mathbf{J}}(k_x, k_y) = \hat{x} I \Delta\ell$. Substituting this into (2.221) through (2.222) and inverse transforming yields

$$E_z(x, y, z > 0) = \\ -\frac{Z_1}{2k_1}\left(\frac{1}{2\pi}\right)^2 (I\Delta\ell) \int_{-\infty}^{\infty}\int_{-\infty}^{\infty} \frac{\partial}{\partial x \partial z}(1 - R^{TM}) \frac{e^{-jk_{1z}^{-} z}}{k_{1z}^{-}} e^{j(k_x x + k_y y)}\, dk_x\, dk_y \tag{2.236}$$

$$H_z(x, y, z > 0) = \\ -\frac{j}{2}\left(\frac{1}{2\pi}\right)^2 (I\Delta\ell) \int_{-\infty}^{\infty}\int_{-\infty}^{\infty} \frac{\partial}{\partial y}(1 + R^{TE}) \frac{e^{-jk_{1z}^{-} z}}{k_{1z}^{-}} e^{j(k_x x + k_y y)}\, dk_x\, dk_y \tag{2.237}$$

where we have also made the replacements $jk_x \to \frac{\partial}{\partial x}$, $jk_y \to \frac{\partial}{\partial y}$, and $-jk_{1z}^{-} = \frac{\partial}{\partial z}$. Since $R^{TM}$ and $R^{TE}$ are functions of $k_\rho^2 = k_x^2 + k_y^2$, the aforementioned double integrals can be reduced to a single integral by invoking Sommerfeld's identity (2.209). In particular, by comparing (2.209) with (2.202), for $z > 0$ we deduce

$$E_z(\rho, \phi, z) = -\frac{Z_1}{k_1}\frac{(I\Delta\ell)}{8\pi}\frac{\partial}{\partial x \partial z}\int_{-\infty}^{\infty} \frac{k_\rho}{k_{1z}^{-}}(1 - R^{TM}) H_o^{(2)}(k_\rho \rho) e^{-jk_{1z}^{-} z}\, dk_\rho \tag{2.238}$$

$$H_z(\rho, \phi, z) = -j\frac{(I\Delta\ell)}{8\pi}\frac{\partial}{\partial y}\int_{-\infty}^{\infty} \frac{k_\rho}{k_{1z}^{-}}(1 + R^{TE}) H_o^{(2)}(k_\rho \rho) e^{-jk_{1z}^{-} z}\, dk_\rho. \tag{2.239}$$

Further, upon carrying out the derivatives we have

$$E_z(\rho, \phi, z) = -\frac{jZ_1}{k_1}\frac{(I\Delta\ell)}{8\pi}(\hat{\rho}\cdot\hat{x})\int_{-\infty}^{\infty} k_\rho^2(1 - R^{TM})H_1^{(2)}(k_\rho\rho)e^{-jk_{1z}^{-}z}\,dk_\rho \tag{2.240}$$

$$H_z(\rho, \phi, z) = j\frac{(I\Delta\ell)}{8\pi}(\hat{\rho}\cdot\hat{y})\int_{-\infty}^{\infty}\frac{k_\rho^2}{k_{1z}^{-}}(1 + R^{TE})H_1^{(2)}(k_\rho\rho)e^{-jk_{1z}^{-}z}\,dk_\rho. \tag{2.241}$$

Far-zone approximations to these integrands can be obtained by introducing the large argument approximation for the Hankel function (see (2.132))

$$H_1^{(2)}(k_\rho\rho) \sim j\sqrt{\frac{2}{\pi k}}e^{j\frac{\pi}{4}}\frac{e^{-jk\rho}}{\sqrt{\rho}} \tag{2.242}$$

and subsequently making use of the stationary phase method. When the result is substituted into (2.214) we again obtain the far-zone field expressions (2.232) and (2.233) with $\widetilde{J}_x(k_{xs}, k_{ys}) = (I\Delta\ell)$ and $\widetilde{J}_y(k_{xs}, k_{ys}) = 0$.

An alternative and more rigorous approach for evaluating the integrals is the steepest descent path method. In this case the integration path illustrated in Figure 2-11 is deformed to one on which the rapidly varying exponential portion of the integrand has constant phase and is decaying. To illustrate this idea let us consider the integral ($\Omega$ is large)

$$I_c(\Omega) = \int_C F(\zeta)e^{\Omega g(\zeta)}d\zeta; \quad \zeta = \zeta_r + j\zeta_i \tag{2.243}$$

where the path $C$ is illustrated in Figure 2-11, and $g(\zeta) = u(\zeta_r, \zeta_i) + jv(\zeta_r, \zeta_i)$. We now define the saddle point, $\zeta_s$, to be such that

$$\left.\frac{dg(\zeta)}{d\zeta}\right|_{\zeta=\zeta_s} = 0 \tag{2.244}$$

where $\zeta_s = \zeta_{rs} + j\zeta_{is}$, and an integration path $C_{SDP}$ on which

$$\begin{aligned} u(\xi_r, \xi_i) &\le u(\xi_{rs}, \xi_{is}) \\ v(\xi_r, \xi_i) &= v(\xi_{rs}, \xi_{is}). \end{aligned} \tag{2.245}$$

Then upon rewriting $I_c(\Omega)$ as

$$I_c(\Omega) = e^{\Omega g(\zeta_s)}\int_C F(\zeta)e^{\Omega[g(\zeta)-g(\zeta_s)]}\,d\zeta \tag{2.246}$$

it is clear from the definitions in (2.244) to (2.245) that the exponent $[g(\zeta) - g(\zeta_s)]$ is real and negative along the path $C_{SDP}$, which is consequently referred to as the steepest descent path. It is thus instructive to rewrite $I_c(\Omega)$ as

$$I_c(\Omega) = 2\pi j\sum Res + I_b(\Omega) + I_{SDP}(\Omega) \tag{2.247}$$

where

$$I_{SDP}(\Omega) = e^{\Omega g(\zeta_s)}\int_{C_{SDP}} F(\zeta)e^{\Omega[g(\zeta)-g(\zeta_s)]}\,d\zeta, \tag{2.248}$$

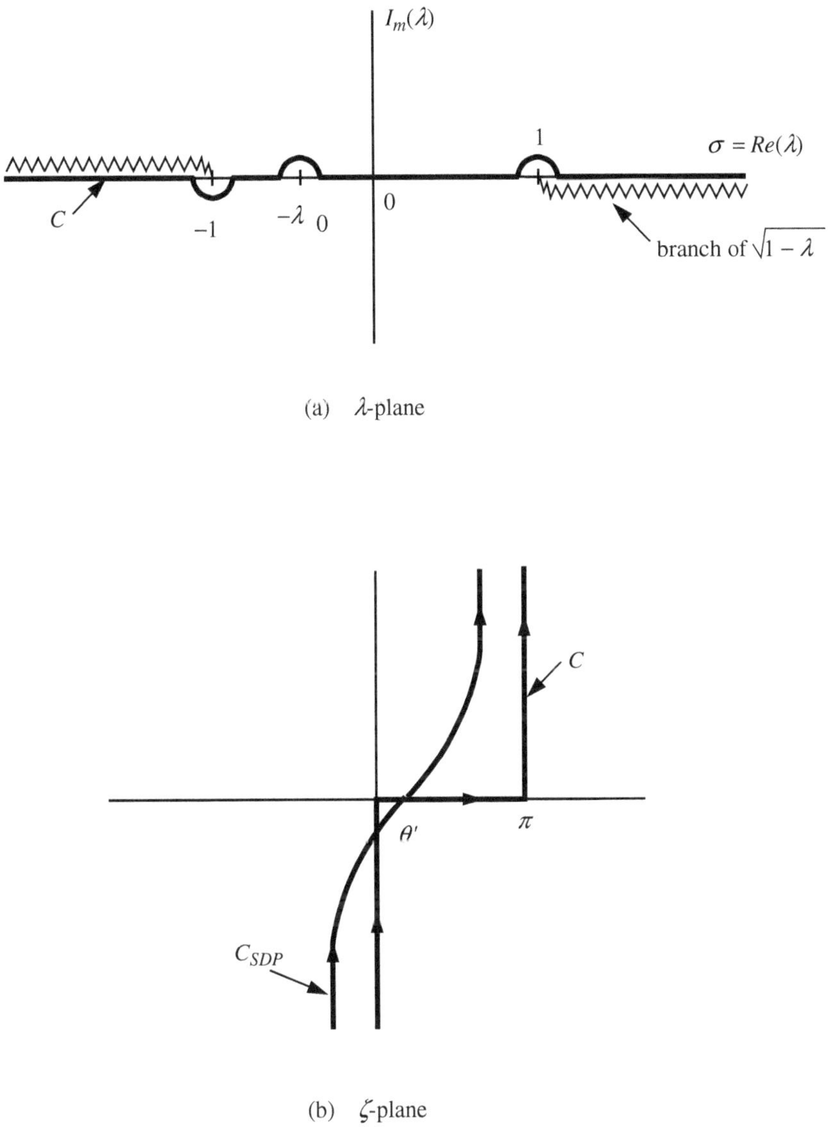


**FIGURE 2-13** Illustration of the $C$ contour in the $\lambda = \cos\zeta$ and $\zeta$ planes for the integral in (2.243). The $C_{SDP}$ path corresponds to $g(\zeta) = j\cos(\zeta - \theta')$.

$I_b(\Omega)$ provides the contribution of any branch points/cuts crossed in the deformation of the contour $C$ to $C_{SDP}$, and $2\pi j \sum Res$ is the residue contribution of the crossed poles. The steepest decent path $C_{SDP}$ corresponding to $g(\zeta) = j\cos(\zeta - \theta')$ is illustrated in Figure 2-13. Following the usual steepest descent path method to evaluate (2.248) (Felsen and Marcuvitz, 1973; Senior and Volakis, 1995) we obtain

$$I_{SDP}(\Omega) \sim e^{\Omega g(\zeta_s)}\sqrt{\frac{-2\pi}{\Omega g''(\zeta_s)}}F(\zeta_s) + O(\Omega) \tag{2.249}$$

and when this formula is employed in conjunction with the pertinent integrals (2.140) and (2.241) we recover the same results obtained via the stationary phase method.

## PROBLEMS

**1.** Given that

$$R = |\mathbf{r} - \mathbf{r}'| = \sqrt{(x - x')^2 + (y - y')^2 + (z - z')^2}$$

and

$$G(R) = \frac{e^{-jk|\mathbf{r}-\mathbf{r}'|}}{4\pi|\mathbf{r} - \mathbf{r}'|} = f(\mathbf{r} - \mathbf{r}')$$

**(a)** Show that

$$\nabla R = \hat{R}$$

$$\nabla \cdot \hat{R} = \frac{2}{R}$$

$$\nabla \hat{R} = \nabla \left( \frac{\mathbf{r} - \mathbf{r}'}{R} \right) = \frac{\bar{\mathbf{I}}}{R} - \frac{\mathbf{R}\hat{R}}{R^2}$$

Note: $\nabla \mathbf{A} = \hat{x}\nabla A_x + \hat{y}\nabla A_y + \hat{z}\nabla A_z$ and $\bar{\mathbf{I}} = \hat{x}\hat{x} + \hat{y}\hat{y} + \hat{z}\hat{z}$.

**(b)** Write out the dyadic operator $\nabla\nabla$ in matrix form and show that

$$\nabla\nabla f(\mathbf{r} - \mathbf{r}') = \nabla\nabla \cdot f(\mathbf{r} - \mathbf{r}')\bar{\mathbf{I}} = -\nabla'\nabla \cdot f(\mathbf{r} - \mathbf{r}')$$

**(c)** Show that

$$\nabla \cdot (\psi \bar{\mathbf{I}}) = \nabla \psi$$

$$\nabla \cdot (\bar{\mathbf{I}} \times \mathbf{A}) = \nabla \times \mathbf{A}$$

$$\bar{\mathbf{I}} \times \mathbf{A} = -\mathbf{A} \times \bar{\mathbf{I}}$$

**2. (a)** Beginning with (2.82) show that, if $\mathbf{E} = \hat{z}u$, the wave equation in the source-free region reduces to

$$\nabla \cdot \left[ \frac{1}{\mu_r} \nabla u \right] + k_o^2 \epsilon_r u = 0$$

provided $u$ is independent of $z$.

**(b)** Multiply this by the weighting functions $W(\overline{\rho})$, and integrate over the region $A$ to obtain

$$\iint_A W(\overline{\rho}) \left\{ \nabla \cdot \left[ \frac{1}{\mu_r(\overline{\rho})} \nabla u(\overline{\rho}) + k_o^2 \epsilon_r u(\overline{\rho}) \right] \right\} ds = 0.$$

Starting from here, show that

$$\iint_A \left\{ \frac{1}{\mu_r} \nabla W \cdot \nabla u - k_o^2 \epsilon_r W u \right\} ds - \oint_\Gamma \frac{1}{\mu_r} W \frac{\partial u}{\partial n} \, d\ell = 0$$

where, as illustrated in Figure 2.P2,

$$\frac{\partial u}{\partial n} = \hat{n} \cdot \nabla u$$

and $\Gamma$ encloses $A$. In carrying out this proof, you must employ a standard identity along with the divergence theorem. Note that this integral relation is the starting point

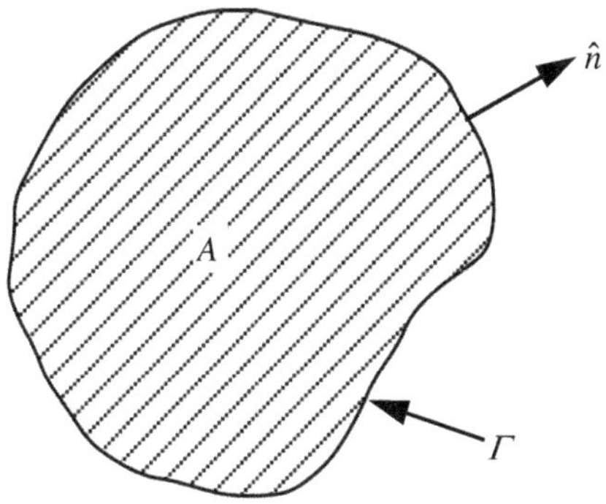


**FIGURE 2.P2** Geometry for Problem 2.

in developing field solutions based on the finite element method for two-dimensional problems.

**3.** **(a)** Show that the $x$-component of the near-zone magnetic field can be written as

$$H_x = \iiint \left[(z - z')J_y(\mathbf{r}') - (y - y')J_z(\mathbf{r}')\right] K(R)\, dv'$$

where

$$K(R) = (1 + jkR)\frac{e^{-jkR}}{4\pi R^3}.$$

**(b)** Show that the far-zone $E_\theta$ spherical component due to the electric current can be written as

$$E_\theta = \frac{j\omega\mu}{4\pi r}e^{-jkr} \iiint [-J_\rho(\mathbf{r}')\cos(\phi - \phi')\cos\theta - J_\phi(\mathbf{r}')\sin(\phi - \phi')\cos\theta$$
$$+ J_z(\mathbf{r}')\sin\theta]e^{jkg}\, dv'$$

with

$$g = \rho'\cos(\phi - \phi')\sin\theta + z'\cos\theta, \quad dv' = \rho'\, d\rho'\, d\phi'\, dz'$$

or as

$$E_\theta = \frac{j\omega\mu}{4\pi r}e^{-jkr} \iiint \big\{[\sin\theta\cos\theta' - \cos\theta\sin\theta'\cos(\phi - \phi')]J_r(\mathbf{r}')$$
$$-[\sin\theta\sin\theta' + \cos\theta\cos\theta'\cos(\phi - \phi')]J_\theta(\mathbf{r}')$$
$$-\cos\theta\sin(\phi - \phi')J_\phi(\mathbf{r}')\big\}e^{jkg}\, dv'$$

with

$$g = [\cos\theta\cos\theta' + \sin\theta\sin\theta'\cos(\phi - \phi')]r'$$

and

$$dv' = r'^2\sin\theta'\, dr'\, d\theta'\, d\phi'.$$

In these $J_\rho$, $J_\phi$ and $J_z$, are the cylindrical components of the current, and likewise $J_r$, $J_\theta$, and $J_\phi$ are the spherical components of $\mathbf{J}$. Depending on the integration volume or surface, one of these expressions is likely to be more convenient than the other.

(c) Obtain similar expressions to those in (b) for the $E_\phi$ component.

(d) Obtain the $E_\theta$ and $E_\phi$ far-zone field components due to the magnetic current using
  (1) Cylindrical coordinates for integration and spherical for observation.
  (2) Spherical coordinates for integration and observation.

4. Consider the plane wave

$$\mathbf{E}^i = \hat{z}E_o e^{-jkx}$$

incident upon a small dielectric sphere of radius $a \ll \lambda$. Based on the volume equivalence principle, the presence of the sphere can be replaced by the radiation of the equivalent volume current

$$\mathbf{J} = j\omega(\epsilon - \epsilon_o)\mathbf{E}$$

where $\mathbf{E} = \mathbf{E}^i + \mathbf{E}^s$, with $\mathbf{E}^s$ being the scattered field radiated by $\mathbf{J}$.

(a) Since the sphere is small, assume that $\mathbf{E} = \hat{z}E_1$ in the sphere with $E_1$ being a constant. Determine $E_1$ (Hint: see (2.69)).

(b) Determine the scattered field in the far zone.

(c) Show that

$$\sigma = \underset{r\to\infty}{4\pi r^2} \frac{|\mathbf{E}^s|^2}{|\mathbf{E}^i|^2} = \frac{4\pi k_o^2(\epsilon_r - 1)^2 a^6 \sin^2\theta}{(\epsilon_r + 2)^2}$$

where $\sigma$ is the bistatic echo area of the sphere. This result is a form of the Rayleigh law of scattering.

(d) Repeat (a) through (c) if the sphere is replaced by a small cube of volume $h^3$, where $h \ll \lambda$. Hint: You must derive a replacement for the last term of (2.69).

(e) Repeat (a) through (c) if the sphere is replaced by a small cylinder having length $h$ and radius $a$, such that $h \ll \lambda$ and $a \ll \lambda$.

5. Evaluate the integral

$$\iiint_{V_o} J(\mathbf{r}) \times \nabla G(\mathbf{r}, \mathbf{r}')\, dv'$$

where $V_o$ is a vanishingly small spherical volume.

6. Give the explicit matrix or components of the free-space dyadic Green's function using spherical coordinates. Do not carry out the derivatives.

7. Give the explicit matrix of the two-dimensional dyadic Green's function in cylindrical and rectangular coordinates. Do not carry out the derivatives.

8. Show that the right-hand side of (2.103) vanishes when $S_c$ is placed at infinity.

9. Find the explicit components of the dyadics $\bar{\mathbf{g}}_{1,2}$ such that

$$\mathbf{A} = \iiint \bar{\mathbf{g}}_1 \cdot \mathbf{J}\, dv'$$

and

$$\mathbf{F} = \iiint \bar{\mathbf{g}}_2 \cdot \mathbf{M}\, dv'$$

where $\mathbf{A}$ and $\mathbf{F}$ denote the magnetic and electric potentials, respectively.

10. Verify equation (2.129), and derive (2.130).

11. Derive appropriate replacements for the last term of (2.147) if $A_o$ in (2.142) is chosen to be the cross section of a small rectangular cylinder having width $w$ and height $h$.

12. Consider the plane wave

$$\mathbf{E}^i = \hat{z}e^{-jkx}$$

incident upon an infinitely long (along the $z$-direction) dielectric cylinder having $\epsilon = \epsilon_o\epsilon_r$ and $\mu = \mu_o$. Given that the echowidth is defined by

$$\sigma = \underset{\rho\to\infty}{4\pi\rho}\frac{|\mathbf{E}^s|^2}{|\mathbf{E}^i|^2}$$

where $\mathbf{E}^s$ denotes the scattered field, and $\rho$ is the usual cylindrical distance, find:

(a) The echowidth of a circular dielectric cylinder of radius $a \ll \lambda$.

(b) The echowidth of a rectangular dielectric cylinder whose dimensions are much smaller than the wavelength in the dielectric.

13. Show that

$$\frac{e^{-jkr}}{4\pi r} = -\frac{j}{8\pi}\int_{-\infty}^{\infty} H_o^{(2)}\left(\sqrt{k_o^2 - k_z^2}\rho\right) e^{\pm jk_z z} dk_z$$

Hint: Take the transform of the wave equations (2.193) with respect to $z$ to obtain an equation similar to (2.116)

14. Consider the planar current distribution $\mathbf{J} = \hat{x}J_x(x, y)\delta(z)$ over the dielectric interface, as shown in Figure 2.P14.

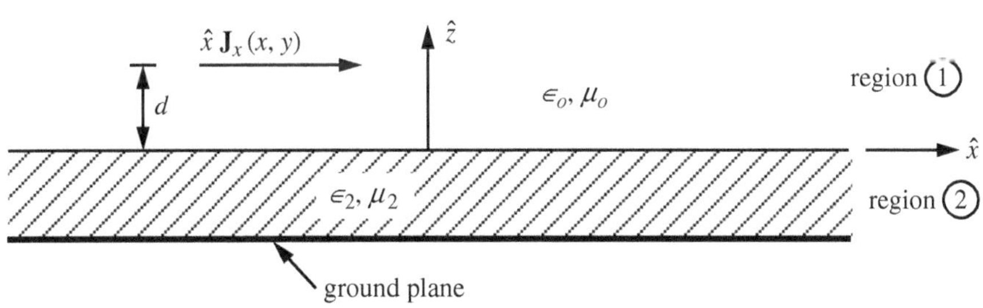


**FIGURE 2.P14** Geometry for Problem 14.

(a) Using the plane wave spectrum representation (2.202) show that for $z > d = 0$

$$H_z = \frac{1}{8\pi^2}\int_{-\infty}^{\infty} \frac{k_y}{k_z^-}\tilde{J}_x(k_x, k_y)e^{-j\mathbf{k}\cdot\mathbf{r}}(1 + R^{TE})\,dk_x\,dk_y$$

Generalize this result to the case when $d \neq 0$, and identify $R^{TE}$.

(b) repeat for $\mathbf{J} = \hat{y}J_y(x, y)\delta(z)$.

15. Assume a $\hat{z}$ directed vertical electric dipole (VED) located a distance $d$ over the interface of Problem 14. Using the Sommerfeld identity show that the radiated field by this dipole in the presence of the interface is

$$E_z = C\int_{-\infty}^{\infty} \frac{(k_\rho)^3}{k_z^-} H_o^{(2)}(k_\rho\rho)\left[e^{-jk_z^-|z|} + R^{TM}e^{-jk_z^-(z+2d)}\right] dk_\rho$$

Identify $R^{TM}$ and the constant $C$.
Note: $k_\rho = \sqrt{k_x^2 + k_y^2}$, $k_z^- = \sqrt{k_o^2 - k_\rho^2}$. (From Sommerfeld, 1909)

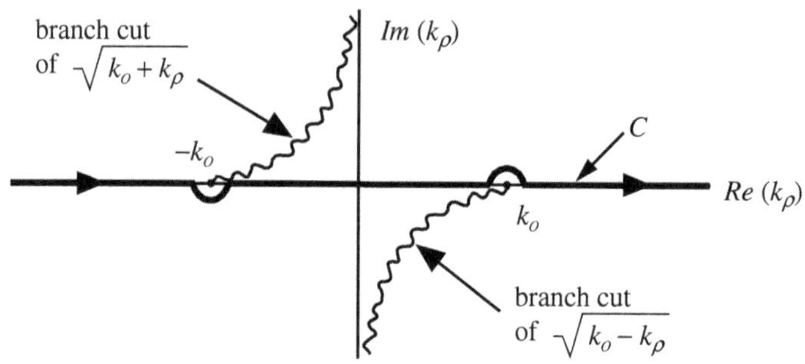


**FIGURE 2.P16a** Geometry for Problem 16(a).

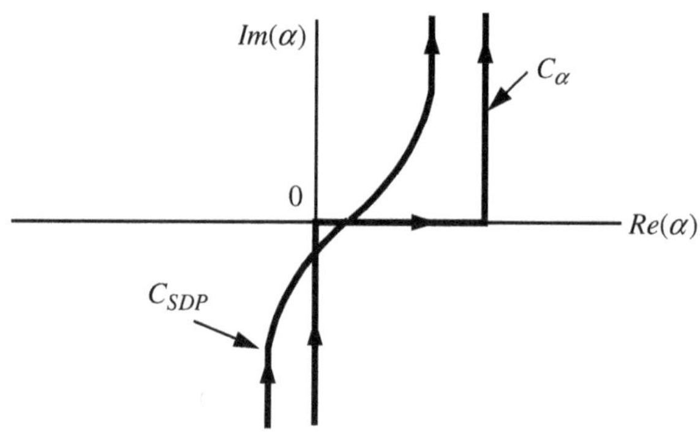


**FIGURE 2.P16b** Geometry for Problem 16(b).

**16.** Consider the integral

$$I = \int_C F(k_\rho) \frac{e^{-j\left(k_\rho \rho + \sqrt{k_o^2 - k_\rho^2}\, z\right)}}{\sqrt{k_o^2 - k_\rho^2}} dk_\rho \tag{P2.1}$$

where $C$ is as shown in Figure 2.P16a, and $F(k_\rho)$ is a slowly varying function. To evaluate this integral via the steepest descent method set $k_\rho = k_o \cos\alpha$ allowing us to rewrite $I$ as

$$I = + \int_{C_\alpha} F(k_o \cos\alpha) e^{-jk_o r \cos(\alpha - \theta')} \, d\alpha \tag{P2.2}$$

**(a)** Show the details in rewriting (P2.1) as given in (P2.2) and define $\theta'$.

**(b)** Find the saddle point in the $\alpha$-plane and give the mathematical definition of the steepest descent path $C_{SDP}$ (see Figure 2.P16b). Where does it cross the real and imaginary axis of $\alpha$?

**(c)** Give the first term of the asymptotic evaluation of (2), and use this result to evaluate the integral

$$E_z = C \int_{-\infty}^{\infty} \frac{k_\rho^3}{k_z^-} H_o^{(2)}(k_\rho \rho) \left[ e^{-jk_z^- z} + R^{TM} e^{-jk_z^- z} \right] dk_\rho$$

encountered earlier. In this evaluation ignore any other singularities of the integral including surface wave poles and branch points.

17. Give explicit expressions for the transform $\widetilde{\mathbf{E}}(k_x, k_y, k_z)$ in terms of $\widetilde{\mathbf{H}}(k_x, k_y, k_z)$ for all rectilinear components.

18. Derive equation (2.214), and find a similar expression when the normal to the dielectric interface is $\hat{x}$ rather than $\hat{z}$.

19. We are interested in the evaluation of the Sommerfeld (1909) integral given by

$$SI = \int_0^{\infty} J_m(k_\rho \rho) h(k_\rho)\, dk_\rho \tag{P2.3}$$

where $J_m$ is the Bessel function of the first kind of order, $m$.

(a) Assuming that $h(k_\rho)$ has a pole at $k_\rho = k_p (0 < k_p < \sqrt{\epsilon} k_o)$, regularize the integral to obtain

$$SI = \int_0^{\sqrt{\epsilon_r} k_o} (J_m(k_\rho \rho) h(k_\rho) - F(k_\rho)) dk_\rho + \int_0^{\sqrt{\epsilon_r} k_o} F(k_\rho)\, dk_\rho + \int_{\sqrt{\epsilon_r} k_o}^{\infty} J_m(k_\rho \rho) h(k_\rho)\, dk_\rho \tag{P2.4}$$

with

$$F(k_\rho) = \frac{A}{k_\rho - k_P} \tag{P2.5}$$

(b) Determine $A$, and evaluate the second integral in closed form.

(c) To make the behavior of the integrands in (P2.4) smoother, apply the following transformation:

$$k_\rho = \sqrt{\epsilon_r} k_o \sin \xi \tag{P2.6}$$

and

$$k_\rho = \sqrt{\epsilon_r} k_o \cosh \xi \tag{P2.7}$$

to the first and third integrals, respectively. Give forms of the integrals involving the new variable, $\xi$.

(d) To evaluate the first and third integrals numerically, first expand the respective integrands using the Chebyshev expansion (Annaert, 1993)

$$f(x) \approx \left[ \sum_{k=1}^{N} c_k T_{k-1}(x) \right] - \frac{1}{2} c_1 \tag{P2.8}$$

Derive expressions for the coefficient $c_k$.

20. Consider the problem of vertical electric dipole placed a distance $d$ above a planar impedance surface satisfying the boundary condition (1.32), as shown in Figure 2.P20.

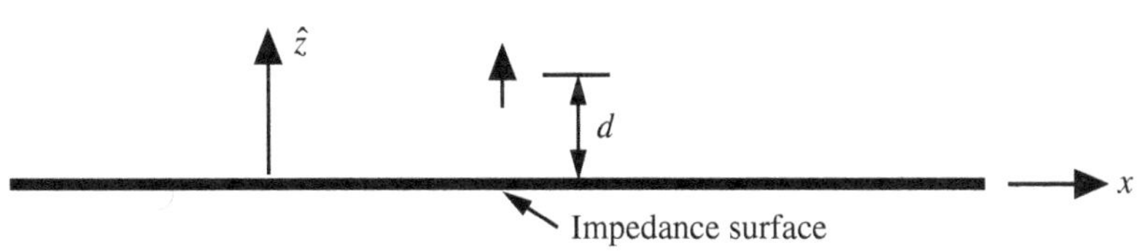


**FIGURE 2.P20** Geometry for Problem 20.

Give an integral representation of $E_z$ similar to that in problem 2.15.

**21.** In the case of an anisotropic medium, Maxwell's equations in the source-free region take the form

$$\nabla \times \mathbf{E} = -jk_o Z_o \overline{\overline{\mu}}_r \cdot \mathbf{H}$$

$$\nabla \times \mathbf{E} = jk_o Y_o \overline{\overline{\epsilon}}_r \cdot \mathbf{E}$$

**(a)** Show that the vector wave equation is given by

$$\nabla \times \left([\overline{\overline{\mu}}_r]^{-1} \cdot \nabla \times \mathbf{E}\right) - k_o^2 \overline{\overline{\epsilon}}_r \cdot \mathbf{E} = 0$$

**(b)** By multiplying the vector wave equation in (a) with the vector weighting function $\mathbf{W}(\mathbf{r})$ and integrating, show that

$$\iiint_V \{\nabla \times \mathbf{W} \cdot [\overline{\overline{\mu}}_r]^{-1} \cdot \nabla \times \mathbf{E} - k_o^2 \mathbf{E} \cdot \overline{\overline{\epsilon}}_r \cdot \mathbf{E}\}\, dv$$

$$- jk_o Z_o \oiint_S \hat{n} \times \mathbf{H} \cdot \mathbf{E}\, ds$$

where $S$ encloses the volume $V$. Note that this integral relation is the starting point in developing solutions based on the finite element method.

**22.** Consider the integral

$$\int_0^\infty \frac{e^{jkR(\xi)}}{R(\xi)} e^{-\beta\xi}\, d\xi = \frac{j}{2} \int_0^\infty \int_{-\infty}^\infty H_0^{(1)}(k\rho) e^{jk_z(z+z'+j\xi)} e^{-j\beta\xi}\, dk_z\, d\xi$$

$$= \frac{j}{2} \int_\infty^\infty \left(\frac{H_0^{(1)}(k_1\rho)}{k_z + \beta}\right) e^{jk_z(z+z')}\, dk_z$$

where $R(\xi) = \sqrt{(x-x')^2 + (y-y')^2 + (z+z'+j\xi)^2 +}$; $k_1^2 = k^2 - k_z^2$. This is an integral that may arise in problem 20. Carry out a steepest descent path evaluation (for $\rho \to \infty$) of the latter integral to obtain

$$\int_0^\infty \frac{e^{jkR(\xi)}}{R(\xi)} e^{\beta\xi}\, d\xi \approx \frac{j}{2} \sqrt{\frac{2}{\pi k\rho}} e^{-j\frac{\pi}{4}} e^{jkr} \frac{\cos\theta_s}{\sin\theta_s + \alpha} \sqrt{2j} \sqrt{\frac{\pi}{kr}} = \frac{e^{jkr}}{r} \frac{1}{k\sin\theta_s + \beta}$$

where $\cos\theta_s = \rho/r$.

**23.** Consider the problem of a plane wave illuminating a submerged circular cylinder as shown in Figure 2.P23. This is a situation that occurs in underground detection of plastic mines. From the standard solution of the wave equation (see Harrington, 1961), the scattered field by a TM plane wave impinging on a dielectric cylinder is

$$E_z = \sum_{n=-\infty}^{\infty} j^{-n} A_n H_n^{(2)}\left(k_1\rho e^{jn(\phi-\phi_i)}\right)$$

where $\phi_i$ denotes the incidence angle, and

$$A_n = \frac{k_2 J_n(k_1 a) J_n'(k_2 a) - k_1 J_n'(k_1 a) J_n(k_2 a)}{k_1 J_n(k_2 a) H_n^{(2)}(k_1 a) - k_2 J_n'(k_2 a) H_n^{(2)}(k_2 a)}$$

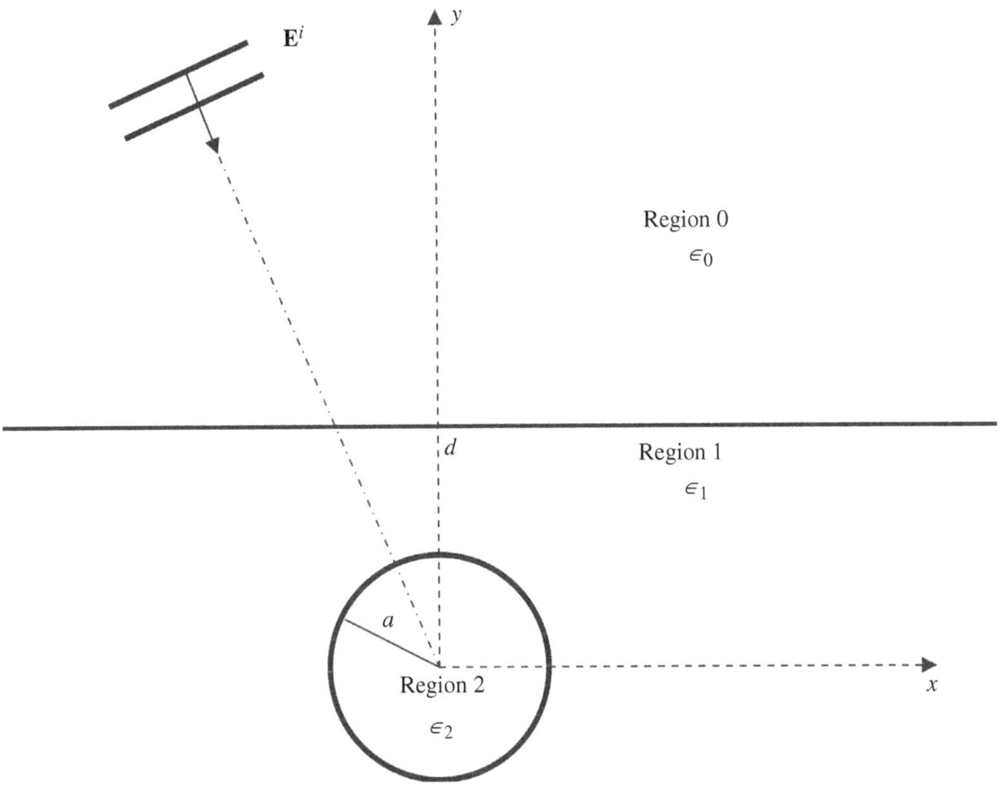


**FIGURE 2.P23** ■ Scattering scenario for a plane wave incident upon a dielectric cylinder buried in a homogeneous dielectric half space.

As usual, $k_{1,2}$ refer to the propagation constants in media 1 and 2 respectively. Invoke the plane wave expansion for the Hankel function to get

$$E_z = \frac{1}{\pi} \int_{-\infty}^{\infty} \frac{e^{-j(k_{1x}x + k_{1y}y)}}{k_1 y} \left( \sum_{-\infty}^{\infty} A_n e^{-jn\phi_i} e^{-jn \tan^{-1}\left(\frac{k_{1y}}{k_{1x}}\right)} \right) dk_{1x}$$

and use this to obtain an expression for the scattered field in the region 0 for the problem illustrated in the figure. Note that the solution is the sum of multiple reflections/transmissions from the dielectric interface.

# BIBLIOGRAPHY


Annaert, G. (1993). Evaluation of Sommerfeld integrals using Chebychev decomposition. *IEEE Trans. Antennas Propagat.*, *41*, 159–164.

Collin, R. (1986a). The dyadic Green's function as an inverse operator. *Radio Sci.*, *21*, 883–890.

Collin, R. E. (1986b). Dyadic Green's function expansion in spherical coordinates. *Electromagnetics*, *6*, 183–207.

Collin, R. E. (1991). *Field Theory for Guided Waves*. New York: IEEE Press, 2nd ed.

Felsen, L. B. and Marcuvitz, N. (1973). *Radiation and Scattering of Waves*. New York: Prentice Hall.

Harrington, R. F. (1961). *Time-Harmonic Electromagnetic Fields*. New York: McGraw-Hill.

Papoulis, A. (1962). *Fourier Integral and Its Applications*. New York: McGraw-Hill.

Senior, T. B. A. and Volakis, J. L. (1995). *Approximate Boundary Conditions in Electromagnetics*. IET Press.

Sommerfeld, A. N. (1909). The propagation of waves in wireless telegraphy. *Ann. Phys. Ser. 4*, *28*, 665.

Tai, C.-T. (1992). *Generalized Vector and Dyadic Analysis*. New York: IEEE Press.

Van Bladel, J. (1985). *Electromagnetic Fields,* New York: Hemisphere Pub. (reprint of the book originally pulished by McGraw-Hill in 1964).

Volakis, J. L., Chatterjee, A., and Kempel, L. (1998). *Finite Element Methods for Electromagnetics*. New York: IEEE Press.

CHAPTER 3

# Integral Equations and Other Field Representations

Chapter Outline



The field integral representations given in the previous chapter, although of sufficient generality, are often inconvenient and possibly inefficient for specific applications. Also, their integrands are highly singular, requiring special treatment when the observation point is in the source region. This difficulty cannot be eliminated, but any reduction in the integrand's singularity is desirable for achieving higher accuracies in numerical computations involving such integrals. Obviously, a variety of field representations, integral equations, and formal solutions could be derived, many of which can be applicable only to a specific situation. This chapter will consider some alternative field representations to construct integral equations that are among the most frequently used. First, we shall develop three-dimensional representations. Many of the two-dimensional representations can then be reduced from the three-dimensional ones. However, for scattering applications a larger variety of two-dimensional representations is available primarily because the topic has been extensively studied.

## 3.1 | THREE-DIMENSIONAL INTEGRAL EQUATIONS

### 3.1.1 Kirchhoff's Integral Equation

Perhaps the simplest integral equation can be derived by considering the wave equations (2.86) in conjunction with Green's second identity (see (2.42)). To proceed, we assume the existence of certain structures whose surfaces will be denoted by $S_1, S_2, \ldots, S_N$ (see Figure 3-1). The collection of these surfaces, henceforth referred to as $S_\Omega$ (enclosing the volume $V_\Omega$), is illuminated by sources enclosed within the volume $V_{is}$. The volume region exterior to $S_\Omega$ shall be denoted by $V_\infty$, which, as seen, is also bounded by the surface $S_\infty$ placed at infinity.

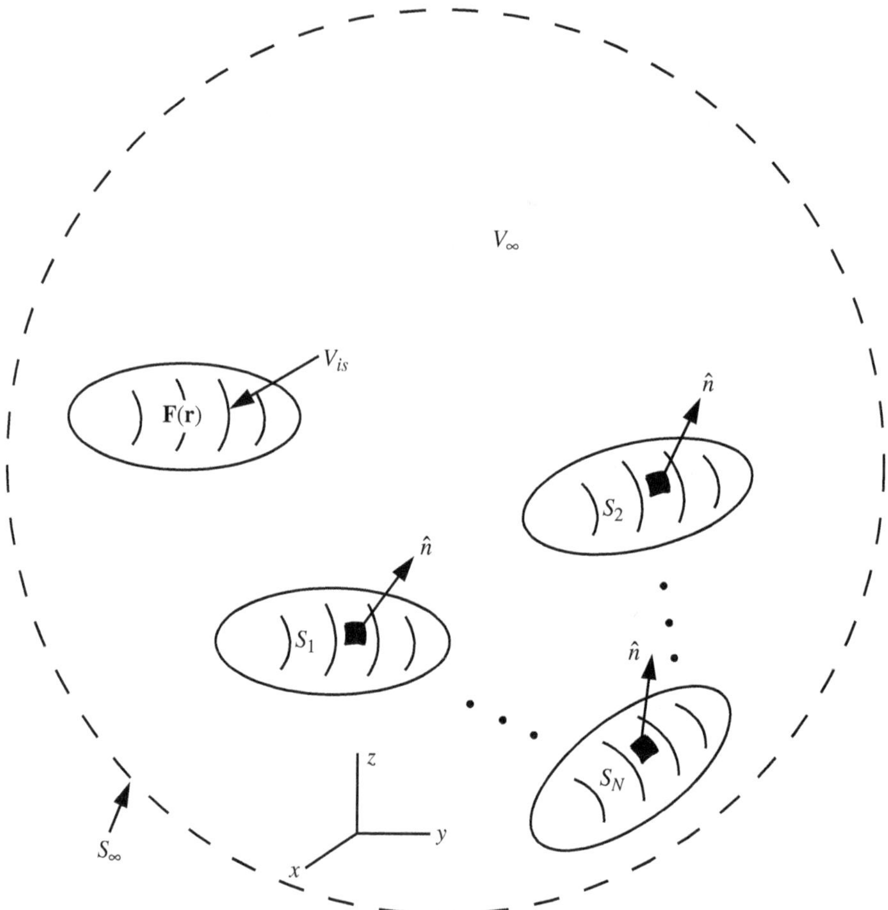


**FIGURE 3-1** Geometry for the application of Green's second identity.

Without loss of generality, let us consider one of the electric field component, say, $E_a$. Then from (2.86)

$$\nabla^2 E_a + k^2 E_a = \begin{cases} F_a(\mathbf{r}) & \mathbf{r} \in V_{is} \\ 0 & \mathbf{r} \notin V_{is} \end{cases} \tag{3.1}$$

in which $F_a(\mathbf{r}) = \mathbf{F}(\mathbf{r}) \cdot \hat{a}$ represents the source terms on the right-hand side of (2.86). Multiplying this by the free-space Green's function and integrating yields

$$\iiint_{V_\infty} G(\mathbf{r}, \mathbf{r}')\left[\nabla^2 E_a(\mathbf{r}) + k^2 E_a(\mathbf{r})\right] dv = \begin{cases} \iiint_{V_{is}} F_a(\mathbf{r}) G(\mathbf{r}, \mathbf{r}')\, dv & \mathbf{r} \in V_{is} \\ 0 & \mathbf{r} \notin V_{is} \end{cases} \tag{3.2}$$

and we remark that $V_\infty$ includes the source volume $V_{is}$. Also, from Green's second identity (see (2.42)) we have

$$\begin{aligned} &\iiint_{V_\infty} \left[E_a(\mathbf{r})\nabla^2 G(\mathbf{r}, \mathbf{r}') - G(\mathbf{r}, \mathbf{r}')\nabla^2 E_a(\mathbf{r})\right] dv \\ &= -\oiint_{S_\Omega} \left[E_a(\mathbf{r})\frac{\partial G(\mathbf{r}, \mathbf{r}')}{\partial n} - G(\mathbf{r}, \mathbf{r}')\frac{\partial E_a(\mathbf{r})}{\partial n}\right] ds \\ &\quad + \oiint_{S_\infty} \left[E_a(\mathbf{r})\frac{\partial G(\mathbf{r}, \mathbf{r}')}{\partial n} - G(\mathbf{r}, \mathbf{r}')\frac{\partial E_a(\mathbf{r})}{\partial n}\right] ds \end{aligned} \tag{3.3}$$

where $\frac{\partial}{\partial n} = \hat{n} \cdot \nabla$, and we remark that the negative sign in front of the integral over $S_\Omega$ was introduced because the unit normal $\hat{n}$ points toward the interior of $V_\infty$. Further, by noting that $G(\mathbf{r}, \mathbf{r}')$ and $E_a(\mathbf{r})$ satisfy the radiation condition (2.39), it follows that the integral over $S_\infty$ in (3.3) vanishes. Thus we have

$$\iiint_{V_\infty} G(\mathbf{r}, \mathbf{r}')\nabla^2 E_a(\mathbf{r})\, dv = \iiint_{V_\infty} E_a(\mathbf{r})\nabla^2 G(\mathbf{r}, \mathbf{r}')\, dv + \oiint_{S_\Omega} \left[ E_a(\mathbf{r}) \frac{\partial G(\mathbf{r}, \mathbf{r}')}{\partial n} - G(\mathbf{r}, \mathbf{r}') \frac{\partial E_a(\mathbf{r})}{\partial n} \right] ds \tag{3.4}$$

and when this is combined with (3.2) we obtain

$$\iiint_{V_\infty} E_a(\mathbf{r}) \left[\nabla^2 G(\mathbf{r}, \mathbf{r}') + k^2 G(\mathbf{r}, \mathbf{r}')\right] dv = - \oiint_{S_\Omega} \left[ E_a(\mathbf{r}) \frac{\partial G(\mathbf{r}, \mathbf{r}')}{\partial n} - G(\mathbf{r}, \mathbf{r}') \frac{\partial E_a(\mathbf{r})}{\partial n} \right] + \iiint_{V_{is}} F_a(\mathbf{r}) G(\mathbf{r}, \mathbf{r}')\, dv. \tag{3.5}$$

We now recall the differential equation (2.38) satisfied by the Green's function, and when this is introduced into (3.5), upon interchanging $\mathbf{r}$ and $\mathbf{r}'$, we obtain

$$\oiint_{S_\Omega} \left[ E_a(\mathbf{r}') \frac{\partial G(\mathbf{r}, \mathbf{r}')}{\partial n'} - G(\mathbf{r}, \mathbf{r}') \frac{\partial E_a(\mathbf{r}')}{\partial n'} \right] ds' - \iiint_{V_{is}} F_a(\mathbf{r}') G(\mathbf{r}, \mathbf{r}')\, dv' = \begin{cases} E_a(\mathbf{r}) & \mathbf{r} \in V_\infty \\ 0 & \mathbf{r} \notin V_\infty \end{cases} \tag{3.6}$$

in which the differentiation is on the primed coordinates and is taken along the normal directed away from $S_1, S_2, \ldots, S_N$.

The result given by (3.6) is often referred to as the *extinction* or *Kirchhoff's integral equation* and is valid for all field components provided these satisfy the radiation condition. No other boundary condition is required to be satisfied by the field, and since $E_a(\mathbf{r})$ is completely arbitrary we can generalize it to the case of vector fields. We have

$$\oiint_{S_\Omega} \left[ \mathbf{E}(\mathbf{r}') \frac{\partial G(\mathbf{r}, \mathbf{r}')}{\partial n'} - G(\mathbf{r}, \mathbf{r}') \frac{\partial \mathbf{E}(\mathbf{r}')}{\partial n'} \right] ds' - \iiint_{V_{is}} \mathbf{F}_E(\mathbf{r}) G(\mathbf{r}, \mathbf{r}')\, dv' = \begin{cases} \mathbf{E}(\mathbf{r}) & \mathbf{r} \in V_\infty \\ 0 & \mathbf{r} \notin V_\infty \end{cases} \tag{3.7a}$$

and by duality

$$\oiint_{S_\Omega} \left[ \mathbf{H}(\mathbf{r}') \frac{\partial G(\mathbf{r}, \mathbf{r}')}{\partial n'} - G(\mathbf{r}, \mathbf{r}') \frac{\partial \mathbf{E}(\mathbf{r}')}{\partial n'} \right] ds' - \iiint_{V_{is}} \mathbf{F}_H(\mathbf{r}) G(\mathbf{r}, \mathbf{r}')\, dv' = \begin{cases} \mathbf{H}(\mathbf{r}) & \mathbf{r} \in V_\infty \\ 0 & \mathbf{r} \notin V_\infty \end{cases} \tag{3.7b}$$

in which

$$\mathbf{F}_E(\mathbf{r}) = j\omega\mu\mathbf{J}(\mathbf{r}) - \frac{\nabla\nabla\cdot\mathbf{J}(\mathbf{r}')}{j\omega\epsilon} + \nabla\times\mathbf{M}(\mathbf{r})$$

and

$$\mathbf{F}_H(\mathbf{r}) = j\omega\epsilon\mathbf{M}(\mathbf{r}) - \frac{\nabla\nabla\cdot\mathbf{M}(\mathbf{r}')}{j\omega\mu} - \nabla\times\mathbf{J}(\mathbf{r}).$$

If no external sources are present or if all sources are located on or within the surfaces (e.g., $S_1$, $S_2$) then $\mathbf{F}_E = \mathbf{F}_M = 0$. In that case $\mathbf{E}$, $\mathbf{H}$, $\frac{\partial\mathbf{E}}{\partial n'}$, or $\frac{\partial\mathbf{H}}{\partial n'}$ when integrated over $S_\Omega$ play the role of equivalent sources of the same type as $\mathbf{F}_E$ and $\mathbf{F}_H$. This will become apparent in later applications. Alternatively, the integrals associated with $\mathbf{F}_E$ and $\mathbf{F}_M$ can be recognized to yield the fields radiated by the sources within $V_\infty$, and we may thus set

$$-\iiint \mathbf{F}_E(\mathbf{r}')G(\mathbf{r},\mathbf{r}')\,dv' \rightarrow \mathbf{E}^i(\mathbf{r})$$

$$-\iiint \mathbf{F}_H(\mathbf{r}')G(\mathbf{r},\mathbf{r}')\,dv' \rightarrow \mathbf{H}^i(\mathbf{r})$$

where $(\mathbf{E}^i, \mathbf{H}^i)$ denote the excitation or incident fields. For scattering computations these are usually plane waves whose source is at infinity.

In practice, additional boundary conditions would be imposed on the fields at the surfaces (e.g., $S_1$, $S_2$). This leads to the construction of integral equations for a unique solution of the fields. However, in their present form, (3.6) and (3.7) are not applicable to the case where $\mathbf{r}$ is on $S_\Omega$, that is, at the boundary of $V_\Omega$ coinciding with $S_\Omega$. To make them applicable to this case we shall consider the limit as the observation point $P$ at $\mathbf{r} = \mathbf{r}_o$ approaches the surface from outside or inside $S_\Omega$. To simulate the latter situation, we distort the surface $S_\Omega$ about the observation point $P$ as shown in Figure 3-2, that is, by adding a hemispherical surface to $S_\Omega$ of radius

**FIGURE 3-2** Geometry for evaluating the field on $S_\Omega$.

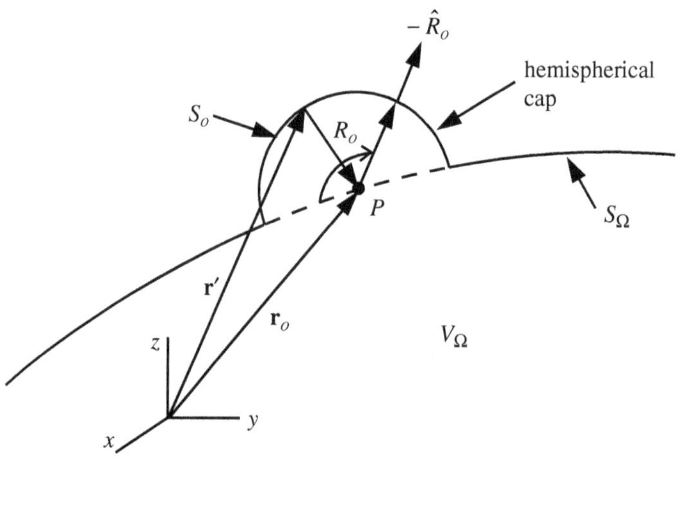

$R_o \to 0$, which has its center at the observation point $P$. Accordingly, from (3.7a)

$$\iint_{S_\Omega - S_o} \left[ \mathbf{E}(\mathbf{r}') \frac{\partial G(\mathbf{r}, \mathbf{r}')}{\partial n'} - G(\mathbf{r}, \mathbf{r}') \frac{\partial \mathbf{E}(\mathbf{r}')}{\partial n'} \right] ds'$$
$$= - \iint_{S_o} \left[ \mathbf{E}(\mathbf{r}') \frac{\partial G(\mathbf{r}, \mathbf{r}')}{\partial n'} - G(\mathbf{r}, \mathbf{r}') \frac{\partial \mathbf{E}(\mathbf{r}')}{\partial n'} \right] ds'.$$

For the integral over the hemispherical surface $S_o$ we have $ds' = R_o^2 \sin\theta_o \, d\phi_o \, d\theta_o$,

$$G(\mathbf{r}, \mathbf{r}') = \frac{e^{-jkR_o}}{4\pi R_o}$$

and

$$\frac{\partial G(\mathbf{r}, \mathbf{r}')}{\partial n'} = -\hat{R}_o \cdot \nabla' G(\mathbf{r}, \mathbf{r}') = \hat{R}_o \cdot \nabla G(\mathbf{r}, \mathbf{r}') = \frac{\partial G}{\partial R_o} = -\left( jk + \frac{1}{R_o} \right) \frac{e^{-jkR_o}}{4\pi R_o}.$$

Substituting these into the integral gives

$$- \iint_{S_o \to 0} [\ ]\, ds' = \mathbf{E}(\mathbf{r}_o) \int_0^{\frac{\pi}{2}} \int_0^{2\pi} \left( jk + \frac{1}{R_o} \right) \frac{e^{-jkR_o}}{4\pi R_o} R_o^2 \sin\theta_o \, d\phi_o \, d\theta_o$$
$$- \int_0^{\pi} \int_0^{2\pi} \frac{\partial \mathbf{E}}{\partial n'} \frac{e^{-jkR_o}}{4\pi R_o} R_o^2 \sin\theta_o \, d\phi_o \, d\theta_o$$

and it is seen that the last integral vanishes as $R_o \to 0$. Also,

$$\mathbf{E}(\mathbf{r}_o) \int_0^{\frac{\pi}{2}} \int_0^{2\pi} \left( jk + \frac{1}{R_o} \right) \frac{e^{-jkR_o}}{4\pi R_o} R_o^2 \sin\theta_o \, d\phi_o \, d\theta_o = \frac{1}{2} \mathbf{E}(\mathbf{r}_o) \tag{3.8}$$

and thus we can write (Kellogg, 1929)

$$\oiint_{S_\Omega} \left[ \mathbf{E}(\mathbf{r}') \frac{\partial G(\mathbf{r}, \mathbf{r}')}{\partial n'} - G(\mathbf{r}, \mathbf{r}') \frac{\partial \mathbf{E}(\mathbf{r}')}{\partial n'} \right] ds' = \frac{1}{2} \mathbf{E}(\mathbf{r}) \tag{3.9}$$

for $\mathbf{r}$ on $S_\Omega$. This simply states that the field on the surface $S_\Omega$ is obtained by averaging its values just inside and just outside $S_\Omega$. Using (3.9) we can now revise the integral expressions (3.7) to read

$$\oiint_{S_\Omega} \left[ \mathbf{E}(\mathbf{r}') \frac{\partial G(\mathbf{r}, \mathbf{r}')}{\partial n'} - G(\mathbf{r}, \mathbf{r}') \frac{\partial \mathbf{E}(\mathbf{r}')}{\partial n'} \right] ds' + \mathbf{E}^i(\mathbf{r})$$
$$= \begin{cases} \mathbf{E}(\mathbf{r}) & \mathbf{r} \text{ in } V_\infty \\ \frac{1}{2}\mathbf{E}(\mathbf{r}) & \mathbf{r} \text{ on } S_\Omega \\ 0 & \mathbf{r} \text{ within } S_\Omega \end{cases} \tag{3.10a}$$

$$\oiint_{S_\Omega} \left[ \mathbf{H}(\mathbf{r}') \frac{\partial G(\mathbf{r}, \mathbf{r}')}{\partial n'} - G(\mathbf{r}, \mathbf{r}') \frac{\partial \mathbf{H}(\mathbf{r}')}{\partial n'} \right] ds' + \mathbf{H}^i(\mathbf{r})$$
$$= \begin{cases} \mathbf{H}(\mathbf{r}) & \mathbf{r} \text{ in } V_\infty \\ \frac{1}{2}\mathbf{H}(\mathbf{r}) & \mathbf{r} \text{ on } S_\Omega \\ 0 & \mathbf{r} \text{ within } S_\Omega \end{cases} \tag{3.10b}$$

and it should be noted that these are valid provided the observation point is not at a corner or an edge formed by $S_\Omega$. They are evocative of Huygens's principle, which states that the fields caused by the presence of the volume enclosed by surface(s) $S_\Omega$ can be determined uniquely everywhere from a knowledge of that field and its normal derivative on $S_\Omega$. Alternatively, it will be shown in the next section that a knowledge of the tangential electric and magnetic fields on $S_\Omega$ is sufficient to uniquely determine the fields exterior to $S_\Omega$ regardless of the volume composition enclosed by $S_\Omega$. These statements are valid even if $(\mathbf{E}^i, \mathbf{H}^i)$ are zero and sources exist within $S_\Omega$. In that case we can state that the fields exterior to $S_\Omega$ can be determined uniquely from a knowledge of the surface tangential electric and magnetic fields or a knowledge of the electric/magnetic field and its normal derivative. By referring to Chapter 1, one concludes that the surface equivalence principle can be thought as another statement of Huygens's principle (Baker and Copson, 1939).

Equations (3.10) can be referred to as the vector form of Kirchhoff's equations who first employed (a scalar form of) these for computing diffraction by apertures. To obtain the standard Kirchhoff's scalar equations the vector field in (3.10) is replaced by a scalar function or a component of the field. Because of their simplicity, Kirchhoff's (1882) (also Rubinowicz, 1917) equations are widely used for obtaining the diffraction by apertures or the scattering by closed surfaces whose surface fields are known or can be reasonably approximated (using, e.g., physical optics).

### 3.1.2 Stratton-Chu Integral Equations

The Stratton-Chu integral formulae for field representations (Stratton and Chu, 1939; Stratton, 1941) are among the most popular in scattering and antenna related problems. Perhaps a primary reason for their popularity is their reduced kernel singularity in comparison with the representations (2.52) or (2.102), which integrate the current sources directly over the volume. The main feature of the Stratton-Chu representations is the transferring of one of the del operators from the Green's function to the current, reducing the kernel singularity from $R^{-3}$ to $R^{-2}$ (see (2.63)). There are several ways to derive the Stratton-Chu equations, but it is instructive to begin their derivation by considering one of the integral expansions given earlier. Let us, for example, begin with equation (2.52a) where our goal is to reduce the singularity of the integrand (or kernel) associated with the last right-hand-side term of this equation. This term can be written as

$$\begin{aligned}-\frac{jZ}{k}\iiint_V \mathbf{J}(\mathbf{r}')\cdot\nabla\nabla G(\mathbf{r},\mathbf{r}')\,dv' &= -\frac{jZ}{k}\nabla\left\{\iiint_V \mathbf{J}(\mathbf{r}')\cdot\nabla G(\mathbf{r},\mathbf{r}')\,dv'\right\}\\ &= \frac{jZ}{k}\nabla\left\{\iiint_V \mathbf{J}(\mathbf{r}')\cdot\nabla' G(\mathbf{r},\mathbf{r}')\,dv'\right\}\end{aligned}$$

and by invoking the identity (2.50) we have

$$\begin{aligned}\iiint_V \mathbf{J}(\mathbf{r}')\cdot\nabla' G(\mathbf{r},\mathbf{r}')\,dv' &= \iiint \nabla'\cdot\{\mathbf{J}(\mathbf{r}')G(\mathbf{r},\mathbf{r}')\}\,dv'\\ &\quad-\iiint_V [\nabla'\cdot\mathbf{J}(\mathbf{r}')]\,G(\mathbf{r},\mathbf{r}')\,dv'.\end{aligned}$$

Next, by employing the divergence theorem we obtain

$$\iiint_V \nabla' \cdot \{\mathbf{J}(\mathbf{r}')G(\mathbf{r},\mathbf{r}')\}\, dv' = \oiint_{S_c} G(\mathbf{r},\mathbf{r}')\left[\mathbf{J}(\mathbf{r}')\cdot\hat{n}(\mathbf{r}')\right] ds' \tag{3.11}$$

where $\hat{n}$ is the unit normal pointing outward of the surface $S_c$ enclosing the volume $V$ containing the source $\mathbf{J}(\mathbf{r}')$. A natural boundary condition is that the current be confined within the volume $V$, implying that the component of $\mathbf{J}$ normal to the surface $S_c$ must be zero. Thus, the integral in (3.11) vanishes, and we can then write

$$\frac{-jZ}{k}\iiint_V \mathbf{J}(\mathbf{r}')\cdot\nabla\nabla G(\mathbf{r},\mathbf{r}')\, dv' = \frac{-jZ}{k}\iiint_V \left[\nabla'\cdot\mathbf{J}(\mathbf{r}')\right]\nabla G(\mathbf{r},\mathbf{r}')\, dv'. \tag{3.12}$$

When this identity and its dual are used in (2.52) we obtain the equations

$$\mathbf{E}(\mathbf{r}) = \iiint_V \Big[\mathbf{M}(\mathbf{r}')\times\nabla G(\mathbf{r},\mathbf{r}') - jkZ\mathbf{J}(\mathbf{r}')G(\mathbf{r},\mathbf{r}') - \frac{jZ}{k}\nabla'\cdot\mathbf{J}(\mathbf{r}')\nabla G(\mathbf{r},\mathbf{r}')\Big]\, dv' \tag{3.13a}$$

$$\mathbf{H}(\mathbf{r}) = \iiint_V \Big[-\mathbf{J}(\mathbf{r}')\times\nabla G(\mathbf{r},\mathbf{r}') - jkY\mathbf{M}(\mathbf{r}')G(\mathbf{r},\mathbf{r}') - \frac{jY}{k}\nabla'\cdot\mathbf{M}(\mathbf{r}')\nabla G(\mathbf{r},\mathbf{r}')\Big]\, dv'. \tag{3.13b}$$

Alternative representations can be obtained by invoking the continuity equations (1.38) and (1.39) to replace the divergence of the current quantities with volume charges. Doing so yields

$$\mathbf{E}(\mathbf{r}) = \iiint_V \Big[\mathbf{M}(\mathbf{r}')\times\nabla G(\mathbf{r},\mathbf{r}') - jkZ\mathbf{J}(\mathbf{r}')G(\mathbf{r},\mathbf{r}') - \frac{\rho(\mathbf{r}')}{\epsilon}\nabla G(\mathbf{r},\mathbf{r}')\Big]\, dv' \tag{3.14a}$$

$$\mathbf{H}(\mathbf{r}) = \iiint_V \Big[-\mathbf{J}(\mathbf{r}')\times\nabla G(\mathbf{r},\mathbf{r}') - jkY\mathbf{M}(\mathbf{r}')G(\mathbf{r},\mathbf{r}') - \frac{\rho_m(\mathbf{r}')}{\mu}\nabla G(\mathbf{r},\mathbf{r}')\Big]\, dv' \tag{3.14b}$$

which are the natural equations that result if we introduce the scalar potentials $\Phi_e$ and $\Phi_m$ in equations (2.19).

When the expressions (3.13) and (3.14) are applied to an antenna or scattering configuration such as that shown in Figure 3-1 it is convenient to employ Love's equivalence principle (see Chapter 1). This allows one to replace the presence of the volume enclosed by the surfaces $S_1, S_2, \ldots, S_N$ (comprising the surface $S_\Omega$) by a set of equivalent sources

$$\mathbf{J} = \hat{n}\times\mathbf{H}, \qquad \mathbf{M} = \mathbf{E}\times\hat{n} \tag{3.15}$$

placed on the surfaces $S_1, S_2, \ldots, S_N$. Also, in accordance with the boundary conditions (1.62) and (1.63) we may set

$$\rho_s = \epsilon(\hat{n}\cdot\mathbf{E}), \qquad \rho_{ms} = \mu(\hat{n}\cdot\mathbf{H}). \tag{3.16}$$

Introducing these into (3.14) yields

$$\mathbf{E}(\mathbf{r}) = \mathbf{E}^i + \oiint_{S_\Omega} \{[\mathbf{E}(\mathbf{r}') \times \hat{n}'] \times \nabla G(\mathbf{r}, \mathbf{r}') - jkZ\,[\hat{n}' \times \mathbf{H}(\mathbf{r}')]\,G(\mathbf{r}, \mathbf{r}') - \hat{n}' \cdot \mathbf{E}(\mathbf{r}')\nabla G(\mathbf{r}, \mathbf{r}')\}\, ds' \tag{3.17a}$$

$$\mathbf{H}(\mathbf{r}) = \mathbf{H}^i + \oiint_{S_\Omega} \{[\mathbf{H}(\mathbf{r}') \times \hat{n}'] \times \nabla G(\mathbf{r}, \mathbf{r}') - jkY\,[\mathbf{E}(\mathbf{r}') \times \hat{n}']\,G(\mathbf{r}, \mathbf{r}') - \hat{n}' \cdot \mathbf{H}(\mathbf{r}')\nabla G(\mathbf{r}, \mathbf{r}')\}\, ds' \tag{3.17b}$$

in which $\hat{n}' = \hat{n}(\mathbf{r}')$, where $\hat{n}(\mathbf{r}')$ denotes the outward unit normal outward to $S_\Omega$ at $\mathbf{r}$. We have also included the incident fields $(\mathbf{E}^i, \mathbf{H}^i)$ to account for any source exterior to $S_\Omega$. We remark that (3.17) give the most common form of the Stratton-Chu equations.

An alternative field representation in terms of the dyadic Green's function can be obtained by substituting (3.15) into (2.102). Since the equivalent sources are only over surfaces $S_\Omega$, we have

$$\mathbf{E}(\mathbf{r}) = \mathbf{E}^i + \iint_{S_\Omega} \{[\nabla \times \overline{\Gamma}(\mathbf{r}, \mathbf{r}')] \cdot [\mathbf{E}(\mathbf{r}') \times \hat{n}'] + jkZ\overline{\Gamma}(\mathbf{r}, \mathbf{r}') \cdot [\hat{n}' \times \mathbf{H}(\mathbf{r}')]\}\, ds' \tag{3.18a}$$

$$\mathbf{H}(\mathbf{r}') = \mathbf{H}^i \iint_{S_\Omega} \{jkY\overline{\Gamma}(\mathbf{r}, \mathbf{r}') \cdot [\mathbf{E}(\mathbf{r}') \times \hat{n}'] - [\nabla \times \overline{\Gamma}(\mathbf{r}, \mathbf{r}')] \cdot [\hat{n}' \times \mathbf{H}(\mathbf{r}')]\}\, ds' \tag{3.18b}$$

and this is equivalent to (3.17) only for closed surfaces in which case the identity (Van Bladel, 1985)

$$\oiint_{S_\Omega} \nabla \cdot [\mathbf{J}(\mathbf{r}')G(\mathbf{r}, \mathbf{r}')]\, ds' = -\oiint_{S_\Omega} \left[\frac{1}{R_1} + \frac{1}{R_2}\right] G(\mathbf{r}, \mathbf{r}')\,[\mathbf{J}(\mathbf{r}') \cdot \hat{n}(\mathbf{r}')]\, ds' = 0 \tag{3.19}$$

holds when $\mathbf{J}$ is replaced by $\hat{n}' \times \mathbf{H}(\mathbf{r}')$ or $\mathbf{E}(\mathbf{r}') \times \hat{n}(\mathbf{r}')$. In (3.19), $R_1$ and $R_2$ denote the principle radii of curvature at the surface point $\mathbf{r}'$ but in the event $S_\Omega$ is not closed (i.e., $S_\Omega$ is the surface of a flat or curved conducting sheet as shown in Figure 1-1) then this identity must be replaced by (Van Bladel, 1985)

$$\begin{aligned}\iint_{S_\Omega} \nabla \cdot [\mathbf{J}(\mathbf{r}')G(\mathbf{r}, \mathbf{r}')]\, ds' &= + \oint_C \hat{b}' \cdot \mathbf{J}(\mathbf{r}')G(\mathbf{r}, \mathbf{r}')\, d\ell' \\ &\quad - \iint_{S_\Omega} \left[\frac{1}{R_1} + \frac{1}{R_2}\right] G(\mathbf{r}, \mathbf{r}')\,[\mathbf{J}(\mathbf{r}') \cdot \hat{n}(\mathbf{r}')]\, ds' \\ &= \oint_C \hat{b}' \cdot \mathbf{J}(\mathbf{r}')G(\mathbf{r}, \mathbf{r}')\, d\ell'\end{aligned} \tag{3.20}$$

in which $C$ denotes the contour defining the outer perimeter of $S_\Omega$ and $\hat{b}' = \hat{\ell}' \times \hat{n}'$, where $\hat{\ell}'$ is the unit tangent to $C$ at $\mathbf{r}'$. Thus, one cannot specialize (3.13) to open surfaces such as curved plates (see Figure 3-3) by simply changing the volume integral to one over the boundary domain of $\mathbf{J}$ and $\mathbf{M}$. Such an interchange of the volume and surface integral is permitted for the dyadic representations (2.102) or

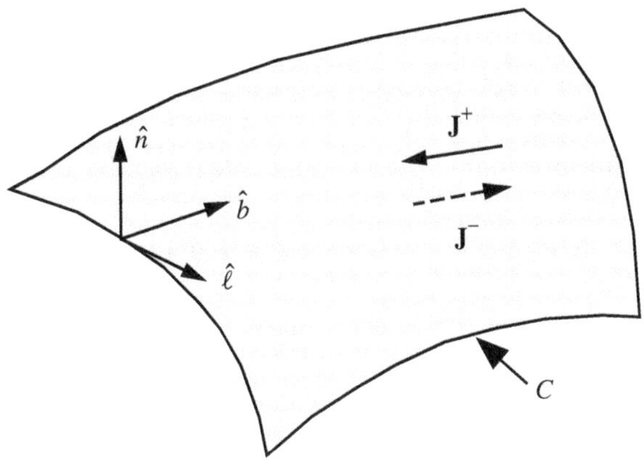


**FIGURE 3-3** Geometry of a curved plate representing an open surface.

(2.42), which are similar to Franz's integral formulas whose precise form is given in one of the problems. However, in the case of (3.13) it is permitted only if the contour integral in (3.20) also vanishes. The last is often referred to as Kottler's boundary line integral, and its presence is necessary to ensure the divergencelessness of the field for all $\mathbf{r}$. If, however, one thinks of $\mathbf{J}$ and $\mathbf{M}$ as representing the net currents on the open surface, then $\mathbf{J} \cdot \hat{b} = (\mathbf{J}^+ - \mathbf{J}^+) \cdot \hat{b} = 0$ and $\mathbf{M} \cdot \hat{b} = (\mathbf{M}^+ - \mathbf{M}^-) \cdot \hat{b} = 0$ at the boundary line $C$. Consequently, the Kottler integral in (3.20) again vanishes, implying that with this interpretation of $\mathbf{J}$ and $\mathbf{M}$, (3.13) remains valid when the volume integrals are replaced by ones over the surface of the curved plate.

We remark that (3.18) are again evocative of Huygens's principle as discussed in the previous section in connection with Kirchhoff's integral equation. In practice, however, the Stratton-Chu equations are more attractive than (3.18) because of the lower singularity of their kernel, leading to a more accurate numerical implementation.

Integral equations such as those in (3.17) can be used for solving the fields on $S_\Omega$ by enforcing the specific boundary conditions associated with the surfaces comprising $S_\Omega$. To enforce these boundary conditions it is necessary to have the observation point directly on $S_\Omega$ leading to singular kernels that must be carefully integrated as done in the previous section. As before, we refer to Figure 3-2 and rewrite (3.17a) as

$$\begin{aligned}\mathbf{E}(\mathbf{r}_o) = \mathbf{E}^i(\mathbf{r}_o) &+ \iint_{S_\Omega - S_o} \{[\mathbf{E}(\mathbf{r}') \times \hat{n}'] \times \nabla G(\mathbf{r}_o, \mathbf{r}') \\ &- jkZ\,[\hat{n}' \times \mathbf{H}(\mathbf{r}')]\, G(\mathbf{r}_o, \mathbf{r}') - \hat{n}' \cdot \mathbf{E}(\mathbf{r}')\nabla G(\mathbf{r}_o, \mathbf{r}')\}\, ds' \\ &+ \iint_{S_o} \{[\mathbf{E}(\mathbf{r}') \times \hat{n}'] \times \nabla G(\mathbf{r}_o, \mathbf{r}') - jkZ\,[\hat{n}' \times \mathbf{H}(\mathbf{r}')]\, G(\mathbf{r}_o, \mathbf{r}') \\ &- \hat{n}' \cdot \mathbf{E}(\mathbf{r}')\nabla G(\mathbf{r}_o, \mathbf{r}')\}\, ds' \end{aligned} \tag{3.21}$$

in which $S_o$ is a vanishingly small hemispherical surface. Noting the identities

$$\begin{aligned}(\mathbf{E} \times \hat{n}') \times \nabla G &= \nabla G \times (\hat{n}' \times \mathbf{E}) \\ &= \hat{n}'(\mathbf{E} \cdot \nabla G) - \mathbf{E}(\hat{n}' \cdot \nabla G)\end{aligned} \tag{3.22}$$

$$-(\hat{n}' \cdot \mathbf{E})\nabla G = -\hat{n}'(\mathbf{E} \cdot \nabla G) + \mathbf{E} \times (\hat{n}' \times \nabla G) \tag{3.23}$$

it follows that

$$(\mathbf{E} \times \hat{n}') \times \nabla G - (\hat{n}' \cdot \mathbf{E})\nabla G = \mathbf{E} \times (\hat{n}' \times \nabla G) - \mathbf{E}(\hat{n}' \cdot \nabla G). \tag{3.24}$$

When the last is substituted into (3.21) the surface integral over $S_o$ becomes

$$\iint_{S_o} \{\mathbf{E}(\mathbf{r}') \times \left[\hat{n}' \times \nabla G(\mathbf{r}_o, \mathbf{r}')\right] - \mathbf{E}(\mathbf{r}') \left[\hat{n}' \cdot \nabla G(\mathbf{r}_o, \mathbf{r}')\right] \\ - jkZ \left[\hat{n}' \times \mathbf{H}(\mathbf{r}')\right] G(\mathbf{r}_o, \mathbf{r}')\} \, ds'.$$

For this integral $\hat{n}' = \hat{R}_o$, $ds' = R_o^2 \sin\theta_o \, d\phi_o \, d\theta_o$, and since $S_o \to 0$ we may set $\mathbf{E}(\mathbf{r}') \approx \mathbf{E}(\mathbf{r}_o)$ and $\mathbf{H}(\mathbf{r}') = \mathbf{H}(\mathbf{r}_o)$. When we substitute for $\nabla G$ as given in (2.55) with $R = R_o$, we find that the first term of the integral vanishes because $\nabla G = \hat{R}_o|\nabla G|$ and $\hat{n} \times \hat{R}_o = 0$. Also the third term goes to zero as $R_o \to 0$. The second term (see (3.8)) when integrated gives $-\frac{1}{2}\mathbf{E}(\mathbf{r}_o)$, and thus we can rewrite (3.21) as

$$\mathbf{E}(\mathbf{r}_o) = 2\mathbf{E}^i(r_o) + 2 \oiint_{S_\Omega} \{\left[\mathbf{E}(\mathbf{r}') \times \hat{n}'\right] \times \nabla G(\mathbf{r}_o, \mathbf{r}') \\ -jkZ \left[\hat{n}' \times \mathbf{H}(\mathbf{r}')\right] G(\mathbf{r}_o, \mathbf{r}') - \hat{n}' \cdot \mathbf{E}(\mathbf{r}')\nabla G(\mathbf{r}_o, \mathbf{r}')\} \, ds'. \tag{3.25}$$

Incorporating this result into (3.17) we have

$$\oiint_{S_\Omega} \{\left[\mathbf{E}(\mathbf{r}') \times \hat{n}'\right] \times \nabla G(\mathbf{r}, \mathbf{r}') - jkZ \left[\hat{n}' \times \mathbf{H}(\mathbf{r})\right] G(\mathbf{r}, \mathbf{r}') \\ - \hat{n}' \cdot \mathbf{E}(\mathbf{r}')\nabla G(\mathbf{r}, \mathbf{r}')\} \, ds' + \mathbf{E}^i(\mathbf{r}) = \begin{cases} \mathbf{E}(\mathbf{r}) & \mathbf{r} \text{ in } V_\infty \\ \frac{1}{2}\mathbf{E}(\mathbf{r}) & \mathbf{r} \text{ on } S_\Omega \\ 0 & \mathbf{r} \text{ within } S_\Omega \end{cases} \tag{3.26a}$$

$$\oiint_{S_\Omega} \{\left[\mathbf{H}(\mathbf{r}') \times \hat{n}'\right] \times \nabla G(\mathbf{r}, \mathbf{r}') - jkY \left[\mathbf{E}(\mathbf{r}) \times \hat{n}'\right] G(\mathbf{r}, \mathbf{r}') \\ - \hat{n}' \cdot \mathbf{H}(\mathbf{r}')\nabla G(\mathbf{r}, \mathbf{r}')\} \, ds' + \mathbf{H}^i(\mathbf{r}) = \begin{cases} \mathbf{H}(\mathbf{r}) & \mathbf{r} \text{ in } V_\infty \\ \frac{1}{2}\mathbf{H}(\mathbf{r}) & \mathbf{r} \text{ on } S_\Omega \\ 0 & \text{within } S_\Omega. \end{cases} \tag{3.26b}$$

For completeness, we note that the impressed fields $(\mathbf{E}^i, \mathbf{H}^i)$ may be replaced by their volume integral representations (3.13) or (2.52). However, when the observation point is within the volume of the impressed or equivalent volume sources $(\mathbf{J}, \mathbf{M})$, we must then revert to the principle-value integral representation given in (2.69). It should also be noted that the Stratton-Chu equations are completely equivalent to the vector Kirchhoff equations (3.10). Notably, both sets of integral equations involve the normal and tangential field components on the surface $S_\Omega$, but Kirchhoff's equations decouple each field component from the others. However, these are unavoidably coupled upon application of the boundary conditions on $S_\Omega$. Nevertheless, in the case of two-dimensional applications where only a $z$-directed electric or magnetic field exists, Kirchhoff's equations are the most simple to use. By setting $\mathbf{E} = \hat{z}E_z$ or $\mathbf{H} = \hat{z}H_z$ in (3.10), a scalar equation is obtained instead of the vector integral equation resulting from (3.26). Consequently, the extinction or Kirchhoff's integral equations are, generally, the preferred choice in formulating two-dimensional problems.

### 3.1.3 Integral Equations for Homogeneous Dielectrics

Man-made structures such as vehicles made of composites and microstrip antennas are typically composed of piecewise homogeneous dielectrics. The effects of these materials must therefore be accounted for in computing the radiated or scattered fields. So far, field representations were given that apply in the presence of structures enclosed within a surface $S_\Omega$ by invoking the equivalence theorem. In this section we will specialize these expressions to the case where the surface $S_\Omega$ encloses a piecewise homogeneous dielectric body. We shall first consider the simplest case, that is, that which is pertinent to a homogeneous dielectric body.

Consider the homogeneous dielectric body enclosed by the surface $S_\Omega = S_d$ as shown in Figure 3-4. The dielectric is immersed in some excitation field $(\mathbf{E}^i, \mathbf{H}^i)$ generated by the sources $(\mathbf{J}^i, \mathbf{M}^i)$, which are exterior to $S_d$, and we are interested in finding a representation of the field in the exterior region (region 1) and perhaps interior to $S_d$, (region 2). One of the simplest integral expressions in this case is obtained by invoking the surface equivalence theorem, and with this in mind we set up the two problems illustrated in Figure 3-4. The set-up in Figure 3-4b assumes zero field interior to $S_d$; thus, the equivalent current $(\mathbf{J}_1, \mathbf{M}_1)$ can be used for computing the fields $(\mathbf{E}_1, \mathbf{H}_1)$ exterior to $S_d$. In contrast, the set-up in Figure 3-4c assumes zero exterior fields and thus the equivalent currents $(\mathbf{J}_2, \mathbf{M}_2)$ can be used for computing the interior fields $(\mathbf{E}_2, \mathbf{H}_2)$.

It should be remarked that the set up assumed here, where the fields are set to zero in the indicated region, is not unique. Any other non-zero field could have been used, and this would result in a different, albeit equivalent, formulation. In fact, certain judicious choices for the interior fields of the set-up in Figure 3-4b or the exterior fields in Figure 3-4c lead to formulations that may involve a single surface equivalent current (Glisson, 1984). An alternative approach will be to eliminate the introduction of the equivalent surface currents altogether and express the scattered fields in terms of the tangential electric and magnetic fields at the dielectric interface. In this case, the representation (3.18) may be used (or some other equivalent expression) to set up integral equations for the tangential fields upon invoking field continuity at the interface. Nevertheless, next we shall consider the solution of the scattered/radiated fields in the presence of a dielectric via the set-up in Figure 3-4 since this appears to be one of the most often used approaches.

The introduced equivalent current illustrated in Figure 3-4 can be substituted into (3.13) to obtain integral expressions for the exterior and interior fields upon changing the volume integral to one over the closed surface, $S_d$. However, before doing so, it is important to note that by enforcing the tangential field continuity equations

$$\hat{n}_1 \times \mathbf{E}_1 = \hat{n}_1 \times \mathbf{E}_2, \quad \hat{n}_1 \times \mathbf{H}_1 = \hat{n}_1 \times \mathbf{H}_2 \tag{3.27}$$

($\hat{n}_1$ denotes the unit normal pointing away from $S_d$) across the surface $S_d$, it follows that

$$\mathbf{J}_1 = -\mathbf{J}_2 = \mathbf{J}, \quad \mathbf{M}_1 = -\mathbf{M}_2 = -\mathbf{M}. \tag{3.28}$$

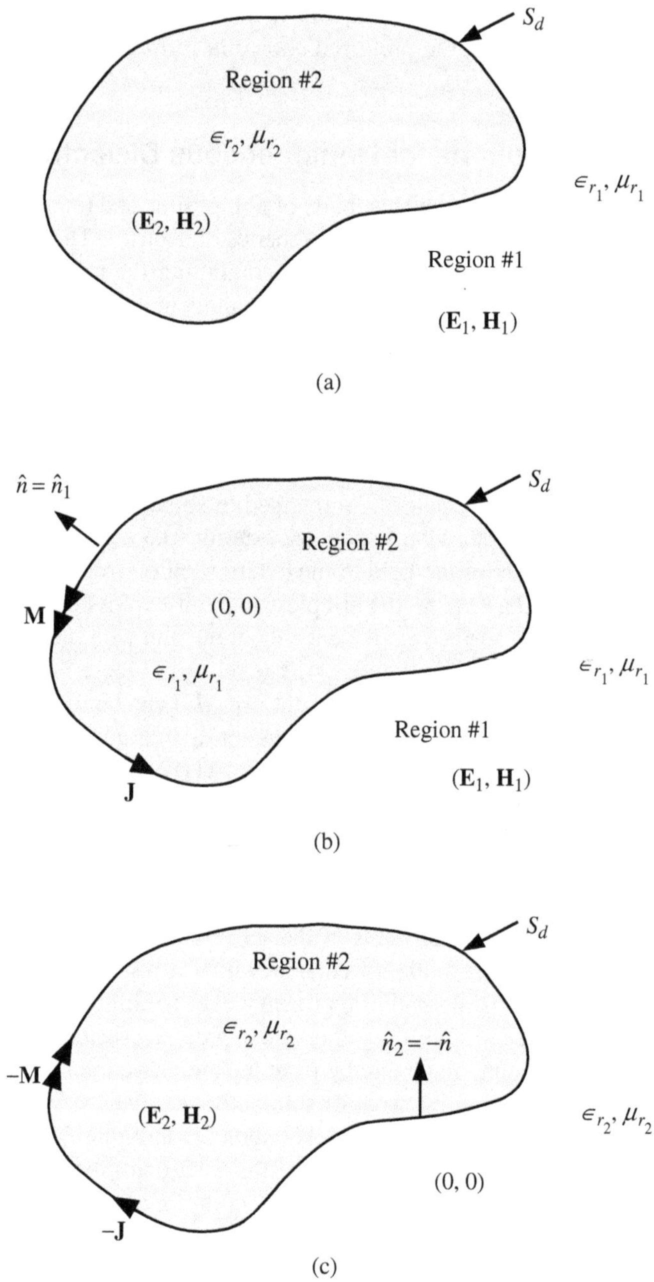


**FIGURE 3-4** Application of the equivalence principle for a dielectric: (a) original problem, (b) equivalent problem for region 1, and (c) equivalent problem for region 2.

In arriving at (3.28) we could have also implied that (see Section 1.10)

$$\begin{aligned} \mathbf{J}_1 &= \hat{n}_1 \times \mathbf{H}_1, & \mathbf{M}_1 &= -\hat{n}_1 \times \mathbf{E}_1 \\ \mathbf{J}_2 &= \hat{n}_2 \times \mathbf{H}_2 = -\hat{n}_1 \times \mathbf{H}_2 & \mathbf{M}_2 &= -\hat{n}_2 \times \mathbf{E}_2 = \hat{n}_1 \times \mathbf{E}_2. \end{aligned} \tag{3.29}$$

However, it is not necessary to introduce these expressions since the surface fields are unknown, and it is thus more convenient to retain $(\mathbf{J}, \mathbf{M})$ as the variable functions to be determined by enforcing the boundary conditions associated with problems defined in Figure 3-4. From Figure 3-4b, since the interior fields have been set to zero, we have that on $S_d$ (actually just inside $S_d$)

$$\begin{aligned} \hat{n}_1 \times \mathbf{E}_1 &= 0 \\ \hat{n}_1 \times \mathbf{H}_1 &= 0. \end{aligned} \tag{3.30}$$

By defining the total fields $(\mathbf{E}_1, \mathbf{H}_1)$ to be the sum of the source fields and those radiated by $(\mathbf{J}, \mathbf{M})$ we may rewrite (3.30) as

$$\begin{aligned} \hat{n}_1 \times \mathbf{E}^i &= -\hat{n}_1 \times \mathbf{E}_1^s \\ \hat{n}_1 \times \mathbf{H}^i &= -\hat{n}_1 \times \mathbf{H}_1^s \end{aligned} \tag{3.31a}$$

where

$$\begin{aligned} \mathbf{E}_1^s = \oiint_{S_d} \Big[ & \mathbf{M}(\mathbf{r}') \times \nabla G_1(\mathbf{r}, \mathbf{r}') - jk_o Z_o \mu_{r_1} \mathbf{J}(\mathbf{r}') G_1(\mathbf{r}, \mathbf{r}') \\ & - j\frac{Z_o}{k_o \epsilon_{r_1}} \nabla_s' \cdot \mathbf{J}(\mathbf{r}') \nabla G_1(\mathbf{r}, \mathbf{r}') \Big] ds' \end{aligned} \tag{3.31b}$$

$$\begin{aligned} \mathbf{H}_1^s = \oiint_{S_d} \Big[ & -\mathbf{J}(\mathbf{r}') \times \nabla G_1(\mathbf{r}, \mathbf{r}') - jk_o Y_o \epsilon_{r_1} \mathbf{M}(\mathbf{r}') G_1(\mathbf{r}, \mathbf{r}') \\ & - j\frac{Y_o}{k_o \mu_{r_1}} \nabla_s' \cdot \mathbf{M}(\mathbf{r}') \nabla G_1(\mathbf{r}, \mathbf{r}') \Big] ds'. \end{aligned} \tag{3.31c}$$

In these $k_o$, $Z_o$, $Y_o$ are the free-space wave number impedance and admittance, respectively, whereas $\epsilon_{r_1}$ and $\mu_{r_1}$ are the relative material constants of the exterior medium and are usually unity. Also,

$$G_1(\mathbf{r}, \mathbf{r}') = \frac{e^{-jk_1|\mathbf{r}-\mathbf{r}'|}}{4\pi|\mathbf{r}-\mathbf{r}'|} = \frac{e^{-jk_o\sqrt{\mu_{r_1}\epsilon_{r_1}}|\mathbf{r}-\mathbf{r}'|}}{4\pi|\mathbf{r}-\mathbf{r}'|} \tag{3.32}$$

is the Green's function associated with the exterior region. The fields $(\mathbf{E}_1^s, \mathbf{H}_1^s)$ are customarily referred to as those scattered by the dielectric body due to the excitation $(\mathbf{E}^i, \mathbf{H}^i)$. Instead of repeatedly using the explicit integral representation (3.31) it is convenient to define the operators

$$L_{1E_m}(\mathbf{M}) = \oiint_{S_d} \mathbf{M}(\mathbf{r}') \times \nabla G_1(\mathbf{r}, \mathbf{r}')\, ds' \tag{3.33}$$

$$L_{1H_m}(\mathbf{M}) = -\oiint_{S_d} \left[ jk_o Y_o \epsilon_{r_1} \mathbf{M}(\mathbf{r}') G_1(\mathbf{r}, \mathbf{r}') + \frac{jY_o}{k_o \mu_{r1}} \nabla_s' \cdot \mathbf{M}(\mathbf{r}') \nabla G_1(\mathbf{r}, \mathbf{r}') \right] ds' \tag{3.34}$$

$$L_{1E_e}(\mathbf{J}) = -\oiint_{S_d} \left[ jk_o Z_o \mu_{r1} \mathbf{J}(\mathbf{r}') G_1(\mathbf{r}, \mathbf{r}') + \frac{jZ_o}{k_o \epsilon_{r1}} \nabla_s' \cdot \mathbf{J}(\mathbf{r}') \nabla G_1(\mathbf{r}, \mathbf{r}') \right] ds' \tag{3.35}$$

$$L_{1H_e}(\mathbf{J}) = -\oiint_{S_d} \mathbf{J}(\mathbf{r}') \times \nabla G_1(\mathbf{r}, \mathbf{r}')\, ds'. \tag{3.36}$$

Then, since (Müller, 1969; Van Bladel, 1985)

$$\hat{n}_1 \times \oiint_{S_d} \mathbf{A}(\mathbf{r}') \times \nabla G_1(\mathbf{r}_o^{\pm}, \mathbf{r}')\, ds' = \mp\frac{1}{2}\mathbf{A}(\mathbf{r}_o) + \hat{n}_1 \times \oiint_{S_d} \mathbf{A}(\mathbf{r}') \times \nabla G_1(\mathbf{r}_o, \mathbf{r}')\, ds' \tag{3.37}$$

where $\mathbf{r}_o^{\pm}$ implies that the observation point is just exterior (+) or interior (−) to $S_d$, we can rewrite (3.31) more explicitly as

$$\begin{aligned} -\frac{1}{2}\mathbf{M}(\mathbf{r}) - \hat{n}_1 \times L_{1E_m}(\mathbf{M}) - \hat{n}_1 \times L_{1E_e}(\mathbf{J}) &= \hat{n}_1 \times \mathbf{E}^i \\ +\frac{1}{2}\mathbf{J}(\mathbf{r}) - \hat{n}_1 \times L_{1H_e}(\mathbf{J}) - \hat{n}_1 \times L_{1H_m}(\mathbf{M}) &= \hat{n}_1 \times \mathbf{H}^i \end{aligned} \tag{3.38}$$

valid for $\mathbf{r}$ on $S_d$. We note that (3.37) can be proven by following a similar procedure to that employed for the derivation of (3.9).

Another set of equations to be coupled with (3.38) can be obtained by enforcing the boundary conditions on $(\mathbf{E}_2, \mathbf{H}_2)$. From Figure 3-4c we have that on $S_d$ (actually just outside $S_d$)

$$\begin{aligned} \hat{n}_1 \times \mathbf{E}_2 = \hat{n}_1 \times \mathbf{E}_2^s = 0 \\ \hat{n}_1 \times \mathbf{H}_2 = \hat{n}_1 \times \mathbf{H}_2^s = 0 \end{aligned} \tag{3.39}$$

and upon making use of (3.37) these can be more explicitly written as

$$\begin{aligned} -\frac{1}{2}\mathbf{M}(\mathbf{r}) - \hat{n}_1 \times L_{2E_m}(-\mathbf{M}) - \hat{n}_1 \times L_{2E_e}(-\mathbf{J}) = 0 \\ +\frac{1}{2}\mathbf{J}(\mathbf{r}) - \hat{n}_1 \times L_{2H_e}(-\mathbf{J}) - \hat{n}_1 \times L_{2H_m}(-\mathbf{M}) = 0. \end{aligned} \tag{3.40a}$$

In these, the integral operators $L_{2E_m}$, $L_{2E_e}$, $L_{2H_e}$, and $L_{2H_m}$ are identical to those defined in (3.33) through (3.36) provided $\epsilon_{r_1}$ and $\mu_{r_1}$ are replaced by $\epsilon_{r_2}$ and $\mu_{r_2}$, respectively. By inspection, it is also seen that the minus sign in the argument of the operators can be factored out, giving

$$\begin{aligned} \frac{1}{2}\mathbf{M}(\mathbf{r}) - \hat{n}_1 \times L_{2E_m}(\mathbf{M}) - \hat{n}_1 \times L_{2E_e}(\mathbf{J}) = 0 \\ -\frac{1}{2}\mathbf{J}(\mathbf{r}) - \hat{n}_1 \times L_{2H_e}(\mathbf{J}) - \hat{n}_1 \times L_{2H_m}(\mathbf{M}) = 0 \end{aligned} \tag{3.40b}$$

valid for $\mathbf{r}$ on $S_d$.

It is apparent that (3.38) and (3.40) are four integral equations involving only two unknowns. This is because we had initially enforced the continuity conditions (3.27) to relate the equivalent currents introduced for representing the exterior and interior fields. It is also a consequence of the fact that only the tangential electric or magnetic fields are needed over a closed surface for determining the fields away from $S_d$. Thus, we are essentially free to use one from each set of equations (3.38) and (3.40) to obtain a pair of them to be solved (usually numerically, and this will be discussed later) for $(\mathbf{J}, \mathbf{M})$. For example, we could select the equation resulting from the pair of conditions

$$\begin{aligned} \hat{n}_1 \times \mathbf{E}^i &= -\hat{n}_1 \times \mathbf{E}_1^s \\ \hat{n}_1 \times \mathbf{E}_2^s &= 0 \end{aligned} \tag{3.41a}$$

or from

$$\begin{aligned} \hat{n}_1 \times \mathbf{H}^i &= -\hat{n}_1 \times \mathbf{H}_1^s \\ \hat{n}_1 \times \mathbf{H}_2^s &= 0. \end{aligned} \tag{3.41b}$$

The integral equations resulting from (3.41a) are usually referred to as the *electric field integral equations* (EFIE), whereas those implied by (3.41b) are referred as the *magnetic field integral equations* (MFIE).

### 3.1.4 Integral Equations for Metallic Bodies

When $S_d$ encloses a conducting surface (i.e., $\epsilon_{r_2} \to 1 - j\infty$) we may then set $\mathbf{M} = 0$ (see Sections 1.4 and 1.10), and in that case the first of (3.41a) gives

$$jk_o Z_o \mu_{r1} \hat{n}_1 \times \oiint_{S_d} \left[ \mathbf{J}(\mathbf{r}') G_1(\mathbf{r}, \mathbf{r}') + \frac{1}{k_o^2 \mu_{r1} \epsilon_{r1}} \nabla_s' \cdot \mathbf{J}(\mathbf{r}') \nabla G_1(\mathbf{r}, \mathbf{r}') \right] ds' = \hat{n}_1 \times \mathbf{E}^i \tag{3.42a}$$

whereas from the first of (3.41b) we have

$$\frac{1}{2}\mathbf{J}(\mathbf{r}') + \hat{n}_1 \times \oiint_{S_d} \mathbf{J}(\mathbf{r}') \times \nabla G_1(\mathbf{r}, \mathbf{r}')\, ds' = \hat{n}_1 \times \mathbf{H}^i. \tag{3.42b}$$

These are, respectively, the well-known EFIE and MFIE for perfectly conducting surfaces. This MFIE is also known as *Maue's integral equation* and is the most common for solving the fields scattered by a closed conducting surface. It will be shown later that Maue's MFIE leads to a better-conditioned matrix than (3.42a), and this is a primary reason for its popularity in simulating closed conducting surfaces. An EFIE of the same form as the MFIE (3.42b), can however be derived from (3.41a) by invoking image theory to eliminate the electric currents (since $S_d$ is perfectly conducting). This gives

$$\frac{1}{2}\mathbf{M}(\mathbf{r}) + \hat{n}_1 \times \oiint_{S_d} \mathbf{M}(\mathbf{r}') \times \nabla G_1(\mathbf{r}, \mathbf{r}')\, ds' = -\hat{n}_1 \times \mathbf{E}^i \tag{3.42c}$$

which is clearly the dual of (3.42b). Since (3.42c) and (3.42b) simulate the same metallic surface, it is not surprising that one can be derived from the other. Specifically, (3.42b) can be derived from (3.42c) by taking the curl of the last and making use of the equivalence relation (see (1.111))

$$\mathbf{J} = \frac{\nabla \times \mathbf{M}}{j\omega\mu_o\mu_{r1}}.$$

The fact that (3.42b) and (3.42c) are equivalent (i.e., they predict the same scattered fields) is a vivid demonstration that, in the case of perfectly conducting surfaces, one could formulate the fields in terms of electric or magnetic currents.

We should remark that neither of (3.42) is valid for open conducting surfaces such as a metallic flat or curved plate (see Figure 3-3). This is because the surface

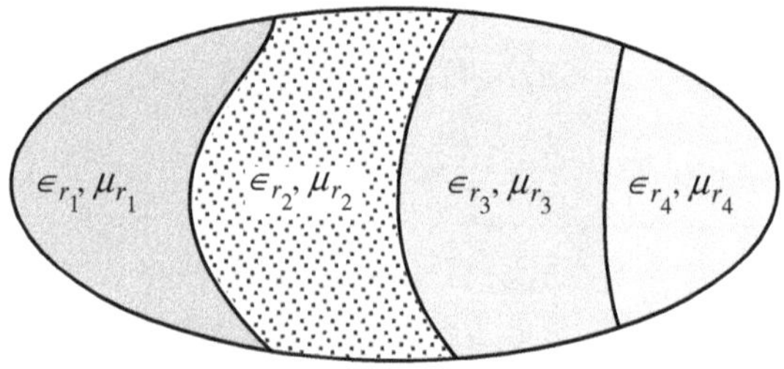


**FIGURE 3-5** Piecewise homogeneous dielectric body.

equivalence principle was used to introduce the equivalent surface currents. To construct an integral equation for the surface currents on a curved plate we may return to the original integral expression (2.52a) or (2.147) and set $\mathbf{M} = 0$. Then upon enforcing the boundary condition $\hat{n} \times (\mathbf{E}^s + \mathbf{E}^i) = 0$, we obtain the integral equation (also an EFIE)

$$+ jk_o Z_o \mu_{r1} \hat{n} \times \iint_S \left[ \mathbf{J}(\mathbf{r}') G_1(\mathbf{r}, \mathbf{r}') + \frac{1}{k_o^2 \epsilon_{r1} \mu_{r1}} \mathbf{J}(\mathbf{r}') \cdot \nabla\nabla G_1(\mathbf{r}, \mathbf{r}') \right] ds' = \hat{n} \times \mathbf{E}^i. \tag{3.43}$$

In contrast to the current appearing in (3.42), the one in this integral equation should be interpreted to represent the net flow between the top and bottom surfaces of the plate as illustrated in Figure 3-3. With this interpretation of $\mathbf{J}$ and from the discussion in Section 3.1.2, it is then seen that (3.43) is equivalent to (3.42a). Nevertheless, (3.43) is more difficult to implement than (3.42a) because of its higher kernel singularity.

### 3.1.5 Combined Field Integral Equations

Returning now to the original integral equation for the dielectric body we must address their uniqueness. Since they were formulated by assuming a null field within certain enclosed volumes, in accordance with the uniqueness theorem (3.41) or (3.42) will fail at those frequencies associated with a resonant mode within $S_d$. Fortunately, the EFIE is associated with different resonant modes than the MFIE, and this has been exploited to construct sets of equations which yield a unique solution. The most obvious approach is to consider various linear combinations of (3.41). For example, we could consider the combination (Mautz and Harrington, 1978)

$$\begin{aligned} \hat{n}_1 \times \left[ \mathbf{E}_1^s + \alpha \mathbf{E}_2^s \right] &= -\hat{n}_1 \times \mathbf{E}^i \\ \hat{n}_1 \times \left[ \mathbf{H}_1^s + \beta \mathbf{H}_2^s \right] &= -\hat{n}_1 \times \mathbf{H}^i \end{aligned} \tag{3.44}$$

where $\alpha$ and $\beta$ are arbitrary non-zero scalars. If we set $\alpha = \beta = 1$ we obtain the Poggio-Miller-Chang-Harrington-Wu (PMCHW) formulation (Poggio and Miller, 1973), while the choice of $\alpha = -\epsilon_{r_2}/\epsilon_{r_1}$ and $\beta = -\mu_{r_2}/\mu_{r_1}$ leads to the Müller (1969) formulation. Another combination proposed (Govind and Wilton, 1979) is

$$\hat{n}_1 \times \left[ \mathbf{H}_1^s + \frac{\alpha}{Z_1} \mathbf{E}_1^s \right] = -\hat{n}_1 \times \left[ \mathbf{H}_1^i + \frac{\alpha}{Z_1} \mathbf{E}^i \right]$$

$$\hat{n}_1 \times \left[ \mathbf{H}_2^s - \frac{\beta}{Z_2} \mathbf{E}_2^s \right] = 0 \tag{3.45}$$

in which $Z_1 = Z_0\sqrt{\mu_{r_1}/\epsilon_{r_1}}$, $Z_2 = Z_0\sqrt{\mu_{r_2}/\epsilon_{r_2}}$, whereas $\alpha$ and $\beta$ are again arbitrary scalars. Finally, a third coupled set of integral was proposed by Yaghjian (1981), who noted that the continuity equations are not necessarily satisfied when resonant modes are present. On this basis, the continuity equations can be combined with (3.41) or (3.42) to yield the conditions

$$\begin{aligned} \mathbf{E}_1^s + \hat{n}\frac{\nabla_s \cdot \mathbf{J}}{-j\omega\epsilon_o\epsilon_{r_1}} &= -\mathbf{E}_1^i \\ \mathbf{H}_1^s + \hat{n}\frac{\nabla_s \cdot \mathbf{M}}{-j\omega\mu_o\mu_{r_1}} &= -\mathbf{H}_1^i. \end{aligned} \tag{3.46}$$

From these we can readily derive integral equations for $(\mathbf{J}, \mathbf{M})$ upon substituting for the fields as given in (3.31). The integral equations based on (3.44) or (3.45) are generally referred to the literature as the *combined field integral equations* (CFIE), whereas the integral equations resulting from (3.46) is referred to as the *augmented field integral equations* (AFIE). They have all been used primarily for scattering computations, and their solution will be considered later. The CFIE have also been used for radiation problems relating to various types of cavity antennas. As can be expected, the CFIE cannot yield unique solutions at those frequencies where the electric and magnetic field integral equations fall concurrently. In addition, for very large structures the spurious resonant modes of $S_d$ are congruent leading to inaccuracies in the solution of CFIE. Further, it has been noted that the AFIE does not ensure the removal of all spurious resonances, and later we will discuss other remedies that can ensure uniqueness at the resonant frequencies of the cavity enclosed by $S_d$.

### 3.1.6 Integral Equations for Piecewise Homogeneous Dielectrics

The formulation presented in the previous section for treating homogeneous dielectrics can be readily extended to bodies composed of various homogeneous dielectric sections as shown in Figure 3-5. Let us for example consider the structure in Figure 3-6 consisting of a dielectric and a perfectly conducting section. We shall denote the surface of the conducting section that borders the exterior region (region 1) as $S_{dc_1}$ and the surface that borders the dielectric region of the body (region 2) as $S_{dc_2}$. Also, the surface of the dielectric that borders the exterior regions will be denoted as $S_{dc}$. The exterior region has relative dielectric constants $(\epsilon_{r_1}, \mu_{r_1})$ and a characteristic impedance $Z_1 = Z_o\sqrt{\frac{\mu_{r1}}{\epsilon_{r_1}}}$. Correspondingly, the interior dielectric region has relative dielectric constants $(\epsilon_{r_2}, \mu_{r_2})$ and a characteristic impedance $Z_2 = Z_o\sqrt{\frac{\mu_{r2}}{\epsilon_{r_2}}}$. We shall assume that the excitation fields $(\mathbf{E}^i, \mathbf{H}^i)$ will be in the exterior region although they can also be placed within the interior dielectric region as is likely the case with cavity type antennas (Arvas, Rahhal-Arabi, and Sadigh, 1991; Shafai, Moheb, and Kishk, 1991)

Following the formulation presented in the previous section, we refer to Figure 3-5 and introduce the equivalent currents $\mathbf{J}_{c_1}$ and $\mathbf{J}_{c_2}$ on the conducting surfaces $S_{dc_1}$ and $S_{dc_2}$, respectively. Since $S_{dc_1}$ and $S_{dc_2}$ border perfect conductors we choose to retain only the electric equivalent currents, although one could also choose to formulate the fields in terms of magnetic currents as discussed in the previous section. On the

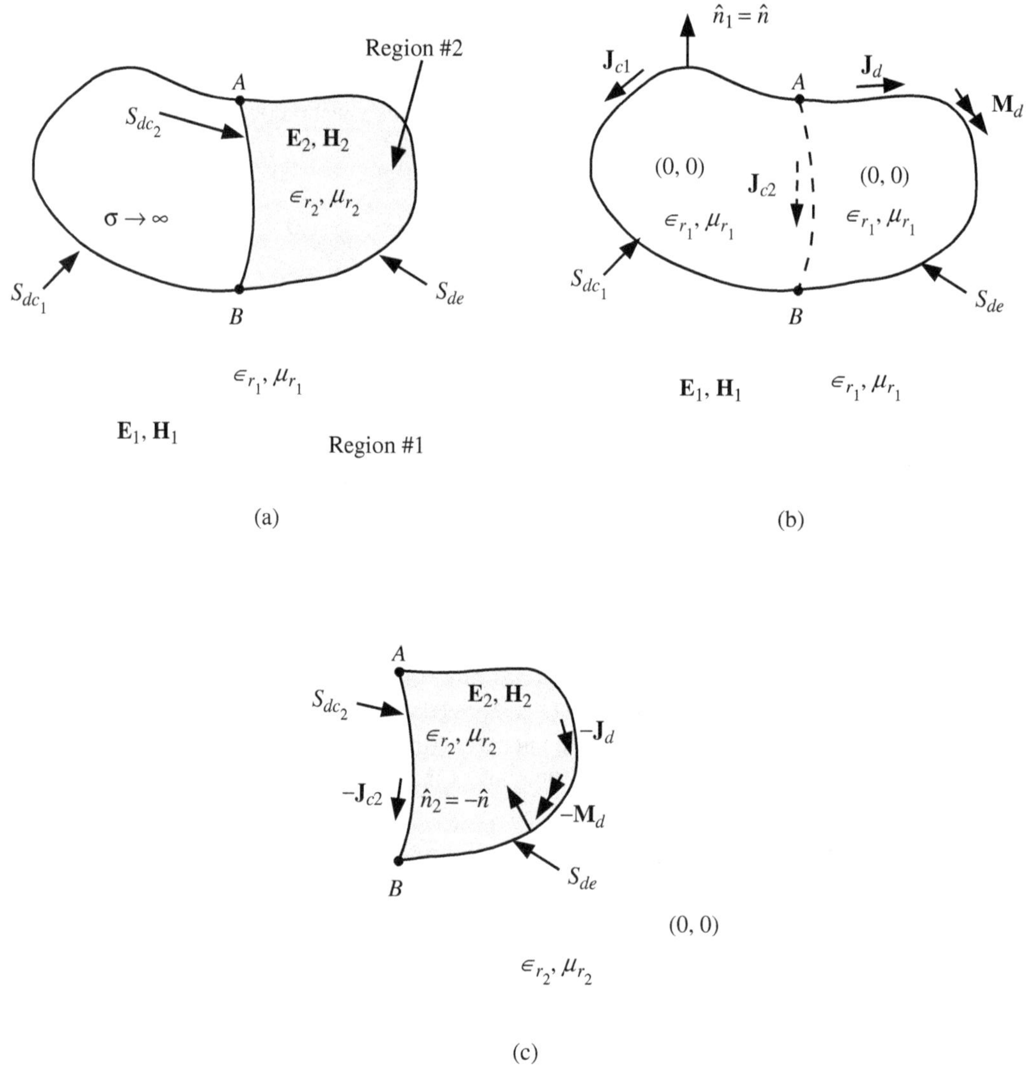


**FIGURE 3-6** Application of equivalence principle for a conductor dielectric body: (a) original problem, (b) equivalent problem for the exterior fields, and (c) equivalent problem for the interior fields.

dielectric surface $S_{de}$, which borders the exterior region, we introduce the equivalent electric and magnetic currents $(\mathbf{J}_d, \mathbf{M}_d)$. In accordance with the model in Figure 3-6b the boundary conditions are

$$
\begin{aligned}
\hat{n} \times \mathbf{E}_1 &= \hat{n} \times \left(\mathbf{E}^i + \mathbf{E}_1^s\right) = 0 \quad &&\text{on } S_{dc_1} \\
\hat{n} \times \mathbf{H}_1 &= \hat{n} \times \left(\mathbf{H}^i + \mathbf{H}_1^s\right) = 0 \quad &&\text{on } S_{dc_1} \\
\hat{n} \times \mathbf{E}_1 &= \hat{n} \times \left(\mathbf{E}^i + \mathbf{E}_1^s\right) = 0 \quad &&\text{on } S_{de} \\
\hat{n} \times \mathbf{H}_1 &= \hat{n} \times \left(\mathbf{H}_1 + \mathbf{H}_1^s\right) = 0 \quad &&\text{on } S_{de}
\end{aligned}
\tag{3.47}
$$

in which the scattered fields are given by

$$\mathbf{E}_1^s = -\iint_{S_{dc_1}} \left[ jk_o Z_o \mu_{r_1} \mathbf{J}_{c_1}(\mathbf{r}') G_1(\mathbf{r},\mathbf{r}') + \frac{jZ_o}{k_o \epsilon_{r_1}} \nabla'_s \cdot \mathbf{J}_{c_1}(\mathbf{r}') \nabla G_1(\mathbf{r},\mathbf{r}') \right] ds'$$
$$+ \iint_{S_{de}} \left[ \mathbf{M}_d(\mathbf{r}') \times \nabla G_1(\mathbf{r},\mathbf{r}') - jk_o Z_o \mu_{r_1} \mathbf{J}_d(\mathbf{r}) G_1(\mathbf{r},\mathbf{r}') \right.$$
$$\left. - \frac{jZ_o}{k_o \epsilon_{r_1}} \nabla'_s \cdot \mathbf{J}_d(\mathbf{r}') \nabla G_1(\mathbf{r},\mathbf{r}') \right] ds' \tag{3.48a}$$

$$\mathbf{H}_1^s = -\iint_{S_{dc_1}} \mathbf{J}_{c_1}(\mathbf{r}') \times \nabla G_1(\mathbf{r},\mathbf{r}')\, ds' + \iint_{S_{de}} \left[ -\mathbf{J}_d(\mathbf{r}') \times \nabla G_1(\mathbf{r},\mathbf{r}') \right.$$
$$\left. - jk_o Y_o \epsilon_{r_1} \mathbf{M}_d(\mathbf{r}') G_1(\mathbf{r},\mathbf{r}') - \frac{jY_o}{k_o \mu_{r_1}} \nabla'_s \cdot \mathbf{M}_d(\mathbf{r}') \nabla G_1(\mathbf{r},\mathbf{r}') \right] ds'. \tag{3.48b}$$

It should be noted that $\mathbf{J}_{c_2}$ does not enter into the representation of the exterior fields $(\mathbf{E}_1, \mathbf{H}_1)$ because it is completely enclosed by the surfaces $S_{dc_1}$ and $S_{de}$. Thus, in accordance with the equivalence principle it does not contribute to $(\mathbf{E}_1, \mathbf{H}_1)$. However, it does radiate in the dielectric region, and its effect in the exterior region is observed through the modifications it causes to $(\mathbf{J}_d, \mathbf{M}_d)$. This will be more apparent when the interior fields $(\mathbf{E}_2, \mathbf{H}_2)$ are formulated. To solve for the current densities, (3.48) are substituted into the boundary conditions (3.47), and upon making use of the identity (3.37) and the operator definitions (3.33) through (3.36), we obtain the integral equations

$$-\hat{n} \times L_{1E_m}(\mathbf{M}_d) - \hat{n} \times L_{1E_e}(\mathbf{J}_d) - \hat{n} \times L_{1E_e}(\mathbf{J}_{c_1}) = \hat{n} \times \mathbf{E}^i \quad \mathbf{r} \text{ on } S_{dc_1} \tag{3.49a}$$

$$+\frac{1}{2}\mathbf{J}_{c_1}(\mathbf{r}) - \hat{n} \times L_{1H_e}(\mathbf{J}_{c_1}) - \hat{n} \times L_{1H_e}(\mathbf{J}_d) - \hat{n} \times L_{1H_m}(\mathbf{M}_d) = \hat{n} \times \mathbf{H}^i \quad \mathbf{r} \text{ on } S_{dc_1} \tag{3.49b}$$

$$-\frac{1}{2}\mathbf{M}_d(\mathbf{r}) - \hat{n} \times L_{1E_m}(\mathbf{M}_d) - \hat{n} \times L_{1E_e}(\mathbf{J}_d) - \hat{n} \times L_{1E_e}(\mathbf{J}_{c_1}) = \hat{n} \times \mathbf{E}^i \quad \mathbf{r} \text{ on } S_{de} \tag{3.49c}$$

$$\frac{1}{2}\mathbf{J}_d(\mathbf{r}) - \hat{n} \times L_{1H_e}(\mathbf{J}_{c_1}) - \hat{n} \times L_{1H_e}(\mathbf{J}_d) - \hat{n} \times L_{1H_m}(\mathbf{M}_d) = \hat{n} \times \mathbf{H}^i \quad \mathbf{r} \text{ on } S_{de}. \tag{3.49d}$$

In addition to these, we can derive another set of four integral equations by referring to Figure 3-6 and enforcing the boundary conditions relating to the interior fields $(\mathbf{E}_2, \mathbf{H}_2)$.

Since we have assumed that no sources exist within the dielectric region, the boundary condition for the interior fields are

$$\begin{aligned} \hat{n} \times \mathbf{E}_2 &= \hat{n} \times \mathbf{E}_2^s = 0 \quad \text{on } S_{dc_2} \\ \hat{n} \times \mathbf{H}_2 &= \hat{n} \times \mathbf{H}_2^s = 0 \quad \text{on } S_{dc_2} \\ \hat{n} \times \mathbf{E}_2 &= \hat{n} \times \mathbf{E}_2^s = 0 \quad \text{on } S_{de} \\ \hat{n} \times \mathbf{H}_2 &= \hat{n} \times \mathbf{H}_2^s = 0 \quad \text{on } S_{de}. \end{aligned} \tag{3.50}$$

By invoking tangential field continuity across the dielectric boundary $S_{de}$, it follows that the interior fields are generated by the negative of the currents used for formulating the exterior fields. Thus, the interior fields are given by

$$
\begin{aligned}
\mathbf{E}_2 = \mathbf{E}_2^s = & \iint_{S_{dc_2}} \Big[ jk_o Z_o \mu_{r_2} \mathbf{J}_{c_2}(\mathbf{r}') G_2(\mathbf{r}, \mathbf{r}') \\
& + \frac{jZ_o}{k_o \epsilon_{r_2}} \nabla_s' \cdot \mathbf{J}_{c_1}(\mathbf{r}') \nabla G_2(\mathbf{r}, \mathbf{r}') \Big] ds' \\
& - \iint_{S_{de}} \Big[ \mathbf{M}_d(\mathbf{r}') \times \nabla G_2(\mathbf{r}, \mathbf{r}') - jk_o Z_o \mu_{r_2} \mathbf{J}_d(\mathbf{r}') G_2(\mathbf{r}, \mathbf{r}') \\
& - \frac{jZ_o}{k_o \epsilon_{r_2}} \nabla_s' \cdot \mathbf{J}_d(\mathbf{r}') \nabla G_2(\mathbf{r}, \mathbf{r}') \Big] ds'
\end{aligned} \tag{3.51a}
$$

$$
\begin{aligned}
\mathbf{H}_2 = \mathbf{H}_2^s = & \iint_{S_{dc_2}} \mathbf{J}_{c_2}(\mathbf{r}') \times \nabla G_2(\mathbf{r}, \mathbf{r}') \\
& + \iint_{S_{de}} \Big[ \mathbf{J}_d(\mathbf{r}') \times \nabla G_2(\mathbf{r}, \mathbf{r}') - jk_o Y_o \epsilon_{r_2} \mathbf{M}_d(\mathbf{r}') G_2(\mathbf{r}, \mathbf{r}') \\
& - \frac{jY_o}{k_o \mu_{r_2}} \nabla_s' \cdot \mathbf{M}_d(\mathbf{r}') \nabla G_2(\mathbf{r}, \mathbf{r}') \Big] ds'
\end{aligned} \tag{3.51b}
$$

in which

$$
G_2(\mathbf{r}, \mathbf{r}') = \frac{e^{-jk_o \sqrt{\mu_{r_2} \epsilon_{r_2}} |\mathbf{r} - \mathbf{r}'|}}{4\pi |\mathbf{r} - \mathbf{r}'|}.
$$

Substituting these into the boundary conditions (3.50) and making use of the identity (3.37) and definitions similar to those in (3.33) through (3.36) we obtain the integral equations

$$
-\hat{n} \times L_{2E_m}(\mathbf{M}_d) - \hat{n} \times L_{2E_e}(\mathbf{J}_d) - \hat{n} \times L_{2E_e}(\mathbf{J}_{c_2}) = 0 \quad \mathbf{r} \text{ on } S_{dc_2} \tag{3.52a}
$$

$$
-\frac{1}{2}\mathbf{J}_{c_2}(\mathbf{r}) - \hat{n} \times L_{2H_e}(\mathbf{J}_{c2}) - \hat{n} \times L_{2H_e}(\mathbf{J}_d) - \hat{n} \times L_{2H_m}(\mathbf{M}_d) = 0 \quad \mathbf{r} \text{ on } S_{dc_2} \tag{3.52b}
$$

$$
\frac{1}{2}\mathbf{M}_d(\mathbf{r}) - \hat{n} \times L_{2E_m}(\mathbf{M}_d) - \hat{n} \times L_{2E_e}(\mathbf{J}_d) - \hat{n} \times L_{2E_e}(\mathbf{J}_{c_2}) = 0 \quad \mathbf{r} \text{ on } S_{de} \tag{3.52c}
$$

$$
-\frac{1}{2}\mathbf{J}_d(\mathbf{r}) - \hat{n} \times L_{2H_e}(\mathbf{J}_d) - \hat{n} \times L_{2H_e}(\mathbf{J}_{c_2}) - \hat{n} \times L_{2H_m}(\mathbf{M}_d) = 0 \quad \mathbf{r} \text{ on } S_{de} \tag{3.52d}
$$

which can be combined with (3.49) for a solution of the surface current densities. We have, of course, eight vector equations, whereas only four are needed to solve for

$\mathbf{J}_{c_1}$, $\mathbf{J}_{c_2}$, $\mathbf{J}_d$, and $\mathbf{M}_d$. We may choose the four EFIE, which result from the boundary conditions

$$\begin{aligned} \hat{n} \times \left(\mathbf{E}^i + \mathbf{E}_1^s\right) &= 0 \quad \text{on } S_{dc_1} \\ \hat{n} \times \mathbf{E}_2 &= 0 \quad \text{on } S_{dc_2} \\ \hat{n} \times \left(\mathbf{E}^i + \mathbf{E}_1^s\right) &= 0 \quad \text{on } S_{de} \\ \hat{n} \times \mathbf{E}_2 &= 0 \quad \text{on } S_{de} \end{aligned} \tag{3.53}$$

or the four MFIE, which result from the boundary conditions

$$\begin{aligned} \hat{n} \times \left(\mathbf{H}^i + \mathbf{H}_1^s\right) &= 0 \quad \text{on } S_{dc_1} \\ \hat{n} \times \mathbf{H}_2 &= 0 \quad \text{on } S_{dc_2} \\ \hat{n} \times \left(\mathbf{H}^i + \mathbf{H}_1^s\right) &= 0 \quad \text{on } S_{de} \\ \hat{n} \times \mathbf{H}_2 &= 0 \quad \text{on } S_{de}. \end{aligned} \tag{3.54}$$

However, as noted in the previous section the integral equation set based on (3.53) or (3.54) fail to yield unique solutions when the excitation frequency coincides with an internal resonance. In this case we may again combine (3.53) and (3.54) to obtain, for example, the CFIE set resulting from the boundary condition

$$\begin{aligned} \hat{n} \times \left[\mathbf{E}_1^s + \alpha \mathbf{E}_2\right] &= -\hat{n} \times \mathbf{E}^i \quad \text{on } S_{de} \\ \hat{n} \times \left[\mathbf{H}_1^s + \beta \mathbf{H}_2\right] &= -\hat{n} \times \mathbf{H}^i \quad \text{on } S_{de} \\ \hat{n} \times \left(\mathbf{E}^i + \mathbf{E}_1^s\right) &= 0 \quad \text{on } S_{dc_1} \\ \hat{n} \times \mathbf{E}_2 &= 0 \quad \text{on } S_{dc_2} \end{aligned} \tag{3.55}$$

in which $\alpha$ and $\beta$ are arbitrary scalars and can be set equal to the values discussed after equation (3.44). Another CFIE set can be obtained by enforcing the boundary conditions

$$\begin{aligned} \hat{n} \times \left[\mathbf{E}^i + \mathbf{E}_1^s\right] + \frac{\alpha}{Z_1} \hat{n} \times \left[\mathbf{H}^i + \mathbf{H}_1^s\right] &= 0 \quad \text{on } S_{dc_1} \\ \hat{n} \times \mathbf{E}_2 + \frac{\beta}{Z_2} \hat{n} \times \mathbf{H}_2 &= 0 \quad \text{on } S_{dc_2} \\ \hat{n} \times \left(\mathbf{E}^i + \mathbf{E}_1^s\right) + \frac{\gamma}{Z_1} \hat{n} \times \left(\mathbf{H}^i + \mathbf{H}_1^s\right) &= 0 \quad \text{on } S_{de} \\ \hat{n} \times \mathbf{E}_2 + \frac{\delta}{Z_2} \hat{n} \times \mathbf{H}_2 &= 0 \quad \text{on } S_{de} \end{aligned} \tag{3.56}$$

where again $\alpha$, $\beta$, $\gamma$, and $\delta$ are arbitrary scalars similar to those appearing in (3.45). As can be realized, several other CFIE could be generated through various combinations of the EFIE and MFIE (Rao and Wilton, 1990; Shafai, Moheb, and Kishk, 1991). Alternatively, one could combine the EFIE and MFIE with the integral equations derived from the continuity equations as discussed in the previous section in connection with the augmented field integral equations (AFIE).

### 3.1.7 Integral Equations for Inhomogeneous Dielectrics

In the previous section we developed integral equations for piecewise homogeneous structures. This was accomplished by treating each homogeneous section separately in a manner permitting us to employ a uniform set of dielectric constants over the entire region of interest. The fields and equations generated from each homogeneous section were then coupled by enforcing tangential field continuity at the transition boundaries of the different dielectric regions. An important aspect of this approach was the use of the unbounded-space Green's function for the treatment of each homogeneous region. Obviously in the case of an inhomogeneous dielectric region (i.e., a region in which the permittivity and permeability are arbitrary functions of $\mathbf{r}$) this approach cannot be used because of a lack of an appropriate Green's function.

A standard approach in modeling inhomogeneous dielectrics is to employ the volume equivalence principle rather than the surface equivalence principle used for sectionally homogeneous dielectrics. Let us consider the inhomogeneous dielectric volume $V_d$ as shown in Figure 3-7 in the presence of some excitation field $(\mathbf{E}^i, \mathbf{H}^i)$ whose source may be within $V_d$ or exterior to it. The total field $(\mathbf{E}, \mathbf{H})$ in the presence of the dielectric volume $V_d$ can then be written as

$$\mathbf{E} = \mathbf{E}^i + \mathbf{E}^s \qquad \mathbf{H} = \mathbf{H}^i + \mathbf{H}^s \tag{3.57}$$

where $(\mathbf{E}^s, \mathbf{H}^s)$ are the scattered fields caused by the presence of the dielectric (i.e., they may comprise the reflected, diffracted fields and their interactions). In

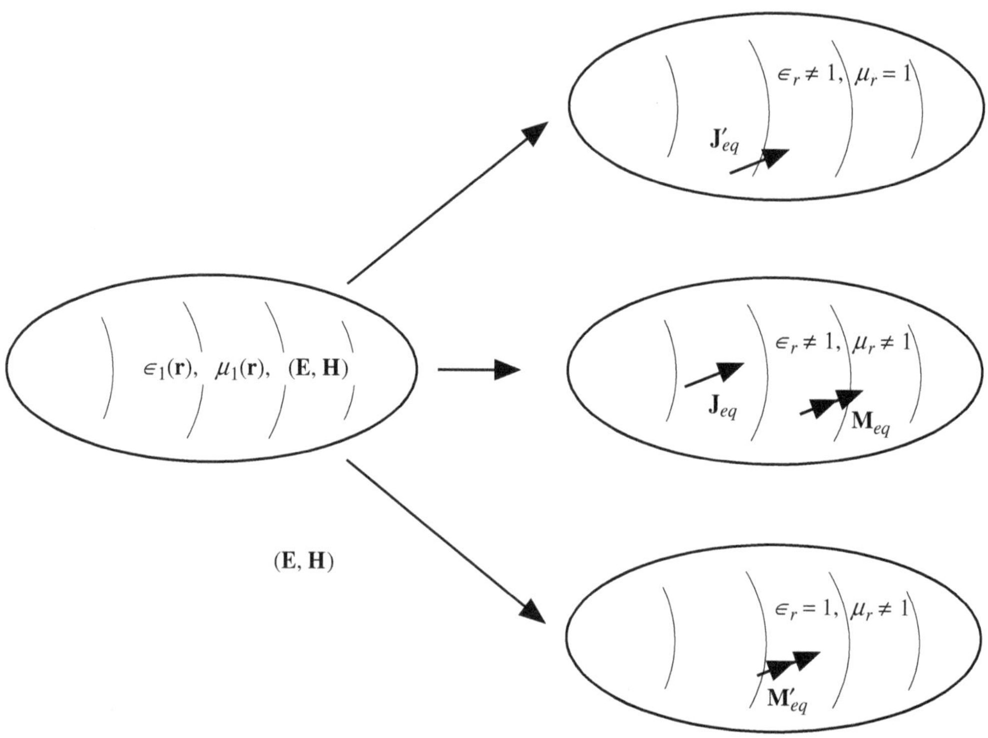


**FIGURE 3-7** ■ Volume equivalences for an inhomogeneous dielectric. The equivalent sources $\mathbf{J}_{eq}$, $\mathbf{M}_{eq}$, $\mathbf{J}'_{eq}$, and $\mathbf{M}'_{eq}$ are, respectively, defined in (3.58), (1.112), and (1.113). (After Volakis, 1992. With permission.)

accordance with the volume equivalence principle (see Section 1.9), the scattered fields may be represented by the field generated from the equivalent volume sources

$$\mathbf{J}_{eq} = j\omega\epsilon_o(\epsilon_r - 1)\mathbf{E}, \quad \mathbf{M}_{eq} = j\omega\mu_o(\mu_r - 1)\mathbf{H} \tag{3.58}$$

where $\epsilon_r$ and $\mu_r$ can be arbitrary functions of position. Obviously, since $\epsilon_r$ and $\mu_r$ are unity exterior to $V_d$, these equivalent currents exist only within the dielectric volume (see Figure 3-7). From (2.52), (2.69), (2.102), or (3.13) the scattered electric field can now be expressed as

$$\begin{aligned}\mathbf{E}^s &= \iiint_{V_d} \Big[\mathbf{M}_{eq}(\mathbf{r}') \times \nabla G_o(\mathbf{r}, \mathbf{r}') - jk_o Z_o \mathbf{J}_{eq}(\mathbf{r}') G_o(\mathbf{r}, \mathbf{r}') \\ &\quad - \frac{jZ_o}{k}\mathbf{J}_{eq}(\mathbf{r}') \cdot \nabla\nabla G_o(\mathbf{r}, \mathbf{r}')\Big]\, dv' + \frac{jZ_o}{3k_o}\mathbf{J}_{eq}(\mathbf{r}) \\ &= \iiint_{V_d} \Big[\mathbf{M}_{eq}(\mathbf{r}') \times \nabla G_o(\mathbf{r}, \mathbf{r}') - jk_o Z_o \mathbf{J}_{eq}(\mathbf{r}') G_o(\mathbf{r}, \mathbf{r}') \\ &\quad - \frac{jZ_o}{k_o}\nabla' \cdot \mathbf{J}_{eq}(\mathbf{r}') \nabla G_o(\mathbf{r}, \mathbf{r}')\Big]\, dv' + \frac{jZ_o}{3k_o}\mathbf{J}_{eq}(\mathbf{r}) \\ &= \iiint_{V_d} \Big[\nabla \times \overline{\Gamma}_o(\mathbf{r}, \mathbf{r}') \cdot \mathbf{M}_{eq}(\mathbf{r}') + jk_o Z_o \overline{\Gamma}_o(\mathbf{r}, \mathbf{r}') \cdot \mathbf{J}_{eq}(\mathbf{r}')\Big]\, dv' \\ &\quad + \frac{jZ_o}{3k_o}\mathbf{J}_{eq}(\mathbf{r})\end{aligned} \tag{3.59}$$

and $\mathbf{H}^s$ can be obtained by duality. In this, $G_o(\mathbf{r}, \mathbf{r}')$ denotes the free-space Green's function given by (2.37) with $k = k_o = \omega\sqrt{\mu_o\epsilon_o}$, that is,

$$G_o(\mathbf{r}, \mathbf{r}') = \frac{e^{-jk_o|\mathbf{r}-\mathbf{r}'|}}{4\pi|\mathbf{r} - \mathbf{r}'|} = \frac{e^{-jk_oR}}{4\pi R} \tag{3.60}$$

and $Z_o = 1/Y_o$ is the free-space intrinsic impedance. Also, $\overline{\Gamma}_o$ is the free-space dyadic Green's function given by (2.95) upon replacing $k$ with $k_o$, that is,

$$\overline{\Gamma}_o(\mathbf{r}, \mathbf{r}') = -\left[\overline{\mathbf{I}} + \frac{\nabla\nabla}{k_o^2}\right] G_o(\mathbf{r}, \mathbf{r}'). \tag{3.61}$$

To obtain an integral equation for the solution of the currents $(\mathbf{J}_{eq}, \mathbf{M}_{eq})$ we substitute (3.57) into (3.58), and this yields

$$\mathbf{J}_{eq} = jk_oY_o(\epsilon_r - 1)[\mathbf{E}^i + \mathbf{E}^s] \tag{3.62a}$$

$$\mathbf{M}_{eq} = jk_oZ_o(\mu_r - 1)[\mathbf{H}^i + \mathbf{H}^s]. \tag{3.62b}$$

By introducing the integral expression for the scattered electric field and its dual, we then obtain a coupled set of integral equations for the solution of $(\mathbf{J}_{eq}, \mathbf{M}_{eq})$. The solution of such an integral set will be considered later.

By this time, it should be obvious to the reader that the introduction of the equivalent currents is a matter of convenience since one could instead formulate the scattered fields without invoking such currents. For example, the Stratton-Chu surface integral equation can be written in terms of the surface fields as given by (3.17) or in terms of equivalent currents as given by (3.13). Also, the integral equations presented in the previous section for homogeneous dielectrics could be rewritten by replacing the equivalent surface currents with tangential surface fields. This amounts to substituting, for example, (3.28) through (3.29) into the integral equations (3.38) or (3.40). These can then be solved to find the unknown surface fields from which we can obtain the fields elsewhere by employing, for example, the Stratton-Chu integral equation (3.17) or some other equivalent expression. Not surprisingly, therefore, we can rewrite the scattered field expressions (3.59) by replacing the equivalent currents with quantities involving the electric and magnetic fields within the volume. For example, (3.58) can be substituted into (3.59), and then (3.57) provide a coupled set of integral equations for the solution of the fields within the dielectric's volume, $V_d$. This integral equation set as well as that implied by (3.62) in conjunction with (3.59) involve a total of six unknown field or current components per volume location for nontrivial $\epsilon_r$ and $\mu_r$. If $\mu_r = 1$ then $\mathbf{M}_{eq} = 0$ or $\mu_r - 1 = 0$, and the number of unknowns reduces to three, involving the three components of the electric field within each point in the volume $V_d$. Alternatively, if $\epsilon_r = 1$ only the three components of magnetic field or the magnetic equivalent current are required for determining the fields everywhere.

As will be seen later, the traditional numerical implementation of the coupled set (3.62) for nontrivial $\epsilon_r$ and $\mu_r$ demands substantial computer memory, and it is thus of interest to reduce the number of unknown quantities as much as possible. One approach is to employ the equivalent current expression (1.112)

$$\mathbf{J}'_{eq} = \frac{\epsilon_r - 1}{\epsilon_r}\nabla \times \mathbf{H} + \nabla \times [(\mu_r - 1)\mathbf{H}]$$

which combines the radiation effects of $\mathbf{J}_{eq}$ and $\mathbf{M}_{eq}$ given by (3.58). The scattered fields can then be expressed as

$$\mathbf{E}^s = jk_o Z_o \iiint_{V_d} \overline{\boldsymbol{\Gamma}}_o(\mathbf{r}, \mathbf{r}') \cdot \left\{\frac{\epsilon_r - 1}{\epsilon_r}\nabla' \times \mathbf{H}(\mathbf{r}') + \nabla' \times \left[(\mu_r - 1)\mathbf{H}(\mathbf{r}')\right]\right\} dv' \tag{3.63a}$$

$$\mathbf{H}^s = -\iiint_{V_d} \nabla \times \overline{\boldsymbol{\Gamma}}_o(\mathbf{r}, \mathbf{r}') \cdot \left\{\frac{\epsilon_r - 1}{\epsilon_r}\nabla' \times \mathbf{H}(\mathbf{r}') + \nabla' \times \left[(\mu_r - 1)\mathbf{H}(\mathbf{r}')\right]\right\} dv' \tag{3.63b}$$

in which the primed del operator $\nabla'$ implies differentiation with respect to the primed variables and in which the integrals $\epsilon_r$ and $\mu_r$ may be functions of $\mathbf{r}'$. By substituting (3.63b) into the second of (3.57), we obtain the integral equation

$$\mathbf{H} = \mathbf{H}^i - \iiint_{V_d} \nabla \times \overline{\boldsymbol{\Gamma}}_o(\mathbf{r}, \mathbf{r}') \cdot \left\{\frac{\epsilon_r - 1}{\epsilon_r}\nabla' \times \mathbf{H}(\mathbf{r}') + \nabla' \times \left[(\mu_r - 1)\mathbf{H}(\mathbf{r}')\right]\right\} dv'. \tag{3.64}$$

We may also substitute (3.63a) into the first of (3.57), and by taking the curl of the resulting equation we recover (3.64). Alternatively, we may formulate an integral equation for the electric field by introducing the magnetic equivalent current (1.113)

$$\mathbf{M}'_{eq} = -\frac{\mu_r - 1}{\mu_r}\nabla \times \mathbf{E} - \nabla \times [(\epsilon_r - 1)\mathbf{E}]$$

to replace $\mathbf{J}_{eq}$ and $\mathbf{M}_{eq}$ of (3.58). The scattered fields are then given by

$$\mathbf{E}^s = -\iiint_{V_d} \nabla \times \overline{\mathbf{\Gamma}}_o(\mathbf{r}, \mathbf{r}') \cdot \left\{ \frac{\mu_r - 1}{\mu_r}\nabla' \times \mathbf{E}(\mathbf{r}') + \nabla' \times \left[(\epsilon_r - 1)\mathbf{E}(\mathbf{r}')\right] \right\} dv' \tag{3.65a}$$

$$\mathbf{H}^s = -jk_oY_o \iiint_{V_d} \overline{\mathbf{\Gamma}}_o(\mathbf{r}, \mathbf{r}') \cdot \left\{ \frac{\mu_r - 1}{\mu_r}\nabla' \times \mathbf{E}(\mathbf{r}') + \nabla' \times \left[(\epsilon_r - 1)\mathbf{E}(\mathbf{r}')\right] \right\} dv' \tag{3.65b}$$

and when these are substituted into (3.57) we obtain the integral equation

$$\mathbf{E} = \mathbf{E}^i - \iiint_{V_d} \nabla \times \overline{\mathbf{\Gamma}}_o(\mathbf{r}, \mathbf{r}') \cdot \left\{ \frac{\mu_r - 1}{\mu_r}\nabla' \times \mathbf{E}(\mathbf{r}') + \nabla' \times \left[(\epsilon_r - 1)\mathbf{E}(\mathbf{r}')\right] \right\} dv' \tag{3.66}$$

which as expected is the dual of (3.64).

From (3.63) through (3.66) it is seen that one could formulate the fields in an inhomogeneous dielectric volume, either in terms of the magnetic field or the electric field. In contrast, the integral expressions (3.59) require knowledge of both the electric and the magnetic fields. Clearly, the numerical solution of (3.64) or (3.66) requires only three unknown components per volume point, but the presence of the curl operations on the unknown volume fields complicates their implementation. It is, thus, instructive to find alternative expressions to (3.63) or (3.65) that do not involve the curl of the unknown field. This is possible, and to develop the resulting integral expression let us begin with (3.59), which upon introducing the volume equivalent current expressions (3.58) becomes

$$\begin{aligned}\mathbf{E}^s &= -k_o^2 \iiint_{V_d} (\epsilon_r - 1)\mathbf{E}(\mathbf{r}') \cdot \overline{\mathbf{\Gamma}}_o(\mathbf{r}, \mathbf{r}')\, dv' \\ &\quad - j\omega\mu_o \nabla \times \iiint_{V_d} (\mu_r - 1)\mathbf{H}(\mathbf{r}')G_o(\mathbf{r}, \mathbf{r}')\, dv'.\end{aligned} \tag{3.67}$$

We can rewrite the second integral in this by introducing Maxwell's equation $\nabla \times \mathbf{E} = -j\omega\mu_o\mu_r\mathbf{H}$, giving

$$\begin{aligned}\mathbf{E}^s_m &= -j\omega\mu_o \nabla \times \iiint_{V_d} (\mu_r - 1)\mathbf{H}(\mathbf{r}')G_o(\mathbf{r}, \mathbf{r}')\, dv' \\ &= \nabla \times \iiint_{V_d} \left(1 - \frac{1}{\mu_r}\right) G_o(\mathbf{r}, \mathbf{r}')\nabla' \times \mathbf{E}(\mathbf{r}')\, dv'.\end{aligned} \tag{3.68}$$

Further by using the identities

$$\nabla \times \left[\nabla' \times (\phi \mathbf{E})\right] = \nabla \times \left[\nabla' \phi \times \mathbf{E}\right] + \nabla \times \left[\phi \nabla' \times \mathbf{E}\right] \tag{3.69}$$

$$\nabla(\phi\psi) = \psi \nabla \phi + \phi \nabla \psi \tag{3.70}$$

we obtain

$$\begin{aligned}
\mathbf{E}_m^s &= \nabla \times \iiint_{V_d} \nabla' \times \left\{\left(1 - \frac{1}{\mu_r}\right) G_o(\mathbf{r}, \mathbf{r}')\mathbf{E}(\mathbf{r}')\right\} dv' \\
&\quad - \nabla \times \iiint_{V_d} \left\{\left(1 - \frac{1}{\mu_r}\right) \nabla' G_o(\mathbf{r}, \mathbf{r}') \times \mathbf{E}(\mathbf{r}')\right\} dv' \\
&\quad + \nabla \times \iiint_{V_d} \left\{G_o(\mathbf{r}, \mathbf{r}')\nabla' \left(\frac{1}{\mu_r}\right) \times \mathbf{E}(\mathbf{r}')\right\} dv' \\
&= \mathbf{F}_{I_1} + \mathbf{F}_{I_2} + \mathbf{F}_{I_3}.
\end{aligned} \tag{3.71}$$

The first of the integrals in (3.71) can be transformed to a surface integral by invoking the vector Stokes identity

$$\iiint_V (\nabla \times \mathbf{F})\, dv = \oiint_{S_c} (\hat{n} \times \mathbf{F})\, ds. \tag{3.72}$$

This gives

$$\begin{aligned}
\mathbf{F}_{I_1} &= \nabla \times \iiint_{V_d} \nabla' \times \left\{\left(1 - \frac{1}{\mu_r}\right) G_o(\mathbf{r}, \mathbf{r}')\mathbf{E}(\mathbf{r}')\right\} dv' \\
&= \nabla \times \oiint_{S_d} \left(1 - \frac{1}{\mu_r}\right) G_o(\mathbf{r}, \mathbf{r}')\hat{n}' \times \mathbf{E}(\mathbf{r}')\, ds' \\
&= - \oiint_{S_d} \left(1 - \frac{1}{\mu_r}\right) \left[\hat{n}' \times \mathbf{E}(\mathbf{r}')\right] \times \nabla G_o(\mathbf{r}, \mathbf{r}')\, ds'
\end{aligned} \tag{3.73}$$

in which $\hat{n}' = \hat{n}(\mathbf{r}')$ is the outward unit surface normal to the surface $S_d$ and $\mu_r = \mu_r(\mathbf{r}')$. For the second integral in (3.71) we have

$$\begin{aligned}
\mathbf{F}_{I_2} &= -\nabla \times \iiint_{V_d} \left(1 - \frac{1}{\mu_r}\right) \nabla' G_o(\mathbf{r}, \mathbf{r}') \times \mathbf{E}(\mathbf{r}')\, dv' \\
&= - \iiint_{V_d} \left(1 - \frac{1}{\mu_r}\right) \nabla \times \left[\nabla' G_o(\mathbf{r}, \mathbf{r}') \times \mathbf{E}(\mathbf{r}')\right] dv'
\end{aligned} \tag{3.74}$$

and by employing the identity (Van Bladel, 1985)

$$\begin{aligned}
\nabla \times \left(\nabla' G_o \times \mathbf{E}\right) &= \nabla' G_o \nabla \cdot \mathbf{E} - \mathbf{E} \nabla \cdot \nabla' G_o + \mathbf{E} \cdot \nabla \nabla' G_o - \nabla' G_o \cdot \nabla \mathbf{E} \\
&= \mathbf{E} \nabla^2 G_o - \mathbf{E} \cdot \nabla \nabla G_o
\end{aligned} \tag{3.75}$$

(since $\nabla G = -\nabla' G$ and $\mathbf{E}$ is a function of $\mathbf{r}'$), it follows that

$$\begin{aligned}
\mathbf{F}_{I_2} &= +k_o^2 \iiint_{V_d} \left(1 - \frac{1}{\mu_r}\right) \mathbf{E}(\mathbf{r}') \cdot \left[\bar{\mathbf{I}} + \frac{\nabla\nabla}{k_o^2}\right] G_o(\mathbf{r}, \mathbf{r}')\, dv' + \left(1 - \frac{1}{\mu_r}\right) \mathbf{E}(\mathbf{r}) \\
&= -k_o^2 \iiint_{V_d} \left(1 - \frac{1}{\mu_r}\right) \mathbf{E}(\mathbf{r}') \cdot \bar{\mathbf{\Gamma}}_o(\mathbf{r}, \mathbf{r}')\, dv' + \left(1 - \frac{1}{\mu_r}\right) \mathbf{E}(\mathbf{r})
\end{aligned} \tag{3.76}$$

for $\mathbf{r}$ in $V_d$. In deriving this expression we also employed the differential equation (2.38) satisfied by the scalar Green's function and the definition for the free-space dyadic Green's function (3.61). Clearly, with the derivation of (3.73) and (3.76), the first two integrals in (3.71) have been rewritten only in terms of undifferentiated electric fields, and we would like to do the same for the last integral of (3.71). We have

$$\begin{aligned}
\mathbf{F}_{I_3} &= \nabla \times \iiint_{V_d} \left\{ G_o(\mathbf{r},\mathbf{r}')\nabla'\left(\frac{1}{\mu_r}\right) \times \mathbf{E}(\mathbf{r}') \right\} dv' \\
&= \iiint_{V_d} \nabla \times \left\{ G_o(\mathbf{r},\mathbf{r}')\nabla'\left(\frac{1}{\mu_r}\right) \times \mathbf{E}(\mathbf{r}') \right\} dv' \\
&= \iiint_{V_d} \nabla G_o(\mathbf{r},\mathbf{r}') \times \left[\nabla'\left(\frac{1}{\mu_r}\right) \times \mathbf{E}(\mathbf{r}')\right] dv' \qquad (3.77)
\end{aligned}$$

where again the last integral involves only the undifferentiated electric field within the volume of the dielectric.

We can now rewrite $\mathbf{E}_m$ by using the simplified expressions (3.73), (3.76), and (3.77). When the result is substituted in (3.66), the complete electric field expression involving undifferentiated volume fields is (Volakis, 1992)

$$\begin{aligned}
&-k_o^2 \iiint_{V_d} \left(\epsilon_r - \frac{1}{\mu_r}\right) \mathbf{E}(\mathbf{r}') \cdot \overline{\boldsymbol{\Gamma}}_o(\mathbf{r},\mathbf{r}')\, dv' \\
&+ \iiint_{V_d} \left[\mathbf{E}(\mathbf{r}') \times \nabla'\left(\frac{1}{\mu_r}\right)\right] \times \nabla G_o(\mathbf{r},\mathbf{r}')\, dv' \\
&- \oiint_{S_d} \left(1 - \frac{1}{\mu_r}\right) [\hat{n}' \times \mathbf{E}(\mathbf{r}')] \times \nabla G_o(\mathbf{r},\mathbf{r}')\, dv' + \mathbf{E}^i \\
&= \begin{cases} \mathbf{E}(\mathbf{r}) & \mathbf{r} \text{ not in } V_d \\ \frac{1}{2}\left(1 + \frac{1}{\mu_r}\right)\mathbf{E}(\mathbf{r}) & \mathbf{r} \text{ on } S_d \\ \frac{1}{3}(\epsilon_r + 2\mu_r)\mathbf{E}(\mathbf{r}) & \mathbf{r} \text{ in } V_d \end{cases} \qquad (3.78a)
\end{aligned}$$

and by duality

$$\begin{aligned}
&-k_o^2 \iiint_{V_d} \left(\mu_r - \frac{1}{\epsilon_r}\right) \mathbf{H}(\mathbf{r}') \cdot \overline{\boldsymbol{\Gamma}}_o(\mathbf{r},\mathbf{r}')\, dv' \\
&+ \iiint_{V_d} \left[\mathbf{H}(\mathbf{r}') \times \nabla'\left(\frac{1}{\epsilon_r}\right)\right] \times \nabla G_o(\mathbf{r},\mathbf{r}')\, dv' \\
&- \oiint_{S_d} \left(1 - \frac{1}{\epsilon_r}\right) [\hat{n}' \times \mathbf{H}(\mathbf{r}')] \times \nabla G_o(\mathbf{r},\mathbf{r}')\, dv' + \mathbf{H}^i \\
&= \begin{cases} \mathbf{H}(\mathbf{r}) & \mathbf{r} \text{ not in } V_d \\ \frac{1}{2}\left(1 + \frac{1}{\epsilon_r}\right)\mathbf{H}(\mathbf{r}) & \mathbf{r} \text{ on } S_d \\ \frac{1}{3}(\mu_r + 2\epsilon_r)\mathbf{H}(\mathbf{r}) & \mathbf{r} \text{ in } V_d. \end{cases} \qquad (3.78b)
\end{aligned}$$

It should be noted that in evaluating the right-hand side of (3.78) we made use of (2.69) and (3.25). Either (3.78a) or (3.78b) can be used for a solution of the fields within the volume of the dielectric. As seen, these integral expressions/equations contain both surface and volume integrals, and we shall therefore refer to them as *volume-surface integral equations* (VSIE). The second integral of (3.78a) involves the gradient of the dielectric constant (which vanishes for a homogeneous dielectric) whose effect is indistinguishable from the equivalent volume magnetic current, $\mathbf{E} \times \nabla\left(\frac{1}{\mu_r}\right)$. Similarly, the surface integral in (3.78a) involves the tangential electric field, which can be thought of as representing equivalent surface magnetic currents.

## 3.2 TWO-DIMENSIONAL REPRESENTATIONS

All integral equations and expressions presented in the previous section can be readily modified for two-dimensional structures, which are invariant with respect to the $z$-variable. As discussed in Section 2.5, in this case the excitation is an electric or a magnetic line source. Typically, this is a $z$-directed source generating $z$-polarized electric or magnetic fields, respectively, which become the excitation fields ($E_z^i$ or $H_z^i$). Consequently, the corresponding total fields and scattered fields will be $z$-directed as well. This fact along with the field's independence on the $z$-variable leads to a substantial simplification of the three-dimensional integral equations and expressions given earlier. In addition, by introducing the two-dimensional Green's function (see (2.113)), the volume integrals reduce to surface integrals over the cross section of the structure, and those over a three-dimensional surface reduce to a line integral over the boundary of the two-dimensional cross section. Next we consider some two-dimensional boundary (surface) and domain (volume) integral equations that are among the most popular in the literature.

### 3.2.1 Boundary Integral Equations

Upon setting $\mathbf{E} = \hat{z}E_z$ and $\mathbf{H} = \hat{z}H_z$, directly from (3.10) we have

$$\oint_C \left[E_z(\rho')\frac{\partial G_{2d}(\overline{\rho},\overline{\rho}')}{\partial n'} - G_{2d}(\overline{\rho},\overline{\rho}')\frac{\partial E_z(\overline{\rho}')}{\partial n'}\right] d\ell' + E_z^i(\overline{\rho}) = \begin{cases} E_z(\overline{\rho}) & \overline{\rho} \text{ in } A_\infty \\ 0 & \overline{\rho} \text{ in } A \end{cases} \tag{3.79a}$$

$$\oint_C \left[H_z(\rho')\frac{\partial G_{2d}(\overline{\rho},\overline{\rho}')}{\partial n'} - G_{2d}(\overline{\rho},\overline{\rho}')\frac{\partial H_z(\overline{\rho}')}{\partial n'}\right] d\ell' + H_z^i(\overline{\rho}) = \begin{cases} H_z(\overline{\rho}) & \overline{\rho} \text{ in } A_\infty \\ 0 & \overline{\rho} \text{ in } A. \end{cases} \tag{3.79b}$$

In these $G_{2d}(\overline{\rho},\overline{\rho}')$ is the two-dimensional Green's function defined in (2.115). Also, as illustrated in Figure 3-8, $C$ is the sum of all contours, $C_1, C_2, \ldots, C_N$ that enclose the regions $A_1, A_2, \ldots, A_N$, respectively, and $A_\infty$ is the region outside $A = \sum_n A_n$.

As usual the contour integrals on the left-hand side of (3.79) contains non-integrable singularities when $\overline{\rho}$ is on $A$ and must thus be evaluated by distorting the contour $C$ at that point and take the limit as this distortion vanishes. In parallel with

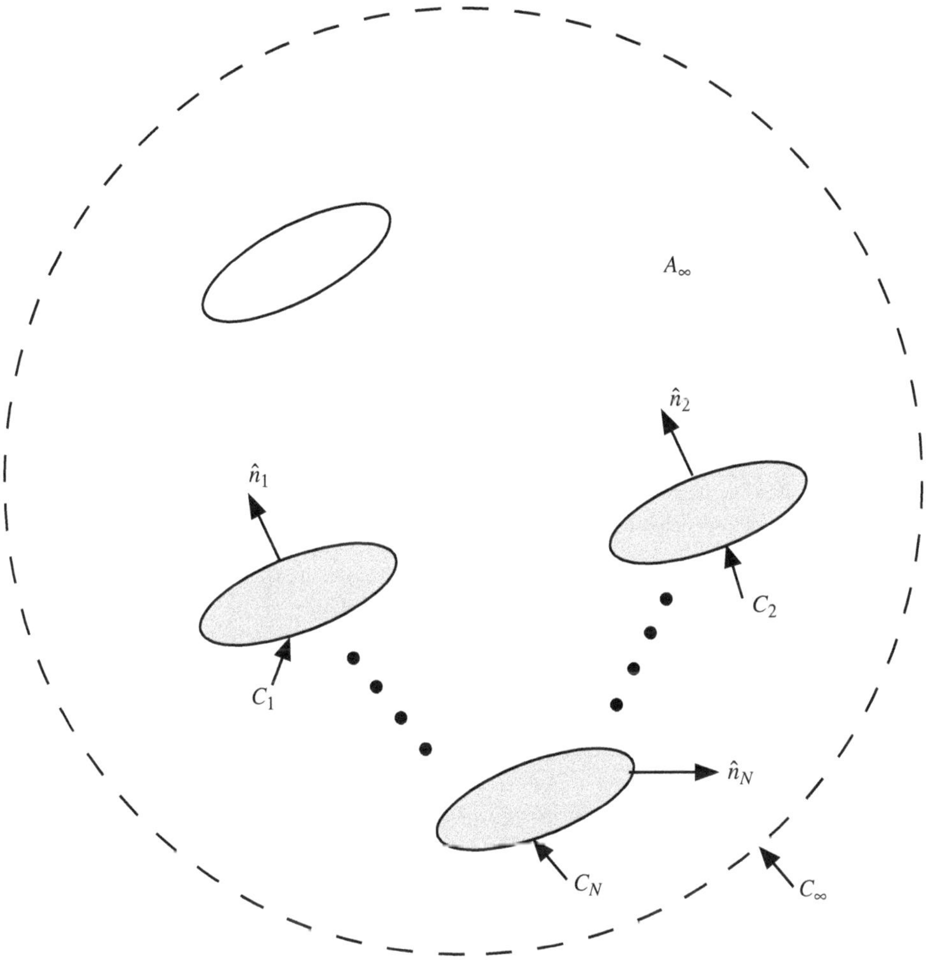


**FIGURE 3-8** ▪ Geometry for the two-dimensional application of Kirchhoff's integral equations.

the previous approaches, we proceed by rewriting (3.79a) as

$$\int_{C-C_o}\left[E_z(\overline{\rho}')\frac{\partial G_{2d}(\overline{\rho}_o,\overline{\rho}')}{\partial n'}-G_{2d}(\overline{\rho}_o,\overline{\rho}')\frac{\partial E_z(\overline{\rho}')}{\partial n'}\right]d\ell'$$

$$=-\int_{C_o}\left[E_z(\overline{\rho}')\frac{\partial G_{2d}(\overline{\rho}_o,\overline{\rho})}{\partial n'}-G_{2d}(\overline{\rho}_o,\overline{\rho})\frac{\partial E_z(\overline{\rho}')}{\partial n'}\right]d\ell' \qquad (3.80)$$

where $C_o$ is a vanishingly small semicircular contour whose center is at the observation point $\overline{\rho}_o$, and this is illustrated in Figure 3-9. To evaluate the right-hand side integral of (3.80) as $C_o \to 0$, we let $d\ell' = a\, d\phi$, where $|\overline{\rho}_o - \overline{\rho}'| = R_o \to 0$, and recall the small argument expression for the Hankel function (2.141) and (2.143). In particular, from (2.143) it follows that the second term of the right-hand side integral

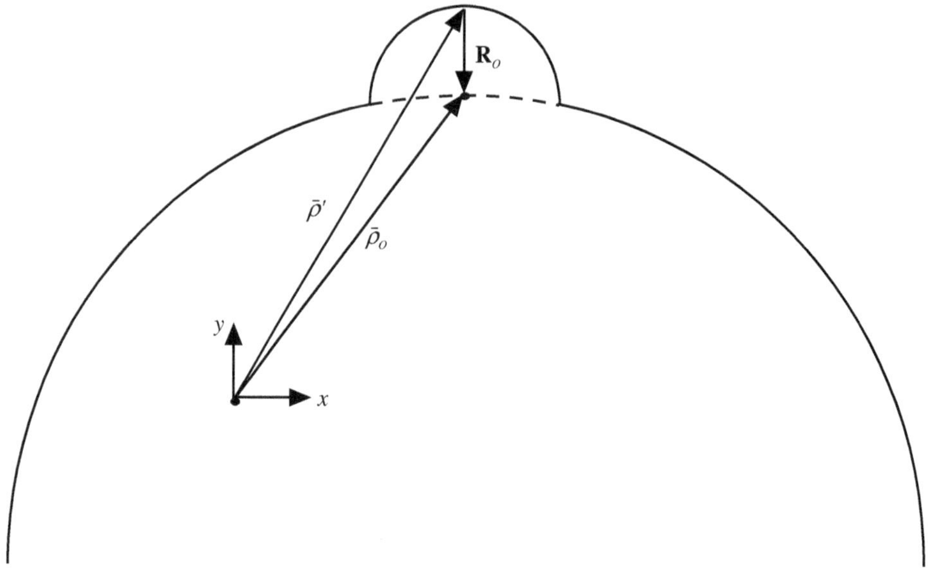


**FIGURE 3-9** ■ Geometry for evaluating the fields on $C$.

in (3.80) vanishes as $a \to 0$, and the first term can be written as

$$
\begin{aligned}
-\int_{C_o} E_z(\overline{\rho}')\frac{\partial G_{2d}(\overline{\rho}_o,\overline{\rho}')}{\partial n'}\,d\ell' &= -\int_{C_o} E_z(\overline{\rho}')\left[\hat{n}\cdot\nabla' G_{2d}(\overline{\rho}_o,\overline{\rho}')\right]d\ell' \\
&= -\int_0^{\pi} E_z(\overline{\rho}')\left[\hat{R}_o\cdot\nabla G_{2d}(\overline{\rho}_o,\overline{\rho}')\right]a\,d\phi' \\
&\approx +E_z(\overline{\rho}_o)\int_0^{\pi}\frac{\hat{R}_o\cdot\hat{R}_o}{2\pi a}a\,d\phi' \\
&= +\frac{1}{2}E_z(\overline{\rho}_o).
\end{aligned}
$$

In general, one can also show that

$$
\oint_C \phi(\overline{\rho}')\frac{\partial G_{2d}(\overline{\rho}_o^{\pm},\overline{\rho}')}{\partial n'}\,d\ell' = \pm\frac{1}{2}\phi(\overline{\rho}_o) + \oint_C \phi(\overline{\rho}')\frac{\partial G_{2d}}{\partial n'}(\overline{\rho}_o,\overline{\rho}')\,d\ell' \quad (3.81)
$$

where $\overline{\rho}_o^{\pm}$ is the observation point just exterior (+) or interior (−) to the contour, $C$. This identity (generalization of the result in (3.8)) can be viewed as the scalar form of the vector identity (3.37), and, as stated earlier, the field discontinuity at the boundary $C$ is due to the implied surface currents responsible for the scattered field.

When the identity (3.81) is incorporated into the integral equations (3.79) we have

$$
\oint_C\left[E_z(\overline{\rho}')\frac{\partial G_{2d}(\overline{\rho},\overline{\rho}')}{\partial n'} - G_{2d}(\overline{\rho},\overline{\rho}')\frac{\partial E_z(\overline{\rho}')}{\partial n'}\right]d\ell' + E_z^i(\overline{\rho}) = \begin{cases} E_z(\overline{\rho}) & \overline{\rho}\text{ in }A_\infty \\ \frac{1}{2}E_z(\overline{\rho}) & \overline{\rho}\text{ on }C \\ 0 & \overline{\rho}\text{ in }A \end{cases}
$$

(3.82a)

$$\oint_C \left[ H_z(\overline{\rho}') \frac{\partial G_{2d}(\overline{\rho}, \overline{\rho}')}{\partial n'} - G_{2d}(\overline{\rho}, \overline{\rho}') \frac{\partial E_z(\overline{\rho}')}{\partial n'} \right] d\ell' + H_z^i(\overline{\rho}) = \begin{cases} H_z(\overline{\rho}) & \overline{\rho} \text{ in } A_\infty \\ \frac{1}{2} H_z(\overline{\rho}) & \overline{\rho} \text{ on } C \\ 0 & \overline{\rho} \text{ in } A. \end{cases} \tag{3.82b}$$

These should be compared with (3.10), which apply to three-dimensional surfaces. Of course, one could have derived (3.82) directly from (3.10) by integrating with respect to $z'$ (see (2.113)). Also, the Stratton-Chu integral equations (3.26) can be shown to reduce to (3.82) upon making use of (2.124) and noting that $\hat{n} \cdot \mathbf{E} = 0$ on the boundary $C$. The first of (3.82) is commonly referred to as the $E$-polarization or transverse magnetic (TM) boundary integral equation since the electric field is transverse to the plane of incidence. Correspondingly, (3.82b) is referred to as the $H$-polarization or transverse electric (TE) integral equation.

### 3.2.2 Homogeneous Dielectrics

The boundary integral equations for two-dimensional homogeneous dielectric cylinders can be derived directly from the corresponding three-dimensional ones given in (3.38) and (3.40). However, it is simpler to consider their derivation beginning with the boundary integral equations (3.82).

Let us consider the dielectric cylinder whose cross section is illustrated in Figure 3-10. The cylinder's constitutive parameters are $(\epsilon_{r2}, \mu_{r2})$, whereas those of the exterior region are $(\epsilon_{r1}, \mu_{r1})$, and $\hat{n}$ denotes the unit normal pointing in the outward direction of the cylinder. We shall denote the total field in medium 1 as $E_{z1}$ or $H_{z1}$ and that in medium 2 as $E_{z2}$ or $H_{z2}$. Then, from (3.82) if we let $\overline{\rho}$ approach $C$ from the exterior (medium 1) we have

$$\oint_C \left[ E_{z1}(\overline{\rho}') \frac{\partial G_{2d_1}(\overline{\rho}, \overline{\rho}')}{\partial n'} - G_{2d_1}(\overline{\rho}, \overline{\rho}') \frac{\partial E_{z1}(\overline{\rho}')}{\partial n'} \right] d\ell' + E_z^i(\overline{\rho}) = \frac{1}{2} E_{z1}(\overline{\rho}) \quad \overline{\rho} \subset C \tag{3.83a}$$

for TM excitation and

$$\oint_C \left[ H_{z1}(\overline{\rho}') \frac{\partial G_{2d_1}(\overline{\rho}, \overline{\rho}')}{\partial n'} - G_{2d_1}(\overline{\rho}, \overline{\rho}') \frac{\partial H_{z1}(\overline{\rho}')}{\partial n'} \right] d\ell' + H_z^i(\overline{\rho}) = \frac{1}{2} H_{z1}(\overline{\rho}) \quad \overline{\rho} \in C \tag{3.83b}$$

for TE excitation with

$$G_{2d_1}(\overline{\rho}, \overline{\rho}') = \frac{-j}{4} H_0^{(2)}\left(k_o \sqrt{\epsilon_{r1}\mu_{r1}}\, |\overline{\rho} - \overline{\rho}'|\right). \tag{3.84}$$

Alternatively, if we let $\overline{\rho}$ approach from the interior of $C$, (3.82) yields

$$-\oint_C \left[ E_{z2}(\overline{\rho}') \frac{\partial G_{2d_2}(\overline{\rho}, \overline{\rho}')}{\partial n'} - G_{2d_2}(\overline{\rho}, \overline{\rho}') \frac{\partial E_{z2}(\overline{\rho}')}{\partial n'} \right] d\ell' = \frac{1}{2} E_{z2}(\overline{\rho}) \quad \overline{\rho} \in C \tag{3.85a}$$

for TM incidence and

$$-\oint_C \left[ H_{z2}(\overline{\rho}') \frac{\partial G_{2d_2}(\overline{\rho}, \overline{\rho}')}{\partial n'} - G_{2d_2}(\overline{\rho}, \overline{\rho}') \frac{\partial H_{z2}(\overline{\rho}')}{\partial n'} \right] d\ell' = \frac{1}{2} H_{z2}(\overline{\rho}) \quad \overline{\rho} \in C \tag{3.85b}$$

for TE incidence with

$$G_{2d_2}(\overline{\rho},\overline{\rho}') = \frac{-j}{4} H_o^{(2)}\left(k_o\sqrt{\epsilon_{r2}\mu_{r2}}\,|\overline{\rho}-\overline{\rho}'|\right). \tag{3.86}$$

We remark that in (3.85) there is no explicit appearance of the excitation field since this is confined in the exterior region of the dielectric (i.e., there are no sources in the dielectric).

To obtain an integral equation set for the unique solution of the boundary fields and their derivatives, it is necessary to couple the respective equations of (3.83) and (3.85). This can be accomplished by enforcing the tangential field continuity conditions

$$\hat{n}\times\mathbf{E}_1 = \hat{n}\times\mathbf{E}_2$$

$$\hat{n}\times\mathbf{H}_1 = \hat{n}\times\mathbf{H}_2.$$

For the TM case, the first of these conditions implies

$$E_{z1} = E_{z2} \tag{3.87a}$$

and from the second condition, in conjunction with (2.124), we have

$$\frac{1}{\mu_{r1}}\frac{\partial E_{z1}}{\partial n} = \frac{1}{\mu_{r2}}\frac{\partial E_{z2}}{\partial n}. \tag{3.87b}$$

Likewise, for the TE case the appropriate boundary conditions are

$$\begin{aligned} H_{z1} &= H_{z2} \\ \frac{1}{\epsilon_{r1}}\frac{\partial H_{z1}}{\partial n} &= \frac{1}{\epsilon_{r2}}\frac{\partial H_{z2}}{\partial n}. \end{aligned} \tag{3.88}$$

Substituting now (3.87) and (3.88) into (3.85) yields

$$-\oint\left[E_{z1}(\overline{\rho}')\frac{\partial G_{2d_2}(\overline{\rho},\overline{\rho}')}{\partial n'} - \frac{\mu_{r2}}{\mu_{r1}}G_{2d_2}(\overline{\rho},\overline{\rho}')\frac{\partial E_{z1}(\overline{\rho})}{\partial n'}\right]d\ell' = \frac{1}{2}E_{z1}(\overline{\rho}) \tag{3.89a}$$

and

$$-\oint\left[H_{z1}(\overline{\rho}')\frac{\partial G_{2d_2}(\overline{\rho},\overline{\rho}')}{\partial n'} - \frac{\epsilon_{r2}}{\epsilon_{r1}}G_{2d_2}(\overline{\rho},\overline{\rho}')\frac{\partial H_{z1}(\overline{\rho}')}{\partial n'}\right]d\ell' = \frac{1}{2}H_{z1}(\overline{\rho}). \tag{3.89b}$$

The integral equations (3.83a) and (3.89a) now form a coupled set of integral equations for the solution of $E_{z1}$ and $\frac{\partial E_{z1}}{\partial n}$ on the boundary $C$. Similarly, the $H$-polarization integral equations (3.83b) and (3.89b) form another set of coupled integral equations that can be solved for $H_{z1}$ and $\frac{\partial H_{z1}}{\partial n}$ on $C$. The solution of these integral equations will be considered in a later chapter.

Upon solution of the boundary field and its derivative, the fields in the exterior (medium 1) and interior (medium 2) regions can be found by returning to (3.82). For example, the exterior field is given by

$$\phi = \phi^s + \phi^i \tag{3.90}$$

where

$$\phi^s = \int_C \left[ \phi(\overline{\rho}') \frac{\partial G_{2d_1}(\overline{\rho}, \overline{\rho}')}{\partial n'} - G_{2d_1}(\overline{\rho}, \overline{\rho}') \frac{\partial \phi(\overline{\rho}')}{\partial n'} \right] d\ell' \tag{3.91}$$

is the scattered field, and $\phi$ represents the $E_{z1}$ or $H_{z1}$ field depending on the polarization of the excitation field. For computational purposes, we may write (see also (2.120))

$$\frac{\partial G_{2d_1}}{\partial n} = \hat{n} \cdot \nabla G_{2d_1} = -\frac{j}{4}(\hat{n} \cdot \hat{R}) H_o^{(2)'}(kR)$$
$$= (\hat{n} \cdot \hat{R}) \frac{jk_o\sqrt{\mu_{r_1}\epsilon_{r_1}}}{4} \cdot H_1^{(2)}\left(k_o\sqrt{\mu_{r_1}\epsilon_{r_1}}\, R\right) \tag{3.92}$$

and when this is substituted in (3.91) along with the large argument expressions for $H_o^{(2)}(kR)$ and $H_1^{(2)}(kR)$ given by (2.132), it follows that in the far zone ($\rho \to \infty$)

$$\phi^s{}_{k_o\rho \to \infty} \simeq -\sqrt{\frac{k_1}{8\pi}} \frac{e^{-j(k\rho+\pi/4)}}{\sqrt{\rho}} \int_C \left[ j(\hat{n}' \cdot \hat{\rho})\phi(\overline{\rho}') + \frac{\psi(\overline{\rho}')}{k_1} \right] e^{jk_1\hat{\rho}\cdot\overline{\rho}'}\, d\ell' \tag{3.93}$$

where we have set

$$\psi(\overline{\rho}) = \frac{\partial \phi(\overline{\rho})}{\partial n}$$

and $k_1 = k_o\sqrt{\mu_{r_1}\epsilon_{r_1}}$. The evaluation of the field in the interior region can be accomplished in a similar manner. It follows that the interior total field is given by the negative of the right-hand side of (3.88), provided $G_{2d_1}$ is also replaced by $G_{2d_2}$ and $\phi$ is identified as $E_{z2}$ or $H_{z2}$. That is

$$E_{z2} = -\int_C \left[ E_{z1}(\overline{\rho}') \frac{\partial G_{2d_2}(\overline{\rho}, \overline{\rho}')}{\partial n'} - \frac{\mu_{r2}}{\mu_{r1}} G_{2d_2}(\overline{\rho}, \overline{\rho}') \frac{\partial E_{z1}}{\partial n'}(\overline{\rho}') \right] d\ell' \tag{3.94a}$$

for the TM case and

$$H_{z2} = -\int_C \left[ H_{z1}(\overline{\rho}') \frac{\partial G_{2d_2}(\overline{\rho}, \overline{\rho}')}{\partial n'} - \frac{\epsilon_{r2}}{\epsilon_{r1}} G_{2d_2}(\overline{\rho}, \overline{\rho}') \frac{\partial H_{z1}}{\partial n'}(\overline{\rho}') \right] d\ell' \tag{3.94b}$$

for the TE case.

### 3.2.3 Metallic Cylinders

A special case of a homogeneous cylinder is the perfectly conducting or metallic cylinder. Then, on the boundary $C$ the tangential electric field vanishes, implying

$$E_{z1} = E_z = 0 \tag{3.95a}$$

for $E$-polarization and

$$\frac{\partial H_{z1}}{\partial n} = \frac{\partial H_z}{\partial n} = 0 \tag{3.95b}$$

for $H$-polarization. Incorporating these into (3.83) we obtain the integral equations

$$\oint \frac{\partial E_z(\overline{\rho}')}{\partial n'} G_{2d_1}(\overline{\rho}, \overline{\rho}')\, d\ell' = E_z^i(\overline{\rho}) \qquad \overline{\rho} \in C \tag{3.96a}$$

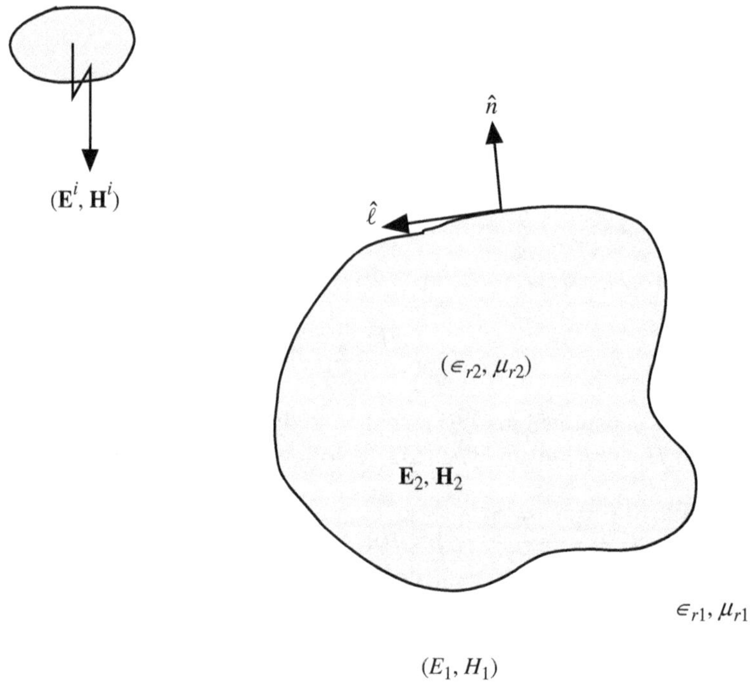


**FIGURE 3-10** Geometry of an inhomogeneous dielectric cylinder.

and

$$\oint_C H_z(\overline{\rho}')\frac{\partial G_{2d_1}(\overline{\rho}, \overline{\rho}')}{\partial n'}\, d\ell' + H_z^i(\overline{\rho}) = \frac{1}{2}H_z(\overline{\rho}) \qquad \overline{\rho} \in C \tag{3.96b}$$

for $E$- and $H$-polarizations, respectively (see Figure 3-10).

As noted earlier, we could have instead formulated the problem by introducing on $C$ the equivalent current

$$\begin{aligned}\mathbf{J} &= \hat{n} \times \mathbf{H}\\ &= \begin{cases} \hat{n} \times \left(\dfrac{-jY_1}{k_1}\hat{z} \times \nabla E_z\right) = -\hat{z}\dfrac{jY_1}{k_1}\dfrac{\partial E_z}{\partial n}, & \text{TM case}\\ -\hat{\ell}H_z, & \text{TE case}\end{cases}\end{aligned} \tag{3.97}$$

where $\hat{\ell} = \hat{z} \times \hat{n}$ denotes the unit vector tangent to $C$, as illustrated in Figure 3-11. Substituting (3.97) into (3.96) yields

$$jk_1 Z_1 \oint_C J_z(\overline{\rho}')G_{2d_1}(\overline{\rho}, \overline{\rho}')\, d\ell' = E_z^i(\overline{\rho}), \qquad \overline{\rho} \in C \tag{3.98a}$$

for the TM case and

$$\oint_C J_\ell(\overline{\rho}')\frac{\partial G_{2d_1}(\overline{\rho}, \overline{\rho}')}{\partial n'}\, d\ell' - H_z^i(\overline{\rho}') = \frac{1}{2}J_\ell(\overline{\rho}), \qquad \overline{\rho} \in C \tag{3.98b}$$

for the TE case, with $\mathbf{J} = \hat{\ell}J_s$. It is now readily recognized that (3.98b) is the two-dimensional counterpart of the MFIE given by (3.42b). As was the case in three dimensions, this is valid only for closed boundaries. In contrast, the EFIE (3.98a),

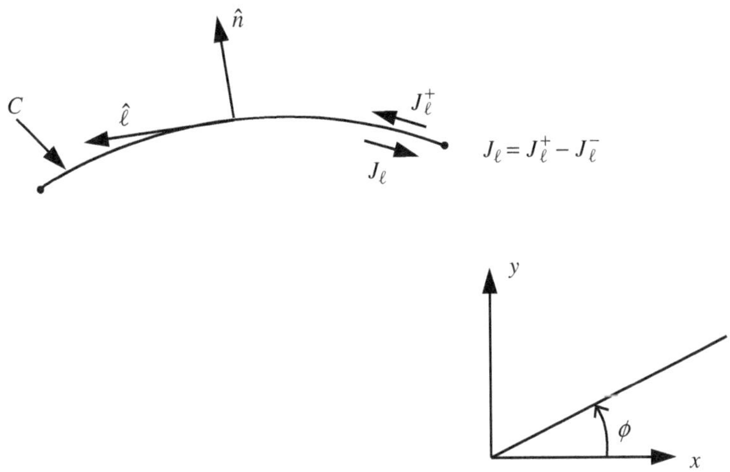


FIGURE 3-11 Geometry of a metallic curved strip.

which is the counterpart of (3.42a) for three dimensions, can be shown to be valid for closed as well as open surfaces such as a curved strip. This can be verified by rederiving it from (2.127a) upon setting $\mathbf{M} = 0$ and then enforcing the boundary condition $\hat{n} \times \mathbf{E} = 0$ on the boundary. Using the same procedure for $H$-polarization, we find directly from (2.127a) the integral equation

$$E_\ell^i = \int_C jk_1 Z_1(\hat{\ell} \cdot \hat{\ell}') J_\ell(\bar{\rho}') G_{2d_1}(\bar{\rho}, \bar{\rho}')\, d\ell' + \frac{jZ_1}{k_1} \frac{\partial}{\partial \ell} \int_C J_\ell(\bar{\rho}') \frac{\partial}{\partial \ell'} G_{2d_1}(\bar{\rho}, \bar{\rho}')\, d\ell' \tag{3.99}$$

in which $\mathbf{E}_\ell^i = \hat{\ell} \cdot \mathbf{E}^i$, and $\frac{\partial}{\partial \ell} = \hat{\ell} \cdot \nabla$ implies differentiation along the direction tangential to the boundary $C$. The three-dimensional counterpart of (3.99) is (3.43), and it is readily seen that the EFIEs (3.98a) and (3.99) can be deduced directly from (3.43) through the application of the identity (2.113). As explained in connection with (3.43), when the EFIEs (3.98) and (3.99) are applied to open surfaces, the current density in those equations represents the net current flow on the curved strip (see Figure 3-11). It should be noted that the non-integrable singularity of the kernel in the second integral of the EFIE (3.99) requires special attention when this integral equation is solved numerically. Thus, although (3.45) is valid for both open and closed surfaces, it is preferable to use the MFIE whenever possible.

### 3.2.4 Piecewise Homogeneous Dielectrics

The treatment of piecewise homogeneous dielectrics is important in many industrial and biomedical applications involving the characterization of the field behavior in materials. The structure in Figure 3-12 is a typical configuration of interest. To compute the interior and scattered fields from this configuration due to an exterior TE or TM excitation, one approach is to formulate appropriate boundary integral equations at each of the dielectric interfaces (Wu and Tsai, 1977a,b) Following the procedure discussed earlier for three-dimensional structures, we can proceed with the construction of the boundary integral equations by setting up the three equivalent problems illustrated in Figures 3-12b and 3-12c. As before, we can then introduce

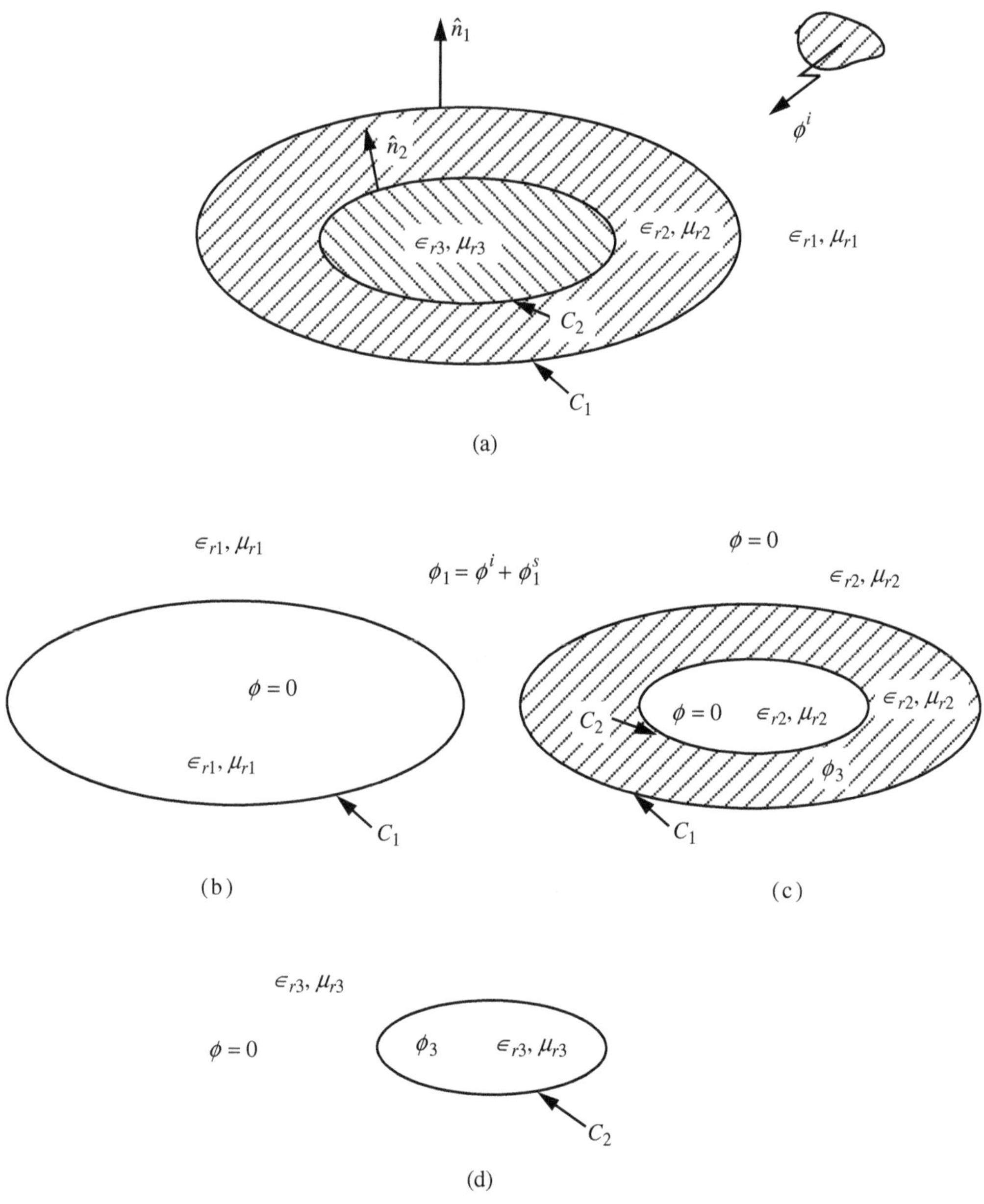


**FIGURE 3-12** ■ Geometry of the layered dielectric cylinder and equivalent problems to formulate the field in each region.

equivalent surface currents at the interface boundaries $C_1$ and $C_2$, satisfying (3.28) and thus ensuring tangential field continuity. Integral equations for the solution of the equivalent current can then be constructed by enforcing conditions such as those in (3.47).

An alternative to the aforementioned procedure is to bypass the introduction of the equivalent currents and to work directly with the $z$-directed field and its normal derivative, as was done in the previous section for homogeneous dielectrics. Such an approach could also be used in the three-dimensional formulation, and in this case use of (3.26) or (3.18) would be required. However, when employing (3.26), one must deal with the presence of $\hat{n} \cdot \mathbf{H}$ and $\hat{n} \cdot \mathbf{E}$. These can be replaced by

$$\hat{n} \cdot \mathbf{H} = -\frac{\nabla_s \cdot \mathbf{M}}{j\omega\mu} = \frac{\nabla_s \cdot (\hat{n} \times \mathbf{E})}{jkZ} \tag{3.100a}$$

$$\hat{n} \cdot \mathbf{E} = -\frac{\nabla_s \cdot \mathbf{J}}{j\omega\mu} = \frac{\nabla_s \cdot (\hat{n} \times \mathbf{H})}{jkY} \tag{3.100b}$$

implying that we could derive integral equations similar to those in Sections 3.1.3 through 3.1.6 which involve only the tangential electric and magnetic fields instead of the equivalent surface currents. Thus, mathematically the two procedures are equivalent, but one could argue that the introduction of the equivalent surface currents is physically more appealing. With this understanding, we now proceed to construct boundary integral equations for the layered dielectric in Figure 3-11a.

Assuming TM excitation and denoting the field exterior to $C_1$ by $E_{z1}$, from (3.83a) we have

$$\oint_{C_1}\left[E_{z1}(\overline{\rho}')\frac{\partial G_{2d_1}(\overline{\rho},\overline{\rho}')}{\partial n_1'} - G_{2d_1}(\overline{\rho},\overline{\rho}')\frac{\partial E_{z1}(\overline{\rho}')}{\partial n_1'}\right] + E_z^i(\overline{\rho}') = \frac{1}{2}E_z(\overline{\rho}) \quad \overline{\rho} \in C. \tag{3.101a}$$

When we approach the observation point on $C_1$ from its interior surface, it is convenient to choose the closed boundary to be the sum of $C_1$ and $C_2$ when making use of (3.82). Then, since there are no sources in region 2, it follows that ($E_{z2}$ is the field between $C_1$ and $C_2$)

$$\begin{aligned}&+\oint_{C_2}\left[E_{z2}(\overline{\rho}')\frac{\partial G_{2d_2}(\overline{\rho},\overline{\rho}')}{\partial n_2'} - G_{2d_2}(\overline{\rho},\overline{\rho}')\frac{\partial E_{z2}}{\partial n_2'}\right]d\ell' \\ &-\oint_{C_1}\left[E_{z1}(\overline{\rho}')\frac{\partial G_{2d_2}(\overline{\rho},\overline{\rho}')}{\partial n_1'} - \frac{\mu_{r2}}{\mu_{r1}}G_{2d_2}(\overline{\rho},\overline{\rho}')\frac{\partial E_{z1}}{\partial n_1'}\right]d\ell' = \begin{cases}\frac{1}{2}E_{z1}(\overline{\rho}), & \overline{\rho} \in C_1 \\ \frac{1}{2}E_{z2}(\overline{\rho}), & \overline{\rho} \in C_2\end{cases}\end{aligned} \tag{3.101b}$$

which, except for the appearance of the integral over $C_2$, is identical to (3.89). Finally, by approaching the observation point on $C_2$ from its interior surface, we obtain

$$-\oint_{C_3}\left[E_{z2}(\overline{\rho}')\frac{\partial G_{2d_3}(\overline{\rho},\overline{\rho}')}{\partial n_2'} - \frac{\mu_{r3}}{\mu_{r2}}G_{2d_3}(\overline{\rho},\overline{\rho}')\frac{\partial E_{z2}(\overline{\rho}')}{\partial n_2'}\right]d\ell' = \frac{1}{2}E_{z2}(\overline{\rho}) \quad \overline{\rho} \in C_2 \tag{3.101c}$$

where

$$G_{2d_3}(\overline{\rho},\overline{\rho}') = -\frac{j}{4}H_o^{(2)}\left(k_o\sqrt{\epsilon_{r3}\mu_{r3}}\,|\overline{\rho}-\overline{\rho}'|\right). \tag{3.102}$$

Together, the integral equations (3.101) form a coupled set for the solution of the fields and their derivatives on the boundaries $C_1$ and $C_2$. The corresponding set of integral equations for the TE case is obtained directly from (3.101) by invoking duality. Once (3.101) are solved via some numerical procedure, the exterior fields can be found from (3.90) through (3.91) with $C = C_1$. The interior fields can be evaluated using one of the equations given in (3.82). For example, the field in medium 2 between the boundaries $C_1$ and $C_2$ is given by

$$E_{z2}(\overline{\rho}) = \oint_{C_2} \left[ E_{z2}(\overline{\rho}') \left( \hat{n}'_2 \cdot \nabla' G_{2d_2}(\overline{\rho}, \overline{\rho}') \right) - G_{2d_2}(\overline{\rho}, \overline{\rho}') \frac{\partial E_{z2}}{\partial n'_2} \right] d\ell' - \oint_{C_1} \left[ E_{z1}(\overline{\rho}') \left( \hat{n}'_1 \cdot \nabla' G_{2d_2}(\overline{\rho}, \overline{\rho}') \right) - \frac{\mu_{r2}}{\mu_{r1}} G_{2d_2}(\overline{\rho}, \overline{\rho}') \frac{\partial E_{z2}}{\partial n'_1} \right] d\ell' \tag{3.103}$$

and the dual of this must be used for TE incidence.

### 3.2.5 Domain Integral Equations

In modeling the cross section of inhomogeneous dielectric cylinders we must resort to a volume/domain formulation as discussed in Section 3.1.7 for three-dimensional dielectrics. Consider the inhomogeneous cylinders shown in Figure 3-13 whose relative permittivity and permeability are $\epsilon_r(\overline{\rho})$ and $\mu_r(\overline{\rho})$. In accordance with the volume equivalence principle, for TM incidence/excitation, we may replace the presence of the cylinder by the two-dimensional currents

$$J_z = jk_o Y_o(\epsilon_r - 1)\left(E_z^s + E_z^i\right) \tag{3.104a}$$

$$\mathbf{M} = \hat{x} M_x + \hat{y} M_y = jk_o Z_o(\mu_r - 1)(\mathbf{H}^s + \mathbf{H}^i). \tag{3.104b}$$

Then, from (2.127) or (2.130) we obtain

$$E_z^s = \frac{jk_o}{4} \iint_A \hat{z} \cdot \left( \mathbf{M}(\overline{\rho}') \times \hat{R} \right) H_1^{(2)}(k_o R)\, ds' - \frac{k_o Z_o}{4} \iint_A J_z(\overline{\rho}') H_o^{(2)}(k_o R)\, ds' \tag{3.105a}$$

$$\mathbf{H}^s = \iint_A \left[ J_z(\overline{\rho}')\hat{z} \times \nabla G_{2d_o}(\overline{\rho}, \overline{\rho}') - jk_o Y_o \mathbf{M}(\overline{\rho}') G_{2d_o}(\overline{\rho}, \overline{\rho}') - \frac{jY_o}{k_o} \mathbf{M}(\overline{\rho}') \cdot \nabla\nabla G_{2d_o}(\overline{\rho}, \overline{\rho}') \right] ds' \tag{3.105b}$$

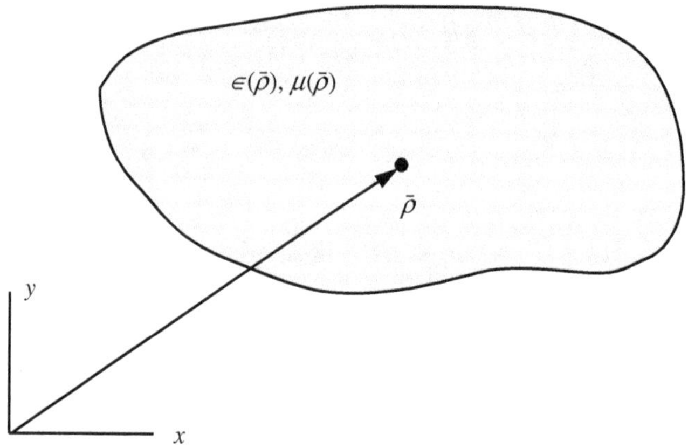


**FIGURE 3-13** Inhomogeneous dielectric cylinder of arbitrary cross section.

with

$$G_{2d_o}(\overline{\rho}, \overline{\rho}') = \frac{-j}{4} H_o^{(2)}(k_o|\overline{\rho} - \overline{\rho}'|). \tag{3.106}$$

Substituting these expressions into (3.104) results in a coupled set of integral equations for the currents $J_z$ and $\mathbf{M}$. The corresponding equations for the TE case are

$$\frac{M_z}{jk_o Z_o(\mu_r - 1)} = H_z^i + H_z^s \tag{3.107a}$$

$$\frac{\mathbf{J}}{jk_o Y_o(\epsilon_r - 1)} = \mathbf{E}^i + \mathbf{E}^s \tag{3.107b}$$

where

$$H_z^s = -\frac{jk_o}{4} \iint_A \hat{z} \cdot (\mathbf{J}(\overline{\rho}') \times \hat{R}) H_1^{(2)}(k_o R)\, ds' - \frac{k_o Y_o}{4} \iint_A M_z(\overline{\rho}') H_o^{(2)}(k_o R)\, ds' \tag{3.108a}$$

and

$$\mathbf{E}^s = -\iint_A M_z(\overline{\rho}') \hat{z} \times \nabla G_{2d_o}(\overline{\rho}, \overline{\rho}')\, ds' - \iint_A \Big[ jk_o Z_o \mathbf{J}(\overline{\rho}') G_{2d_o}(\overline{\rho}, \overline{\rho}') + \frac{jZ_o}{k_o} \mathbf{J}(\overline{\rho}') \cdot \nabla\nabla G_{2d_o}(\overline{\rho}, \overline{\rho}') \Big]\, ds' \tag{3.108b}$$

are the associated scattered fields.

When $\mu_r = 1$ in (3.104), $\mathbf{M}$ vanishes, and thus the solution of (3.104) through (3.105) involves a single unknown. However, for nonunity $\mu_r$, the integral equations involve three unknowns that can be reduced in number by following the procedure discussed in Section 3.1.7. That is, the effect of the currents (3.104) may be replaced by the single magnetic equivalent current

$$\begin{aligned} \mathbf{M}' &= -\frac{\mu_r - 1}{\mu_r} \nabla \times (\hat{z} E_z) - \nabla \times [\hat{z}(\epsilon_r - 1) E_z] \\ &= \frac{\mu_r - 1}{\mu_r} \hat{z} \times \nabla E_z + \hat{z} \times \nabla [(\epsilon_r - 1) E_z] \end{aligned} \tag{3.109}$$

which can be used in the first integral (3.105a) to obtain the scattered field expression

$$\hat{z} E_z^s = -\frac{jk_o}{4} \iint_A \hat{R} \times \Big\{ \frac{\mu_r - 1}{\mu_r} \hat{z} \times \nabla E_z(\overline{\rho}') + \hat{z} \times \nabla \big[ (\epsilon_r - 1) E_z(\overline{\rho}') \big] \Big\} H_1^{(2)}(k_o R)\, ds'. \tag{3.110}$$

When this is substituted into (3.57) we have

$$\hat{z} E_z(\overline{\rho}) = \hat{z} E_z^i(\overline{\rho}) - \frac{jk_o}{4} \iint_A \hat{R} \times \Big\{ \frac{\mu_r - 1}{\mu_r} \hat{z} \times \nabla E_z(\overline{\rho}') + \hat{z} \times \nabla \big[ (\epsilon_r - 1) E_z(\overline{\rho}') \big] \Big\} H_1^{(2)}(k_o R)\, ds' \tag{3.111}$$

which is an integral equation for the solution of the field $E_z$ across the cross section of the cylinder. Similarly, by invoking duality the corresponding integral equation for the TE case is

$$\hat{z}H_z(\overline{\rho}) = \hat{z}H_z^i(\overline{\rho}) - \frac{jk_o}{4}\iint_A \hat{R} \times \Big\{\frac{\epsilon_r - 1}{\epsilon_r}\hat{z} \times \nabla H_z(\overline{\rho}')$$
$$+\hat{z} \times \nabla\left[(\mu_r - 1)H_z(\overline{\rho}')\right]\Big\} H_1^{(2)}(k_o R)\, ds' \tag{3.112}$$

and the implied equivalent source for this field is the electric current

$$\mathbf{J}' = -\frac{\epsilon_r - 1}{\epsilon_r}\hat{z} \times \nabla H_z - \hat{z} \times \nabla\left[(\mu_r - 1)H_z\right] \tag{3.113}$$

which can be deduced from (1.112).

Special forms of (3.111) and (3.112) have been solved by Peterson and Klock (1988) and Peterson (1989) (see also Michielssen, Peterson, and Mittra (1991)). Because they involve derivatives of the unknown quantities $E_z$ or $H_z$, their solution requires use of higher-order expansion functions. However, these can be replaced by domain-boundary integral equations, which again involve a single undifferentiated unknown quantity. Such a domain-boundary integral equation can be deduced directly from the volume-surface integral formulation described in Section 3.1.7. In particular, by making use of (2.113) and setting $\mathbf{E} = \hat{z}E_z$ in (3.67)-(3.77), it follows that (Jin, Liepa, and Tai, 1988)

$$E_z = E_{ze} + E_{zm} + E_z^i \tag{3.114a}$$

where

$$E_{ze}(\overline{\rho}) = k_o^2 \iint_A (\epsilon_r - 1)E_z(\overline{\rho}')G_o(\mathbf{r}, \mathbf{r}')\, ds' \tag{3.114b}$$

$$E_{zm}(\overline{\rho}) = k_o^2 \iint_A \left(1 - \frac{1}{\mu_r}\right) E_z(\overline{\rho}')G_o(\overline{\rho}, \overline{\rho}')\, ds'$$
$$- \iint_A \nabla'\left(\frac{1}{\mu_r}\right) \cdot \left[E_z(\overline{\rho}')\nabla G_o(\overline{\rho}, \overline{\rho}')\right] ds'$$
$$+ \oint_C \left(1 - \frac{1}{\mu_r}\right) E_z(\overline{\rho}')\frac{\partial G_o(\overline{\rho}, \overline{\rho}')}{\partial n'}\, d\ell'$$
$$+ \begin{cases} (1 - \frac{1}{\mu_r})E_z(\overline{\rho}) & \overline{\rho} \in A \\ \frac{1}{2}(1 - \frac{1}{\mu_r})E_z(\overline{\rho}) & \overline{\rho} \in C \\ 0 & \text{elsewhere} \end{cases} \tag{3.114c}$$

which is an integral equation for the solution of $E_z(\overline{\rho})$ in the domain $A$ and on the boundary $C$. Clearly, $E_{ze} + E_{zm} = E_z^s$ is the scattered field from the dielectric cylinder. To derive the corresponding expression for $H$-polarization we follow the same procedure, which yields the dual of (3.114).

We remark that an alternative and equivalent integral equation to (3.114) was derived by Ricoy, Kilberg, and Volakis (1989) who chose to work with equivalent currents rather than the fields themselves. In this case, the scattered field by the cylinder is expressed as

$$E_z^s = -jk_o Z_o \iint_A J_z'(\overline{\rho}')G_o(\overline{\rho}, \overline{\rho}')\, ds' + Z_o \int_C M_z'(\overline{\rho}')G_o(\overline{\rho}, \overline{\rho}') \qquad (3.115)$$

where $J_z'$ and $M_z'$ are some unknown equivalent current density quantities. They can be found by solving the integral equation

$$\left[-\nabla\left(\frac{\mu_r - 1}{\mu_r}\right) \times \mathbf{H}^i(\overline{\rho}) - \left(\frac{\epsilon_r\mu_r - 1}{\mu_r}\right)\nabla \times \mathbf{H}^i(\overline{\rho})\right] \cdot \hat{z}$$
$$= -\frac{1}{\mu_r}J_z'(\overline{\rho}) - jk_o \iint_A J_z'(\overline{\rho}')K(\overline{\rho}, \overline{\rho}')\, ds' + \int_C M_z'(\overline{\rho}')K(\overline{\rho}, \overline{\rho}')\, d\ell' \qquad (3.116)$$

in which the kernel $K(\overline{\rho}, \overline{\rho}')$ is given by

$$K(\overline{\rho}, \overline{\rho}') = \left[\frac{jk_o(\epsilon_r\mu_r - 1)}{\mu_r} - \frac{j}{k_o}\nabla\left(\frac{\mu_r - 1}{\mu_r}\right) \cdot \nabla\right] G_o(\mathbf{r}, \mathbf{r}'). \qquad (3.117)$$

Both integral equations (3.114) and (3.116) involve the same (undifferentiated) number of unknown quantities and have the same kernel singularity. As a result, there is no advantage between the two integral equations.

## PROBLEMS

1. Modify the right-hand side of (3.10) to account for the case when the observation point $\mathbf{r}$ is at $P$, where $P$ coincides with the tip of a cone. See Figure 3.P1.

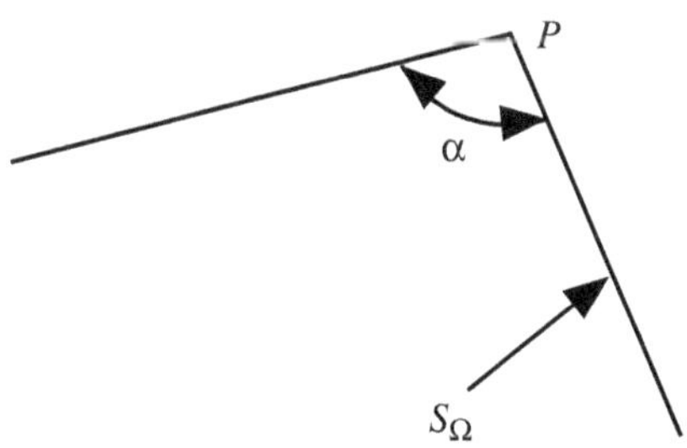


**FIGURE 3.P1** Geometry for Problem 1.

2. Using (2.21), we obtain the integral representation (Franz, 1948)

$$\mathbf{E} = -\nabla \times \iint_S \mathbf{M}(\mathbf{r}')G(\mathbf{r}, \mathbf{r}')\, ds' - \frac{jZ}{k}\nabla \times \iint_S \mathbf{J}(\mathbf{r}') \times \nabla G(\mathbf{r}, \mathbf{r}')\, ds'$$

and its dual. Using the identities (Tai, 1972)

$$\nabla \times \iint (\hat{n} \times \mathbf{H}) \times \nabla' G\, ds' = \nabla \times \nabla \times \iint (\hat{n} \times \mathbf{H})G\, ds'$$

and

$$\nabla \times \iint (\hat{n} \cdot \mathbf{H})\nabla' G\, ds' = 0.$$

Show that Franz's formula is equivalent to (3.26). Discuss the advantages and disadvantages of the two integral expressions.

3. Prove equations (3.37) and (3.81).

4. Beginning with (3.104) show that an alternative TM integral expression for the scattered field by a dielectric cylinder is

$$E_z^s = -\frac{jk_o^2}{4}\iint_A (\epsilon_r - 1)E_z(\overline{\rho}')H_o^{(2)}\left(k_o|\overline{\rho} - \overline{\rho}'|\right) ds'$$
$$+\frac{j}{4}\iint_A \hat{z}\cdot\nabla\times\mathbf{M}(\overline{\rho}')H_o^{(2)}\left(k_o|\overline{\rho} - \overline{\rho}'|\right) ds'$$

where

$$\mathbf{M} = -\frac{\mu_r - 1}{\mu_r}\nabla\times(\hat{z}E_z).$$

For a discretization of the resuting integral equation by enforcing $E_z^i + E_z^s = E_z$, refer to Baucke (1991) and Van Bladel (1993). The dual of this expression is applicable for the TE excitation.

5. Consider the problem of radiation or scattering in the presence of a dielectric boundary as illustrated in Figure 3.P5a. The fields $(\mathbf{E}^i, \mathbf{H}^i)$ denote the excitation fields that are known, $(\mathbf{E}_e, \mathbf{H}_e)$ are the fields scattered by the dielectric in the external region, and $(\mathbf{E}_d, \mathbf{H}_d)$ are the total fields within the dielectric volume. To find an integral equation for the solution of $(\mathbf{E}_d, \mathbf{H}_d)$ and $(\mathbf{E}_e, \mathbf{H}_e)$, set up the following equivalent subproblems.

As seen in Figure 3.P5b, problem A recovers the original fields exterior to the dielectric, whereas problem B provides the original fields interior to the dielectric. Our goal is to choose the arbitrary fields $(\mathbf{E}_{ei}, \mathbf{H}_{ei})$ and $(\mathbf{E}_{do}, \mathbf{H}_{do})$ in such a way so that the resulting integral equations have the least number of unknowns. Choosing

$$\hat{n}\times\mathbf{E}_{ei} = \hat{n}\times\mathbf{E}_e$$

implies that

$$\mathbf{M}_e = 0, \quad \mathbf{J}_e = \hat{n}\times(\mathbf{H}_e - \mathbf{H}_{ei}).$$

Also, we may set $(\mathbf{E}_{do}, \mathbf{H}_{de}) = (0, 0)$, implying

$$\mathbf{J}_d = -\hat{n}\times\mathbf{H}_d, \quad \mathbf{M}_d = -\mathbf{E}_d\times\hat{n}.$$

(a) Based on the previously given assumptions show that

$$\mathbf{J}_d = -\hat{n}\times(\mathbf{H}_e + \mathbf{H}^i), \quad \mathbf{M}_d = -(\mathbf{E}_e + \mathbf{E}^i)\times\hat{n} \qquad \mathbf{r}\in S^+.$$

(b) Use the equalities in (a) to write an integral equation involving $\mathbf{J}_e$ and $\mathbf{J}_d$ and another involving $\mathbf{J}_e$ and $\mathbf{M}_d$. Employ these in the field continuity conditions to obtain the

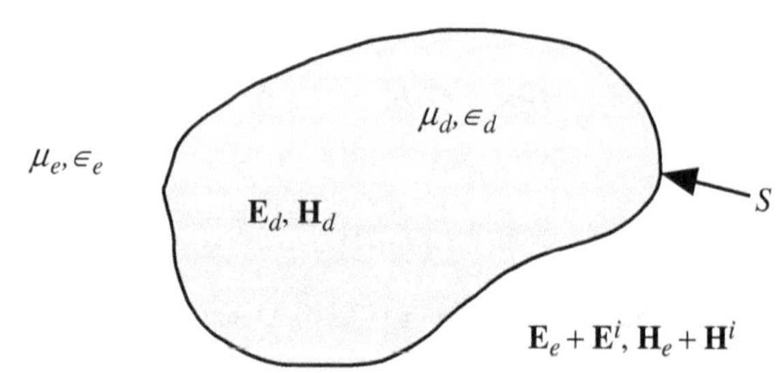


**FIGURE 3.P5a** Geometry for Problem 5.

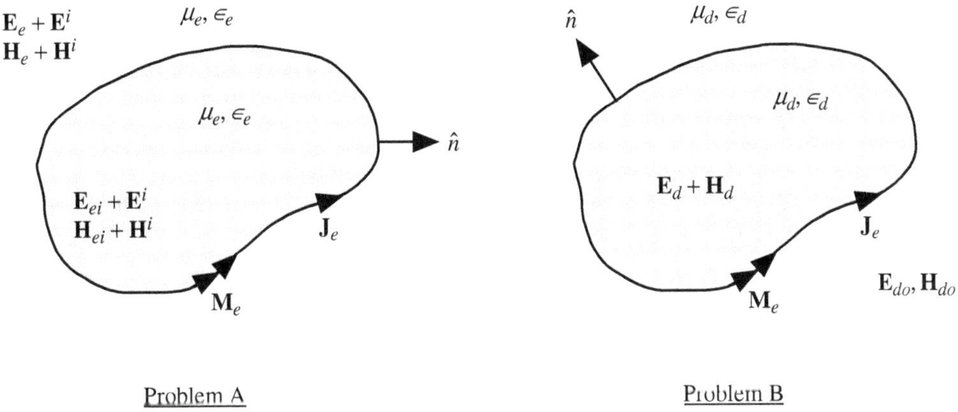


**FIGURE 3.P5b** ▪ Geometry for Problem 5.

integral equation

$$\hat{n} \times \left[\frac{1}{2}\mathbf{E}_e(\mathbf{J}_e, 0) + \frac{1}{2}\mathbf{E}_d(\mathbf{J}_e, 0) + \mathbf{E}_d(\hat{n} \times L_e(\mathbf{J}_e), 0) + L_d(\hat{n} \times \mathbf{E}_e(\mathbf{J}_e, 0)\right]$$

$$= \hat{n} \times \left[\frac{1}{2}\mathbf{E}^i + L_d(\hat{n} \times \mathbf{E}^i) + \mathbf{E}_d(\hat{n} \times \mathbf{H}^i, 0)\right]$$

where

$$L_{e,d}(J) = \iint \nabla \times (\mathbf{J}G_{e,d})\, ds' = \iint (\mathbf{J} \times \nabla G_{e,d})\, ds'$$

and $\mathbf{E}_{e,d}(\mathbf{J}, 0)$ imply the Stratton-Chu integral equation of the electric field in terms of the electric current with $\mathbf{M} = 0$. Also, $G_e$ and $G_d$ represent the Green's functions of a homogeneous space having the properties of the exterior and interior regions, respectively.

6. Consider the orthogonal direction pairs $(\hat{\ell}, \hat{n})$ and $(\hat{\ell}', \hat{n}')$ illustrated in Fig. 3.P6.

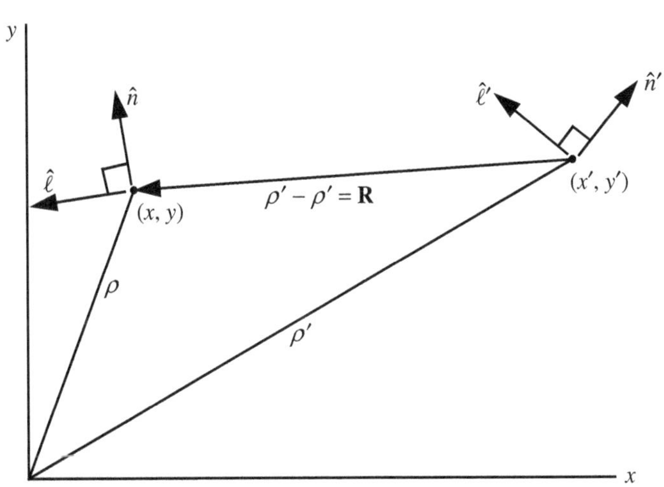


**FIGURE 3.P6** ▪ Illustration of the directions and parameters for Problem 6.

Show that

$$\frac{\partial G_{2d}(\bar{\rho},\bar{\rho}')}{\partial n} = +\frac{jk_o}{4}\left[(\hat{n}'\cdot\hat{n})(\hat{n}'\cdot\hat{R}) + (\hat{\ell}'\cdot\hat{n})(\hat{\ell}'\cdot\hat{R})\right] H_1^{(2)}(k_o R)$$

$$\frac{\partial G_{2d}(\bar{\rho},\bar{\rho}')}{\partial \ell} = +\frac{jk_o}{4}\left[(\hat{n}'\cdot\hat{\ell})(\hat{n}'\cdot\hat{R}) + (\hat{\ell}'\cdot\hat{\ell})(\hat{\ell}'\cdot\hat{R})\right] H_1^{(2)}(k_o R).$$

7. Consider the problem of scattering by an impedance cylinder satisfying the boundary condition (1.132) (see Figure 3.P7):

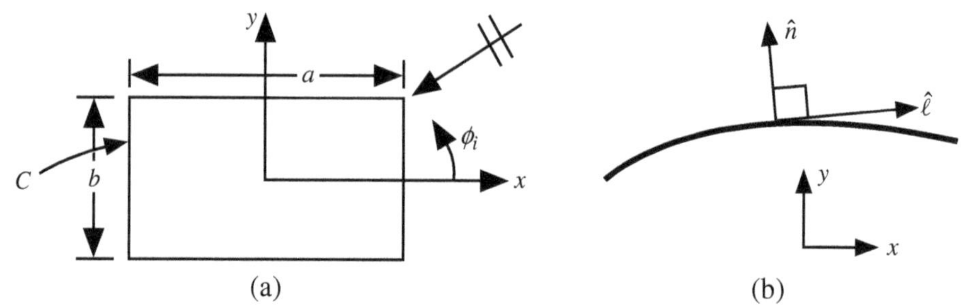


**FIGURE 3.P7** (a) Geometry of a square impedance cylinder. (b) Geometry of an impedance strip.

(a) Develop an integral equation for the $H_z$ field component (on the cylinder's surface) using Kirchhoff's integral equation method. This integral equation should involve only the $H_z$ component and not its derivative.

(b) If the cylinder's normalized impedance is $\eta$, what should be the value of $\eta$ so that the integral equation in (a) reduces to (3.96b)?

(c) Use (2.127a) along with (1.132) to derive an integral equation for the impedance strip. Hint: Use (1.132) to relate **J** to **M**.

8. Several approximations have been introduced to bypass the solution of the integral equations found in Problem 3.1, with the most popular of them being the physical optics (PO) approximation.

(a) Derive from (1.132) that

$$\frac{\partial H_z}{\partial n} = \hat{n}\cdot\nabla H_z = jk_o\eta H_z$$

and use this to show that the far-zone field for an impedance cylinder is given by

$$H_z^s\Big|_{\rho\to\infty} \sim -\sqrt{\frac{k_o}{8\pi}}\frac{e^{-jk_o\rho}}{\sqrt{\rho}}e^{j\frac{\pi}{4}}\int_C(\eta+\hat{\rho}\cdot\hat{n}')H_z(\rho')e^{jk_o\bar{\rho}'\cdot\hat{\rho}}\,d\ell'$$

(b) The PO approximation amounts to setting

$$H_z = (1+R)H_z^i$$

in the lit region of the cylinder's surface, where $R = \frac{\sin\phi_o-\eta}{\sin\phi_o+\eta}$ is the plane wave reflection coefficient for an impedance surface (with $\phi_o$ measured from the dielectric interface), and $H_z^i$ is the incident plane wave

$$H_z^i = e^{jk_o(x\cos\phi_i + y\sin\phi_i)}.$$

Use this approximation to derive a closed-form expression for the far-zone scattered field of a square impedance cylinder (see Figure 3.P7) for $0 < \phi < 180°$ and $\phi_i = 0°$ or $90°$.

(c) Another approximation is to instead enforce the first- or second-order absorbing boundary conditions (ABCs) on the cylinder's surface. The first-order ABC is given by

$$\frac{\partial H_z^s}{\partial n} = -\left(jk_o + \frac{\kappa}{2}\right) H_z^s$$

where $\kappa$ denotes the curvature at each point on $C$ (equal to 1/radius), and $H_z^s$ is the scattered field. Use this ABC in conjunction with the impedance boundary condition $\frac{\partial H_z}{\partial n} = jk_o\eta H_z$ to find the far-zone scattered field as in (b).

(d) The second-order ABC is given by

$$\frac{\partial H_z^s}{\partial n} = -\left[jk_o + \frac{\kappa}{2} - \frac{j\kappa^2}{8(j\kappa - k_o)}\right] H_z^s + \frac{j}{2(j\kappa - k_o)} \frac{\partial^2 H_z^s}{\partial \ell^2}$$

Use this in conjunction with the impedance boundary condition to derive a second-order differential equation for $H_z^s$. Give a closed-form solution of this for the square cylinder and the corresponding far-zone scattered field expression.

(e) Compare the far-zone bistatic scattered field expressions derived in (b), (c), and (d) for a square cylinder having $a = 2\lambda_o$, $b = 1\lambda_o$ and with $\phi_i = o°$.

9. Consider the problem of an antenna placed in a rectangular cavity and radiating in the presence of a substructure as shown in Figure 3.P9.

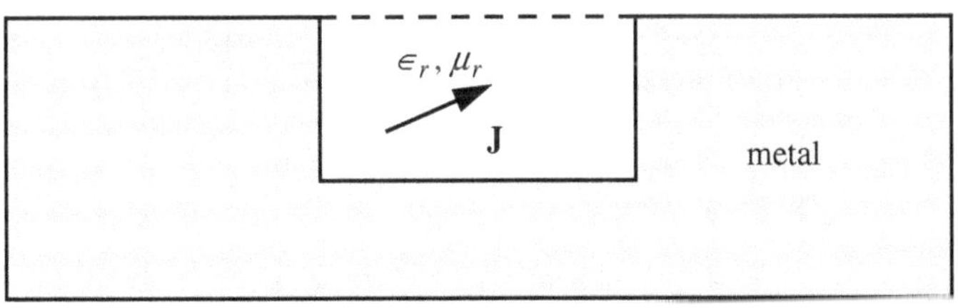


**FIGURE 3.P9** Illustration of an antenna radiating in a rectangular cavity.

Construct the necessary interior and exterior MFIEs and EFIEs for this problem. Give the final equations that must be solved on the basis of the PMCHW formulation.

10. By using the Green's function of the rectangular cavity, we can limit the support of the surface equivalent currents only across the aperture as shown in Figure 3.P10. Construct the integral equation for the interior fields by using the field expressions

$$\mathbf{E}^d = -jk_c Z_d \left[1 + \frac{1}{k_c^2}\nabla\nabla\cdot\right](\hat{z}\Psi^{TM}) - \nabla \times (\hat{z}\Psi^{TM})$$

$$\mathbf{H}^d = \nabla \times (\hat{z}\Psi^{TM}) - jk_d Y_d \left(1 + \frac{1}{k_d^2}\nabla\nabla\cdot\right)(\hat{z}\Psi^{TE})$$

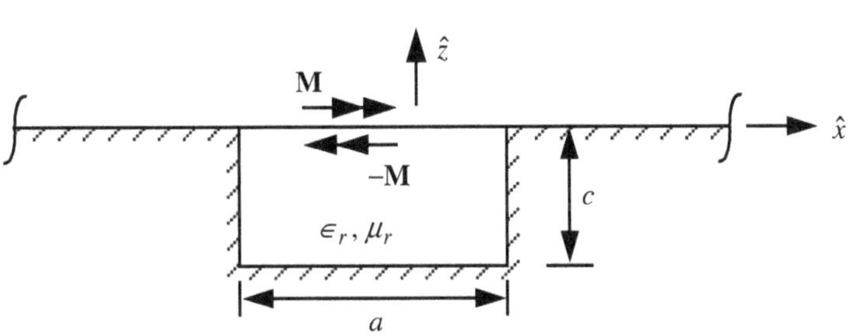


**FIGURE 3.P10** By using the Green's function of the rectangular cavity, we can limit the support of the surface equivalent currents only across the aperture.

where $k_d = \sqrt{\epsilon_r \mu_r}\, k_o$, $Z_d = \sqrt{\frac{\mu_r}{\epsilon_r}}$, $Z_o = \frac{1}{Y_c}$

$$\Psi^{TM} = \sum_{m=1}^{\infty}\sum_{n=1}^{\infty} A_{mn} \sin\left(\frac{m\pi}{a}x\right) \sin\left(\frac{n\pi}{b}y\right) \cos\left[k_z(z+c)\right]$$

$$\Psi^{TE} = \sum_{m=0}^{\infty}\sum_{n=0}^{\infty} B_{mn} \cos\left(\frac{m\pi}{a}x\right) \cos\left(\frac{n\pi}{b}y\right) \sin\left[k_z(z+c)\right]$$

and

$$\left(\frac{m\pi}{a}\right)^2 + \left(\frac{n\pi}{b}\right)^2 + k_z^2 = k_d^2.$$

Hint: Show that $E_x^c$, for example, can be written as

$$E_x^c = \sum_m \sum_n \left[ jZ_c \left(\frac{m\pi}{a}\right) \frac{k_{mn}}{k_d} A_{mn} + \left(\frac{n\pi}{b}\right) B_{mn}\right]$$

$$\cdot \cos\left(\frac{m\pi}{a}x\right) \sin\left(\frac{n\pi}{b}y\right) \sin\left[k_{mn}(z+c)\right]$$

where $k_{mn} = k_z$ with $Im(k_{mn}) < 0$

$$A_{mn} = -\frac{2jk_d Y_d}{k_{mn} ab \sin(k_{mn} c)} \left\{ \left(\frac{m\pi}{a}\right)^2 + \left(\frac{n\pi}{b}\right)^2 \right\}^{-1}$$

$$\cdot \left[ \epsilon_n \left(\frac{n\pi}{b}\right) I_x^{mn} - \epsilon_m \left(\frac{m\pi}{a}\right) I_y^{mn} \right]$$

$$B_{mn} = -\frac{2}{ab \sin(k_{mn} c)} \left\{ \left(\frac{m\pi}{a}\right)^2 + \left(\frac{n\pi}{b}\right)^2 \right\}^{-1}$$

$$\cdot \left[ \epsilon_n \left(\frac{m\pi}{a}\right) I_x^{mn} + \epsilon_m \left(\frac{n\pi}{b}\right) I_y^{mn} \right]$$

$$I_{\substack{x\\y}}^{mn} = \iint_{\text{Aperture}} M_{\substack{x\\y}}(x',y') \left\{ \begin{matrix}\sin\\ \cos\end{matrix} \left(\frac{m\pi}{a}x'\right) \begin{matrix}\cos\\ \sin\end{matrix} \left(\frac{n\pi}{b}y'\right)\right\} dx'\,dy'$$

and $\epsilon_m = 1$ for $m = 0$ or $\epsilon_m = 2$ for $m \geq 1$.

**11.** Consider a patch radiating on a finite size substrate with a superstrate placed over the patch (Figure 3.P11). Construct an integral equation set for the net electric current density flowing on the infinitesimally thin patch at the dielectric interface. The excitation field is due to a dipole placed below the patch. Hint: Introduce the electric current $\mathbf{J}_p(x, y)$ in place of the patch and treat the field due to this current in the same manner as $(\mathbf{E}^i, \mathbf{H}^i)$. After constructing the volume integral equations, generate a last integral equation for $\mathbf{J}_p$ by enforcing $\hat{n} \times \mathbf{E}_{total} = 0$ on the patch. Assume the patch is just below the superstrate.

**12.** Consider the problem of two-dimensional scattering by the illustrated configuration in Figure 3.P12:

**(a)** Construct the necessary boundary integral equations. To model the presence of the strip, use the approach described in Problem 10. The strip is just inside the superstrate and is infinitesimally thin as is the case with the ground plane.

**(b)** Instead of the boundary integral equation approach, use the appropriate volume integral equations to model the structure along with a boundary integral equation over the perfectly electric conducting (PEC) sections. For thin substrates this would lead to a numerical system with less unknowns.

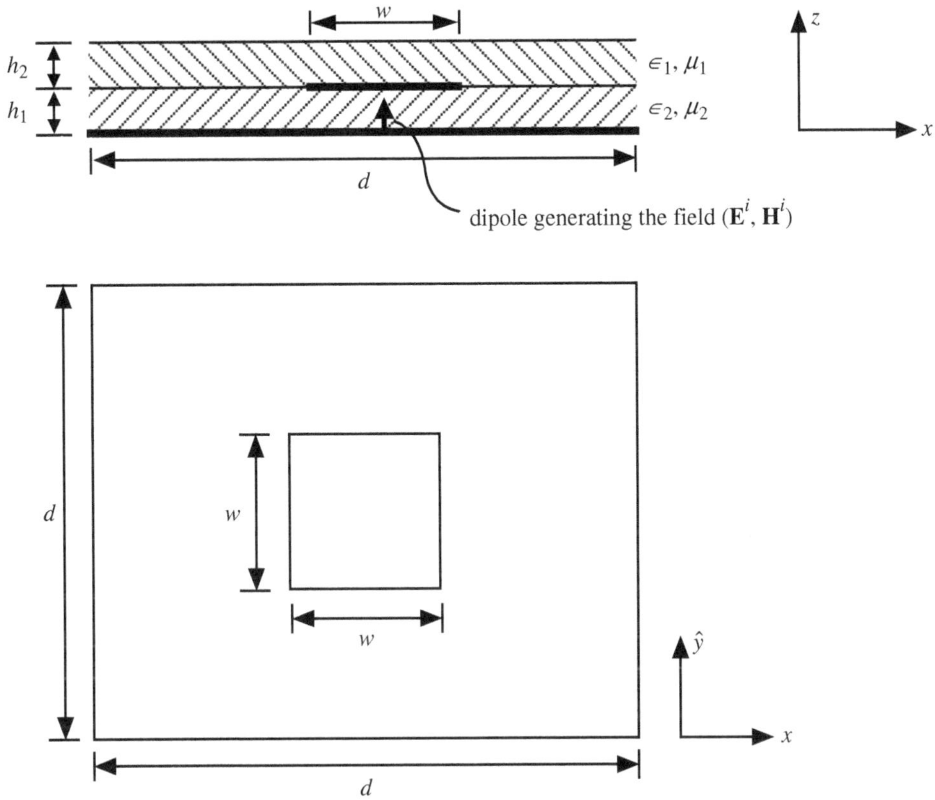


**FIGURE 3.P11** ■ A patch radiating on a finite size substrate with a superstrate placed over the patch.

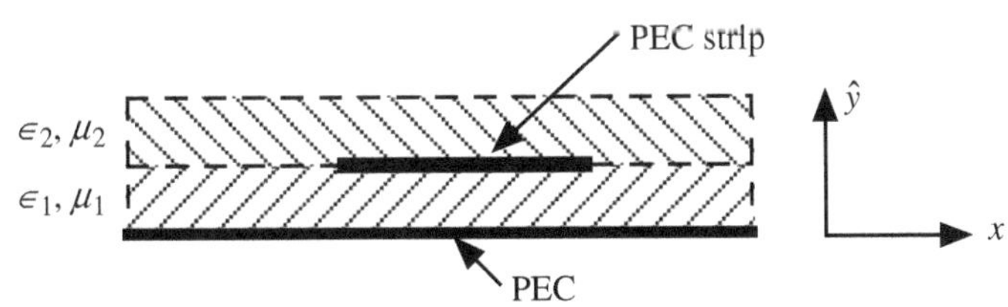


**FIGURE 3.P12** ■ Geometry for Problem 12.

## BIBLIOGRAPHY


Arvas, E., Rahhal-Arabi, A., and Sadigh, A. (1991). Scattering from multiple conducting and dielectric bodies of arbitrary shape. *IEEE Antennas Propagation Society Magazine*, *33*, 29–36.

Baker, B. B. and Copson, E. T. (1939). *The Mathematical Theory of Huygens' Principle.* Oxford: Oxford University Press.

Baucke, R. C. (1991). Scattering by two-dimensional lossy inhomogeneous dielectric and magnetic cylinders using linear pyramidal basis functions and point matching. *IEEE Trans. Antennas Propagat.*, *39*, 255–259.

Franz, W. (1948). Zur formulierung des huygensschen prinzips. *Z. für Naturforshung*, *3a*, 500–506.

Glisson, A. W. (1984). An integral equation for electromagnetic scattering from homogeneous dielectric bodies. *IEEE Trans. Antennas Propagat.*, *AP-32*, 173–175.

Govind, S. and Wilton, D. R. (1979). Numerical computation of electromagnetic scattering by inhomogeneous penetrable bodies. *Rome Air Dev. Center, U.S. Air Force, report RADC-TR-79-172.*

Jin, J.-M., Liepa, V. V., and Tai, C. T. (1988). A volume-surface integral equation for electromagnetic scattering by inhomogeneous cylinders. *J. of Electromagnetic Waves Appl.*, *2*, 573–588.

Kellogg, D. D. (1929). *Foundations of Potential Theory*. Springer (republished by Dover, New York, 1953).

Kirchhoff, G. (1882). *Berlin Ber., XX, 641.*

Mautz, J. R. and Harrington, R. F. (1978). H-field, E-field and combined field solutions for conducting bodies of revolution. *Arch. Elekron. Ubertragungstech (AEU)*, *32*, 159–164.

Michielssen, E., Peterson, A. F., and Mittra, R. (1991). Oblique scattering from inhomogeneous cylinders using a coupled integral equation formulation with triangular cells. *IEEE Transactions on Antennas and Propagation*, *39*(4), 485–490.

Müller, C. (1969). *Foundations of the Mathematical Theory of Electromagnetic Waves*. Berlin: Springer-Verlag.

Peterson, A. F. (1989). Comparison of integral, differential and hybrid methods for TE scattering from inhomogeneous dielectric cylinders. *J. Electromagnetic Wave Appl.*, *3*, 87–106.

Peterson, A. F. and Klock, P. W. (1988). An improved MFIE formulation for TE-wave scattering from lossy inhomogeneous dielectric cylinders. *IEEE Trans. Antennas Propagat.*, *36*, 45–49.

Poggio, A. and Miller, E. (1973). *"Integral equation solutions of three-dimensional scattering problems"* (chap. in *Computer Techniques for Electromagnetics,* R. Mittra, ed.), New York: Pergamom Press, chap. 4.

Rao, S. M. and Wilton, D. R. (1990). E-field, H-field, and combined field solution for arbitrary shaped three dimensional dielectric bodies. *Electromagnetics*, *10*, 407–421.

Ricoy, M. A., Kilberg, S. M., and Volakis, J. L. (1989). Simple integral equations for two-dimensional scattering with further reduction in unknowns. *IEE Proc.-H*, *136*, 298–304.

Rubinowicz, A. (1917). Die beugungswelle in der Kirchhoffshen theorie der beugungserschei nungum. *Ann. Physik*, *53*, 257–278.

Shafai, L., Moheb, H., and Kishk, A. A. (1991). Multiple formulations for solution verification in numerical handling of antenna analysis and design problems. *IEEE Antennas and Propagation Society Magazine*, *33*, 7–16.

Stratton, J. A. (1941). *Electromagnetic Theory*. New York: McGraw-Hill.

Stratton, J. A. and Chu, L. J. (1939). Diffraction theory of electromagnetic waves. *Phys. Rev.*, *56*, 99–107.

Tai, C. T. (1972). Kirchhoff theory: scalar, vector or dyadic. *Proc. IEEE*, *20*, 114–115.

Van Bladel, J. G. (1985). *Electromagnetic Fields*. New York: Hemisphere Pub. Co. (Original printing in 1964 by McGraw Hill).

Van Bladel, J. G. (1993). Three-dimensional operators acting on surface sources. *Radio Science*, *28*, 841–845.

Volakis, J. L. (1992). Alternative field representations and integral equations for modeling inhomogeneous dielectrics. *IEEE Trans. Microwave Theory and Techn.*, 604–608.

Wu, T.-K. and Tsai, L. L. (1977a). Electromagnetic fields induced inside arbitrary cylinders of biological tissue. *IEEE Trans. Microwave Theory and Techn.*, *25*, 61–65.

Wu, T.-K. and Tsai, L. L. (1977b). Scattering by arbitrarily cross-sectioned layered lossy cylinders. *IEEE Trans. Antennas Propagat.*, *25*, 518–523.

Yaghjian, A. D. (1981). Augmented electric and magnetic field integral equations. *Radio Sci.*, *16*, 987–1001.

CHAPTER 4

# Solution of Integral Equations for Wire Radiators and Scatterers

**Chapter Outline**



## 4.1 FORMULATION

Let us consider a wire of length $2\ell$ and radius $a$ ($\ell \gg a$), as shown in Figure 4-1. The wire is excited by an incident field $\mathbf{E}^i$, and we are interested in computing the current generated on the wire due to this excitation. Upon determination of the current we can then compute the radiated field in the usual manner.

To solve for the wire surface currents, we must enforce the boundary condition demanding that the total tangential electric field vanishes on the surface of the perfectly conducting wire. That is,

$$E_z^{tot} = E_z^i + E_z^r = 0 \tag{4.1}$$

where $E_z^r$ is the field radiated by the wire surface current density $\mathbf{J}(\phi, z) = \hat{z} J_z(\phi, z) + \hat{\phi} J_\phi(\phi, z)$. However, on the assumption of a very thin wire, that is, $k_o a \ll 1$, where $k_o = 2\pi/\lambda_o$ is the free-space wavenumber, $J_\phi(\phi, z)$ will either be negligible or not affect the radiated field. Thus, from (2.52a), (2.109c), or (2.102a) in conjunction with (2.101) we may express the wire-radiated field as

$$\mathbf{E}^r(\rho, \phi, z) = -jk_o Z_o \hat{z} \int_{-\ell}^{\ell} \int_0^{2\pi} J_z(\phi', z') \left(1 + \frac{1}{k_o^2} \frac{\partial^2}{\partial z^2}\right) \frac{e^{-jk_o R}}{4\pi R} a \, d\phi' \, dz' \tag{4.2}$$

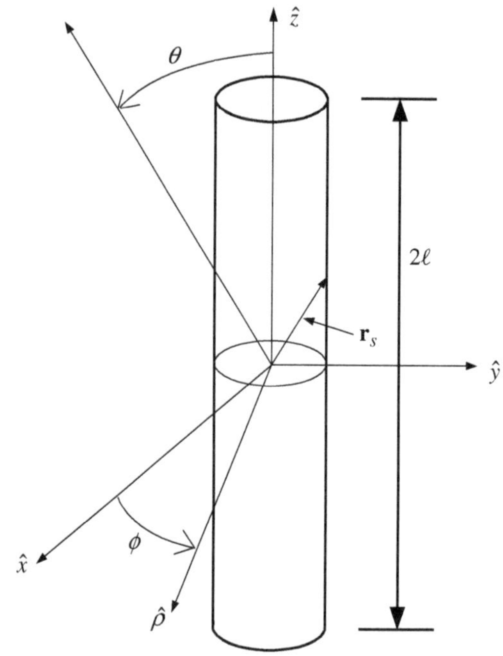


**FIGURE 4-1** Cylindrical wire geometry.

in which $Z_o = 1/Y_o$ denotes the free-space intrinsic impedance and

$$R = |\mathbf{r} - \mathbf{r}'| = \sqrt{\rho^2 + a^2 - 2\rho a \cos(\phi - \phi') + (z - z')^2} \tag{4.3}$$

since $\mathbf{r} = \rho\hat{\rho} + z\hat{z}$ and $\mathbf{r}' = a\hat{\rho}' + z'\hat{z}$. This expression can be further simplified by assuming that $J_z(\phi, z)$ is symmetric with respect to $\phi$, a reasonable assumption since the wire is very thin and is typically part of a transmission line fed by a voltage source at its center. The surface current $J_z(\phi, z)$ can then be equivalently replaced by a filamentary line current $I(z)$ placed at the center of the tubular conductor. For the two currents to generate the same field when $\rho \gg a$, it is necessary that they satisfy the relation

$$I(z) = \int_0^{2\pi} J_z(\phi, z) a \, d\phi = 2\pi a J_z(z). \tag{4.4}$$

Introducing this into (4.2) yields

$$E_z^r(\rho, \phi = 0, z) = E_z^r(\rho, z) = -jk_o Z_o \int_{-\ell}^{\ell} I(z') \left(1 + \frac{1}{k_o^2} \frac{d^2}{dz^2}\right) G_w(z - z') \, dz' \tag{4.5}$$

where

$$G_w(z - z') = \frac{1}{2\pi} \int_0^{2\pi} \frac{e^{-jk_o\sqrt{\rho^2 + a^2 - 2\rho a \cos\phi' + (z - z')^2}}}{4\pi\sqrt{\rho^2 + a^2 - 2\rho a \cos\phi' + (z - z')^2}} \, d\phi' \tag{4.6}$$

and we have arbitrarily set $\phi = 0$ since by symmetry the radiated field is expected to be independent of $\phi$.

To construct the integral equation for the solution of the current $I(z)$ we set $\rho = a$ in (4.6) and substitute (4.5) into (4.1). This gives

$$E_z^i(\rho = a, z) = +jk_o Z_o \int_{-\ell}^{\ell} I(z') \left(1 + \frac{1}{k_o^2} \frac{d^2}{dz^2}\right) G_{wu}(z - z')\, dz'. \qquad (4.7)$$

The kernel $G_{wu}(z - z')$ is now given by

$$G_{wu}(z - z') = \frac{1}{2\pi} \int_0^{2\pi} \frac{e^{-jk_o R_u}}{4\pi R_u}\, d\phi' \qquad (4.8)$$

with

$$R_u = \sqrt{(z - z')^2 + 4a^2 \sin^2(\phi'/2)}. \qquad (4.9)$$

$G_{wu}$ is often referred to as the unreduced thin-wire kernel. In practice, though, to avoid the integration over $\phi'$, $G_{wu}(z - z')$ is replaced by the reduced kernel

$$G_{wr}(z - z') = \frac{e^{-jk_o\sqrt{(z-z')^2+a^2}}}{4\pi\sqrt{(z - z')^2 + a^2}} = \frac{e^{-jk_o R_o}}{4\pi R_o} \qquad (4.10)$$

which is obtained by letting $\mathbf{r}' = z'\hat{z}$. That is, the reduced kernel refers to the problem where the filamentary current is introduced from the start of the analysis. Substituting (4.10) into the integral equation (4.7) gives

$$E_z^i(\rho = a, z) = jk_o Z_o \int_{-\ell}^{\ell} I(z') \left(1 + \frac{1}{k_o^2} \frac{d^2}{dz^2}\right) \frac{e^{-jk_o R_o}}{4\pi R_o}\, dz' \qquad (4.11)$$

with $R_o$ as defined in (4.10). One readily observes that the right-hand side of this equations is simply the negative of the field radiated by the filamentary current $I(z)$ and evaluated at $\rho = a$, that is, on the surface of the perfectly conducting wire as shown in Figure 4-2. Obviously, (4.11) could have been derived in a more direct manner by first invoking the approximation (4.4) and then referring to the integral representation (2.109c). Nevertheless, the previously outlined steps should serve to

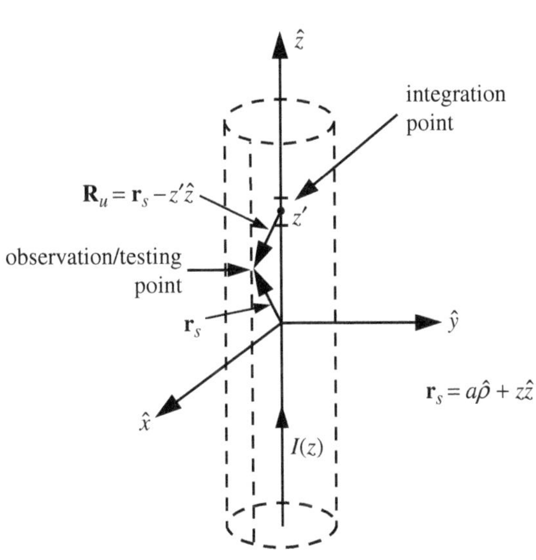


**FIGURE 4-2** Geometry for testing on the wire surface.

clarify the implied approximations. As will be shown later, (4.7) and (4.11) can be solved with nearly equal efforts when an iterative solution scheme is employed.

The thin-wire integral equation (4.11) is commonly referred to as Pocklington's (1897) integral equation. More generally, it belongs to the general class of Fredholm integral equations of the first kind. These are characterized by the presence of the unknown function only under the integral whose limits are constant. Integral equations that have the unknown quantity both under and outside the integral are of the second kind, and we shall consider them at the end of this chapter. Also, if the integral limits are not constant, then the corresponding integral equations are of the Voltera type, which are the typical equations for nonharmonic (time-dependent) field quantities.

An analytical solution of (4.11) is not possible unless the wire is semi-infinite, in which case function theoretic techniques such as the Weiner-Hopf method (Noble, 1958) can be employed for their solution in the transform domain. However, Pocklington's integral equation can be numerically solved without difficulty, particularly because the integral's kernel is never singular since $R_o > a$ for all values of $z$ and $z'$. Nevertheless, to reduce the kernel's singularity, it is still instructive to transfer one of the derivatives from the Green's function to the current, as was done in Section 3.1.2 in conjunction with the Stratton-Chu integral equation. In particular, from the one-dimensional form of (3.12) via integration by parts we have (note $\frac{d}{dz}G_{wr} = -\frac{d}{dz'}G_{wr}$)

$$\int_{-\ell}^{\ell} I(z')\frac{d^2}{dz^2}G_{wr}(z-z')\,dz' = \int_{-\ell}^{\ell} \frac{dI(z')}{dz'}\frac{d}{dz}G_{wr}(z-z')\,dz' - \frac{d}{dz}\left[G_{wr}(z-z')I(z')\right]_{z'=-\ell}^{z'=\ell}. \tag{4.12}$$

Since the current at the wire ends must vanish, we observe that the last term of (4.12) is zero, and thus Pocklington's integral equation can be rewritten as

$$E_z^i(\rho = a, z) = jk_oZ_o\int_{-\ell}^{\ell}\left[I(z')G_{wr}(z-z') + \frac{1}{k_o^2}\frac{d}{dz'}I(z')\frac{d}{dz}G_{wr}(z-z')\right]dz'. \tag{4.13}$$

An alternative way to derive Pocklington's equation is through the use of the vector and scalar potentials. Accordingly, from (2.4) $E_z^r$ can be expressed as

$$E_z^r(\rho, z) = -jk_oZ_oA_z - \frac{\partial\Phi_e}{\partial z} \tag{4.14}$$

where

$$A_z(\rho = a, z) = \int_{-\ell}^{\ell} I(z')\frac{e^{-jk_oR_o}}{4\pi R_o}\,dz' \tag{4.15}$$

and

$$\Phi_e(\rho = a, z) = \int_{-\ell}^{\ell}\frac{\rho(z')}{\epsilon_o}\frac{e^{-jk_oR_o}}{4\pi R_o}\,dz'. \tag{4.16}$$

From the continuity equation (1.38) we have

$$\frac{\rho(z)}{\epsilon_o} = -\frac{Z_o}{jk_o}\frac{dI(z)}{dz} \tag{4.17}$$

and, thus, when (4.14) along with (4.15) through (4.17) is substituted into (4.1) we obtain (4.13).

The standard procedure for solving the previous integral equation amounts to first expanding the currents in terms of a class of basis functions. That is, $I(z)$ is approximately expressed as a linear sum of $N$ known expansion functions. Upon substitution of this expansion into (4.13) we obtain an equation for the coefficients of the expansion that is a function of the surface observation point $z$. The second step in the numerical solution process is the enforcement of the integral equation at specific values of $z$. In this manner we obtain a single linear equation for each enforcement point. If we have $N$ expansion coefficients, a total of $N$ linear equations must then be generated by changing the location of the testing point. These comprise a system that can be solved for the unknown expansion coefficients. Depending on the type of expansion functions or enforcement scheme, different linear systems will be obtained. The procedure of expanding the current in terms of a finite set of functions and then of enforcing the boundary condition is referred to as the discretization of the integral equation. Discretization is therefore the procedure that generates the linear system. In turn, the resulting system can be solved through various direct or iterative methods to obtain the coefficients of the expansion. A knowledge of these provides an approximation for the current distribution, and once the current is known we can proceed with computing the radiated field, input impedance, radiated power, and antenna gain using standard formulae.

Before proceeding with the discretization of the integral equation (4.11), we first present the most commonly used expansion basis functions for the current distributions.

## 4.2 | BASIS FUNCTIONS

A first step in discretizing (4.13) is to expand the current distribution as

$$I(z) = \sum_{n=0}^{N-1} I_n f_n(z) = \sum_{n=0}^{N-1} I_n f(z - z_n) \tag{4.18}$$

where $f_n(z)$ are the basis functions of the expansion, and $I_n$ are unknown expansion coefficients. Referring to Figure 4-3, some of the most popular choices for $f_n(z)$ are as follows:

1. Pulse-basis functions/piecewise constant (PWC):

$$f_n(x) = P_{\Delta x}(x - x_n) = \begin{cases} 1 & x_n - \frac{\Delta x}{2} < x < x_n + \frac{\Delta x}{2} \\ 0 & \text{elsewhere.} \end{cases} \tag{4.19}$$

2. Triangular function/piecewise linear:

$$f_n(x) = T_n(x) = \left(1 - \frac{|x - x_n|}{\Delta x}\right) P_{2\Delta x}(x - x_n). \tag{4.20}$$

3. Piecewise sinusoidal (PWS):

$$f_n(x) = S_n(x) = \frac{\sin k_o(\Delta x - |x - x_n|)}{\sin k_o \Delta x} P_{2\Delta x}(x - x_n) \tag{4.21}$$

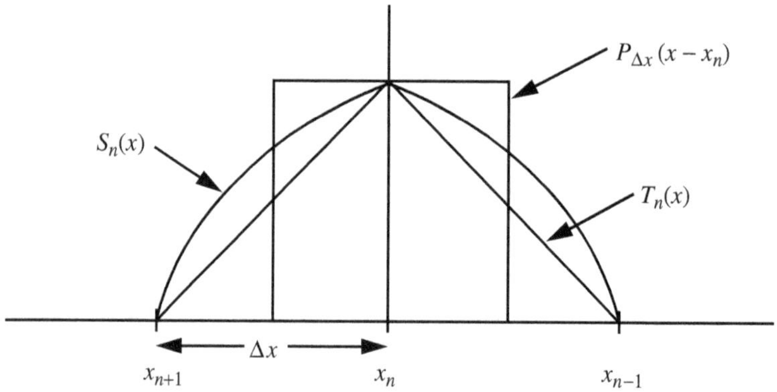


**FIGURE 4-3** Three subsectional expansion functions.

where $\Delta x$ is usually small (of the order $\lambda_o/10$) and $N = 2\ell/\Delta x$. Because their domain is confined to a small section of the wire, they are commonly referred to as subsectional or subdomain basis functions. A major reason for their popularity is owed to their capability to model any arbitrary function provided $\Delta x$ is sufficiently small.

As illustrated, however, in Figure 4-4, they cause artificial discontinuities in the current or its derivatives at the transition between two consecutive expansion functions. Specifically, the current expansion with the PWC basis function is inherently discontinuous at the junction of two adjacent segments, and from (4.17) this implies the existence of a fictitious charge at that point. Nevertheless, in spite of this deficiency when the segments are sufficiently small, they provide a reasonable approximation to the current distribution. In that case, the constant value over the segment should be interpreted to represent the average of the true current over that segment. Because of their simplicity, and this will soon be apparent in the next section, they have been used extensively in electromagnetics but more so for scattering than antenna parameter computations. In the latter case, excessive sampling may be required for the correct evaluation of the antenna's input impedance.

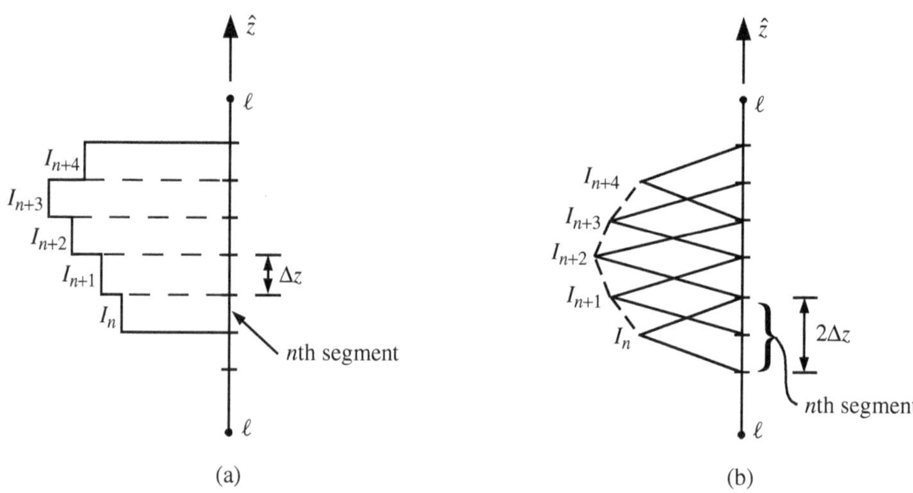


**FIGURE 4-4** Illustration of wire segmentation and current approximation with subdomain basis: (a) pulse-basis expansion, and (b) triangular basis expansion.

The piecewise linear basis functions are seen to generate continuous current distributions. This is because the adjacent basis functions are overlaid as shown in Figure 4-4b. Thus, the current at any point on the wire is obtained by summing the overlaid basis. From their definition, though, when one of the overlaying expansion functions is at a maximum, the left and right adjacent expansion functions are zero. Further, because each expansion is normalized, the coefficients correspond to the current's value at the middle of the $n$th segment. The PWS expansion functions are very similar to the linear basis functions in nearly all respects. One difference between the two is that the PWS basis functions can be differentiated any arbitrary number of times within its range without vanishing. Nevertheless, similar to the piecewise linear basis functions, PWS basis functions also yield a current expansion that has a discontinuous first derivative at the middle of each wire segment. The only advantage of the PWS basis functions is drawn from their property to yield potential integrals that can be evaluated analytically once $S_n(x)$ is expressed as a sum of two exponentials.

Instead of using the aforementioned subsectional or subdomain basis to represent the wire current, one could alternatively employ the usual full basis expansions such as $\cos nx$ and $\sin nx$. For example, noting that $I(\pm\ell) = 0$, an appropriate expansion for the wire current would be

$$I(z) = \sum_{n=1}^{N} C_n \cos\left[\frac{(2n-1)\pi z}{2\ell}\right] \tag{4.22}$$

or

$$I(z) = \sum_{n=1}^{N} C_n \sin\left[\frac{(2n-1)\pi z}{\ell}\right]. \tag{4.23}$$

In contrast to the expansions (4.19) through (4.21), the coefficients of these expansions do not coincide with specific values of the current $I(z)$. More importantly, $N$ may have to be quite large in case $I(z)$ is rapidly varying or not sinusoidal in form. However, for wire antennas $I(z)$ is generally sinusoidal, particularly when the wire is excited by an external incident field. In this case only a few terms of the full basis expansions (4.22) or (4.23) may be required, making them attractive. Generally, though, (4.22) through (4.23) cannot be effectively used for curved wires or other complex wire structures on which the current's distribution is much more irregular. In the following, we shall therefore concentrate on the discretization and solution of Pocklington's integral equation using subdomain/basis functions since such a solution is less specific to the straight wire.

## 4.3 | PULSE-BASIS–POINT-MATCHING SOLUTION

For simplicity, let us first consider the pulse-basis expansion to represent the wire current distribution. This results in a summation of shifted pulses over the total length of the wire, that is,

$$I(z) = \sum_{n=0}^{N-1} I_n P_{\Delta z}(z - z_n) \tag{4.24}$$

where

$$N = \frac{2\ell}{\Delta z} \tag{4.25}$$

are the number of pulses used to approximate the current distribution on the wire and

$$z_n = -\ell + \left(n + \frac{1}{2}\right)\Delta z; \qquad n = 0, 1, 2, \ldots \tag{4.26}$$

Substituting (4.24) into (4.13) yields

$$E_z^i = \frac{jk_o Z_o}{4\pi}\left\{\sum_{n=0}^{N-1} I_n \int_{z_n-\frac{\Delta z}{2}}^{z_n+\frac{\Delta z}{2}} \frac{e^{-jk_o R_o}}{R_o}\,dz' + \frac{1}{k_o^2}\sum_{n=0}^{N-1} I_n \frac{d}{dz}\cdot\left[\frac{e^{-jk_o R_{1n}}}{R_{1n}} - \frac{e^{-jk_o R_{2n}}}{R_{2n}}\right]\right\} \tag{4.27}$$

where

$$R_{1n} = \sqrt{\left(z - z_n + \frac{\Delta z}{2}\right)^2 + a^2}\,, \qquad R_{2n} = \sqrt{\left(z - z_n - \frac{\Delta z}{2}\right)^2 + a^2} \tag{4.28}$$

and we have invoked the expression

$$\frac{dI(z)}{dz} = \sum_{n=-N/2}^{N/2} I_n\left[\delta\left(z - z_n + \frac{\Delta z}{2}\right) - \delta\left(z - z_n - \frac{\Delta z}{z}\right)\right] \tag{4.29}$$

in deriving (4.27). After differentiating the last term of (4.27) with respect to $z$, we obtain

$$E_z^i(\rho = a, z) = \frac{jZ_o}{2\lambda_o}\sum_{n=0}^{N-1} I_n\left[\Psi_n(z) + \Phi_n(z)\right] \tag{4.30}$$

where

$$\Psi_n(z) = \int_{z_n-\frac{\Delta z}{2}}^{z_n+\frac{\Delta z}{2}} \frac{e^{-jk_o R_o}}{R_o}\,dz' \tag{4.31}$$

and

$$\begin{aligned}\Phi_n(z) = -k_o\Bigg[&\left(z - z_n + \frac{\Delta z}{2}\right)\frac{(jkR_{1n} + 1)}{(k_o R_{1n})^3}e^{-jk_o R_{1n}} \\ &- \left(z - z_n - \frac{\Delta z}{2}\right)(jk_o R_{2n} + 1)\frac{e^{-jk_o R_{2n}}}{(k_o R_{2n})^3}\Bigg].\end{aligned} \tag{4.32}$$

Equation (4.30) can now be solved for $I_n$ by demanding that it be satisfied (matched) at $N$ points on the surface of the wire. A convenient set of such points is

$$z = z_m = -\ell + \left(m + \frac{1}{2}\right)\Delta z, \qquad m = 0, 1, 2, 3, \ldots$$

with $\phi = 0$, that is, along the line formed by the wire surface and the $x$-$z$ plane. This results in a set of matrix equations

$$\begin{bmatrix} Z_{00} & Z_{01} & Z_{02} & \cdots & Z_{0n} & \cdots & Z_{0,N-1} \\ Z_{10} & Z_{11} & Z_{12} & \cdots & Z_{1n} & \cdots & Z_{1,N-1} \\ \vdots & & \vdots & & \vdots & & \vdots \\ Z_{m0} & Z_{m1} & Z_{m2} & \cdots & Z_{mn} & \cdots & Z_{m,N-1} \\ \vdots & & \vdots & & \vdots & & \vdots \\ Z_{(N-1),0} & Z_{(N-1),1} & Z_{(N-1),2} & \cdots & Z_{(N-1),n} & \cdots & Z_{(N-1),(N-1)} \end{bmatrix} \begin{bmatrix} I_o \\ I_1 \\ I_2 \\ \vdots \\ I_m \\ \vdots \\ I_{N-1} \end{bmatrix} = \begin{bmatrix} V_0 \\ V_1 \\ V_2 \\ \vdots \\ V_m \\ \vdots \\ V_{N-1} \end{bmatrix}$$

which are commonly written as

$$[Z_{mn}]\{I_n\} = \{V_m\} \tag{4.33}$$

Obviously,

$$\{I_n\} = \left\{I_0, I_1, I_2, \ldots, I_{N/2}, \ldots, I_{N-1}\right\}^T \tag{4.34}$$

is a column matrix, and $[Z_{mn}]$ is a square matrix referred to as the impedance matrix since in this case the excitation column $\{V_m\}$, where

$$V_m = -E_z^i(\rho = a, z_m) \tag{4.35}$$

has units of volts. The corresponding elements of the impedance matrix can be obtained directly from (4.30) through (4.32). We find that

$$Z_{mn} = -\frac{jZ_o}{2\lambda_o}\left[\Psi_n(z_m) + \Phi_n(z_m)\right] \tag{4.36}$$

where $\Psi_n(z_m)$ can be rewritten as

$$\Psi_n(z_m) = \int_{(z_m-z_n)-\frac{\Delta z}{2}}^{(z_m-z_n)+\frac{\Delta z}{2}} \frac{e^{-jk_o\sqrt{t^2+a^2}}}{\sqrt{t^2+a^2}}\, dt.$$

It is seen that $Z_{nm} = Z_{mn}$, indicating that the impedance matrix is symmetric. It is also observed that $[Z_{mn}]$ is completely independent of the excitation.

The integral $\Psi_n(z_m)$ cannot be evaluated analytically but can be approximated in closed form with sufficient accuracy. For $m \neq n$, $\sqrt{t^2 + a^2}$ is not very small, and we may therefore employ midpoint integration to approximately express it as

$$\Psi_n(z_m) \approx \Delta z \frac{e^{-jk_o\sqrt{(z_m-z_n)^2+a^2}}}{\sqrt{(z_m - z_n)^2 + a^2}}; \qquad m \neq 0.$$

When $m = n$, $\sqrt{t^2 + a^2}$ is nearly zero over the midrange of integration. In this case we can employ the two-term expansion

$$e^{-jk_oR} \cong 1 - jk_oR$$

allowing us to approximate $\Psi_n(z_n)$ as

$$\Psi_n(z_n) \approx \int_{\frac{-\Delta z}{2}}^{\frac{\Delta z}{2}} \left[\frac{1}{\sqrt{t^2+a^2}} - jk_o\right] dt = \ln\left(\frac{\sqrt{\left(\frac{\Delta z}{2}\right)^2 + a^2} + \frac{\Delta z}{2}}{\sqrt{\left(\frac{\Delta z}{2}\right)^2 + a^2} - \frac{\Delta z}{2}}\right) - jk_o\Delta z$$

and for $\Delta z \gg a$ this can be further simplified to give

$$\Psi_n(z_n) \approx 2\ln\left(\frac{\Delta z}{a}\right) - jk_o\Delta z; \qquad \Delta z \gg a.$$

An alternative way for computing the integral $\Psi_n(z_m)$ is to *regularize* its near singular integrand with the addition and subtraction of the term $\frac{1}{\sqrt{t^2+a^2}}$, which can be integrated analytically. This gives

$$\Psi_n(z_m) = \int_{(z_m-z_n)-\frac{\Delta z}{2}}^{(z_m-z_n)+\frac{\Delta z}{2}} \left[\frac{e^{-jk_o\sqrt{t^2+a^2}}}{\sqrt{t^2+a^2}} - \frac{1}{\sqrt{t^2+a^2}}\right] dt$$

$$+ \ln\left(\frac{\sqrt{\left(z_m - z_n + \frac{\Delta z}{2}\right)^2 + a^2} + z_m - z_n + \frac{\Delta z}{2}}{\sqrt{\left(z_m - z_n - \frac{\Delta z}{2}\right)^2 + a^2} + z_m - z_n - \frac{\Delta z}{2}}\right).$$

The new integrand is now slowly varying and can thus be evaluated numerically without difficulty.

To compute the current coefficients we must solve the system (4.33), and a number of routines are commercially available to perform this operation in a user-friendly manner. Commonly used software libraries such as IMSL, LINPACK, and NAG include a variety of subroutines for a solution of (4.33). These are based on solution methods such as Gauss-Jordan elimination, Gaussian elimination, Crout, or lower-upper (LU) decomposition, most of which are discussed in numerical analysis textbooks.

If we choose to solve $\{I_m\}$ by inverting the matrix $[Z_{mn}]$, the required central processing unit (CPU) time will be approximately

$$t \approx AN^2 + BN^3 + CN^2N_i \tag{4.37}$$

where $N$, of course, denotes the number of unknowns or the length of the column $\{I_n\}$, and $N_i$ is the number of different excitations for which $\{I_n\}$ must be computed. In addition,

$$A = \text{time required to compute each value of } Z_{mn}$$

$$BN^3 = \text{time required to invert } [Z_{mn}]$$

$$CN^2 = \text{time required to perform the matrix multiplication } [Z_{mn}]^{-1}\{V_m\}.$$

The actual values of the constants $A$, $B$, and $C$ are machine dependent. Expression (4.37) holds regardless of the procedure used to obtain the inverse, but, clearly, for large $N$ the second term of (4.37) dominates. However, a solution for $\{I_n\}$ can be obtained without a need to complete the inverse. In this case the Gauss-Jordan elimination requires $N^3$ ($\frac{N^2}{2}$ if the inverse is not returned) operations to complete the solution, whereas the Gaussian elimination needs $5N^3/6$ operations. In contrast, the LU decomposition approach requires $N^3/3$ operations and is thus much faster. The LU decomposition scheme is also preferred because it results in better accuracy and stability compared with other methods, particularly when $N$ is large. Nevertheless, when $N$ becomes very large, a direct solution of the linear system (4.33) may yield an inaccurate result due to machine round-off errors. An alternative in this case is to use an iterative solution scheme allowing some control on the solution error, and such a scheme is discussed at the end of this chapter.

Often, as is the case with the linear wire discussed here, the impedance matrix will posses certain symmetries that can be exploited in the solution of (4.33). It is easy to observe from (4.28), (4.31), (4.32), and (4.36) that $Z_{nm} = Z_{mn} = Z_{m-n} = Z_{|m-n|}$. Matrices of this type are referred to as symmetric Toeplitz and require order $N^2$ operations to complete the solution. Also, since the elements of $[Z_{mn}]$ can be generated from those in one row or a column, the fill time of the matrix can be reduced to only order $N$ operations. Note, that if we were to consider a solution of the currents on a curved wire, then $Z_{mn} \neq Z_{|m-n|}$, but $Z_{nm} = Z_{mn}$ as a consequence of reciprocity (i.e., the matrix is still symmetric).

## 4.4 | SOURCE MODELING

### 4.4.1 Delta Gap Excitation

The wire antenna is usually center fed by a transmission line whose voltage can be measured at the terminals of the antenna. Assuming, the transmission line voltage at the wire terminals is $V_i$ (see Figure 4-5), we may then write (Collin, 1985)

$$V_i = -\int_{-\delta/2}^{\delta/2} \mathbf{E}^r \cdot \hat{z}dz = +\int_{-\delta/2}^{\delta/2} E_z^i dz = +E_z^i \delta \tag{4.38}$$

from small $\delta$. Consequently

$$E_z^i = \begin{cases} \frac{+V_i}{\delta} & z = 0 \\ 0 & \text{elsewhere} \end{cases} \tag{4.39}$$

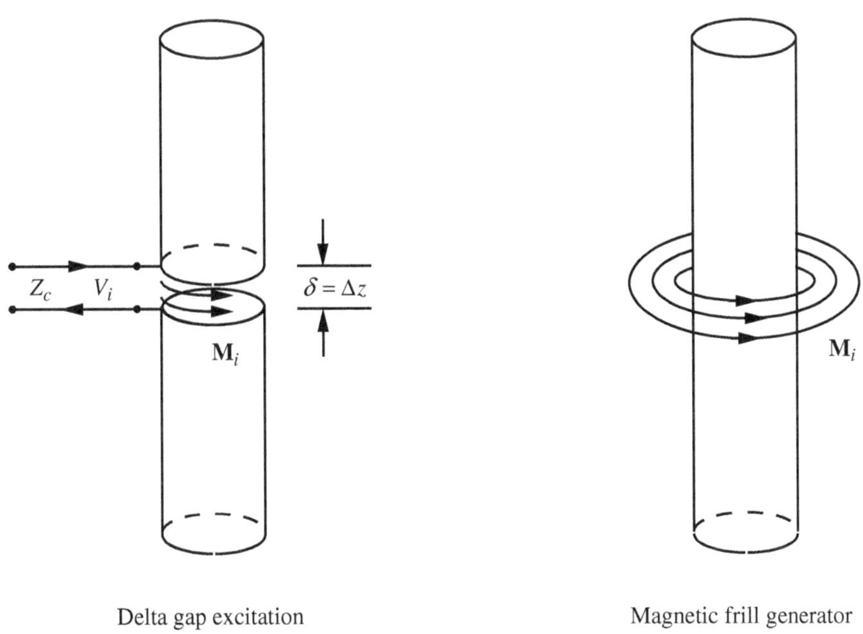


**FIGURE 4-5** Source modeling for the center-fed cylindrical dipole.

and this is referred to as the delta gap excitation model for the source field $E_z^i$. Note that (4.39) is equivalent to having a magnetic current loop

$$\mathbf{M}^i = -\hat{\rho} \times \mathbf{E}^i = \hat{\phi}\frac{V_i}{\delta} \tag{4.40}$$

of radius $a$ as the excitation. In fact, the derivation of (4.38) requires that the delta gap is first closed, making the conductor continuous. The excitation field $E_z^i$, which is confined over the original gap length, can then be replaced by the equivalent magnetic current loop $\mathbf{M}^i$. This, in turn generates a scattered field $E_z^r$ at the conductor's surface so that the total field $E_z^i + E_z^r$ vanishes as required, a condition imposed in deriving (4.38). Inherently, the presence of the magnetic current generates discontinuous electric fields across its surface, and for this particular case the electric field is zero in the interior side of $\mathbf{M}_i$ and equal to $\hat{z}E_z^i$ at its exterior side.

Wire current distributions obtained by solving the system (4.33) in conjunction with a delta gap modeling of the source are illustrated in Figures 4-6 and 4-7. (Figures 4-6–4-7, 4-10–4-12, 4-14–4-18 and 4-19 are courtesy of K. Barkeshli. Some of the figures can be found in Barkeshli and Volakis, 1988) The curves in each figure correspond to $\lambda_o/2$ and $\lambda_o$ long dipoles, respectively, of radius $a = 0.005\lambda_o$. It is seen that a rather large number of pulse basis are required for the current to converge to its final value. Generally, (i.e., provided the system has an acceptable condition number), the correct distribution is obtained if the computed values of $I(z)$ do not change appreciably as $N$ is increased. Having the correct value of $I(z)$ is extremely important for input impedance computations, but the radiation pattern can be predicted with sufficient accuracy once $I(z)$ is known approximately. As expected, the computed current is sinusoidal in form except near the feed point, and, thus, it is not surprising that the often assumed sinusoidal behavior of the wire current is sufficient for pattern prediction but much less so for input impedance computations. This is more apparent for the $\lambda_o$ long dipole in which case the sinusoidal distribution will predict zero current at the feed.

Perhaps one of the reasons for the large number of expansion pulses required to reach convergence is the difficulty of the point matching procedure in satisfying the boundary condition at all $z$. As seen from Figure 4-8, the wire surface fields obtained by integrating the numerically computed current given in Figure 4-6 do not vanish except at the match points $z_m$. Nevertheless, on the average, the surface field is zero as can be attested from the oscillatory behavior of the computed surface field given in Figure 4-8. Later, it will be discussed that higher-order expansion functions and more robust testing procedures yield more satisfactory results with less unknowns.

In implementing (4.33) we were careful to maintain $\frac{\Delta z}{a}$ small in accordance with the thin wire approximation. Studies (e.g., Burke and Poggio, 1980) have shown that the thin-wire approximation is less than 1% in error if $\frac{\Delta z}{a} < 8$. Since typically $\Delta z \leq \frac{\lambda_o}{10}$, this implies that $k_o a \leq 0.08$ to limit the error to 1%.

### 4.4.2 Magnetic Frill Generator

As can be expected, (4.39) is not as accurate (particularly as the wire radius becomes greater than $0.007\lambda$ (Imbriale and Ingersong, 1963) since the field is unlikely to be concentrated only within the gap. An alternative source model giving a smoothly varying excitation field around the gap is the magnetic frill generator. In this case the gap is equivalently replaced by a circumferentially directed surface magnetic

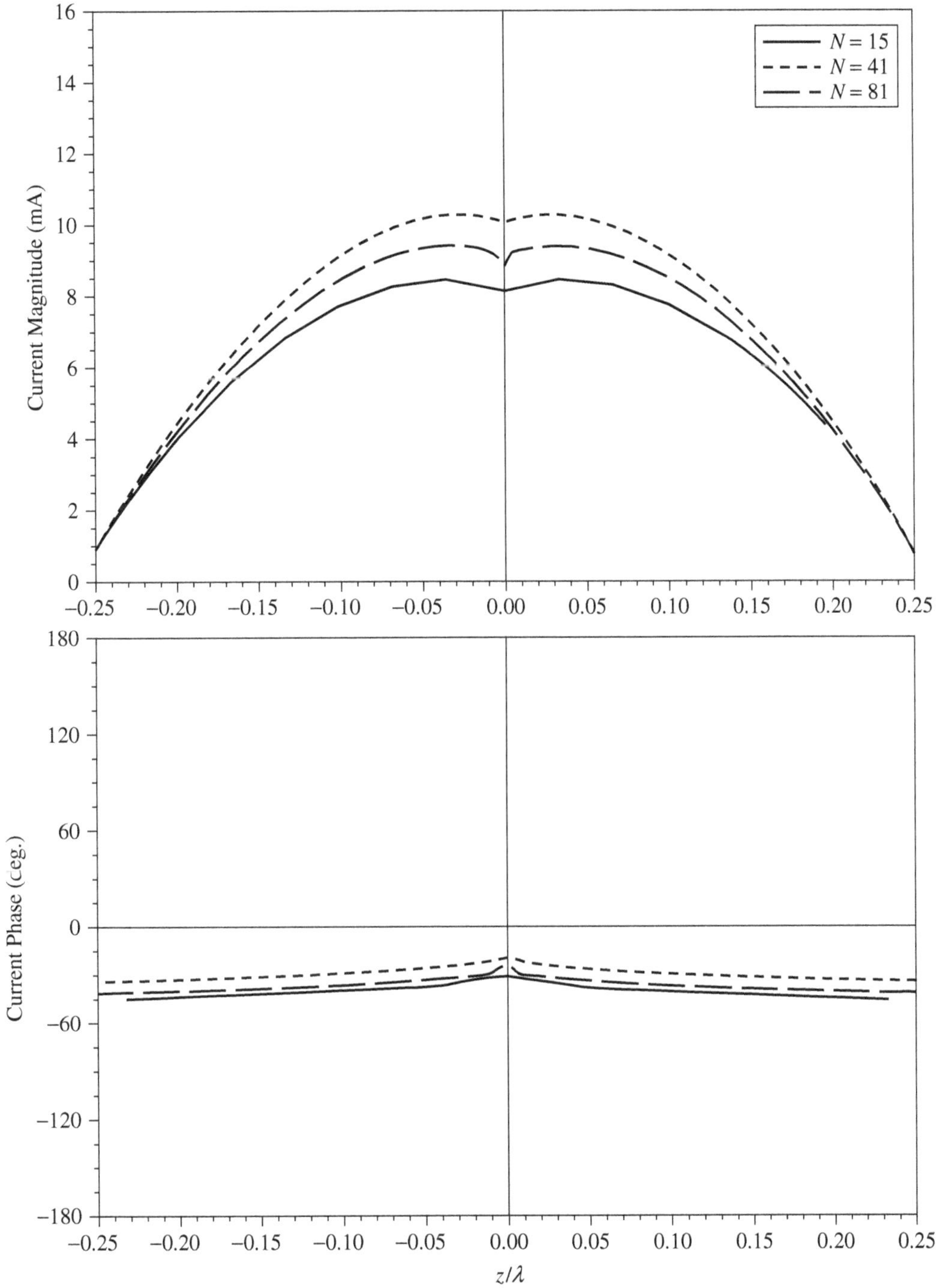


**FIGURE 4-6** ■ Computed current on a center-fed $\lambda_o/2$ dipole of radius $a = .005\lambda_o$ via the pulse-basis–point-matching solution method as a function of the sampling density. The source/excitation is a delta gap as given in (4.39) (Barkeshli and Volakis, 1988).

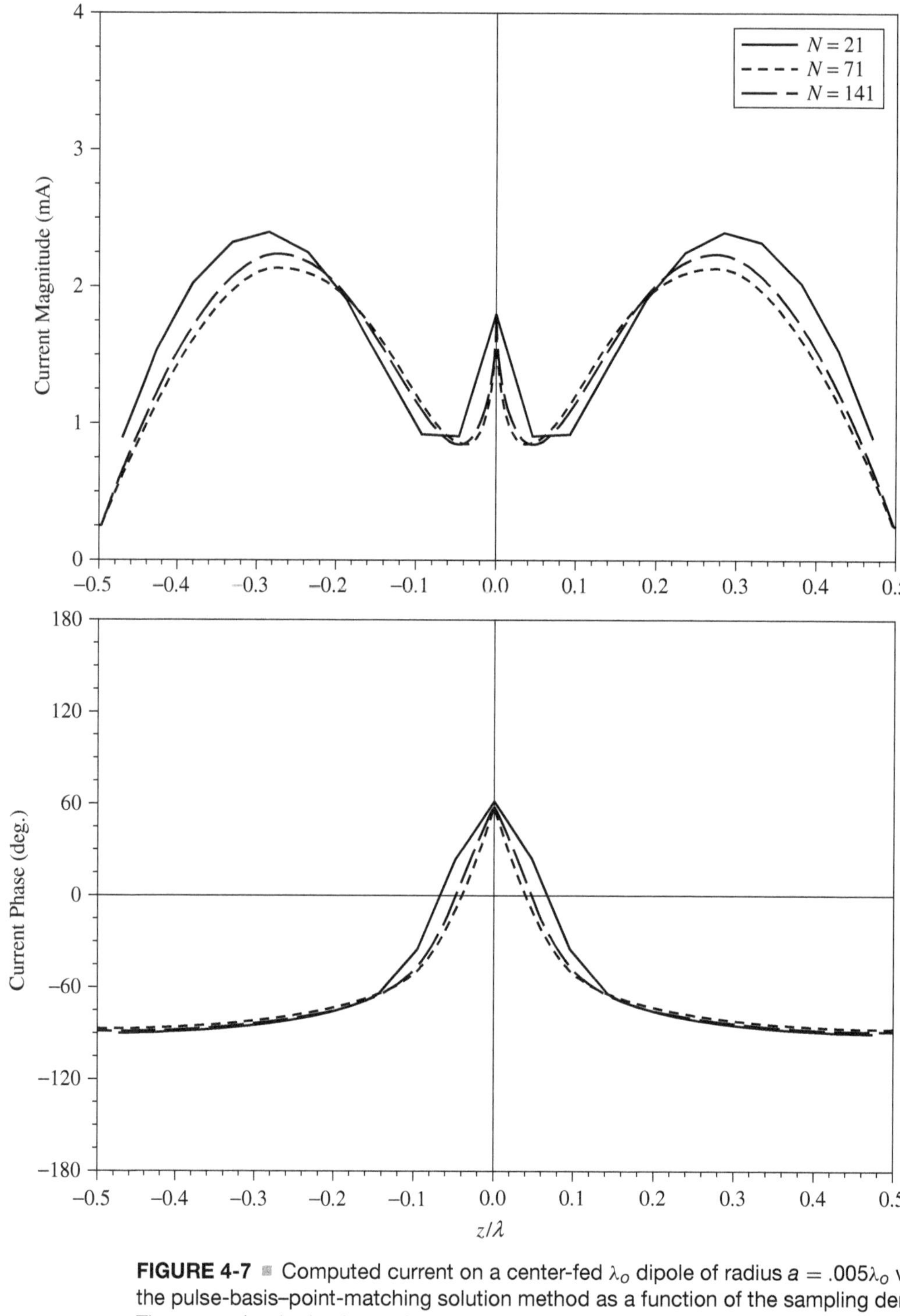


**FIGURE 4-7** ■ Computed current on a center-fed $\lambda_o$ dipole of radius $a = .005\lambda_o$ via the pulse-basis–point-matching solution method as a function of the sampling density. The source/excitation is a delta gap as given in (4.39) (Barkeshli and Volakis, 1988).

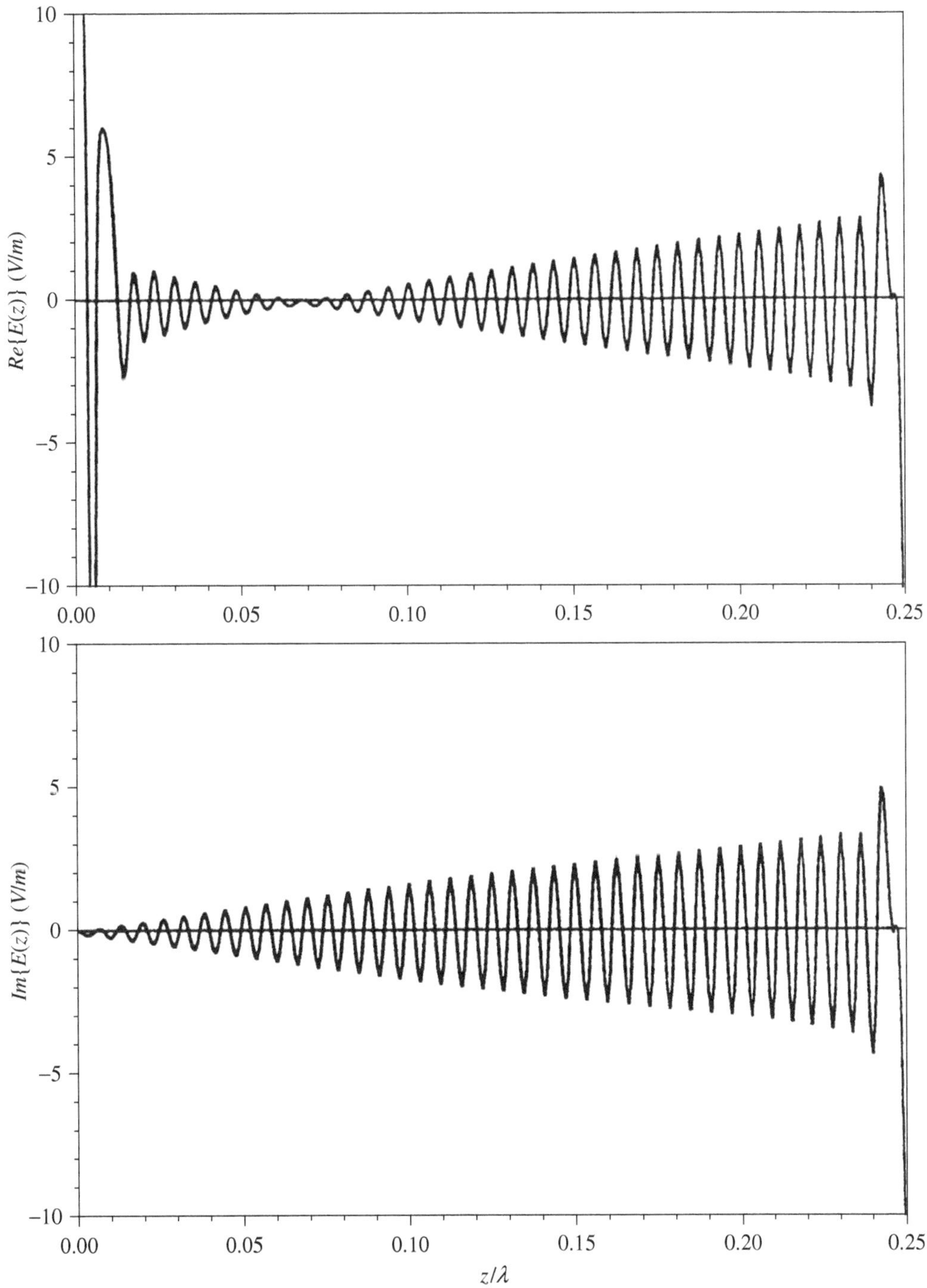


**FIGURE 4-8** ■ Value of the total field on the surface of the dipole computed by integrating the current obtained from a pulse-basis–point-matching solution ($2\ell = 0.5\lambda$, $a = 0.005\lambda$, and $N = 81$). A value of 5 V/m corresponds to a 3.1% error (Barkeshli and Volakis, 1988).

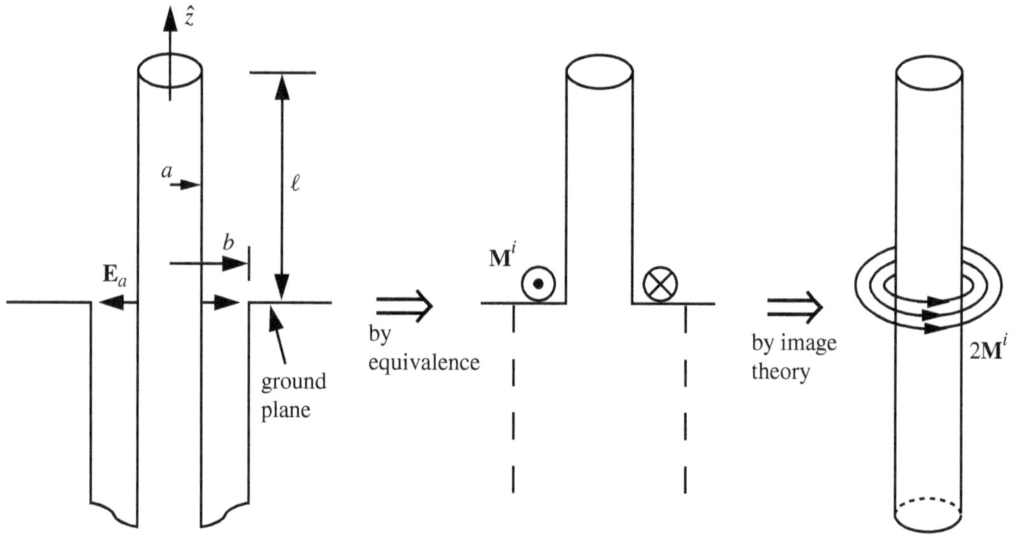


**FIGURE 4-9** Magnetic frill model for a coaxially fed monopole/dipole.

current density existing in the region between $\rho = a$ and $\rho = b$, as shown. The value of the outer radius $b$ is computed from a knowledge of the transmission lines characteristic impedance $Z_c$. When the wire antenna is fed by a coaxial cable the equivalent magnetic frill current can be computed in terms of the aperture fields in the usual manner, as shown in Figure 4-9.

Using the equivalence principle, the aperture is closed and replaced by the surface magnetic current

$$\mathbf{M}^i = \mathbf{E}^a \times \hat{z} \tag{4.41}$$

where

$$\mathbf{E}^a(\rho) = \hat{\rho}\frac{V_i}{2\rho \ln b/a} \qquad (V_i \text{ is a constant}) \tag{4.42}$$

as dictated by the lowest-order mode supported in the coaxial transmission line. The radiated field by $\mathbf{M}^i$ may now be evaluated after invoking image theory to double its strength and the length of the monopole to that of a dipole. From (2.63)

$$\mathbf{E}^i(\rho, z) = -\int_a^b \int_0^{2\pi} \left[\mathbf{M}^i(\rho') \times \hat{R}\right] \left(jk_o + \frac{1}{R}\right) \frac{e^{-jkR}}{4\pi R} \rho' \, d\phi' \, d\rho' \tag{4.43}$$

where

$$\mathbf{R} = |\mathbf{r} - \mathbf{r}'|$$

$$\mathbf{r} = \rho\hat{\rho} + z\hat{z}, \qquad \mathbf{r} = \rho'\hat{\rho} = \rho'(\hat{x}\cos\phi' + \hat{y}\sin\phi').$$

For $\rho = 0$ (observation at the center of the wire)

$$R = \sqrt{(\rho')^2 + z^2}$$

$$\hat{R} = \frac{\mathbf{r} - \mathbf{r}'}{R} = \left[-\hat{\rho}'\rho' + z\hat{z}\right]/R$$

$$\mathbf{M}^i(\rho') = -\hat{\phi}'\frac{V_i}{\rho' \ln b/a} = (\hat{x}\sin\phi' - \hat{y}\cos\phi')\frac{V_i}{\rho' \ln b/a}$$

$$\mathbf{M}^i(\rho') \times \hat{R} = \left[-\hat{z}\rho' - \hat{\rho}'z\right]\frac{V_i}{R\rho' \ln b/a}.$$

Substituting these expressions into (4.43) yields

$$\begin{aligned} E_z^i(\rho = 0, z) &= \frac{V_i}{\ln b/a} \int_a^b \int_o^{2\pi} \left( jk_o + \frac{1}{R} \right) \frac{e^{-jk_o R}}{4\pi R^2} \rho' \, d\phi' \, d\rho' \\ &= \frac{V_i}{2 \ln b/a} \int_a^b \left( jk_o + \frac{1}{R} \right) \frac{e^{-jk_o R}}{R^2} \rho' \, d\rho'. \end{aligned} \tag{4.44}$$

Noting that

$$-\frac{d}{d\rho'} \left\{ \frac{e^{-jk_o R}}{R} \right\} = \left( jk_o + \frac{1}{R} \right) \frac{e^{-jk_o R}}{R} \frac{\rho'}{R}$$

(4.44) may be written as

$$E_z^i(\rho = 0, z) = -\frac{V_i}{2 \ln b/a} \int_a^b \frac{d}{d\rho'} \left\{ \frac{e^{-jk_o R}}{R} \right\} d\rho'$$

to yield

$$E_z^i(\rho = 0, z) = +\frac{V_i}{2 \ln b/a} \left[ \frac{e^{-jk_o \sqrt{z^2+b^2}}}{\sqrt{z^2 + b^2}} - \frac{e^{-jk_o \sqrt{z^2+a^2}}}{\sqrt{z^2 + a^2}} \right]. \tag{4.45}$$

For simplicity, we may assume

$$E_z^i(\rho = a, z) \approx E_z^i(\rho = 0, z)$$

to be substituted into (4.30) and (4.33) for the solution of the wire currents. Alternatively, we may pursue a direct evaluation of (4.43) to find (Tsai, 1972; Thiele, 1973)

$$\begin{aligned} E_z^i(\rho = a, z) = +V_i \frac{k_o(b^2 - a^2)}{8 \ln b/a} \frac{e^{-jk_o R_o}}{R_o^2} & \left\{ 2 \left[ \frac{1}{k_o R_o} + j \left( 1 - \frac{b^2 + a^2}{2R_o^2} \right) \right] \right. \\ & + \frac{a^2}{R_o} \left[ \left( \frac{1}{k_o R_o} + j - j \frac{b^2 + a^2}{2R_o^2} \right) \left( -jk_o - \frac{2}{R_o} \right) \right. \\ & \left. \left. + \left( -\frac{1}{k_o R_o^2} + j \frac{b^2 + a^2}{R_o^3} \right) \right] \right\} \end{aligned} \tag{4.46}$$

where now

$$R_o = \sqrt{z^2 + a^2}$$

with

$$-\ell < z < \ell.$$

Figure 4-10 illustrates the current on a $1\lambda_o$ dipole computed with a magnetic frill model excitation. The current near the feed is now smoother than that obtained with the delta gap model. However, more samples are required to reach convergence, and this is owed to the near singular behavior of the excitation field in (4.45).

### 4.4.3 Plane Wave Incidence

If the cylindrical wire is considered as a scatterer, then $\mathbf{E}^i$ represents the incident field. The simplest form of this is a plane wave given by

$$\mathbf{E}^i = e^{jk_o \mathbf{r}_s \cdot \hat{r}_i} \tag{4.47}$$

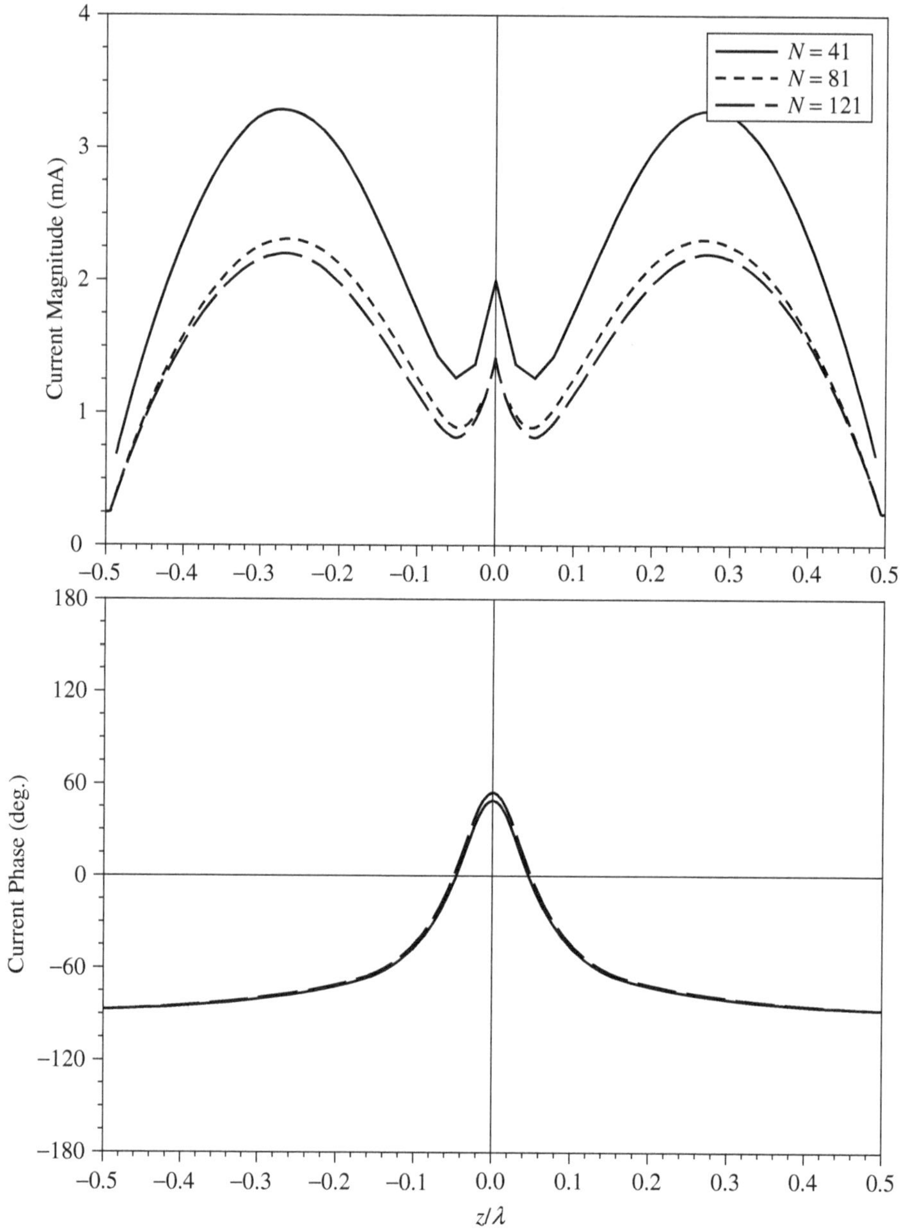


**FIGURE 4-10** ■ Computed current on a center-fed $\lambda_o$ dipole of radius $a = .005\lambda_o$ via the pulse-basis–point-matching technique as a function of the sampling density. The source/excitation is the magnetic frill equivalent current as given in (4.45) (Barkeshli and Volakis, 1988).

where

$$\mathbf{r}_s = a\hat{\rho} + z\hat{z} = a\hat{x} + z\hat{z}|_{\phi=0}$$

if measured on the surface of the wire and

$$\hat{r}_i = \hat{x}\cos\phi_i \sin\theta_i + \hat{y}\sin\phi_i \sin\theta_i + \hat{z}\cos\theta_i \tag{4.48}$$

with $(\theta_i, \phi_i)$ being the usual spherical angles denoting the direction of incidence.

Figures 4-11 and 4-12 show the current on the $\lambda_o/2$ and $1\lambda_o$ wire dipoles due to a plane wave incidence excitation. In contrast to the current on a center-fed dipole, this current has no discontinuous derivatives throughout the length of the dipole. Its form on the $\lambda_o/2$ dipole is clearly sinusoidal with its amplitude depending on the incidence angle. The same holds for longer wires with the exception of having a more complex lobing structure, which can be explained by invoking the traveling wave theory.

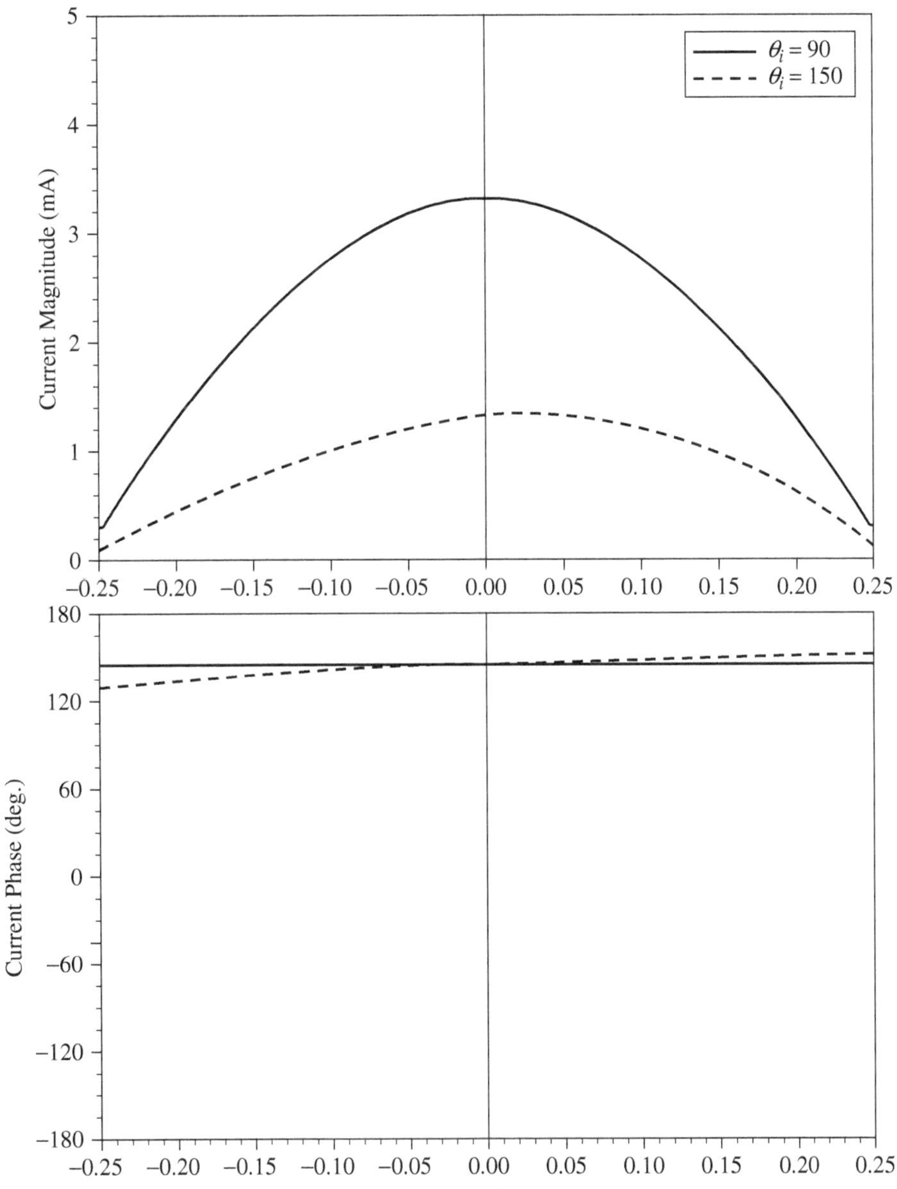


**FIGURE 4-11** ■ Current on a $\lambda_o/2$ wire of radius $a = .005\lambda$ generated by an incident plane wave at $\theta_i = 90°$ and $\theta_i = 150°$ as computed by the pulse-basis–point-matching technique ($a = 0.005\lambda_o$, $N = 101$) (Barkeshli and Volakis, 1988).

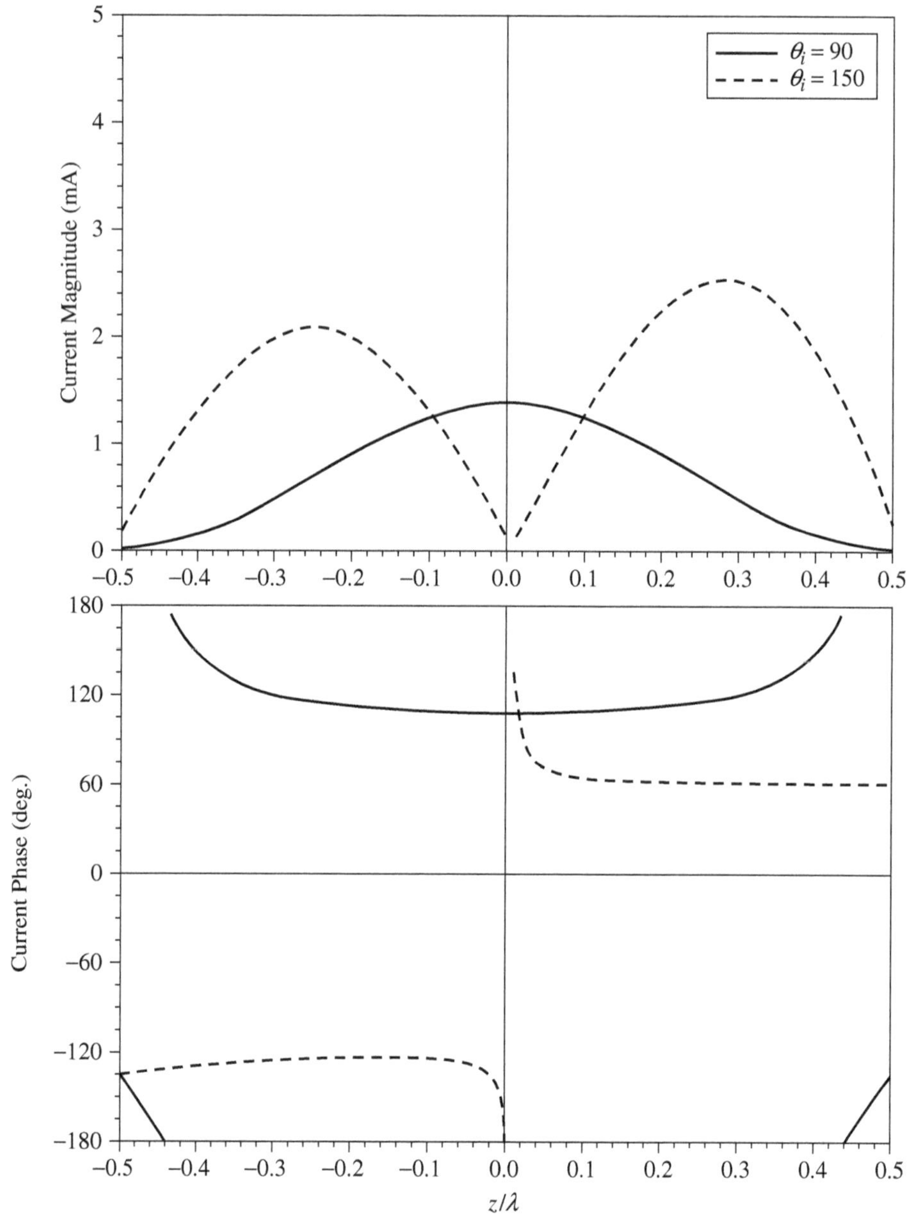


**FIGURE 4-12** Current on a $\lambda_o$ wire of radius $a = .005\lambda$ generated by an incident plane wave at $\theta_i = 90°$ and $\theta_i = 150°$ as computed by the pulse-basis–point-matching technique ($a = 0.005\lambda_o$, $N = 151$) (Barkeshli and Volakis, 1988).

## 4.5 | CALCULATION OF THE FAR-ZONE FIELD AND ANTENNA CHARACTERISTICS

Upon solution of the system (4.33), one can proceed with the evaluation of the radiation or the scattering patterns if **E** is given by (2.48). Geometry for computing the linear antenna's radiated field is illustrated in Figure 4-13. From (2.77)

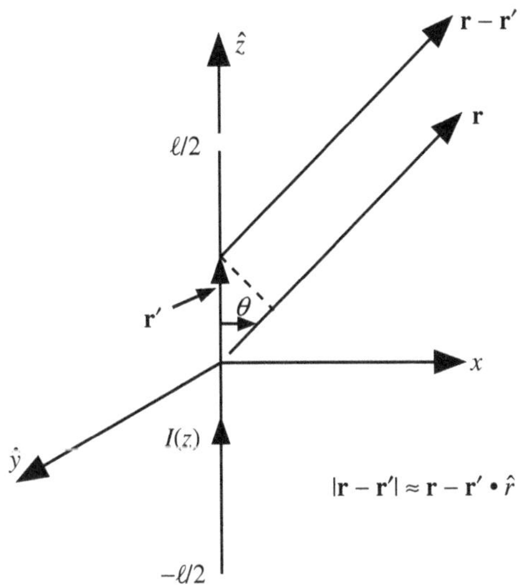


**FIGURE 4-13** Geometry for computing the linear antenna's radiated field.

we have

$$E_\theta^r \approx jk_o Z_o \sin\theta \int_{-\ell}^{\ell} I(z') \frac{e^{-jk_o|\mathbf{r}-\mathbf{r}'|}}{4\pi|\mathbf{r}-\mathbf{r}'|} dz' \\ \approx jk_o Z_o \sin\theta \frac{e^{-jk_o r}}{4\pi r} \int_{-\ell}^{\ell} I(z') e^{jk_o z' \cos\theta} dz'. \tag{4.49}$$

Using (4.14), we get

$$E_\theta^r(r,\theta) \approx jk_o Z_o \sin\theta \frac{e^{-jk_o r}}{4\pi r} \sum_{n=0}^{N-1} I_n \int_{z_n - \frac{\Delta z}{2}}^{z_n + \frac{\Delta z}{2}} e^{jk_o z' \cos\theta} dz' \tag{4.50}$$

and upon performing the trivial integration we have

$$\mathbf{E}_\theta^r(r,\theta) = jk_o Z_o \frac{e^{-jk_o r}}{4\pi r} \Delta z \sin\theta \frac{\sin\left(\frac{k_o \Delta z \cos\theta}{2}\right)}{\left(\frac{k_o \Delta z \cos\theta}{2}\right)} \sum_{n=0}^{N-1} I_n e^{jk_o z_n \cos\theta} \\ = jk_o Z_o \frac{e^{-jk_o r}}{4\pi r} \Delta z \sin\theta \,\mathrm{sinc}\left(k_o \frac{\Delta z}{2} \cos\theta\right) \sum_{n=0}^{N-1} I_n e^{jk_o z_n \cos\theta}. \tag{4.51}$$

The radiation intensity of the antenna is given by

$$U(\theta,\phi) = U(\theta) = \frac{r^2}{2Z_o} \left|E_\theta^r(\theta)\right|^2 \\ = \frac{Z_o}{2} \left(\frac{k_o \Delta z \sin\theta}{4\pi}\right)^2 \mathrm{sinc}^2\left(k_o \frac{\Delta z}{2} \cos\theta\right) \left(\sum_{n=1}^{N} I_n e^{jk_o z_n \cos\theta}\right)^2 \tag{4.52}$$

and the radiated power can be computed from

$$P_{\mathrm{rad}} = \int_0^{2\pi} \int_0^{\pi} U(\theta,\phi) \sin\theta \, d\theta \, d\phi = 2\pi \int_0^{\pi} U(\theta) \sin\theta \, d\theta \tag{4.53}$$

with the integral to be evaluated numerically.

Given the radiated power, the directivity is found from

$$D = \frac{4\pi U_{\max}}{P_{\text{rad}}} = \frac{4\pi U(\theta = \pi/2)}{P_{\text{rad}}}. \tag{4.54}$$

Finally, the gain of the antenna can be easily computed from

$$G(\theta, \phi) = \left.\frac{4\pi U(\theta, \phi)}{P_{\text{in}}}\right|_{\theta=\pi/2} = \frac{Z_o}{2P_{\text{in}}} \frac{(k_o \Delta z)^2}{4\pi} \left(\sum_{n=1}^{N} I_n\right)^2 \tag{4.55}$$

where $P_{\text{in}}$ denotes the input power from the generator.

A parameter of crucial importance in controlling the efficiency of the antenna is its input impedance. This is given by

$$Z_{\text{in}} = \frac{V_i}{I_{\text{in}}} = \frac{E_z^i \Delta z}{I_{\text{in}}} \tag{4.56}$$

where $I_{\text{in}} = I_{\frac{N-1}{2}+1}$ is the value of the current element at the terminal under the obvious assumption that $N$ is odd. However, the accuracy of (4.56) depends on the accuracy of $I_{\text{in}}$ as computed from the solution of the system (4.33). Since we discretized the actual current distribution, $I_{\text{in}}$ is only an approximation to the input current and is often not of acceptable accuracy unless $N$ is very large. To avoid this difficulty when employing (4.56) we may instead use a stationary expression for the input impedance based on power relations. From Poynting's theorem we have

$$\begin{aligned} \frac{1}{2} I_{in} I_{in}^* Z_{in} &= \frac{1}{2} \oiint \mathbf{E}^r \times \mathbf{H}^{r*} \cdot d\mathbf{s} = -\int_{-\ell}^{\ell} \int_0^{2\pi} E_z^r H_\phi^{r*} a \, d\phi \, dz' \\ &= -\frac{1}{2} \int_{-\ell}^{\ell} E_z^r(a, z') I^*(z') \, dz'. \end{aligned} \tag{4.57}$$

Since $\mathbf{J} = \hat{\rho} \times \hat{\phi} H_\phi$ giving $H_\phi = J_z = \dfrac{I(z)}{2\pi a}$. Thus,

$$Z_{in} = +\frac{1}{|I_{\text{in}}|^2} \int_{-\ell}^{\ell} E_z^i(a, z') I^*(z') \, dz \tag{4.58}$$

where we have set $E_z^r(a, z) = -E_z^i(a, z)$ as required by the boundary condition on the surface of the wire. Substituting (4.24) into (4.58) we obtain

$$Z_{in} = \frac{\Delta z}{|I_{\text{in}}|^2} \sum_{n=0}^{N-1} E_z^i(a, z_n) I_n^*. \tag{4.59}$$

It is observed, that for a delta gap excitation (see (4.39)) (4.59) again reduces to (4.56). Note also that

$$Re(Z_{in}) = R_{in} = \frac{2P_{\text{rad}}}{|I_{\text{in}}|^2}. \tag{4.60}$$

The input impedance as computed from (4.56) is shown in Figure 4-14 as a function of the wire's length and for various wire radii. As can be concluded from the presented current computations using the pulse-basis–point-matching solutions, up to 120 segments per wavelength may be required to accurately sample the current

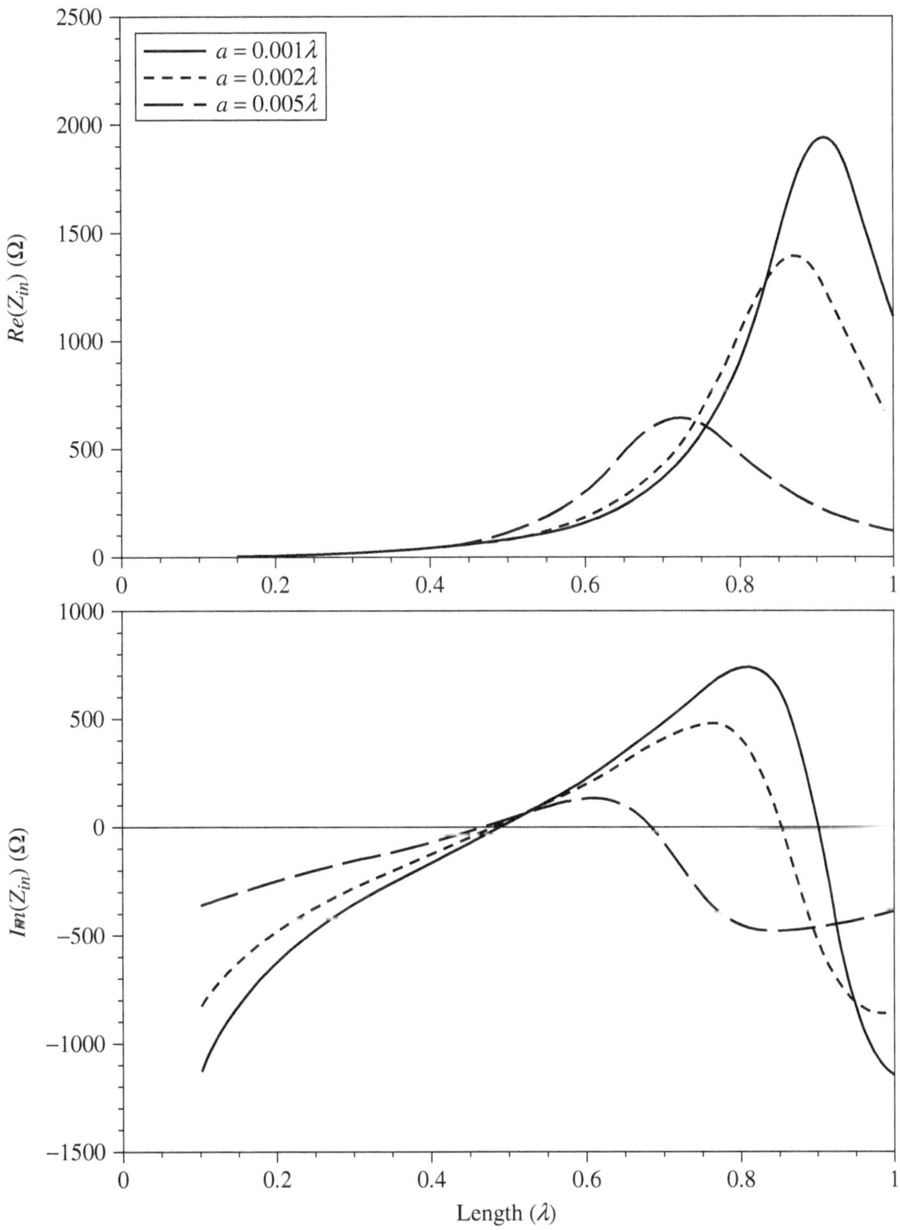


**FIGURE 4-14** ■ Input impedance of a dipole as a function of its length $2\ell$ for three different wire radii: (a) resistive, and (b) reactance. The dipole is resonant when the reactance is zero (Barkeshli and Volakis, 1988).

near the feed. When $Im(Z_{in}) = 0$, the dipole is said to be at resonance, and its first resonance occurs when $2\ell$ is just less than $0.5\lambda_o$, depending on the value of its radius. The bandwidth of an antenna is related to the slope of $Z_{in}$ as a function of frequency, and it is seen from Figure 4-14 that thicker dipoles have a larger bandwidth (the thin-wire solution should not be used for $a > \frac{\lambda_o}{50}$ (Lin and Richmond, 1975)).

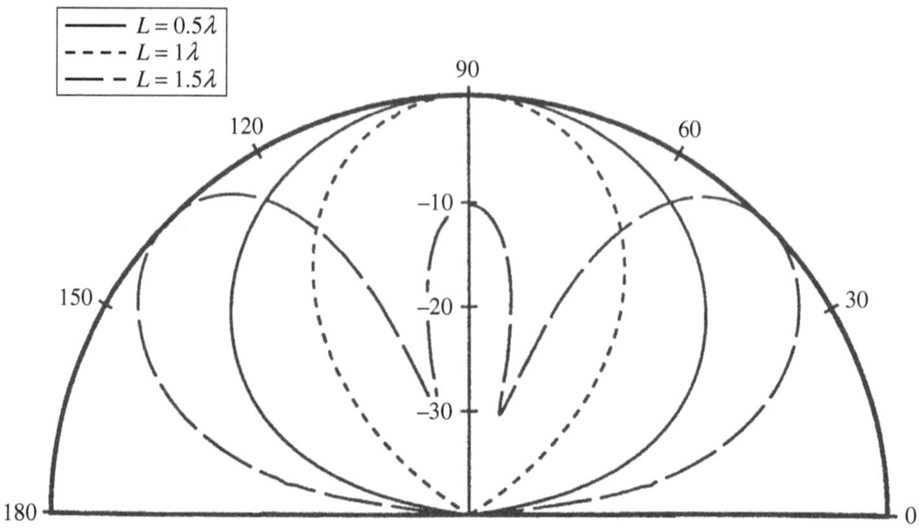


**FIGURE 4-15** Radiation power patterns for the $\lambda_o/2$, $1\lambda_o$, and $1.5\lambda_o$ dipoles computed from the numerical solution of the dipole currents ($a = 0.005\lambda_o$) (Barkeshli and Volakis, 1988).

Radiation patterns for the $\lambda_o/2$, $\lambda_o$ and $3\lambda_o/2$ dipoles are given in Figure 4-15. We note, however, that these are identical to those predicted with the assumed sinusoidal distribution that follows from the transmission line model.

When the excitation is a plane wave, we are generally interested in the echo area or radar cross section (RCS) of the wire structure. The RCS is measured in units of length squared and is given by

$$\sigma = \lim_{r\to\infty} 4\pi r^2 \frac{|\mathbf{E}^r|^2}{|\mathbf{E}^i|^2}. \tag{4.61}$$

If the wire length is measured in wavelengths, then the units of $\sigma$ are square wavelengths ($\lambda_o^2$), and if the wire length is measured in meters then $\sigma$ will be given in $m^2$. The RCS of the $\lambda/2$, $\lambda$, and $3\lambda$ long wires are shown in Figures 4-16 and 4-17. The effect of wire thickness on the wire's RCS is predicted in Figure 4-18, where the value of broadside ($\theta_i = 90^\circ$) $\sigma$ is plotted as a function of the wire's length for three different radii. This is a characteristic curve for the wire scatterer and displays its resonant behavior when $2\ell \approx (n+1)\lambda_o/2$ for odd $n$. Basically, the RCS of the wire at those lengths reaches a local peak with each successive peak becoming larger as $2\ell$ is increased. This property of the wire has been explored in many practical situations, and we remark that the location of the RCS peaks should correspond to the wire length at which $Im(Z_{in}) \approx 0$.

We observe that the echo area pattern of the longer wire as given in Figure 4-17 has a very strong lobe near $\theta = \pi$ (near grazing). This is a lobe characteristic to all thin-wire scattering patterns and is always the one closest to $\theta = 0^\circ$ or $\theta = \pi$. It is often referred to as the traveling wave lobe, and to explain its presence let us assume that the wire is infinite in length. The incident plane wave (4.47) will then generate a current of the form $I_1 e^{jk_o z\cos\theta_i}$, where $I_1$ is a complex constant proportional to the incident wave's strength and can be computed analytically. This

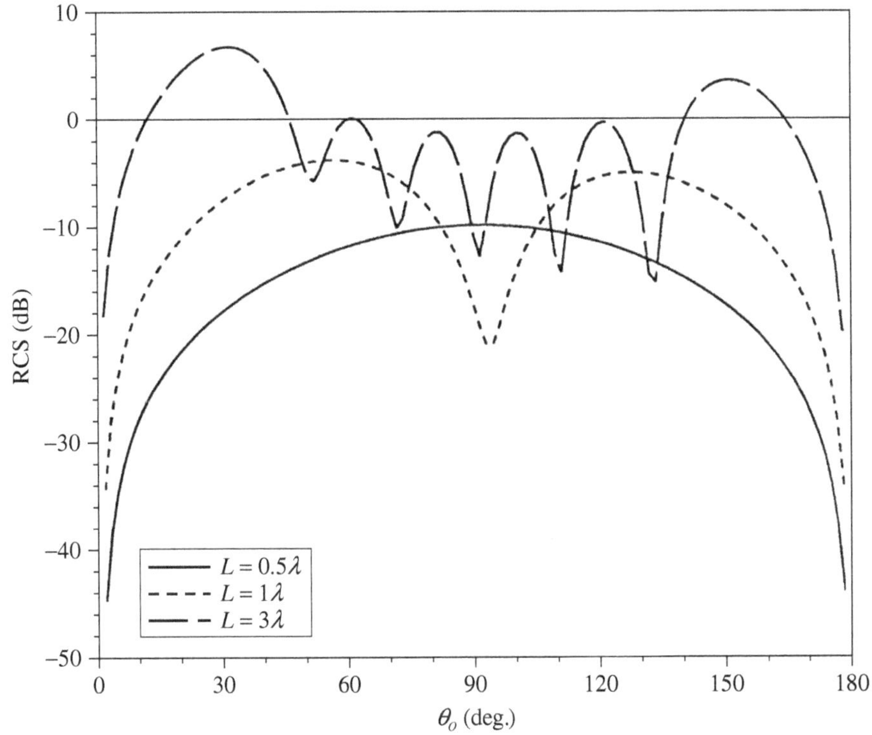


**FIGURE 4-16** Bistatic radar cross section of three straight wires of length $2\ell = \lambda_o/2$, $\lambda_o$, and $3\lambda_o$. The wires have a radius of $a = 0.005\lambda_o$, and the incident plane wave is illuminating the wire at an angle of $\theta_i = 150^\circ$ (PWS basis solution) (Barkeshli and Volakis, 1988).

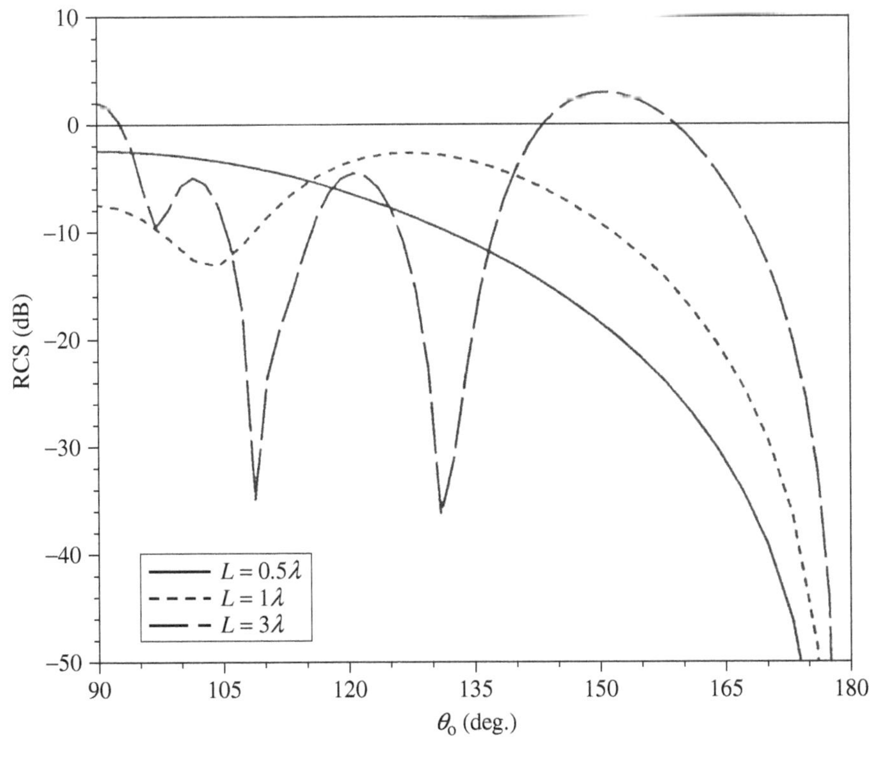


**FIGURE 4-17** Backscatter radar cross section ($\theta = \theta_i$) for the three straight wires whose bistatic patterns are given in Figure 4-16 (Barkeshli and Volakis, 1988).

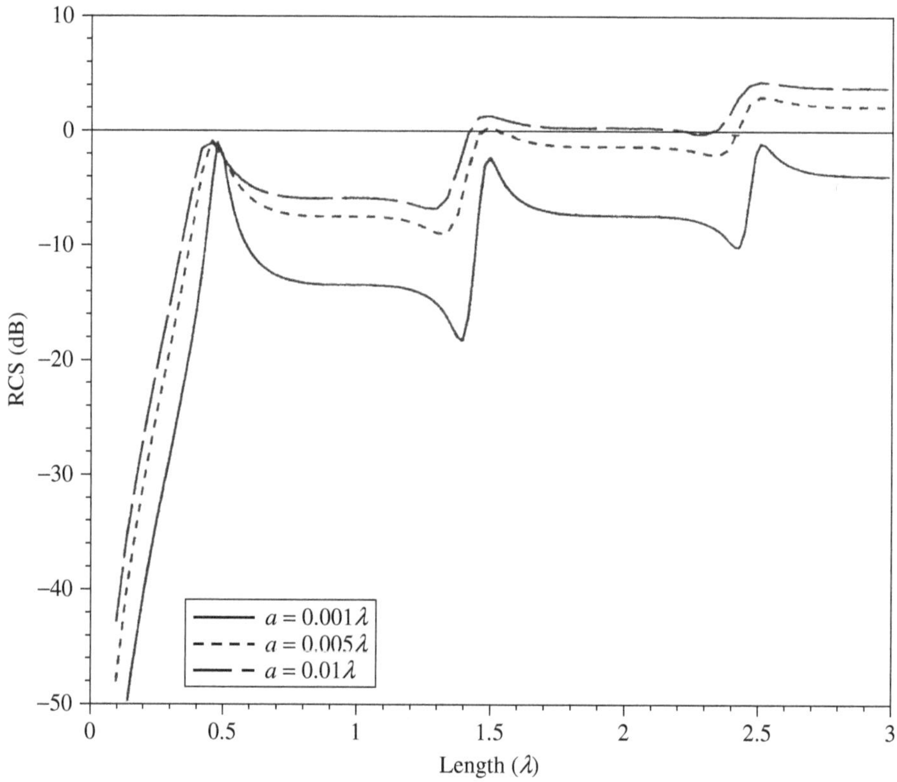


**FIGURE 4-18** Backscatter RCS of three thin wires as a function of length ($2\ell$) illuminated by a plane wave at normal incidence ($\theta_i = 90°$). (PWS point-matching solution with $\Delta Z = 0.01\lambda_o$) (Barkeshli and Volakis, 1988).

is, of course, a traveling current (whose propagation constant matches that of the incident wave), and if the wire is of finite extent, when it reaches the wire ends it generates additional reflected currents of the form $I_2 e^{jk_o z}$ and $I_3 e^{-jk_o z}$, where $I_2$ and $I_3$ are again complex constants. Thus the current on the wire due to a plane wave excitation can be approximately represented as

$$I(z) = \sum_{n=1}^{3} I_n e^{jk_n z} \tag{4.62}$$

with $k_1 = k_o \cos\theta_o$, $k_2 = +k_o$, and $k_3 = -k_o$. From this representation it is not difficult to observe from the radiation integral (4.49) that the scattered field would peak at $\theta = \pi - \theta_i$ and at $\theta = 0$ or $\pi$ if the coefficients $I_n$ were comparably weighted. However, this is not the case, and it turns out that the traveling wave lobe peak occurs when $I_{2,3}$ are at maximum.

The expansion (4.62) is, of course, a linear sum of three full-wave basis functions, similar to those given by (4.22) through (4.23). It was constructed on the basis of the physical phenomena that take place on the straight wire and is thus most efficient for computational purposes. However, as noted earlier, this expansion (which may be referred to as a *solution wave* expansion) is specific to the straight wire scatterer and cannot be employed for other wire shapes or arbitrary multiple wire structures.

## 4.6 | PIECEWISE SINUSOIDAL-BASIS–POINT-MATCHING SOLUTION

The piecewise sinusoidal basis expansion renders a continuous current distribution and is thus more representative of the actual solution. This usually translates in fewer subsections/zones to reach convergence.

Substituting (4.21) into (4.18) yields

$$I(z) = \sum_{n=0}^{N} I_n \frac{\sin k_o(\Delta z - |z - z_n|)}{\sin k_o \Delta z} \tag{4.63}$$

where $z_n = -\ell + (n+1)\Delta z$. When this is substituted into (4.13), we obtain (see Chapter 10 in Jordan and Balmain (1968))

$$E_z^i(\rho = a, z) = \frac{+j30}{\sin k_o \Delta z} \sum_{n=0}^{N-1} I_n \left[ \frac{e^{-jk_o R_{1n}}}{R_{1n}} - 2\cos(k_o \Delta z) \frac{e^{-jk_o R_{2n}}}{R_{2n}} + \frac{e^{-jk_o R_{3n}}}{R_{3n}} \right] \tag{4.64}$$

in which

$$R_{1n} = \sqrt{(z - z_{n-1})^2 + a^2}$$

$$R_{2n} = \sqrt{(z - z_n)^2 + a^2}$$

and

$$R_{3n} = \sqrt{(z - z_{n+1})^2 + a^2}.$$

The fact that the radiated field by a sinusoidal source can be evaluated in a closed form is the principal advantage of $S_n(z)$ over $T_n(z)$.

A point-matching solution of (4.64) follows the same procedure as discussed previously in connection with the pulse-basis expansion. Upon evaluation of the coefficients $I_n$, the radiation pattern is again given by (4.49). From (4.63) we obtain

$$E_\theta(r, \theta) \approx \frac{jk_o Z_o}{\sin\theta} \frac{e^{-jk_o r}}{4\pi r} \sum_{n=0}^{N-1} I_n \int_{z_n - \Delta z}^{z_n + \Delta z} \frac{\sin k_o(\Delta z - |z' - z_n|)}{\sin(k_o \Delta z)} e^{jk_o z' \cos\theta} dz' \tag{4.65}$$

and on carrying out the integration we find

$$E_\theta(r, \theta) \approx j60 \frac{e^{-jkr}}{r} \underbrace{\frac{\cos(k_o \Delta z \cos\theta) - \cos(k_o \Delta z)}{\sin\theta \sin(k_o \Delta z)}}_{\text{element pattern}} \sum_{n=0}^{N-1} I_n e^{jk_o z_n \cos\theta}. \tag{4.66}$$

The evaluation of other parameters such as radiated power, gain, directivity, and input impedance can be performed in a straightforward manner. Not surprisingly, the PWS representation can be shown to yield more accurate results for input impedance computations. This can be attested by examining the wire surface fields generated by the PWS point-matching solution. In contrast to the results in Figure 4-8, it is found that the surface field of this solution is now practically zero without even resorting to the more robust weighted residual method discussed next.

## 4.7 | METHOD OF WEIGHTED RESIDUALS/METHOD OF MOMENTS

The previously described point-matching technique for solving integral equations ensures that the boundary condition is satisfied only at the match points $z_m$. In general, however, the boundary condition is not necessarily satisfied elsewhere, unless the sampling or testing interval $\Delta z$ is extremely small, and this fact was illustrated in Figure 4-8. This is, of course, not cost-effective since the CPU time is proportional to $N^3$ as given by (4.37).

An alternative testing procedure is to satisfy the boundary condition on an average sense over the length of the segment from $z_n$ to $z_{n+1}$. To express this mathematically, let us first define the interproduct (see e.g., Harrington, 1968)

$$\langle R(z), W_m(z)\rangle = \int_{-\ell}^{\ell} R(z) W_m^*(z)\, dz \tag{4.67}$$

and we will hereafter refer to $R(z)$ as the residual and to $W_m(z)$ as the weighting/test basis functions. Setting

$$R(z) = E_z^i(\rho = a, z) + E_z^r(\rho = a, z) \tag{4.68}$$

choosing

$$W_m(z) = \begin{cases} 1 & z_m - \frac{\Delta z}{2} < z < z_m + \frac{\Delta z}{2} \\ 0 & \text{elsewhere} \end{cases} \tag{4.69}$$

and demanding that

$$\langle R(z), W_m(z)\rangle = 0 \tag{4.70}$$

leads to the integral equation

$$-\int_{z_m - \frac{\Delta z}{2}}^{z_m + \frac{\Delta z}{2}} E_z^i(\rho = a, z)\, dz = \int_{z_m - \frac{\Delta z}{2}}^{z_m + \frac{\Delta z}{2}} E_z^r(\rho = a, z)\, dz. \tag{4.71}$$

Upon substitution of the expression for $E_z^r$ as extracted from (4.30) yields,

$$-\int_{z_m - \frac{\Delta z}{2}}^{z_m + \frac{\Delta z}{2}} E_z^i(\rho = a, z) = -\frac{jZ_o}{2\lambda_o}\sum_{n=0}^{N-1} I_n \int_{z_m - \frac{\Delta z}{2}}^{z_m + \frac{\Delta z}{2}} [\Psi_n(z) + \Phi_n(z)]\, dz \tag{4.72}$$

from which we obtain the system

$$[Z_{mn}][I_n] = [V_m] \tag{4.73}$$

where now

$$V_m = -\int_{z_m-\frac{\Delta z}{2}}^{z_m+\frac{\Delta z}{2}} E_z^i(\rho = a, z)\, dz \tag{4.74}$$

$$Z_{mn} = \frac{-jZ_o}{2\lambda_o}\int_{z_m-\frac{\Delta z}{2}}^{z_m+\frac{\Delta z}{2}} [\Psi_n(z) + \Phi_n(z)]\, dz. \tag{4.75}$$

These can be evaluated numerically using, for example, Simpson's, midpoint, or Gaussian rules of integration (Abramowitz and Stegun, 1964). Clearly, (4.70) along with (4.68) and (4.69) demand that the boundary conditions be satisfied on an average sense over the wire subintervals. When the weighting functions are piecewise constant, each current value over the subinterval is given equal weighting in this averaging process. A variety of other choices for $W_m(z)$ have, though, been employed successfully in the past. When $W_m(z)$ are chosen to be the same as the current expansion basis function, the procedure for deriving the resulting system of equations is referred to as Galerkin's method (Jones, 1956; Kantorovich and Krylov, 1959). We also note that when

$$W_m(z) = \delta(z - z_m) \tag{4.76}$$

(4.70) reduces to the system (4.30) derived by the point-matching technique. The previous procedure for discretizing the integral equation is formally referred to as the *weighted residual method* but is most often called the *method of moments* (MoM)[1]. Also, the pulse-basis–point-matching procedure is more formally referred to as the *collocation method*.

The application of Galerkin's technique has been studied extensively and has been found quite robust for many applications. It is indeed pleasing to know that the method minimizes the residual in the least squares sense. It can be expected that the Galerkin's implementation leads to more robust and efficient computer codes. For example, in the case of a pulse-basis expansion, the Galerkin's solution converges to the correct current using 30% to 50% fewer segments as shown in Figure 4-19. The solution convergence improves even further if higher-order basis functions are used and next we describe the Galerkin's formulation and derive the resulting system of equations for the subsectional sinusoidal basis.

From (4.70), with

$$W_m(z) = \begin{cases} \frac{\sin(k_o(\Delta z - |z - z_m|))}{\sin k_o \Delta z} & z_{m-1} < z < z_{m+1} \\ 0 & \text{elsewhere} \end{cases} \tag{4.77}$$

---

[1]According to R. F. Harrington, the term "method of moments" was first used by Kantorovich and Akilov (1964).

and (see (4.64))

$$R(z) = E_z^i(\rho = a, z) - \frac{j30}{\sin(k_o \Delta z)} \times \sum_{n=0}^{N-1} I_n \left[ \frac{e^{-jk_o R_{1n}}}{R_{1n}} - 2\cos(k_o \Delta z) \frac{e^{-jk_o R_{2n}}}{R_{2n}} + \frac{e^{-jk_o R_{3n}}}{R_{3n}} \right] \tag{4.78}$$

we obtain the usual system (4.73) with

$$V_m = -\int_{z_m - \Delta z}^{z_m + \Delta z} E^i(\rho = a, z) \frac{\sin[k_o(\Delta z - |z - z_m|)]}{\sin(k_o \Delta z)} dz \tag{4.79}$$

$$Z_{mn} = -\frac{j30}{\sin k_o \Delta z} \int_{z_m - \Delta z}^{z_m + \Delta z} \frac{\sin[k_o(\Delta z - |z - z_m|]}{\sin(k_o \Delta z)} \times \left[ \frac{e^{-jk_o R_{1n}}}{R_{1n}} - 2\cos(k_o \Delta z) \frac{e^{-jk_o R_{2n}}}{R_{2n}} + \frac{e^{-jk_o R_{3n}}}{R_{3n}} \right] dz. \tag{4.80}$$

The impedance matrix elements may be easily evaluated numerically as given in (4.80) since the integrand is nonsingular. However, after some manipulation, the integral expression can be simplified and written in terms of the exponential integral that is tabulated (Abramowitz and Stegun, 1964, p. 228). A compact expression for $Z_{mn}$ is (King, 1957; Richmond and Geary, 1970)

$$Z_{mn} = \frac{+15}{\sin^2(k_o \Delta z)} \sum_{p=-2}^{2} \sum_{q=-1,1} A(p+3) e^{-jk_o q[|z_m - z_n| + p\Delta z]} E(k_o \beta_{pq}) \tag{4.81}$$

where

$$A(1) = A(5) = 1$$

$$A(2) = A(4) = -4\cos(k_o \Delta z)$$

$$A(3) = 2 + 4\cos^2(k_o \Delta z)$$

$$\beta_{pq} = \sqrt{a^2 + [|z_m - z_n| + p\Delta z]^2} - q[|z_m - z_n| + p\Delta z]$$

and $E(\alpha)$ is the exponential integral. It can be defined in terms of the cosine and sine integrals as

$$E(\alpha) = Ci(\alpha) - jSi(\alpha) \tag{4.82}$$

where

$$Ci(\alpha) = -\int_{\alpha}^{\infty} \frac{\cos x}{x} dx \tag{4.83}$$

and

$$Si(r) = \int_0^{\alpha} \frac{\sin x}{x} dx. \tag{4.84}$$

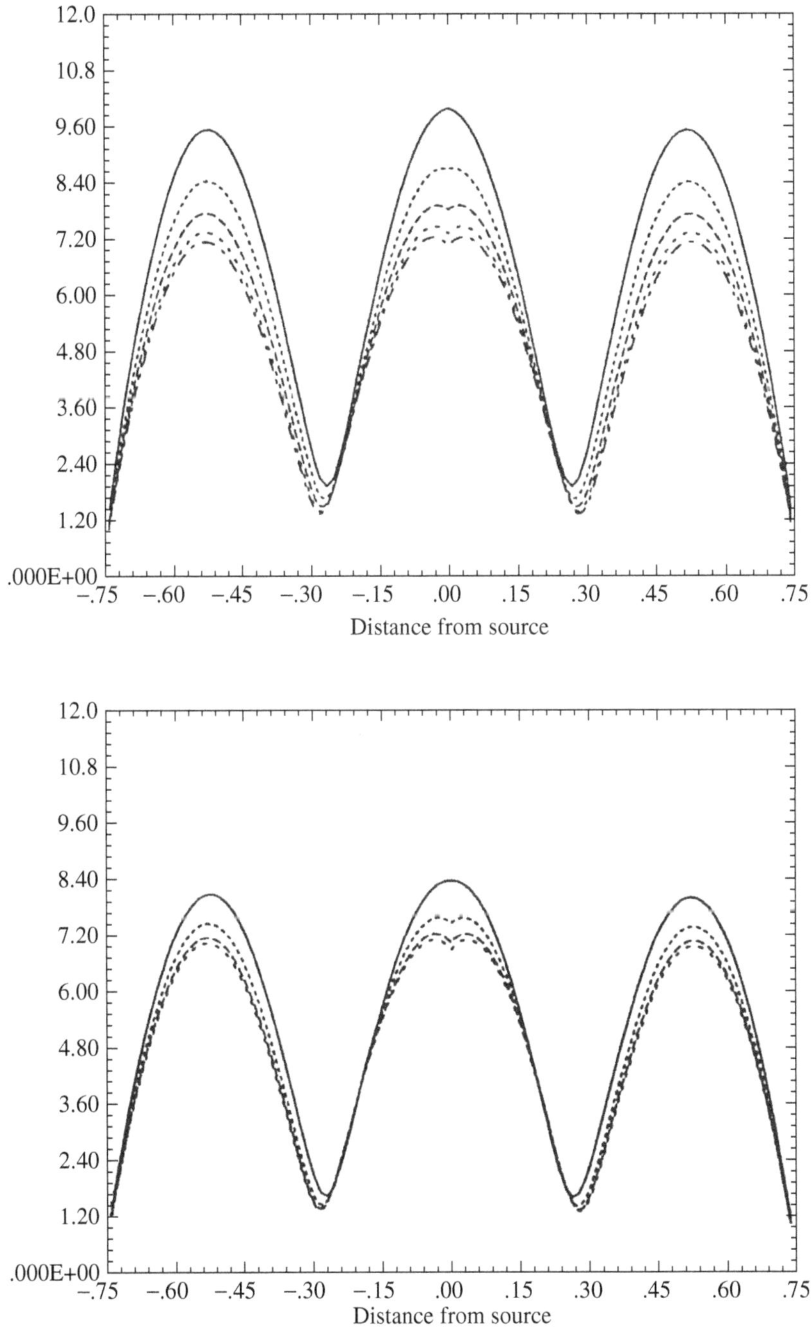


**FIGURE 4-19** Convergence of the current distribution on a center-fed dipole of radius $a = 0.005\lambda_o$ with a magnetic frill source: (a) pulse-basis–point-matching solution using $N = 81, 101, 181, 221$, and 261 points, and (b) Galerkin's pulse-basis solution using $N = 101, 141, 181$, and 221 points (Barkeshli and Volakis, 1988).

A FORTRAN subroutine (one page long) for the numerical evaluation of $E(\alpha)$ is given in (Press, Teukolsky, Vetterling, and Flannery, 1992). Figure 4-19 illustrates that, as can be expected, a smaller number of PWS is required to converge to the correct value of the wire current when compared with the pulse-basis implementation.

## 4.8 | METHOD OF MOMENTS FOR NONLINEAR WIRES

Typically, the antenna or scatterer will be composed of curved wire elements. Also, it is possible to model continuous metallic surfaces with a wire grid of sufficient density as shown in Figure 4-26. Acceptable grid densities are often 10 or more wires per linear wavelength on the surface.

To develop a method of moments formulation applicable to curved wires, let us consider the following curved wire geometry and assume for simplicity a constant wire thickness equal to $2a \gg \lambda$. As usual, we are interested in determining the wire surface currents or more specifically the equivalent line current through the center of the wire.

In proceeding with a numerical solution, it is first necessary to discretize the wire as shown in Figure 4-20. This amounts to generating a model of the curved wire composed of a set of linear segments. Denoting the unit vector along the direction of the $m$th element as $\hat{\ell}_m$, the boundary condition to be satisfied on its surface is

$$(\mathbf{E}^i + \mathbf{E}^r) \cdot \hat{\ell}_m = 0. \tag{4.85}$$

If the curved wire is divided into $N$ straight segments, then $\mathbf{E}^r$ can be expressed as

$$\mathbf{E}^r = \sum_{n=1}^{N} \mathbf{E}_n^r = \sum_{n=1}^{N} \hat{\ell}_n \mathbf{E}_{\ell n}^r \tag{4.86}$$

where $\mathbf{E}_n^r$ is the field radiated by each linear segment. When employing PWC basis to expand the current on each element, $\mathbf{E}_n^r$ can be found from (2.52) or (3.13) upon performing the necessary coordinate transformations. To see this, we begin with

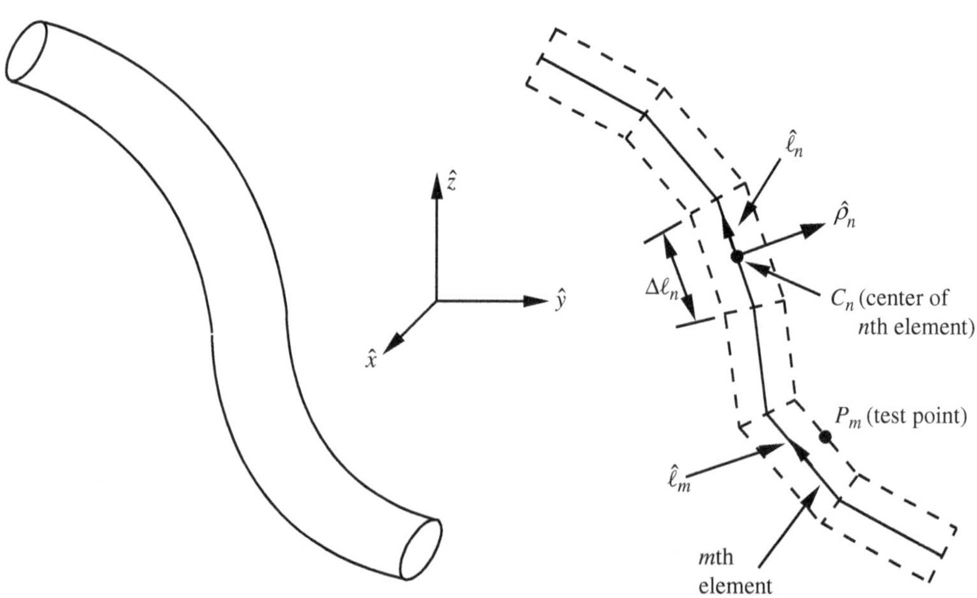


**FIGURE 4-20** ■ Segmentation of a curved wire for numerical modeling.

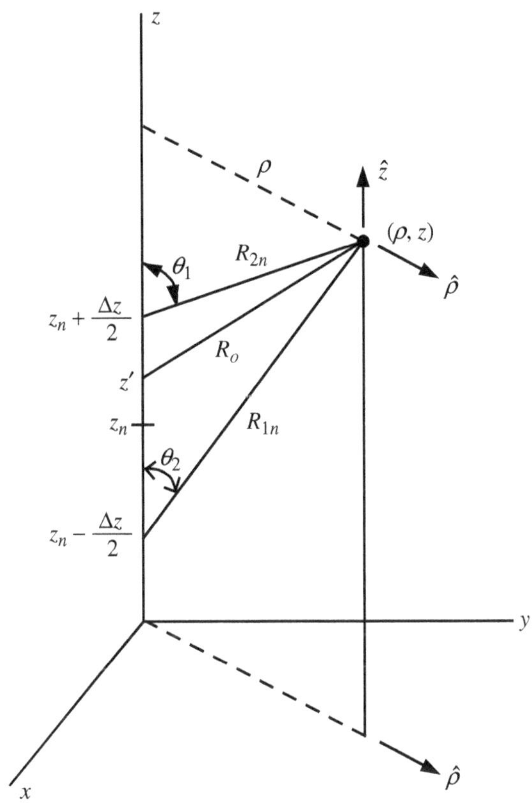


**FIGURE 4-21** Geometry of a monopole situated on the z-axis.

the derivation of the fields radiated by a straight segment ($n$th segment) carrying a constant current $I_n$ as shown in Figure 4-21. This element is often referred to as a monopulse, and we find from (3.13) that it radiates the field

$$\mathbf{E}_n^r = \hat{z}E_{nz}^r(\rho, z) + \hat{\rho}E_{np}^r(\rho, z) \tag{4.87}$$

where

$$E_{nz}^r = -\frac{jZ_o}{2\lambda_o}I_n\left[\Psi_n(\rho, z) + \Phi_n(\rho, z)\right] \tag{4.88}$$

and

$$E_{np}^r = j\frac{Z_o k_o}{2\lambda_o}I_n\rho\left[\frac{(1 + jk_oR_{1n})}{(k_oR_{1n})^3}e^{-jk_oR_{1n}} - \frac{(1 + jk_oR_{2n})}{(k_oR_{2n})^3}e^{-jk_oR_{2n}}\right]. \tag{4.89}$$

In these expressions,

$$R_{1n} = \sqrt{\left(z - z_n + \frac{\Delta z}{2}\right)^2 + \rho^2}, \quad R_{2n} = \sqrt{\left(z - z_n - \frac{\Delta z}{2}\right)^2 + \rho^2} \tag{4.90}$$

are simply generalizations of those given in (4.28), whereas $\Psi_n(\rho, z)$ and $\Phi_n(\rho, z)$ are the same functions as those defined in (4.31) and (4.32), respectively, except that

$$R_o = \sqrt{(z - z')^2 + \rho^2}. \tag{4.91}$$

FIGURE 4-22 Geometry of an arbitrary skewed monopulse.

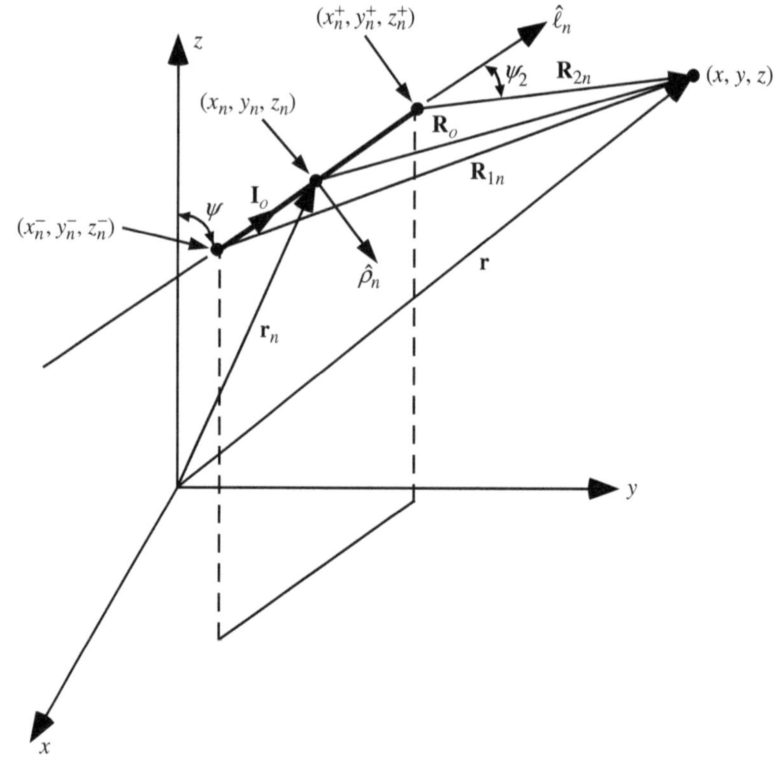


Also, $R_{1n}$ and $R_{2n}$ in (4.31) and (4.32) must be replaced by the more general expressions (4.90).

Given the fields due to the $z$-directed monopole, it is now a straightforward task to derive the corresponding field due to an arbitrarily oriented monopole. Specifically, for the monopole shown in Figure 4-22 we have that

$$\mathbf{E}_n^r(x, y, z) = \hat{\ell}_n E_{n\ell}^r(x, y, z) + \hat{\rho}_n E_{n\rho}^r(x, y, z) \tag{4.92}$$

where $E_{n\ell}$ and $E_{n\rho}$ are given by (4.88) and (4.89) with $R_o$, $R_{1n}$, and $R_{2n}$ redefined as

$$\begin{aligned} R_o &= \sqrt{(x - x_n)^2 + (y - y_n)^2 + (z - z_n)^2} \\ R_{\substack{1n \\ 2n}} &= \sqrt{\left(x - x_n^{\mp}\right)^2 + \left(y - y_n^{\mp}\right)^2 + \left(z - z_n^{\mp}\right)^2}. \end{aligned} \tag{4.93}$$

Also, in (4.89) $\rho$ must be replaced by $R_{2n} \cos \psi_2 = \mathbf{R}_{2n} \cdot \hat{\ell}_n$.

Having an expression for the field radiated by an arbitrarily oriented wire segment carrying a constant current, we may now proceed with the construction of the system of equations for a pulse-basis–point-matching solution. On enforcing (4.85) at $P_m$ as illustrated in Figure 4-20, for $m = 0, 1, 2, \ldots, N$, we get the usual system (4.33). The impedance matrix elements are now given by

$$Z_{mn} = +\left[(\hat{\ell}_n \cdot \hat{\ell}_m) E_{n\ell}(x_m, y_m, z_m) + (\hat{\rho}_n \cdot \hat{\ell}_m) E_{n\rho}^r(x_m, y_m, z_m)\right] \tag{4.94}$$

where $\mathbf{r} = \hat{x}\tilde{x}_m + \hat{y}\tilde{y}_m + \hat{z}\tilde{z}_m = (\hat{x}x_m + \hat{y}y_m + \hat{z}z_m) + a\hat{\rho}_m$ with $(x_m, y_m, z_m)$ as the center point of the $m$th segment, and $a$ its radius. Also,

$$V_m = -\hat{\ell}_m \cdot \mathbf{E}_n^r(\tilde{x}_m, \tilde{y}_m, \tilde{z}_m) \tag{4.95}$$

are the elements of the excitation vector. Note that $Z_{mn}$ in (4.94) reduces to (4.36) for $m = n$.

One of the first implementations of the pulse-basis–point-matching solution for wire structures was carried out by Richmond (1965, 1966). In Figures 4-23 and 4-24

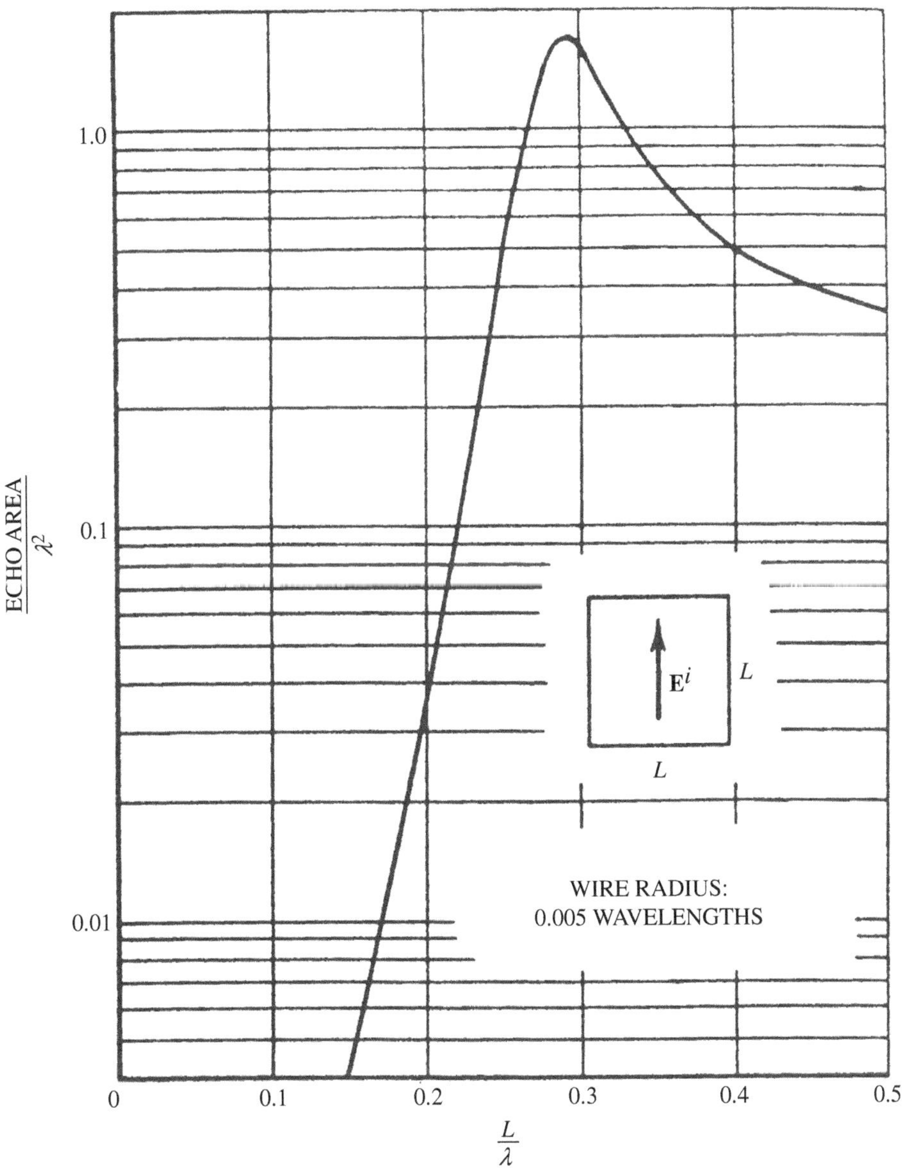


**FIGURE 4-23** ■ Calculated backscatter echo area of square wire loops at the broadside aspect. (After Richmond, 1966. With permission.)

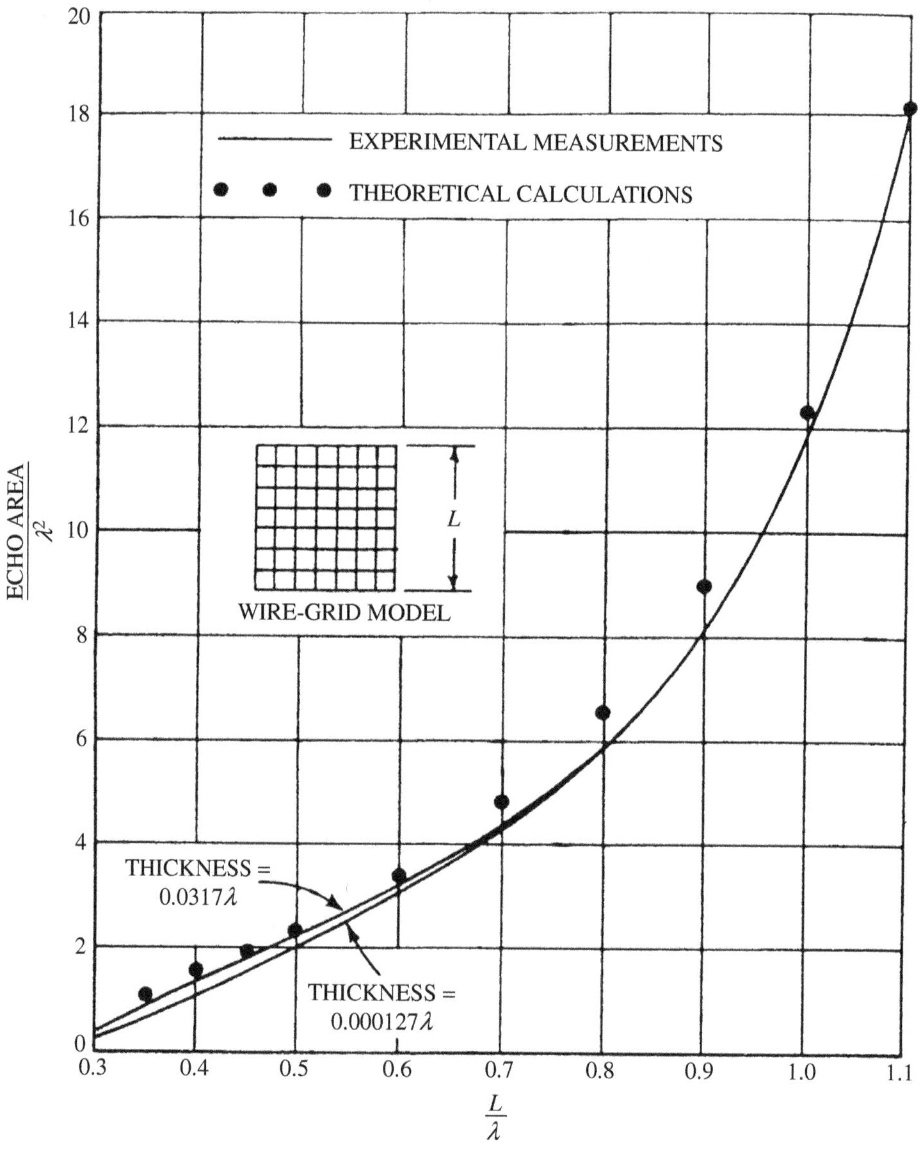


**FIGURE 4-24** ■ Echo area of perfectly conducting square plates at the broadside aspect (wire diameter $d = L/50$). (After Richmond, 1966. With permission.)

we present some RCS calculations from Richmond's (1966) paper, which should serve for validating implementations based on the given pulse-basis–point-matching solution. Figure 4-23 shows the broadside RCS (incident plane wave is impinging along a direction normal to the plane containing the wire loop) of a square loop as a function of $L/\lambda$, where $L^2$ denotes the area enclosed by the loop. Of importance is the observation that solid structures can be modeled by a grid of wires. For example, the square metallic plate shown in Figure 4-24 was modeled by a grid of eight vertical and eight horizontal wires with the wire radii set equal to $L/100$, where $L$ is the side length of the plate. For an accurate simulation of solid surfaces (see e.g. Figure 4-25)), Lin and Richmond (1975) recommended that the wire separation

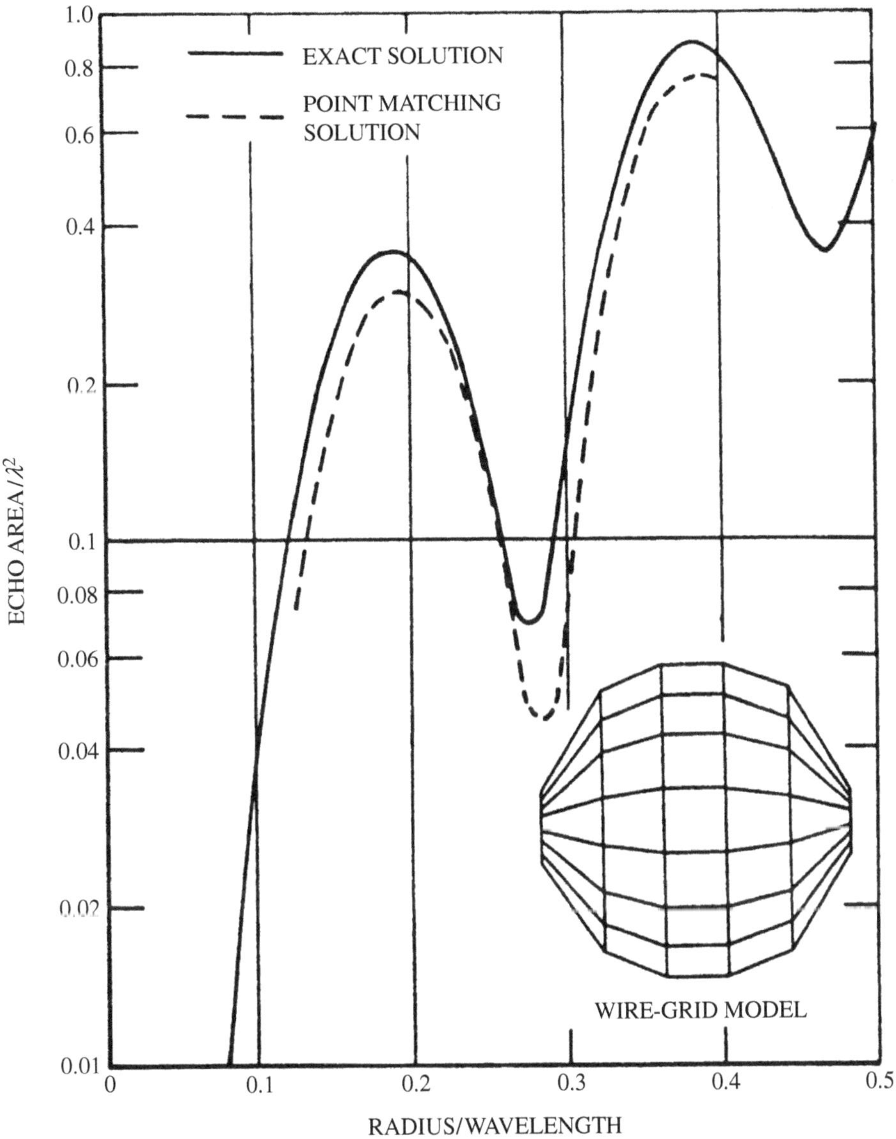


**FIGURE 4-25** ■ Backscatter echo area of perfectly conducting spheres. (The wire grid was generated by rotating a 20-sided regular polygon. The number of polygons was $\frac{70R}{\lambda}$, where $R$ is the sphere radius. The wire radius was set to $a = 0.005\lambda$.) (After Richmond, 1966. With permission.)

be no greater than $\lambda_o/4$ and that the wire radius be chosen to be about 1/25 of this separation distance. Others have used wire grid modeling (see Figure 4-26) to evaluate the radiation performance of reflector antennas (Poggio and Miller, 1973), or of wire antennas on aircraft (Diaz, 1970) or on some solid metallic structure (Thiele, 1973; Trueman, Mishra, Kubina, and Larose, 1993).

To construct the method of moments equations for a nonlinear wire using the Galerkin's procedure with a PWS basis function, we must first obtain the radiated

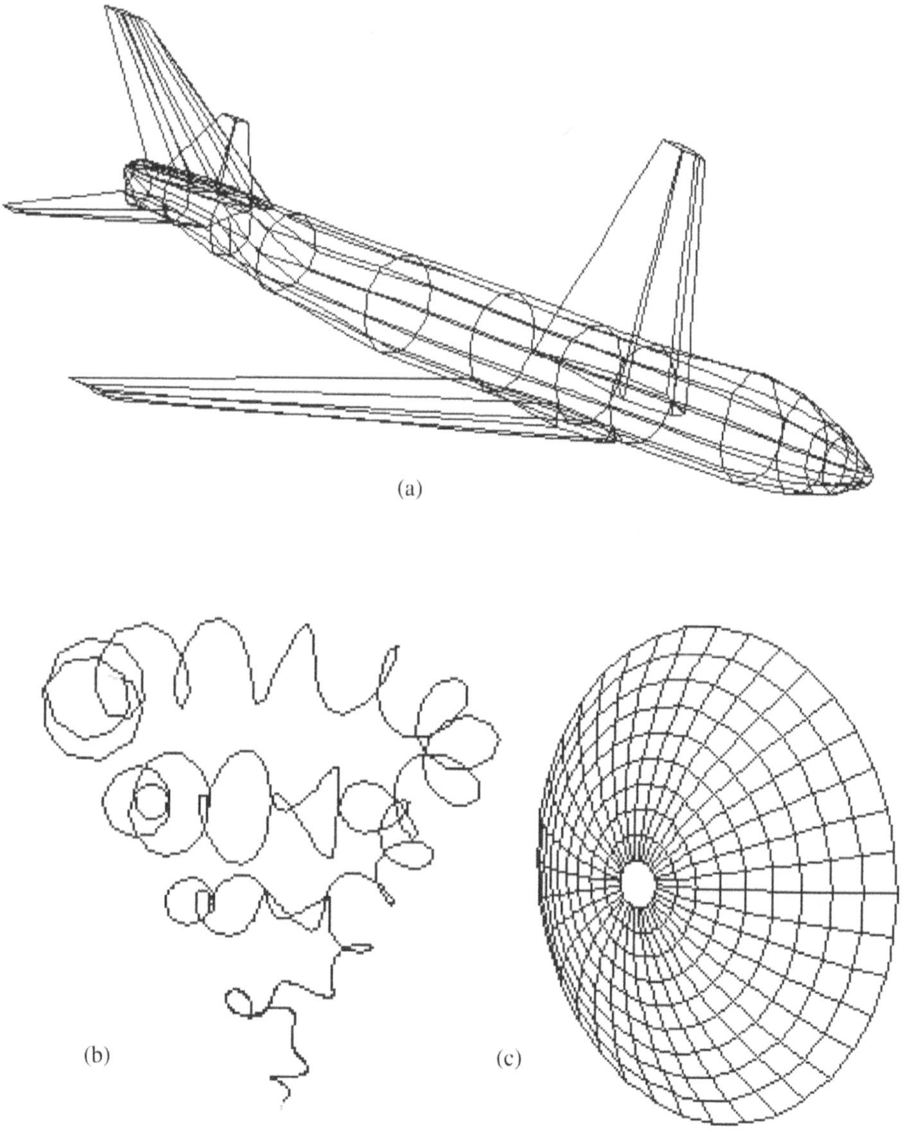


**FIGURE 4-26** ■ Illustration of wire grid models of solid surfaces: (a) wire model of a Boeing 747, (b) coiled conical helical antenna, and (c) wire grid of a parabolic reflector.

fields due to an arbitrarily oriented dipole. Referring to Figure 4-27 and generalizing the $z$-directed wire radiated fields, we obtain that

$$\mathbf{E}_n^r(x, y, z) = \hat{\ell}_n E_{n\ell}^r(x, y, z) + \hat{\rho}_n E_{n\rho}^r(x, y, z) \tag{4.96}$$

where

$$E_{n\ell}^r(x, y, z) = -\frac{j30}{\sin k_o \Delta\ell} I_n \left[ \frac{e^{-jk_o R_{1n}}}{R_{1n}} - 2\cos(k_o \Delta\ell) \frac{e^{-jk_o R_{2n}}}{R_{2n}} + \frac{e^{-jk_o R_{3n}}}{R_{3n}} \right] \tag{4.97}$$

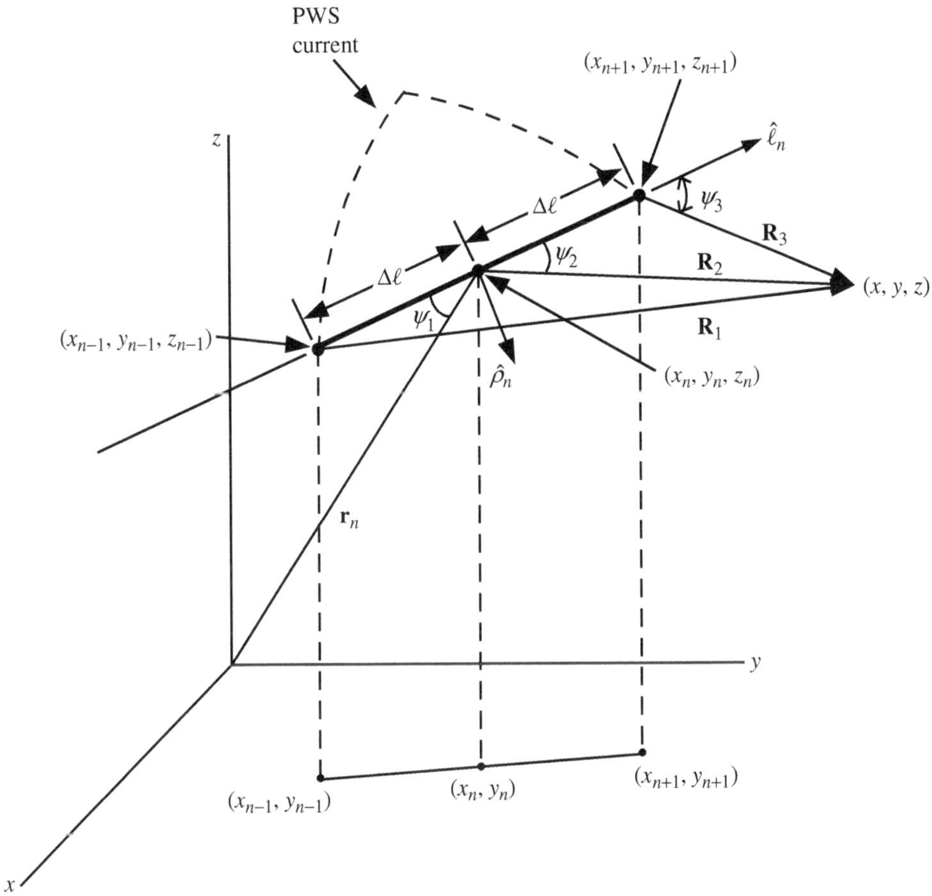


**FIGURE 4-27** Geometry of an arbitrarily oriented dipole.

and

$$E^r_{n\rho}(x, y, z) = \frac{j30}{R_{2n} \sin \psi_{2n} \sin(k_o \Delta\ell)} I_n \left[\cos \psi_{1n} e^{-jk_o R_{1n}} - 2\cos(k_o \Delta\ell) \cos \psi_{2n} e^{-jk_o R_{2n}} + \cos \psi_{3n} e^{-jk_o R_{3n}}\right]. \tag{4.98}$$

In these, $I_n$ denotes the value of the current at $(x_n, y_n, z_n)$,

$$\begin{aligned} R_{1n} &= \sqrt{(x - x_{n-1})^2 + (y - y_{n-1})^2 + (z - z_{n-1})^2} \\ R_{2n} &= \sqrt{(x - x_n)^2 + (y - y_n)^2 + (z - z_n)^2} \\ R_{3n} &= \sqrt{(x - x_{n+1})^2 + (y - y_{n+1})^2 + (z - z_{n+1})^2} \end{aligned} \tag{4.99}$$

have the same geometrical definitions as those given earlier in Section 4.6, and $\cos \psi_{in} = (\hat{\ell}_n \cdot \mathbf{R}_{in})/R_{in}$, for $i$ = 1, 2, 3, as illustrated in Figure 4-27. We are now ready to proceed with the application of Galerkin's method to construct the system of $N$ equations. The equations are constructed from ($m = 0, 1, 2, \ldots, N-1$)

$$\sum_{n=0}^{N-1} I_n \int_{C_m} \mathbf{E}_n^r(x, y, z) \cdot \hat{\ell}_m W(\ell)\, d\ell = -\int_{C_m} \mathbf{E}^i \cdot \hat{\ell}_m W(\ell)\, d\ell \tag{4.100}$$

where the weighting function is now given by

$$W(\ell) = \frac{\sin[k_o(\Delta\ell - |\ell|)]}{\sin(k_o \Delta\ell)} \qquad |\ell| < \Delta\ell \tag{4.101}$$

and the contour $C_m$ specifies integration over the $m$th straight segment of the curved wires. When (4.100) is put in the usual matrix form (4.33), the associated impedance matrix elements are therefore given by

$$Z_{mn} = \int_{C_m} \mathbf{E}_n^r(x, y, z) \cdot \hat{\ell}_m W(\ell)\, d\ell \tag{4.102}$$

and, to perform the implied integration, it is necessary to replace $(x, y, z)$ with their appropriate parametric representations. Since (4.100) is enforced on the wire surface, it follows that the appropriate parametric representations are

$$\begin{aligned} x &= x_m + a(\hat{\rho}_m \cdot \hat{x}) + \ell(\hat{\ell}_m \cdot \hat{x}) \\ y &= y_m + a(\hat{\rho}_m \cdot \hat{y}) + \ell(\hat{\ell}_m \cdot \hat{y}) \\ \hat{z} &= z_m + a(\hat{\rho}_m \cdot \hat{z}) + \ell(\hat{\ell}_m \cdot \hat{z}) \end{aligned} \tag{4.103}$$

and the integration is then carried out from $-\Delta\ell$ to $\Delta\ell$. As expected, when $m = n$, $Z_{mn}$ in (4.102) reduces to (4.80).

A general purpose computer program based on the PWS Galerkin's method has been written by Richmond (1974). This program was used (Lin and Richmond, 1975) to construct wire grid models and aircraft for RCS analysis (see Figure 4-26). As can be realized, because of the importance of wire antennas and their earlier utility for modeling solid structures, the interested reader will find that the literature is very rich on different formulations for wire analyses (see e.g., Miller and Deadrick, 1975, for a survey). Also, general purpose computer programs are readily available. Perhaps the most widely used program is the numerical electromagnetics code (NEC) developed by Burke and Poggio (1980) at Lawrence Livermore Laboratories. The basis functions employed in this program are (Yeh and Mei, 1967)

$$f_n(z) = A_n + B_n \sin[k_o(z - z_n)] + C_n \cos[k_o(z - z_n)] \tag{4.104}$$

for $|z - z_n| < \frac{\Delta z}{2}$, where $z_n$ denotes the center of the wire segment, and $A_n$, $B_n$, $C_n$ are constants to be determined. Two of these constants are eliminated by simply enforcing current continuity at the two ends of the $n$th segment or via application of Kirchhoff's current law (charge conservation) at wire junctions (the other constant is determined by solving the system resulting from point matching). This is necessary, since the previously given basis functions do not guarantee continuity across the wire junctions as is the case with the overlapping PWS and piecewise linear expansions. If a wire segment has one of its ends "free" (i.e., not connected to any other segment), the NEC code enforces the condition

$$|I(z)|_{\substack{at\\end}} = \mp\frac{1}{k_o}\frac{J_1(k_o a)}{J_o(k_o a)}\left.\frac{dI(z)}{dz}\right|_{\substack{at\\end}} \tag{4.105}$$

where $J_{0,1}(k_o a)$ denote the Bessel functions of order 0 and 1, and the + sign is selected if the current is flowing toward the segment's termination. This condition accounts for current leakage onto the end of the finite thickness wire.

## 4.9 | WIRES OF FINITE CONDUCTIVITY

When the wire (or a portion of it) is of finite conductivity, the boundary condition to be satisfied is

$$\mathbf{J}_v = \sigma \mathbf{E}^{tot} \tag{4.106}$$

where $\mathbf{E}^{tot}$ is the total field within the wire, $\mathbf{J}_v$ is the volume current in $A/m^2$, and $\sigma$ denotes the wire conductivity. For $a \ll \ell$ we can again replace $\mathbf{J}_v$ by an equivalent filamentary current at the center of the wire given by

$$I = \pi a^2 J_v. \tag{4.107}$$

Incorporating this into (4.106) yields the condition

$$\mathbf{E}^{tot} = \hat{\ell} R_w I(\ell) \quad \text{or} \quad E^i_\ell + E^r_\ell = R_w I(\ell) \tag{4.108}$$

where

$$R_w = \frac{1}{\pi a^2 \sigma} \tag{4.109}$$

can be referred to as the resistivity of the wire.

The boundary condition (4.108) must now replace the one given in (4.1). This amounts to a modification of the impedance elements from $Z_{mn}$ to $Z'_{mn}$, where

$$Z'_{mn} = \begin{cases} Z_{mn} & n \neq m \\ -R_w(\ell_m) \int_{\Delta \ell_m} f_m(\ell) W_m(\ell)\, d\ell + Z_{mm} & n = m \end{cases} \tag{4.110}$$

where $R_w(\ell_m)$ denotes the resistivity of the wire at the $m$th element, and $\mathbf{E}^r_n$ is the field radiated by the $n$th element. The wire current is expanded in the usual manner to yield the system

$$[Z'_{mn}]\{I_n\} = \{V_m\} \tag{4.111}$$

for a solution of the element amplitude coefficients $I_n$.

Often the wire antenna or scatterer with distributed loads is characterized with a surface impedance $Z_s$. The boundary condition satisfied on the surface of the wire then is

$$\mathbf{E}^{tot} = Z_s \mathbf{J}_s \tag{4.112}$$

where $\mathbf{J}_s$ denotes the surface current. Since

$$\mathbf{J}_s = \hat{\ell}\frac{I(\ell)}{2\pi a} \tag{4.113}$$

(4.112) may be rewritten as

$$\mathbf{E}^{tot} = \hat{\ell}\frac{Z_s}{2\pi a}I(\ell). \tag{4.114}$$

This is similar to (4.108), and thus a solution for the wire currents is found by setting the impedance elements equal to

$$Z'_{mn} = -\frac{Z_s}{2\pi a}\int_{\Delta\ell_m} f_n(\ell)W_m(\ell)\,d\ell + Z_{mn} \tag{4.115}$$

where $Z_{mn}$ are the corresponding elements for the perfectly conducting wire. The integral in (4.115) is over only the common domain of the weighting and expansion functions. Thus, if $W_m(\ell)$ and $f_n(\ell)$ are among those in (4.19) through (4.21), $Z'_{mn}$ will be equal to $Z_{mn}$ except for the matrix elements $Z_{m(m\pm1)}$ and $Z_{(m\pm1)m}$.

## 4.10 | CONSTRUCTION OF INTEGRAL EQUATIONS VIA THE REACTION/RECIPROCITY THEOREM

The integral equations derived earlier via the application of the method of moments procedure can also be derived by invoking the reaction or reciprocity theorem discussed in Chapter 1. The reaction theorem is a mathematical relationship between two sets of sources and their generated fields. Assuming $(\mathbf{J}, \mathbf{M})$ generate the fields $(\mathbf{E}, \mathbf{H})$ and that $(\mathbf{J}_t, \mathbf{M}_t)$ generate the fields $(\mathbf{E}_t, \mathbf{H}_t)$, the reaction theorem states

$$\iiint (\mathbf{E}\cdot\mathbf{J}_t - \mathbf{H}\cdot\mathbf{M}_t)\,dv = \iiint (\mathbf{E}_t\cdot\mathbf{J} - \mathbf{H}_t\cdot\mathbf{M})\,dv. \tag{4.116}$$

Let us now set

$$(\mathbf{E}, \mathbf{H}) = (\mathbf{E}^r + \mathbf{E}^i, \mathbf{H}^r + \mathbf{H}^i)$$

where $(\mathbf{E}^r, \mathbf{H}^r)$ are the fields radiated by the wire current $J_s$ in the presence of the incident field $(\mathbf{E}^i, \mathbf{H}^i)$ having their source at infinity. Expression (4.116) then becomes

$$\iiint (\mathbf{E}^r\cdot\mathbf{J}_t - \mathbf{H}^r\cdot\mathbf{M}_t)\,dv + \iiint (\mathbf{E}^i\cdot\mathbf{J}_t - \mathbf{H}^i\cdot\mathbf{M}_t)\,dv$$
$$= \oiint (\mathbf{E}_t\cdot\mathbf{J}_s - \mathbf{H}_t\cdot\mathbf{M}_s)\,ds. \tag{4.117}$$

We now choose $\mathbf{M}_t = 0$ and set

$$\mathbf{J}_t = \hat{\ell}_m W_m(\ell) = \mathbf{J}_m(\ell) \tag{4.118}$$

concentrated at the center of the $n$th element of the perfectly conductive wire, where $W_m(\ell)$ is usually chosen to be equal (Galerkin's method) to the equivalent line current of the wire's $m$th element. The field generated by this source is now zero (essentially $\mathbf{J}_t$ radiates inside a closed hollow conductor) and (4.117) further reduces to

$$\oiint_{\substack{\text{wire}\\ \text{surface}}} \mathbf{E}^r \cdot \mathbf{J}_m \, ds = -\oiint_{\substack{\text{wire}\\ \text{surface}}} \mathbf{E}^i \cdot \mathbf{J}_m \, ds \tag{4.119}$$

when $\mathbf{J}_m$ is replaced by its equivalent line current at the center of the wire we have

$$\iint \mathbf{E}^r \cdot \hat{\ell}_m W_m(\ell) \, d\ell = -\iint \mathbf{E}^i \cdot \hat{\ell}_m W_m(\ell) \, d\ell \tag{4.120}$$

which is the same as (4.71) derived by the method of weighted residuals.

## 4.11 | ITERATIVE SOLUTION METHODS: THE CONJUGATE GRADIENT METHOD

Instead of inverting the impedance matrix $[Z_{mn}]$ for a solution of the system (4.33) or (4.73), one could employ an iterative solution scheme. Among the numerous iterative solution schemes for such a system, the conjugate gradient (CG) method is most attractive because it guarantees convergence in a maximum of $N$ iterations for an $N$-dimensional system (ignoring round-off errors). The CG method is a nonlinear, semidirect iterative scheme and was introduced by Hesteness and Steifel (1952) independently more than 40 years ago. Beginning with a random initial guess of the solution (including the zero) vector $\{I_n\}$, convergence is accomplished via a systematic orthogonalization of the solution vector with respect to the residual vector defined as the difference between the left- and right-hand sides of the system. The residual vector is computed at the end of each iteration and is used to find the next correction to the solution vector. The correction vectors are chosen to be orthogonal to the residual vectors, which are linearly independent. This is an essential condition for guaranteeing the convergence of the algorithm since at the $N$th iteration the solution vector would have been constructed by $N$ independent vectors (conjugate directions) which form a basis set of the $N$-dimensional space. Moreover, the algorithm will first proceed with corrections that will most greatly impact the minimization of the next residual vector. Consequently, convergence to within a reasonable degree of accuracy can be achieved after only a few (normally less than $N/3$) iterations.

The CG algorithm for the pertinent system proceeds as follows (Volakis *et al.*, 1998; Wilton and Wheeler, 1991):

Initialize the residual vector and conjugate direction:

$$\{r_1\} = \{V\} - [Z]\{I^1\}$$

$$\beta_o = \frac{1}{\|[Z]^a\{r_1\}\|^2}$$

$$p_1 = \beta_o[Z]^a\{r_1\}$$

For $k = 1, \ldots, n$ do

$$\alpha_k = \frac{1}{|[Z]\{p_k\}|^2}$$

$$\{I^{k+1}\} = \{I^k\} + \alpha_k\{p_k\}$$

$$\{r_{k+1}\} = \{r_k\} - \alpha_k[Z]\{p_k\}$$

$$\beta_k = \frac{1}{|[Z]^a\{r_{k+1}\}|^2}$$

$$\{p_{k+1}\} = \{p_k\} + \beta_k[Z]^a\{r_{k+1}\}$$

terminate loop when normalized residual error

$$\frac{|r_{k+1}|}{|[Z]^a\{V\}|} < \text{tolerance}$$

or when $k = N$. In this algorithm, the columns or vectors $\{I^k\}$ represent the current expansion coefficients after the $(k-1)$th iteration, $\{r_k\}$ are the residual vectors, and $\{p_k\}$ are the conjugate directions previously discussed. Also, $[Z]^a$ denotes the adjoint of the impedance matrix, which is equal to the complex conjugate transpose of $[Z_{mn}]$, and

$$|I^k|^2 = \sum_{n-1}^{N} I_n^k (I_n^k)^* \tag{4.121}$$

is the square norm of the vector $\{I^k\}$. Typical values for the tolerance range from .01 to $10^{-4}$.

Excluding initialization, the previous CG algorithm requires $2N^2 + 5N + 2$ multiplications and divisions (i.e., operations) per iteration. Thus, the CPU time required to reach convergence is of order $N^2 N_I$, if $N_I$ is the number of iterations required to satisfy the tolerance condition. Thus, if $N_I$ is not a small faction of $N$, the required CPU time to solve the system will again be of order $N^3$. However, the major advantage of the CG method is realized when the $[Z]$ matrix is Toeplitz, as is the case for the straight wire. In this case, the fast Fourier transform (FFT) can be combined with the CG method to reduce the storage requirements and to speed up the solution. To see how this is accomplished let us first return to the original integral equation (4.11). By invoking the one-dimensional Fourier transform pair defined in (2.171) and the convolution theorem we can rewrite (4.11) as

$$\{E_z^i(\rho = a, z)\} = \frac{jZ_o}{k_o}\mathcal{F}^{-1}\{\widetilde{I}(k_z)(k_o^2 - k_z^2)\widetilde{G}_{wr}(k_z)\} \tag{4.122}$$

where

$$\widetilde{I}(k_z) = \mathcal{F}\{I(z)\} = \int_{-\infty}^{\infty} I(z)P_{2\ell}(z)e^{-jk_z z}\,dz = \int_{-\ell}^{\ell} I(z)e^{-jk_z z}\,dz \tag{4.123}$$

and $\widetilde{G}_{wr}(k_z)$ is correspondingly the Fourier transform of $G_{wr}(z)$ defined in (4.10). It is given by

$$\widetilde{G}_{wr}(k_z) = \int_{-\infty}^{\infty} \frac{e^{-jk_o\sqrt{z^2+a^2}}}{4\pi\sqrt{z^2+a^2}}\,dz = \frac{1}{2\pi}K_o\left(a\sqrt{k_z^2-k_o^2}\right) \tag{4.124}$$

where $K_o$ is the modified Bessel function of the second kind, and from (2.170) $\sqrt{k_o^2-k_z^2} = -j\sqrt{k_z^2-k_o^2}$. If we were to use the integral equation (4.7) involving the unreduced wire kernel, then $\widetilde{G}_{wr}(k_z)$ needs to be replaced by the transform of $G_{wu}(z)$ given by

$$\widetilde{G}_{wu}(k_z) = \frac{1}{2\pi}I_o\left(a\sqrt{k_z^2-k_o^2}\right)K_o\left(a\sqrt{k_z^2-k_o^2}\right) \tag{4.125}$$

in which $I_o$ is the modified Bessel function of the first kind. Note that, although $K_o$ in (4.124) and (4.125) becomes infinite when $k_o = k_z$, the argument of the inverse transform operator in (4.122) vanishes at that point because of the multiplying factor $(k_o^2 - k_z^2)$.

The importance of the algebraic expression (4.122) is apparent because its right-hand side gives the value of the entire column (i.e., for all $z_m$) resulting from the operation $[Z]\{I\}$ without having to actually generate and store the square matrix $[Z]$ or perform the matrix multiplication. However, before we can make practical use of this advantage, it is necessary to rewrite (4.122) in terms of the discrete Fourier transform (DFT) to permit its implementation on a computer. As a first step toward this, we define the discrete transform pair

$$\hat{I}_p = \hat{I}(p\Delta k_z) = \sum_{n=0}^{N-1} I(n\Delta z)W^{np} \tag{4.126a}$$

$$I_n = I(n\Delta z) = \frac{1}{N}\sum_{p=0}^{N-1} \hat{I}^*(p\Delta k_z)(W^{np})^* \tag{4.126b}$$

where $W = e^{-2\pi/N}$, $\Delta k_z$ is the subinterval in the spectral domain given by $\Delta k_z = 1/N\Delta z$ and $\hat{I}_p = \hat{I}(p\Delta f)$ is the discrete transform of the sequence $I_n$.

Upon rewriting the expansion (4.18) as

$$I(z) = \sum_{n=0}^{N-1} I_n f_n(z - z_n) = f(z) * \sum_{n=0}^{N-1} I_n \delta(z - z_n) \tag{4.127}$$

where the * indicates convolution and taking its Fourier transform it is seen that the discrete form of $\widetilde{I}(k_z)$ is given by

$$\widetilde{I}(k_z) = \tilde{f}(k_z)\hat{I}_n \tag{4.128}$$

provided we set $k_z = n\Delta k_z$ for calculating $\tilde{f}(k_z)$. If pulse (PWC) basis functions are employed as expansion functions then

$$\tilde{f}(k_z) = \widetilde{P}_{\Delta z}(k_z) = \Delta z\frac{\sin(k_z\Delta z/2)}{k_z\Delta z/2} = \Delta z \,\mathrm{sinc}\,(k_z\Delta z/2) \tag{4.129}$$

and for PWS basis functions,

$$\tilde{f}(k_z) = \tilde{S}(k_z) = \frac{2k_o[\cos(k_z\Delta z) - \cos(k_o\Delta z)]}{(k_o^2 - k_z^2)\sin(k_o\Delta z)}. \tag{4.130}$$

We note that as $\Delta z \to 0$ then (4.129) and (4.130) reduce to a value equal to $\Delta z$, and thus

$$\tilde{I}(k_z) \approx \Delta z \hat{I}_n$$

which is an expected result implying the basis function $f(z) = \Delta z \delta(z)$.

The result (4.128) can now be substituted into (4.122), and the required DFT can be performed by setting $k_z = n\Delta k_z = n/N\Delta z$. However, before doing so, it is instructive to also replace the transform of the derivative $\frac{\partial}{\partial z}$ by its discrete counterpart. To obtain it, we observe that

$$\mathcal{F}\left\{\frac{\Delta G(z)}{\Delta z}\right\} = \mathcal{F}\left\{\frac{G\left(z+\frac{\Delta z}{2}\right) - G\left(z-\frac{\Delta z}{2}\right)}{\Delta z}\right\} = j\frac{2\sin\left(k_z\frac{\Delta z}{2}\right)}{\Delta z}\tilde{G}(k_z) \tag{4.131}$$

which is the transform of the discrete derivative based on the two-point formula. It simply implies that in (4.122) we must make the replacement

$$k_z \to \frac{2\sin\left(k_z\frac{\Delta z}{2}\right)}{\Delta z} = k_z \operatorname{sinc}\left(k_z\frac{\Delta z}{2}\right)$$

where $\operatorname{sinc}\left(k_z\frac{\Delta z}{2}\right) \to 1$ as $\Delta z \to 0$, an expected result. Finally, to obtain the discrete counterpart of $\tilde{G}_{wr}(k_z)$ the simplest approach is to replace it by the sample train $\tilde{G}_{wr}(p\Delta k_z)$ with $p = -(N-1), \ldots, 0, 1, \ldots, N$ since the DFT must be of finite length equal to $2N$ to satisfy the convolution requirements. However, unless $\tilde{G}_{wr}(k_z)$ is of negligible value for $|k_z| > N\Delta k_z = 1/\Delta z$, this truncation will cause aliasing errors that will affect the convergence of the CG algorithm and the accuracy of the solution. To avoid aliasing, one approach is to increase the size of the DFT. Generally, though, the DFT must be an integer power of 2 to take advantage of the available FFT algorithms. If we then set $M = 2^\gamma$, where $M > 2N - 1$, in accordance with the previous discussion we must have

$$\gamma = \text{Integer}\left\{\log_2(2N-1) + \rho\right\} \tag{4.132}$$

where $\rho \geq 1$ is an integer and determines the order of the FFT's dimension or pad. The minimum value of $\rho$ is unity and can be increased as required to reduce aliasing. In this case all arrays must be increased accordingly, and except for $\tilde{G}_{wr}(n\Delta k_z)$ the rest must be padded with zeros. In particular, the first $N$ points of the array $\{I_n\}$ are filled with the initial guess, and the array is then transformed using a length of $M = 2^\gamma$ to obtain $\{\hat{I}_n\}$. To recover the next $\{I_n\}$ after inverse transformation only the first $N$ points are again kept, and the others are zeroed.

In accordance with the previous discussion, the discrete counterpart of (4.79) is

$$\{V_m\} = -\frac{jZ_o}{k_o}DFT^{-1}\left\{\left[k_o^2 - n^2(\Delta k_z)^2\operatorname{sinc}^2\left(\frac{n\Delta k_z}{2}\Delta z\right)\right]\hat{I}_n\tilde{f}(n\Delta k_z)\tilde{G}_{wr}(n\Delta k_z)\right\} \tag{4.133}$$

The left- and right-hand sides of this equality should be interpreted as columns or vectors of length $N$ with the one from the right side obtained after truncating the padded array. More specifically $V_m = -E_z^i(\rho = a, z_m)$, whereas the right side should be equal to the column generated from the operation $[Z]\{I_n\}$. Thus, in the CG algorithm we should set

$$[Z]\{I_n\} = -\frac{jZ_o}{k_o} DFT^{-1}\left\{\left[k_o^2 - n^2(\Delta k_z)^2 \operatorname{sinc}^2\left(\frac{n\Delta k_z \Delta z}{2}\right)\right] \cdot \hat{I}_n \tilde{f}(n\Delta k_z)\widetilde{G}_{wr}(n\Delta k_z)\right\}. \tag{4.134}$$

This eliminates a need to generate the matrix, thus reducing the storage to $O(N)$ instead of $O(M^2)$ required with the direct solution. Moreover, because the DFT can be computed by performing only $M \log_2 M$ operations (provided $M = 2^\gamma$), the required CPU time per iteration is reduced to $4M(1 + \log_2 M)$. Thus, the total solution CPU time becomes $4MN_I(1+\log_2 M)$, and as before $N_I$ denotes the number of iterations to reach convergence. Actual CPU times for computing the current of a $1\lambda_o$ dipole are given in Figure 4-28. As seen, the CPU time in conjunction with the FFT (usually referred to as the CGFFT method) is nearly a linear function of

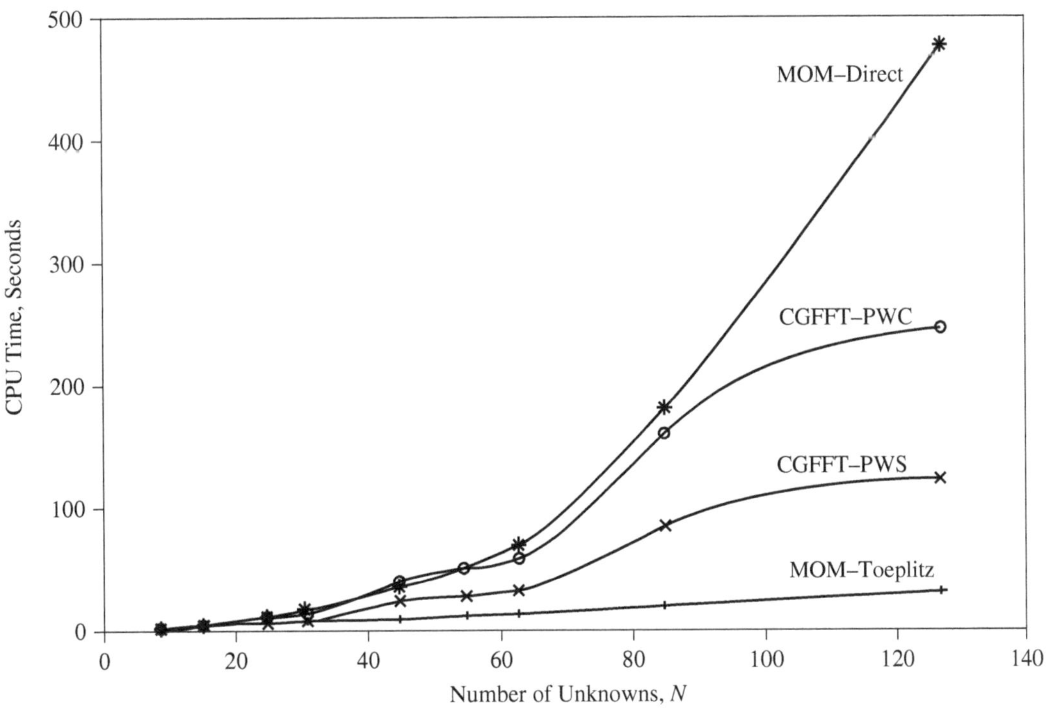


**FIGURE 4-28** ■ A comparison of the CPU time required by the MOM and the CGFFT formulations for the solution of the $1\lambda_o$ wire dipole problem (Barkeshli and Volakis, 1989).

the number of unknowns, whereas the CPU time associated with the direct (matrix inversion) solution is a quadratic function of the unknowns. Also, we observe from Figure 4-19 that the use of higher-order basis functions leads to better convergence rates.

We close this section by noting that an alternative and more appropriate way to compute the column $[Z]\{I_n\}$ is to consider the discretized equation

$$V_m = \sum_{n=0}^{N-1} I_n Z_{mn} = \sum_{n=0}^{N-1} I_n Z(m-n) = \sum_{n=0}^{N-1} I_n Z_{m-n}.$$

Then by application of the discrete convolution theorem it follows that

$$\{V_m\} = DFT^{-1}\{\hat{Z}\hat{I}_n\}$$

where $\hat{Z}$ denotes the discrete transform of the sequence $Z_{0n}$ or the sequence $Z_{m-n} = Z_p$ with $p = -(N-1), \ldots, 0, 1, 2, \ldots, N-1$. Because of the periodicity of the discrete FFT, aliasing is eliminated once the FFT length is set equal to $2N-1$ to accommodate spreading due to the convolution. The sequence $Z_{0n}$ can be obtained from the expressions given earlier by (4.36), (4.75), or (4.80). If $2N-1$ is not a power of 2 then the values of the $Z_p$ sequence should be arranged as shown in Figure 4-29.

In general the computation of the $Z_{mn}$ elements may be difficult due to kernel singularities and the requirement to perform rather involved integrations. In this case, a third alternative would be to return to (4.134) and replace $\widetilde{G}_{wr}(k_z)$ by the discrete transform of the sequence

$$G_{wr}(z_n) = \frac{e^{-jk_o\sqrt{(n\Delta z)^2+a^2}}}{4\pi\sqrt{(n\Delta z)^2+a^2}}$$

with $n = -N-1, \ldots, 0, 1, \ldots, N-1$. This procedure should substantially reduce aliasing errors and is equivalent to setting

$$\hat{Z}_n = -\frac{jZ_o}{k_o}\left\{\left[k^2 - n^2(\Delta k_z)^2 \operatorname{sinc}\left(\frac{n\Delta k_z \Delta z}{2}\right)\right]\tilde{f}(n\Delta k_z)\hat{G}_{wr}\right\} \tag{4.135}$$

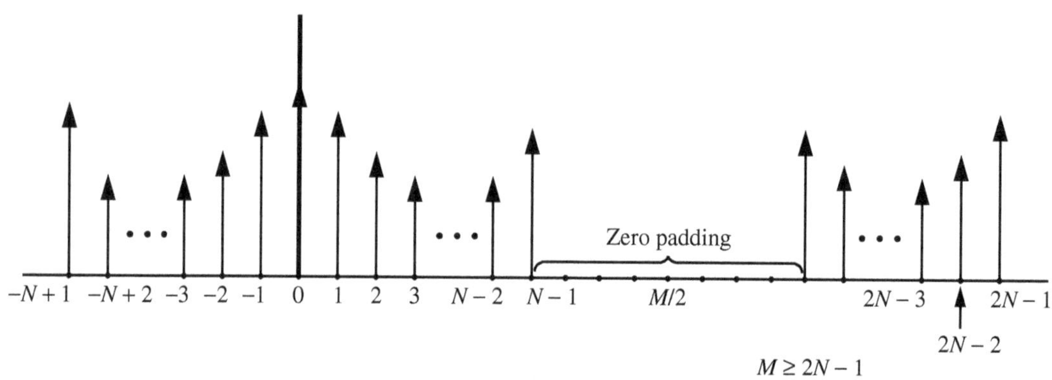


**FIGURE 4-29** ■ Arrangement of the $Z_{m-n} = Z_p$ sequence before inverse Fourier transformation.

where $\hat{G}_{wr}$ corresponds to the transform of the discrete sequence $G_{wr}(z_n)$. By taking the inverse DFT of this $\hat{Z}_n$ we will then (approximately) recover the original sequence $Z_n$. Any aliasing errors will be due to the truncation of $\tilde{f}(k_z)$, but these should be negligible.

## PROBLEMS

1. Show that Pocklington's integral equation can be rewritten as

$$\int_{-\ell}^{\ell} I_z(z')\frac{e^{-jk_oR}}{4\pi R^5}\left[(1+jk_oR)(2R^2-3a^2)+(k_oaR)^2\right]dz' = -jk_oY_oE_z^i(z)$$

where $R=\sqrt{a^2+(z-z')^2}$ and $a$ denotes the wire radius.

2. Consider a small $0.1\lambda_o$ long center-fed dipole of radius $a = 10^{-4}\lambda_o$. On the basis of the pulse-basis–point-matching solution it is found that the impedance matrix (with five segments) elements are given by

$$Z_{12}=Z_{21}=-15.7538-j42613.9$$
$$Z_{13}=Z_{31}=-15.6791-j3487.71$$
$$Z_{14}=Z_{41}=-15.5555-j995.97$$
$$Z_{15}=Z_{51}=-15.3835-j429.1$$
$$Z_{11}=Z_{22}=Z_{33}=Z_{44}=Z_{55}=-15.7834+j93605.74$$

   (a) Assuming a delta gap excitation, determine the input impedance using a five-segment tessellation.

   (b) Increase the number of subdivisions to reach a converged value for the input impedance of the $0.1\lambda_o$ dipole using the pulse basis point-matching solution. What do you observe regarding the solution convergence?

3. Implement the pulse-basis–point-matching solution of Pocklington's integral equation, and determine the current distribution and input impedance as a function of the feed's distance from the center for (a) a resonant ($\lambda/2$) dipole, and (b) a $3\lambda/2$ long dipole. Assume a wire radius of $a = 0.001\lambda$, and choose $N$ sufficiently large for a converged solution.

4. Consider a small $0.1\lambda_o$ long center-fed dipole of radius $a = 10^{-4}\lambda_o$. Using the Galerkin's subsectional sinusoidal basis solution, it is found that the impedance matrix (with 4 segments – 3 PWS) is given by

$$Z_{12}=Z_{21}=Z_{23}=Z_{32}=-.493889-j1577.1$$
$$Z_{13}=Z_{31}=-.490242-j132.354$$
$$Z_{11}=Z_{22}=Z_{33}=-0.495107+j3426.99$$

   (a) For a delta gap excitation determine the input impedance using the four-segment tessellation.

   (b) How many subdivisions are needed to reach the converged value of $Z_{in} = 1.83769 - j1895$?

   (c) Find the converged value of $Z_{in}$ for a magnetic frill excitation.

5. Using the Galerkin's subsectional sinusoidal basis solution,

   (a) Determine the shortest dipole resonant length for $a = 0.001\lambda_o$, $a = 0.005\lambda_o$, $a = 0.01\lambda_o$, and $a = 0.05\lambda_o$. In each case give the corresponding input impedance.

   (b) Determine the input impedance and radiation patterns of the dipoles ($a = 10^{-4}\lambda_o$) having lengths $2\ell = 0.2\lambda_o$, $0.3\lambda_o$, and $0.45\lambda_o$. How many segments are required in each case to reach convergence?

   (c) Determine the backscatter radar cross section patterns of the dipoles given in part (b) of this problem.

6. Consider two identical resonant $\lambda/2$ wire dipoles ($a = 0.001\lambda$) separated by a distance $d$. One of these is fed by a voltage source at its center. Using a Galerkin's solution with sinusoidal subsectional basis, find (a) the input impedance of the excited dipole (driving point impedance) in the presence of the second collinear dipole for $0 < d \le 1.0\lambda_o$, and (b) the mutual impedance $Z_{21} = \frac{V_2}{I_1}$ with the second dipole open circuited at its center (i.e., with the center segment of the second dipole removed) for $0 < d < 1.0\lambda_o$. The voltage $V_2$ is that measured across the open terminal of the second dipole, and $I_1$ is the current at the center of the driven dipole.

7. Consider two identical resonant $\lambda_o/2$ wire dipoles ($a = 0.001\lambda_o$) placed side by side as shown in Figure 4.P7

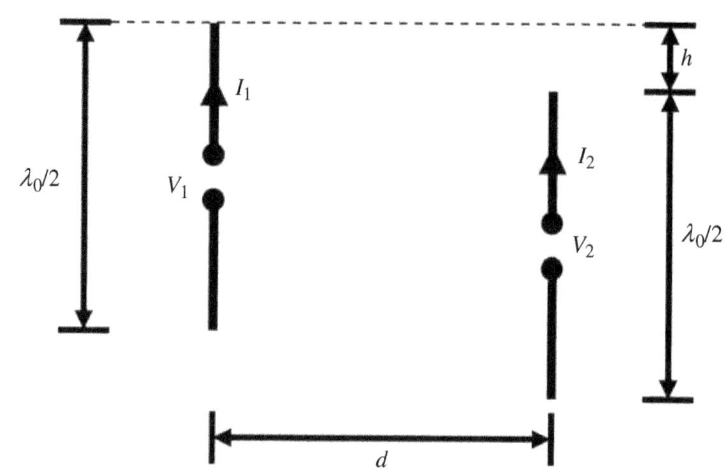


**FIGURE 4.P7** Geometry for Problem 7.

Using a Galerkin's solution with sinusoidal subsectional basis, plot the following:

   (a) The input (driving point) impedance $Z_{in} = \frac{V_1}{I_1}$ with $V_2 = 0$ for $0 < d < 1.0\lambda_o$ and $h = 0$.

   (b) The mutual impedance $\frac{V_2}{I_1}$ with the second dipole open-circuited (i.e., $I_2 = 0$ at its center) for $0 < d < 1.0\lambda_o$ and $h = 0$.

   (c) Repeat (a) and (b) for $h = 0.25\lambda_o$ and $h = 0.5\lambda_o$.
   (Mutual impedances of collinear and side-by-side dipoles have been studied extensively. See e.g., Weeks, 1968)

8. Hallen's (1938) integral equation is specific to a symmetrically fed linear antenna. To derive it, we begin with the assumption that the wire current is symmetric about the feed point as is the case with a center-fed dipole illustrated in Figure 4.P8

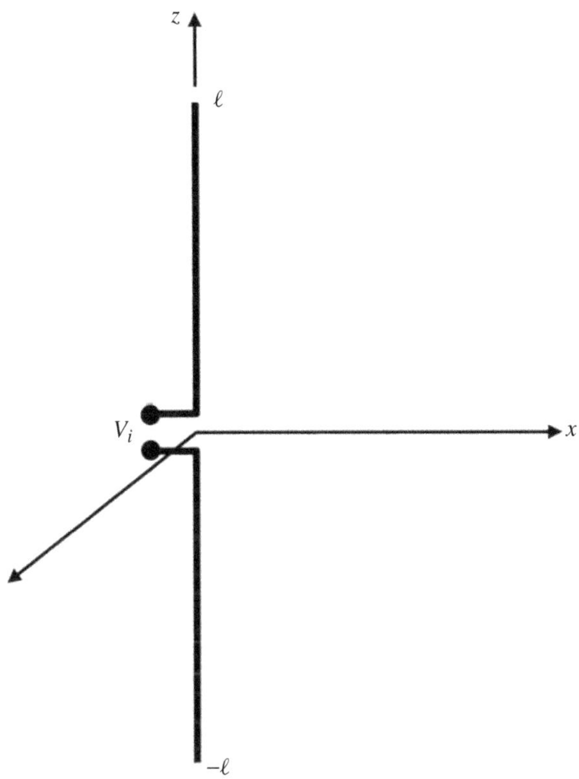


**FIGURE 4.P8** Geometry of the center-fed dipole for Problem 8.

**(a)** Use the differential equation

$$\frac{d^2 A_z}{dz^2} + k_o^2 A_z = 0$$

($A_z$ is the vector potential) along the current symmetry condition to derive Hallen's integral equation

$$\int_{-\ell}^{\ell} I(z') \frac{e^{-jk_o R} dz'}{4\pi R} = -\frac{j}{Z_o} [A_1 \cos(k_o z) + B_1 \sin(k_o |z|)]$$

where $R = \sqrt{a^2 + (z - z')^2}$, $a$ is the wire radius, and $A_1$ and $B_1$ are constants. Note that this integral equation has lower kernel singularity than Pocklington's integral equation. On the other hand, the latter has no restriction on the current shape.

**(b)** Discretize Hallen's integral equations using the pulse-basis–point-matching and pulse-basis Galerkin's methods. It can be shown that $B_1 = 0.5V_i$ and that $A_1$ is determined by introducing an additional equation that enforces a vanishing current at $z = \pm\ell$.

**(c)** Write a computer program implementing Hallen's integral equation pulse-basis–point-matching solution. Compute the current distribution for $2\ell = 0.5\lambda_o$, $a = 0.005\lambda$ and $2\ell = \lambda_o$, $a = 0.01\lambda_o$.

**9.** Consider the dipole shown in Figure 4-27.

**(a)** Derive the near-zone field expressions given by (4.96) through (4.98).

**(b)** Derive the far-zone field expressions radiated by the dipole.

**10.** Consider a wire segment along the $z$-axis. This wire segment carries the current

$$I(z) = \frac{I_1 \sin[k_o(z_2 - z)] + I_2 \sin[k_o(z - z_1)]}{\sin[k_o(z_2 - z_1)]}$$

where $z_1$ and $z_2$ denote the endpoint of the wire with $z_2 > z_1$. Determine the near- and far-zone fields of this monopole. Note that the constants $I_{1,2}$ denote the current values at $z = z_1$ and $z = z_2$, respectively.

**11.** Using the expansion

$$I(z) = \sum_{n=1}^{N} I_n \cos\left[(2n-1)\frac{\pi z}{2\ell}\right]$$

where $2\ell$ is the wire's length:

**(a)** Discretize Pocklington's integral equation in the form

$$[Z_{mn}]\{I_n\} = \{V_n\}$$

and determine $Z_{mn}$. Use point matching at $N - 1$ equidistant locations.

**(b)** Sketch the $n = 1, 2$, and $3$ terms of the expansion.

**(c)** Determine the radar cross section pattern of $2\lambda_o$ and $10\lambda_o$ long dipoles ($a = 10^{-4}\lambda_o$) due to a plane wave incident at an angle 30° with respect to the $z$- axis. How many terms are required to reach convergence?

**12.** Consider the dipoles shown in Figure 4.P12
Their mutual impedance is given by

$$Z_{12} = \left.\frac{V_1}{I_2}\right|_{I_1=0}$$

**FIGURE 4.P12** Geometry for Problem 12.

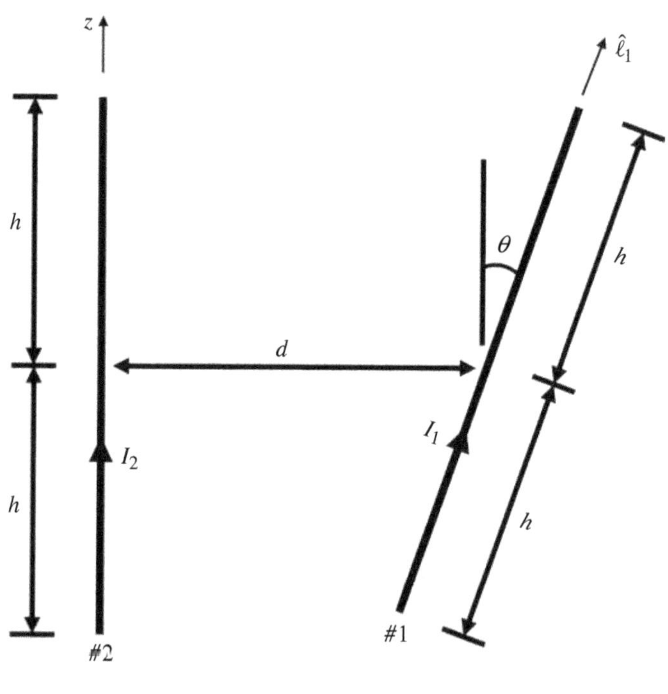

where

$$V_1 = -\frac{1}{I_{1t}} \int_{-h}^{h} (\hat{\ell}_1 \cdot \mathbf{E}_2) I_1(\ell)\, d\ell$$

where $I_{1t}$ denotes the current at the terminal of element 1, and $\mathbf{E}_2$ is the field due to element 2 in the absence of element 1. For $d = 1\lambda_o$ and $h = 0.1\lambda_o$ find the following:

**(a)** $Z_{12}$ as a function of $\theta$ when $I_{1,2}(\ell)$ are assumed constant over the entire length of the elements.

**(b)** $Z_{12}$ as a function of $\theta$ when $I_{1,2}(\ell)$ are assumed to be sinusoidal in form,

$$I_{1,2}(\ell) = \frac{\sin[k_o(h - |\ell|)]}{\sin k_o h}$$

where $\ell$ is measured from the element's center (as a reference for $\theta = 0°$, $Z_{12} = 0.337 + j1.952$).

**13.** Consider the problem of scattering by a coated wire as shown in Figure 4.P13. It can be shown (Chatterjee and Volakis, 1992) that the presence of the coating can be simulated by enforcing the boundary condition

$$E_z = -\left(\frac{\epsilon_r - 1}{jk_o Y_o \epsilon_r}\right) \frac{1}{2\pi} \ln \frac{b}{a} \frac{d^2 I(z)}{dz^2}$$

at $\rho \approx a$, where $a$ is the wire radius, and $\epsilon_r$ is the relative permittivity of the coating of thickness $t = (b - a)$.

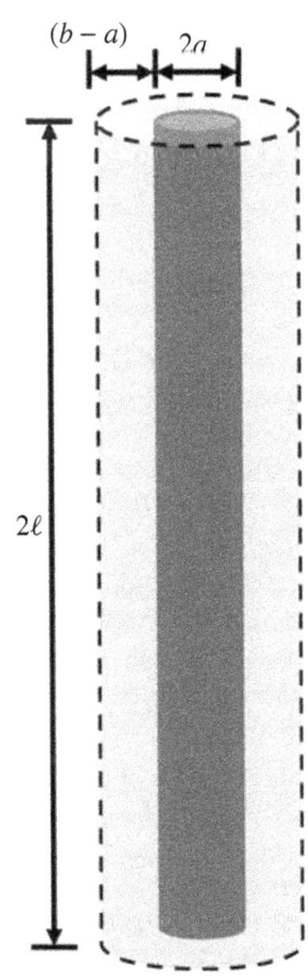


**FIGURE 4.P13** Geometry for Problem 13.

**(a)** Represent the current on the coated wire as

$$I(z) = C_1 e^{jk_o z \cos\theta} + C_2 e^{j\beta z} + C_3 e^{-j\beta z}$$

where $C_{1,2,3}$ are constants to be found, and $\beta$ can be found by solving the transcendental equation.

$$\epsilon_r\left(\beta^2 - k_o^2\right) b \ln\left(0.89\sqrt{\beta^2 - k_o^2}\, b\right) = -(\epsilon_r - 1)k_o^2(b - a)$$

Using this current representation, construct the $3 \times 3$ impedance matrix on the basis of Galerkin's method (i.e., the weighting functions should be $e^{-jk_o z\cos\theta}$, $e^{-j\beta z}$, and $e^{+j\beta z}$).

**(b)** Determine the bistatic radar cross section as a function of $\theta$ for a coated wire having length $16.66\lambda_o$, $b = 0.0067\lambda$, $a = 0.0033\lambda$. The incidence angle is $\theta_i = 10°$.

**(c)** Repeat (b) for a coated wire of length $3\lambda_o$.

**(d)** Repeat (a) and (b) for a bare wire of the same length, and compare the obtained RCS patterns. What can you say about the traveling wave lobe? When should the proposed three-term expansion fail?

## BIBLIOGRAPHY


Abramowitz, M. and Stegun, I. A. (1964). *Handbook of Mathematical Functions with Formulas, Graphs, and Mathematical Tables*. New York: Dover.

Barkeshli, K. and Volakis, J. L. (1988). Applications of the conjugate gradient FFT method to a class of large radiating and scattering problems, University of Michigan Radiation Laboratory Report 389604-3-T.

Barkeshli, K. and Volakis, J. L. (1989). Improving the convergence rate of the conjugate-gradient fft method using subdomain basis functions. *IEEE Trans. Antennas Propagat.*, *37*(7), 893–900.

Burke, G. J. and Poggio, A. J. (1980). Numerical electromagnetic code (NEC) - Method of Moments, Parts I, II and III. *Naval Ocean Systen Center, San Diego, CA, NOSC TD 116 (revised)*.

Chatterjee, A. and Volakis, J. L. (1992). Scattering by perfectly conducting and coated thin wire using a physical basis model. *IEEE Trans. Antennas Propagat.*, *40*, 761–769.

Collin, R. E. (1985). *Antennas and Radiowave Propagation*. New York: McGraw-Hill.

Diaz, M. (1970). Computer techniques for electromagnetics and antennas. *Short course notes, University of Illinois*.

Hallen, E. (1938). Theoretical investigations into the transmitting and receiving qualities of antennae. *N. A. R. Soc. Sci. Upsal., Series IV*, *11*, 1–44.

Harrington, R. F. (1968). *Field Computations by Moment Method*. New York: MacMillan Co.

Hesteness, M. R. and Steifel, E. (1952). Method of conjugate gradients for solving linear systems. *J. Res. Nat. Bur. Standards*, *49*(6), 409–436.

Imbriale, W. A. and Ingersong, P. G. (1963). On numerical convergence of moment method solutions of moderately thick wire antennas using sinusoidal basis functions. *IEEE Trans. Antennas Propagat.*, *21*(3), 363–366.

Jones, D. S. (1956). A critique of the variational method in scattering problems. *IRE Trans. Antennas Propagat.*, *AP-4*(3), 297–301.

Jordan, E. C. and Balmain, K. C. (1968). *Electromagnetic Waves and Radiating Systems*. New York: Prentice-Hall, 2nd ed.

Kantorovich, L. V. and Akilov, G. P. (1964). *Functional Analysis in Normal Spaces*. (Trans. D. E. Brown). Oxford: Pergamon Press.

Kantorovich, L. V. and Krylov, V. I. (1959). *Approximate Methods of Higher Analysis.* (4th ed., Trans. C. D. Benster). New York: John Wiley & Sons.

King, H. E. (1957). Mutual impedance of unequal length in echelon. *IEEE Trans. Antennas Propagat., AP-5*, 306–313.

Lin, Y. T. and Richmond, J. H. (1975). EM modeling of aircraft at low frequencies. *IEEE Trans. Antennas Propagat., AP-23*, 53–56.

Miller, E. K. and Deadrick, F. J. (1975). *"Some computational aspects of thin wire modeling"* (chap. in *Numerical and Asymptotic Techniques in Electromagnetics,* R. Mittra, ed.), New York: Springer-Verlag.

Noble, B. (1958). *Method Based on the Wiener-Hopf Technique*. New York: Pergamon Press.

Pocklington, H. C. (1897). Electrical oscillations in wire. *Cambridge Phil. Proc., 9*, 324–332.

Poggio, A. and Miller, E. (1973). *"Integral equation solutions of three-dimensional scattering problems"* (chap. in *Computer Techniques for Electromagnetics,* R. Mittra, ed.), New York: Pergamom Press, chap. 4.

Press, W. H., Teukolsky, S. A., Vetterling, W. T., and Flannery, B. P. (1992). *Numerical Recipes in Fortran.* (2nd ed.). New York: Cambridge Univ. Press.

Richmond, J. H. (1965). Digital computer solution of the rigorous equations for scattering problems. *Proc. of the IEEE, 53.*

Richmond, J. H. (1966). A wire-grid model for scattering by conducting bodies. *IEEE Trans. Antennas Propagat., AP-14*(6), 782–786.

Richmond, J. H. (1974). Radiation and scattering by thin-wire structures in a homogeneous conducting medium. *IEEE Trans. Antennas Propagat., AP-22*, 365.

Richmond, J. H. and Geary, N. H. (1970). Mutual impedance between coplanar-skew dipoles. *IEEE Trans. Antennas Propagat., AP-18*, 414–416.

Thiele, G. A. (1973). *"Wire Antennas"* (chap. in *Computer Techniques for Electromagnetics,* R. Mittra, ed.), Hemisphere Pub. Co., chap. 2.

Trueman, C. W., Mishra, S. R., Kubina, S. J., and Larose, C. L. (1993). RCS of resonant scattering with attached wires. *IEEE Trans. Antennas Propagat., AP-41*, 351–254.

Tsai, L. L. (1972). A numerical solution for the near and far fields of an annular ring of magnetic current. *IEEE Trans. Antennas Propagat.*, 596–576.

Volakis, J. L., Chatterjee, A., and Kempel, L. (1998). *Finite Element Methods for Electromagnetics*. New York: IEEE Press.

Weeks, W. L. (1968). *Antenna Engineering*. New York: McGraw-Hill.

Wilton, D. R. and Wheeler, J. E. I. (1991). *"Comparison of the convergence rates of the conjugate gradient method applied to various integral equation formulations"* (chap. in *PIER 5,* J.A. Kong and T.K. Sarkar, eds.), New York: Elsevier, chap. 5.

Yeh, Y. S. and Mei, K. K. (1967). Theory of conical equiangular spiral antennas, Part I - Numerical techniques. *IEEE Trans. Antennas Propagat., AP-15*, 634.

CHAPTER 5

# Two-Dimensional Scattering

**Chapter Outline**



The wire structure considered in the previous chapter is perhaps the simplest version of a three-dimensional geometry because it can be modeled by a filamentary current. In the case of other antenna structures such as reflectors, horns, or microstrip patches, the current will be a three-dimensional function, and this leads to more involved kernels for the resulting integral equations. The complexity of the integral equation is increased further when we consider the radar scattering by an aircraft structure, a portion of which may be nonmetallic or coated with a dielectric. Thus, before proceeding with the treatment of three-dimensional structures, it is instructive to consider solutions of simpler structures that can serve to understand and validate the methods. It turns out that, in the case of scattering, we can obtain information from two-dimensional treatments, which under certain assumptions can be extrapolated to three dimensions.

In this chapter we consider the numerical solution of the scattering by (a) a flat and a curved resistive strip, (b) a dielectric strip, (c) a homogeneous dielectric cylinder, and (d) an inhomogeneous dielectric cylinder.

## 5.1 FLAT RESISTIVE STRIP

Consider the flat resistive strip of width $w$ as shown in Figure 5-1. From (1.143), this strip is mathematically characterized by the transition condition

$$\hat{y} \times \hat{y} \times \mathbf{E} = -R_e Z_o \hat{y} \times [\mathbf{H}^+ - \mathbf{H}^-] \tag{5.1}$$

where $R_e$ is the resistivity of the strip in Ohms per square, and the superscript $\pm$ denotes the field above and below the strip. If $R_e$ is real, the strip represents a thin

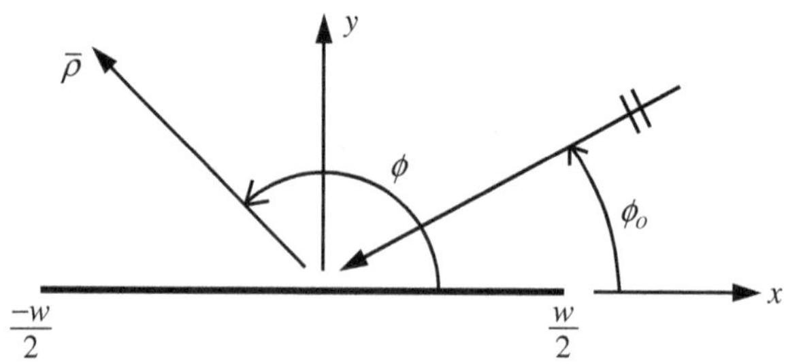


**FIGURE 5-1** Geometry of the flat strip.

conducting layer having a finite conductivity $\sigma$ and a thickness $\tau$. Alternatively, if $R_e$ is complex or imaginary and given by (1.149), it can then be interpreted to represent a thin dielectric layer.

From the natural boundary conditions (1.64) and (1.65), we rewrite (5.1) as

$$\mathbf{J} = -\frac{1}{R_e Z_o}\hat{y} \times \hat{y} \times \mathbf{E} \tag{5.2}$$

where $\mathbf{J}$ is identified as the net surface current density (in A/m) on the surface of the strip. This current is responsible for the scattered field due to an excitation $\mathbf{E}^i$. To determine it, we set $\mathbf{E} = \mathbf{E}^i + \mathbf{E}^s$ in (5.2), and upon replacing $\mathbf{E}^s$ with its integral representation for the scattered field we obtain an integral equation for $\mathbf{J}$. For a perfectly conducting strip ($R_e = 0$) we expect the resulting integral equation to be identical to (3.98) or (3.99). Nevertheless, the resulting equations for $E$- and $H$-polarized excitations are different, and it is thus necessary to consider these two cases separately.

### 5.1.1 *E*-Polarization

#### 5.1.1.1 Pulse-Basis–Point-Matching Solution

Assume a plane wave of the form

$$\mathbf{E}^i = \hat{z}e^{jk_o(x\cos\phi_o + y\sin\phi_o)} \tag{5.3}$$

where $\phi_o$ denotes the incidence angle. Since $\mathbf{E}^i$ is $z$-polarized, $\mathbf{E}^s$ will be polarized in a like manner; consequently, from (2.124)

$$\mathbf{H} = \frac{-jY_o}{k_o}\hat{z} \times \nabla E_z = \frac{jY_o}{k_o}\left(\hat{x}\frac{\partial E_z}{\partial y} - \hat{y}\frac{\partial E_z}{\partial x}\right) = \hat{x}H_x + \hat{y}H_y.$$

This implies that $\mathbf{J} = \mathbf{J}^+ + \mathbf{J}^- = \hat{y} \times (\mathbf{H}^+ - \mathbf{H}^-) = \hat{z}J_x$ allowing us to express the scattered field as

$$\begin{aligned} E_z^s &= -jk_o Z_o \int_{-w/2}^{w/2} J_z(x')G_{2d}(\overline{\rho}, \overline{\rho}')\, dx' \\ &= \frac{-k_o Z_o}{4}\int_{-w/2}^{w/2} J_z(x')H_o^{(2)}(k_o|\overline{\rho} - \overline{\rho}'|)\, dx' \end{aligned} \tag{5.4}$$

with $\overline{\rho} = x\hat{x} + y\hat{y}$ and $\overline{\rho}' = x'\hat{x} + y'\hat{y}$. Substituting this into (5.2) and setting $E_z = E_z^i + E_z^s$, it follows that

$$Y_o E_z^i(x) = R_e J_z(x) + \frac{k_o}{4}\int_{-w/2}^{w/2} J_z(x')H_o^{(2)}(k_o|x - x'|)\, dx' \tag{5.5}$$

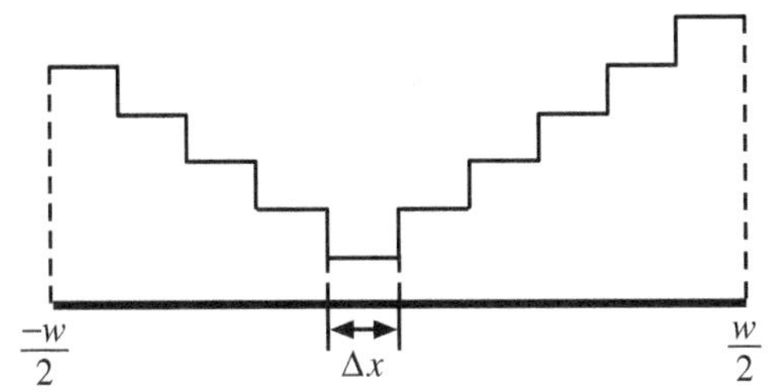


**FIGURE 5-2** Discretization of the flat strip current.

which is a Fredholm integral equation of the second kind. As expected, when $R_e = 0$, it reduces to (3.98a).

The simplest way to solve (5.5) is to expand $J_z(x)$ in terms of a pulse-basis function as (see Figure 5-2)

$$J_z(x) = \sum_{n=0}^{N-1} J_n P_{\Delta x}(x - x_n) \tag{5.6}$$

where $x_n = \frac{-w}{2} + \left(n + \frac{1}{2}\right)\Delta x$, and $N = \frac{w}{\Delta x}$ is the number of segments used to discretize the strip and the integral in (5.5). Substituting this expansion into (5.5) and point matching at $x = x_m$, yields the linear set of equations[1]

$$Y_o E_z^i(x_m) = R_e J_z(x_m) + \frac{k_o}{4} \sum_{n=0}^{N-1} J_n \int_{x_n - \frac{\Delta x}{2}}^{x_n + \frac{\Delta x}{2}} H_o^{(2)}(k_o|x_m - x'|)\, dx' \qquad m = 0, 1, 2, \ldots, N-1. \tag{5.7}$$

This set of equations can be written in the usual matrix form

$$\{V_m\} = [Z_{mn}]\{J_n\} \tag{5.8}$$

where the elements of $\{V_m\}$ are given by

$$V_m = Y_o E_z^i(x_m) \tag{5.9}$$

and the elements of the square matrix $Z_{mn}$ are formally given by

$$\begin{aligned} Z_{mn} &= R_e \delta(n-m) + \frac{k_o}{4} \int_{x_n - \frac{\Delta x}{2}}^{x_n + \frac{\Delta x}{2}} H_o^{(2)}(k_o|x_m - x'|)\, dx' \\ &= R_e \delta(n-m) + \frac{k_o}{4} \int_{\frac{-\Delta x}{2}}^{\frac{\Delta x}{2}} H_o^{(2)}(k_o|x_m - x_n + t|)\, dt \end{aligned} \tag{5.10}$$

in which

$$\delta(n-m) = \begin{cases} 1 & n = m \\ 0 & n \neq m. \end{cases} \tag{5.11}$$

For $x_m = x_n$, we observe that the integrand of (5.10) is singular at $t = 0$ (but the integral is convergent) and must thus be evaluated analytically to avoid numerical

[1] Mei and Van Bladel (1963) were among the first to employ this solution scheme for metallic rectangular schemes.

difficulties. To do so, a standard approach is to employ the small argument expansion of the Hankel function

$$H_o^{(2)}(z) \cong 1 - j\frac{2}{\pi}\ln\frac{\gamma z}{2} + O(z^2, z^2\ln z) \tag{5.12}$$

where $\gamma = 1.781$ is Euler's constant. When this is substituted in (5.10), we obtain

$$\begin{aligned} Z_{nn} &= R_e + \frac{k_o}{4}\int_{\frac{-\Delta x}{2}}^{\frac{\Delta x}{2}}\left[1 - \frac{j2}{\pi}\ln\left(\frac{k_o\gamma|t|}{2}\right)\right]dt \\ &= R_e + \frac{k_o\Delta x}{4}\left[1 - \frac{j2}{\pi}\left(\ln\frac{k_o\gamma\Delta x}{4} - 1\right)\right]. \end{aligned} \tag{5.13}$$

For $m \neq n$ we may employ midpoint integration, since the integrand will likely be slowly varying, to find

$$Z_{mn} \cong \frac{k_o\Delta x}{4}H_o^{(2)}(k_o|x_m - x_n|), \qquad m \neq n. \tag{5.14}$$

In summary, for the case of the flat resistive strip the impedance matrix elements are approximately given by

$$Z_{mn} = \begin{cases} R_e + \frac{k_o\Delta x}{4}\left[1 - \frac{j2}{\pi}\left(\ln\left(\frac{k_o\gamma\Delta x}{4}\right) - 1\right)\right] & n = m \\ \frac{k_o\Delta x}{4}H_o^{(2)}(k_o|x_m - x_n|) & m \neq n. \end{cases} \tag{5.15}$$

This approximation should remain valid provided $\Delta x$ is not extremely small, in which case the midpoint approximation (5.14) is not valid for the matrix elements near the diagonal. Typically, $\frac{\lambda_o}{20} < \Delta x < \frac{\lambda_o}{10}$, but for smaller values of $\Delta x$ it may be necessary to resort to a more accurate integration for those elements using, for example, a higher-order midpoint integration or Gaussian quadrature. Alternatively, the near-diagonal terms can be evaluated analytically by replacing the Hankel function in (5.10) with several terms of its small argument expansion. The results of integration on using such an expansion lead to (Jin and Volakis, 1990; Barkeshli and Volakis, 2004)

$$\int_{x_n-\frac{\Delta x}{2}}^{x_m+\frac{\Delta x}{2}} H_o^{(2)}(k_o|x_m - x'|)\,dx' = 2\int_{\epsilon\to 0}^{\alpha} H_o^{(2)}(k_o\sqrt{t^2})\,dt$$

$$\begin{aligned} 2\int_o^{\alpha} H_o^{(2)}(k_o|t|)\,dt = 2\alpha&\left[1 - \frac{2j}{\pi}\left(\ln\left(\frac{k_o\gamma|\alpha|}{2}\right) - 1\right)\right] \\ &- \alpha\left[1 - \frac{2j}{\pi}\left(\ln\left(\frac{k_o\gamma|\alpha|}{2}\right) - \frac{5}{3}\right)\right]\frac{(k_o\alpha)^2}{6} \\ &+ j\frac{\alpha(k_o\alpha)^2}{9\pi} \end{aligned} \tag{5.16}$$

with $\alpha = (x_m - x_n) + \frac{\Delta x}{2}$.

We observe that, similar to the straight wire, the impedance matrix for the flat strip is also symmetric Toeplitz since $Z_{mn} = Z_{nm} = Z_{|m-n|}$. Consequently we can resort to a conjugate gradient solution to reduce the central processing unit (CPU) time and memory requirements. The matrix product $[Z]\{J_n\}$ appearing in the algorithm must then be computed as

$$[Z]\{J_n\} = DFT^{-1}\{\hat{Z}_n\hat{J}_n\} \tag{5.17}$$

where $\hat{Z}_n$ is the discrete Fourier transform of the sequence

$$Z_n = R_e\delta(n) + \frac{k_o}{4}\int_{x_n+\frac{\Delta x}{2}}^{x_n-\frac{\Delta x}{2}} H_o^{(2)}(k_o|x_o - t|)\, dt \tag{5.18}$$

for $n = -(N-1), \ldots, 0, 1, \ldots, (N-1)$. If the fast Fourier transform (FFT) pad is of dimension $M > 2N - 1$ the sequence $Z_n$ must be arranged as shown in Figure 4-29. That is, since $Z_n = Z_{-n}$ then for $q = M - n > \frac{M}{2}$, $Z_q = Z_n$ and for $N - 1 < q < M/2$, $Z_q = 0$.

The current distributions depicted in Figure 5-3 correspond to a metallic strip and were obtained from a direct solution of the system (5.8). From these plots it is seen that the current is consistently very large near the edges of the metallic strip

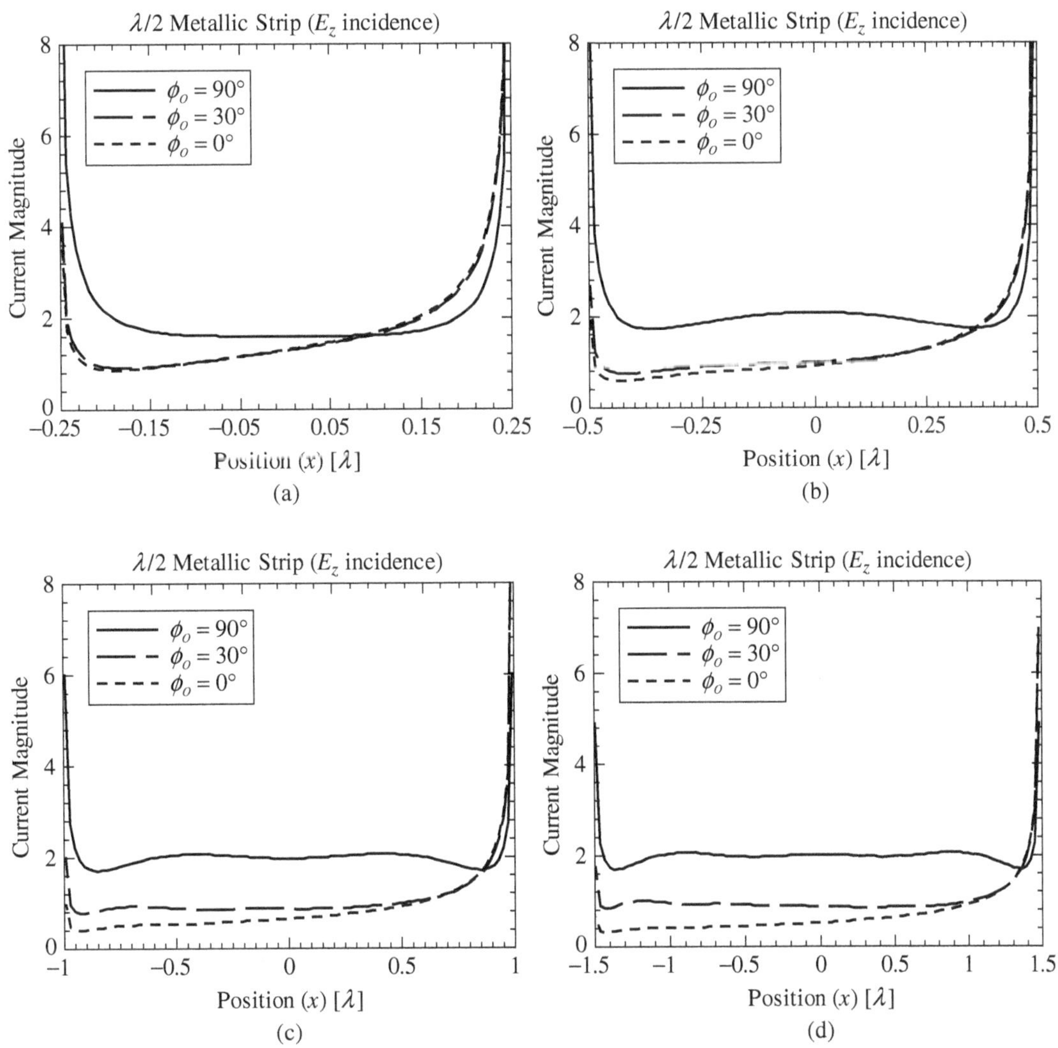


**FIGURE 5-3** Currents on a flat metallic strip excited by an $E$-polarized plane wave incident at $\phi_o = 0°$, 30° and 90°: (a) $w = \lambda/2$, (b) $w = \lambda$, (c) $w = 2\lambda$, and (d) $w = 3\lambda$.

regardless of incidence angle and the strip's width. This is a particular characteristic of the $E$-polarization current near any metallic edge, and it can be shown that at this location $J_z \sim 1/\sqrt{d}$, where $d$ is the distance away from either edge of the strip. As we move away from the edge, the current decreases rapidly, and for near-normal incidences it approximately reaches its physical optics value $2\hat{y} \times \mathbf{H}^i = 2Y_o \sin\phi_o E_z^i$. For nonmetallic strips, that is, for $R_e \neq 0$, the current is not singular at the edges but finite, and for near-normal incidence it reaches its physical optics (PO) value of

$$J_{zPO} = -2\Gamma_e Y_o \sin\phi_o e^{jk_o x \cos\phi_o} = \frac{2Y_o \sin\phi_o}{1 + 2R_e \sin\phi_o} e^{jk_o x \cos\phi_o} \tag{5.19}$$

at a shorter distance from the edge ($\Gamma_e$ is identified as the plane wave reflection coefficient of the planar resistive sheet). This is displayed in Figure 5-4, where we

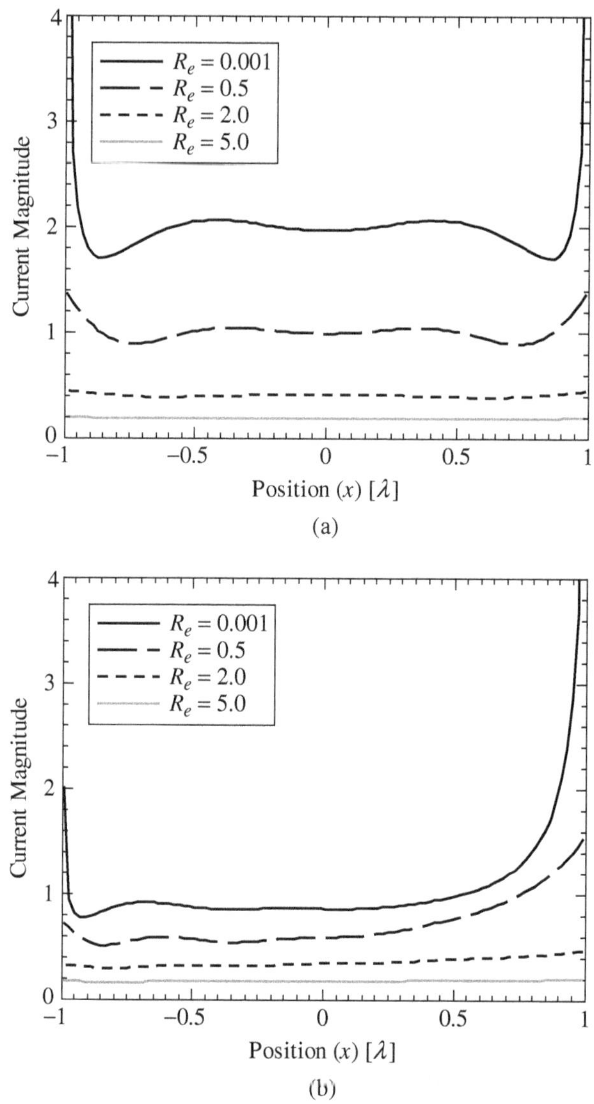


**FIGURE 5-4** Currents on a 2λ flat resistive strip excited by an $E$-polarized plane wave with $R_e = 0.001, 0.5, 2,$ and 5: (a) $\phi_o = 90^\circ$, and (b) $\phi_o = 30^\circ$.

plot the normal incidence ($\phi_o = 90°$) current for four different strip resistivities. As is well known, the PO approximation becomes less accurate as the angle of incidence moves away from broadside, and this can be deduced from the results in Figures 5-3 and 5-4 [2]. The amplitude of the previous PO expressions is independent of $x$, but the exact (numerical) results given in Figures 5-3 and 5-4 are increasingly more dependent on $x$, as $\phi_o$ decreases (approaches grazing).

Given the current distribution $J_z$, the far-zone scattered field is computed from (2.133), and for this case we have

$$E^s_{z\,ff} \sim \sqrt{\frac{k_o}{8\pi}} \frac{e^{-j(k_o\rho-\frac{\pi}{4})}}{\sqrt{\rho}} \int_{-w/2}^{w/2} Z_o e^{jk\hat{\rho}\cdot\hat{x}x'} \hat{\rho}\times\hat{\rho}\times\hat{z}J_z(x')\,dx'$$

$$\sim -\frac{e^{-jk_o\rho}}{\sqrt{\rho}} e^{j\frac{\pi}{4}} \frac{k_o Z_o}{4} \sqrt{\frac{2}{\pi k_o}} \int_{-w/2}^{w/2} J_z(x') e^{jk_o x'\cos\phi}\,dx'$$

since $\hat{\rho} = \hat{x}\cos\phi + \hat{y}\sin\phi$. Introducing the expansion (5.6), we can rewrite this as

$$\begin{aligned} E^s_{z\,ff} &\sim \frac{-e^{-jk_o\rho}}{\sqrt{\rho}} e^{j\frac{\pi}{4}} \frac{k_o Z_o}{4} \sqrt{\frac{2}{\pi k_o}} \sum_{n=0}^{N-1} J_n \int_{x_n-\frac{\Delta x}{2}}^{x_n+\frac{\Delta x}{2}} e^{jk_o x'\cos\phi}\,dx' \\ &= \frac{-e^{-jk_o\rho}}{\sqrt{\rho}} e^{j\frac{\pi}{4}} \frac{k_o Z_o}{4} \sqrt{\frac{2}{\pi k_o}} \Delta x \operatorname{sinc}\left(\frac{k_o \Delta x}{2}\cos\phi\right) \sum_{n=0}^{N-1} J_n e^{jk_o x_n \cos\phi}. \end{aligned} \tag{5.20}$$

Generally, of interest is the echowidth of the strip defined by

$$\sigma_E = \lim_{\rho\to\infty} 2\pi\rho \frac{|\mathbf{E}^s|^2}{|\mathbf{E}^i|^2} \tag{5.21}$$

and measured in units of length (wavelengths or meters). Using this formula, the strip's echowidth is found to be

$$\sigma_E = \frac{k_o}{4} Z_o (\Delta x)^2 \operatorname{sinc}^2\left(\frac{k_o \Delta x}{2}\cos\phi\right) \left|\sum_{n=0}^{N-1} J_n e^{jk_o x_n \cos\phi}\right|^2. \tag{5.22}$$

Plots of $\sigma_E$ for a metallic strip are given in Figure 5-5 as a function of $\phi$ (bistatic) for different incidence angles and strip widths. Corresponding plots for the resistive strip are shown in Figure 5-6. These patterns are very close to the physical optics results near the specular direction (i.e., near $\phi = \pi - \phi_o$), and in fact $\sigma_E = |\Gamma_e|^2 k_o w^2$ at $\phi = \phi_o$, which is equal to the value obtained by integrating the physical optics current given in (5.19). This is always the maximum value in the pattern that to a great degree resembles the sinc $\left[\frac{k_o w}{2}(\cos\phi + \cos\phi_o)\right]$ function.

For radar applications, the receiver and transmitter are at the same location, and, thus, often of interest is the echowidth pattern with $\phi = \phi_o$. Such backscatter patterns of $\sigma_E$ are shown in Figure 5-7. A particular observation is that $\sigma_E \approx -8$ dB $> \lambda = 0.1585\lambda_o$ when $\phi_o = \phi \approx \pi$ for a metallic strip (edge-on

[2]Figures 5-3–5-6 were originally created by Leo C. Kempel. They were later recreated by the authors.

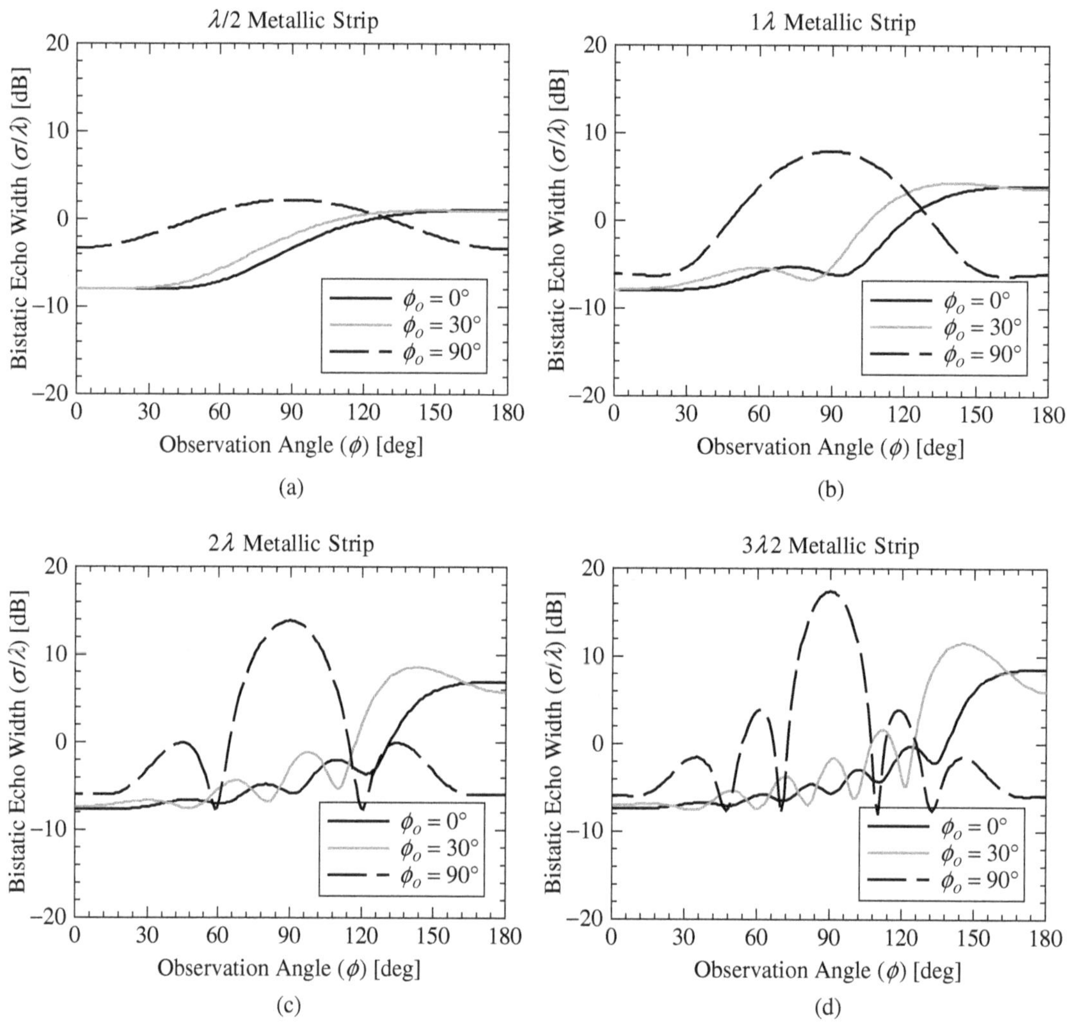


**FIGURE 5-5** ■ Bistatic echowidth of a flat metallic strip excited by an $E$-polarized plane wave incident at an angle $\phi_o = 0^\circ$, $30^\circ$, and $90^\circ$: (a) $w = \lambda/2$, (b) $w = \lambda$, (c) $w = 2\lambda$, and (d) $w = 3\lambda$.

or grazing incidence), provided the strip's width is at least $\lambda/4$. This is because at grazing the rear edge of the strip is completely shielded by the front edge. Thus, for grazing incidences, the strip behaves more like a metallic half plane whose echowidth at edge-on is $-8$ dB $> \lambda$ based on analytical results from diffraction theory. Another way of explaining the shielding of the rear strip edge is to note that at grazing incidence the total field arriving at the rear edge is zero (as a result of the boundary condition on the strip's surface). Thus, the rear edge is not excited for incidences near grazing.

### 5.1.1.2 Narrow Strips

For very narrow metallic strips it is possible to obtain an analytical solution of the strip current by introducing certain approximations in the exact integral equation (5.5)

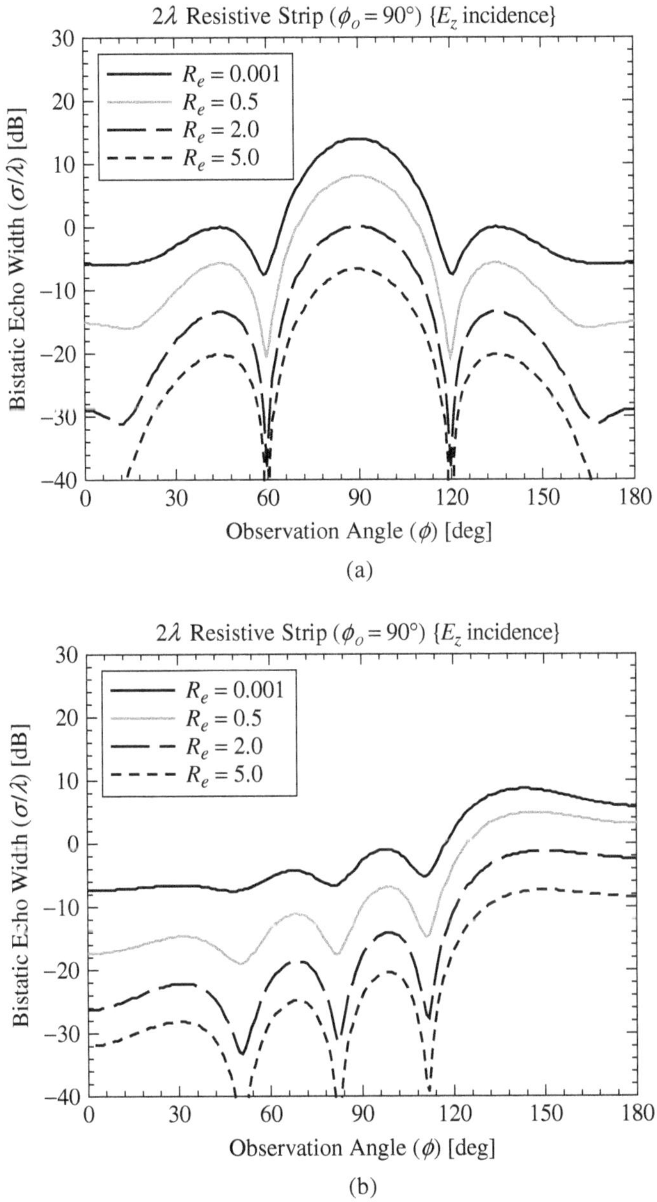


**FIGURE 5-6** $E$-polarization bistatic echowidth of a flat $2\lambda$ wide resistive strip with $R_e = 0.001, 0.5, 2,$ and $5$: (a) $\phi_o = 90°$, and (b) $\phi_o = 30°$.

with $R_e = 0$. In particular, assuming that $k_o w \ll 1$, it follows that the argument of the Hankel function under the integral will also be very small (actually its max value is $2k_o w$). Consequently, we can replace it by the first two terms of its series expansion, given in (5.12). Also, to $O(k_o w)$ the incident plane wave may be replaced by unity which along with (5.12) permits us to rewrite (5.5) as

$$\int_{-w/2}^{w/2} J_z(x') \ln|x - x'|\, dx' = j\frac{2\pi}{k_o} Y_o - \left[\ln\left(\frac{k_o\gamma}{2}\right) + j\frac{\pi}{2}\right] \int_{-w/2}^{w/2} J_z(x')\, dx'. \tag{5.23}$$

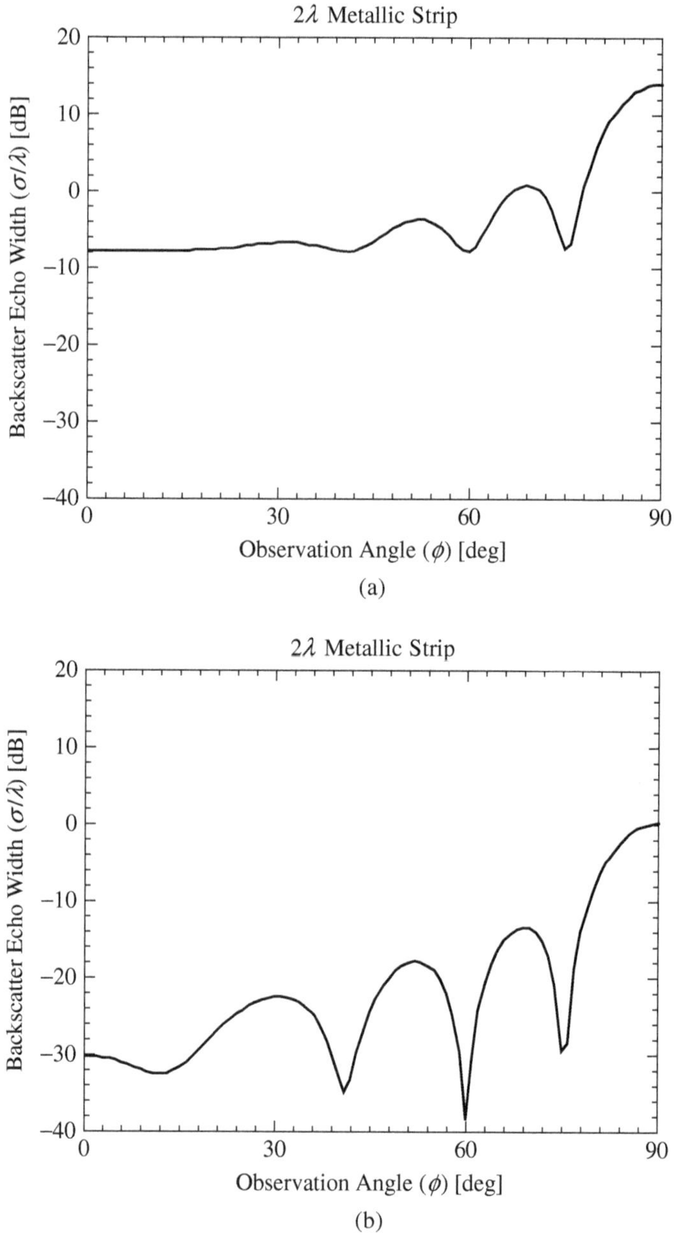


**FIGURE 5-7** Backscatter ($\phi_o = \phi$) $E$-polarization echowidth patterns for a flat strip: (a) $2\lambda$ metallic strip, and (b) resistive strip of width $w = 2\lambda$ and having $R_e = 2$.

This integral equation can be simplified further by introducing the change of variables

$$\xi = \frac{2x}{w}, \qquad \xi' = \frac{2x'}{w}$$

to obtain

$$\int_{-1}^{1} J_z(\xi') \ln|\xi - \xi'|\, d\xi' = j\frac{4\pi}{k_o w} Y_o - \left[\ln\left(\frac{k_o w \gamma}{4}\right) + j\frac{\pi}{2}\right] \int_{-1}^{1} J_z(\xi')\, d\xi'. \tag{5.24}$$

To solve for $J_z(\xi)$ we note the identity (Tricomi, 1985)

$$\int_{-1}^{1} \frac{\ln|x - x'|}{\sqrt{1 - x'^2}}\, dx' = -\pi \ln 2 \qquad x \in (-1, 1) \tag{5.25}$$

which can be applied to (5.24) whose right-hand side is independent of $\xi$ and can thus be considered as a constant in this process. Consequently, we set

$$J_z(\xi) = \frac{\chi_e}{Z_o\sqrt{1 - \xi^2}} = \frac{\chi_e}{Z_o\sqrt{1 - \left(\frac{2x}{w}\right)^2}} \tag{5.26}$$

where $\chi_e$ is a constant to be determined by back-substitution into (5.24). Doing so, in conjunction with (5.25), we obtain

$$\chi_e = \frac{4j}{k_o w\left[\ln\left(\frac{k_o w \gamma}{8}\right) + j\frac{\pi}{2}\right]}.$$

It is seen that (5.26) displays the familiar behavior of the current density near the edge of the strip as discussed already. That is, near the edges, the current of the metallic strips grows as $1/\sqrt{d}$, where $d$ is a measure of the distance from the edge. Equation (5.26) obviously states that the current density goes to infinity at the edge of the strip, implying that the magnetic field must also be infinite at the edge. One should therefore question its physical meaning and significance since in practice we cannot expect such behavior. To explain the validity of the result, we should note that any physical requirement should be imposed on the energy carried by the field. Thus, we can demand only that the power is finite at the edge, and in this particular case it can be shown that the power propagating away from the edge is indeed finite. Moreover, in practice the strip will have a finite conductivity, in which case it must be modeled as a resistive strip.

From the current expression (5.26) we readily find that the far-zone scattered field from the narrow metallic strip is given by

$$E_z^s \sim \frac{e^{-j(k_o\rho - \pi/4)}}{\sqrt{\rho}}\left\{-\frac{k_o}{4}\sqrt{\frac{2}{\pi k_o}}\chi_e \int_{-w/2}^{w/2} \frac{e^{jk_o x'\cos\phi}}{\sqrt{1 - \left(\frac{2x'}{w}\right)^2}}\, dx'\right\}$$

$$\sim -\frac{e^{-j(k_o\rho - \pi/4)}}{\sqrt{\rho}}\sqrt{\frac{k_o}{8\pi}}\left(\frac{w}{2}\chi_e\right)\int_{-1}^{1} \frac{1}{\sqrt{1 - t^2}}\, dt$$

$$\sim -\frac{e^{-j(k_o\rho - \pi/4)}}{\sqrt{\rho}}\sqrt{\frac{k_o}{8\pi}}\left(\frac{\pi w}{2}\right)\chi_e.$$

When this is substituted in the echowidth formula (5.21) we obtain

$$\sigma_E = k_o\left|\frac{\pi w}{4}\chi_e\right|^2 = \frac{\pi\lambda_o}{2\left|\ln\left(\frac{k_o w\gamma}{8}\right) + j\frac{\pi}{2}\right|^2}. \tag{5.27}$$

Plots of $\sigma_E$ as a function of $w$ are displayed in Figure 5-8. As seen, (5.27) is in good agreement with the rigorous solution described earlier provided $k_o w < 1.0$. For larger $k_o w$ the dependence of $\sigma_E$ on the incidence angle becomes noticeable.

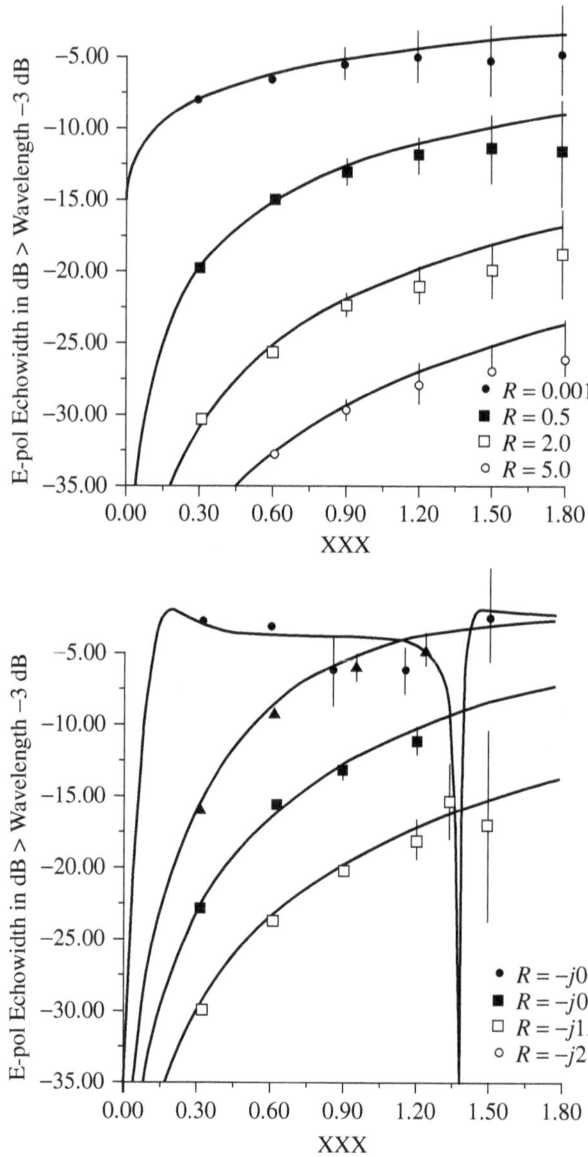


**FIGURE 5-8** $E$-polarization backscatter echowidth plots for a narrow resistive strip based on (5.27). The method of moments data are displayed by the symbols, and the bar lengths indicate the angular field variation from $\phi = 0$ to $\phi = 90^\circ$ as computed by the method of moments. (After Senior and Volakis, 1989. With permission.)

### 5.1.2 *H*-Polarization

The plane wave illuminating the strip is now of the form

$$\mathbf{H}^i = \hat{z}e^{jk_o(x\cos\phi + y\sin\phi_o)} \tag{5.28}$$

with $\phi_o$ again denoting the incidence angle. The scattered field will likewise be $z$-directed, and from (2.124)

$$\mathbf{E} = \frac{jZ_o}{k_o}\hat{z} \times \nabla H_z = -\frac{jZ_o}{k_o}\left(\hat{x}\frac{\partial H_z}{\partial y} - \hat{y}\frac{\partial H_z}{\partial x}\right) = \hat{x}E_x + \hat{y}E_y \tag{5.29}$$

where $H_z = H_z^i + H_z^s$ is the total field. Consequently, the net current on the flat strip will be $x$-directed, and from (5.2) or from a modification of (3.99) we obtain the integral equation

$$\sin\phi_o e^{jk_o x\cos\phi_o} = R_e J_x(x) + \frac{k_o}{4}\int_{-w/2}^{w/2} J_x(x')\left(1+\frac{1}{k_o^2}\frac{\partial^2}{\partial x^2}\right)H_o^{(2)}(k_o|x-x'|)dx'. \tag{5.30}$$

Since $H_o^{(2)}(k_o|x-x'|)$ satisfies the wave equation, this can be alternatively written as

$$\sin\phi_o e^{jk_o x\cos\phi_o} = R_e J_x(x) - \frac{1}{4k_o} \times\left[\frac{\partial^2}{\partial y^2}\int_{-w/2}^{w/2} J_x(x')H_o^{(2)}\left(k_o\sqrt{(x-x')^2+y^2}\right)dx'\right]_{y\to 0} \tag{5.31}$$

so that the integral is now proper (convergent) prior to differentiation.

#### 5.1.2.1 Pulse-Basis–Point-Matching Solution

The solution of (5.30) using pulse-basis expansion and point matching is not generally a suitable approach because of the kernel's singularity, which is of order $R^{-2}$ as $R \to 0$. Consequently, the resulting integral is improper unless at least one of the derivatives is transferred to the current. Nevertheless, such a solution can be meaningful if we retain the second derivative outside the integrand and then carry out the differentiation after the integration is completed. In this manner all integrals remain proper and can be evaluated in the usual manner.

In view of this, let us substitute in (5.31) the expansion

$$J_x(x) = \sum_{n=0}^{N-1} J_n P_{\Delta x}(x - x_n) \tag{5.32}$$

with

$$x_n = -\frac{w}{2} + \left(n + \frac{1}{2}\right)\Delta x \tag{5.33}$$

and $\Delta x = w/N$. Upon point matching at $x = x_m$, $m = 0, 1, \ldots, N-1$ we obtain the system

$$\{V_m\} = [Z_{mn}]\{J_n\} \tag{5.34}$$

where

$$V_m = \sin\phi_o e^{jk_o x_m\cos\phi_o} \tag{5.35}$$

and, formally, $Z_{mn}$ are given by

$$Z_{mn} = R_e\delta(n-m) - \frac{1}{4k_o}\left[\frac{\partial^2}{\partial y^2}\int_{x_n-\frac{\Delta x}{2}}^{x_n+\frac{\Delta x}{2}} H_o^{(2)}\left(k_o\sqrt{(x_m-x')^2+y^2}\right)dx'\right]_{y\to 0}. \tag{5.36}$$

For $m \neq n$, we can move the second derivative with respect to $y$ under the integral and carry out the differentiation to obtain

$$Z_{mn} = \frac{1}{4}\int_{x_n-\frac{\Delta x}{2}}^{x_n+\frac{\Delta x}{2}} \frac{H_1^{(2)}(k_o|x_m-x'|)}{|x_m-x'|}dx', \qquad n \neq m. \tag{5.37a}$$

This result can be readily deduced from (2.129) by first dotting with the dyad $\hat{y}\hat{y}$ and then setting $y' = y = 0$. Alternatively, we can revert to the integral equation (5.30) and obtain

$$
\begin{aligned}
Z_{mn} &= \frac{k_o}{4} \int_{x_n - \frac{\Delta x}{2}}^{x_n + \frac{\Delta x}{2}} H_o^{(2)}(k_o|x_m - x'|)\, dx' \\
&\quad - \frac{1}{4k_o} \int_{x_n - \frac{\Delta x}{2}}^{x_n + \frac{\Delta x}{2}} \frac{\partial^2}{\partial x' \partial x} \{ H_o^{(2)}(k_o|x - x'|) \} \big|_{x = x_m} dx' \\
&\simeq \frac{k_o \Delta x}{4} H_o^{(2)}(k_o|x_m - x_n|) + \frac{1}{4}\left[ \pm H_1^{(2)} \left( k_o \left| x_m - x_n - \frac{\Delta x}{2} \right| \right) \right. \\
&\quad \left. \mp H_1^{(2)} \left( k_o \left| x_m - x_n + \frac{\Delta x}{2} \right| \right) \right], \qquad n \neq m, \quad x_m \gtrless x_n \pm \frac{\Delta x}{2}
\end{aligned}
\tag{5.37b}
$$

since $\frac{\partial}{\partial x} H_o^{(2)}(k_o|x - x'|) = \mp k_o H_1^{(2)}(k_o|x - x'|)$ for $x \gtrless x'$. Obviously, this result is easier to evaluate in addition to being more accurate.

To evaluate the self-cell, the usual procedure is to replace the Hankel function with the first few terms of its small argument expansion. The integration must in this case be carried out with $y \neq 0$. After integration, the result is differentiated before setting $y = 0$. Consequently, since we must at the end differentiate with respect to $y$, we replace $H_o^{(2)}$ in (5.36) with the first four terms of its small argument expansion, that is,

$$
H_o^{(2)}(z) = c_1 + c_2 z^2 - j\frac{2}{\pi} \ln z + j\frac{z^2}{2\pi} \ln z + O(z^4, z^4 \ln z) \tag{5.38}
$$

where $c_1 = 1 - j\frac{2}{\pi}(\ln \gamma - \ln 2)$ and $c_2 = -0.5 + j\frac{1}{2\pi}(\ln \gamma - 1 - \ln 2)$. Then, upon making use of the integral identities

$$
\int \ln(a^2 + x^2)\, dx = x \ln(a^2 + x^2) - 2x + 2a \tan^{-} 1(x/a) \tag{5.39}
$$

$$
\begin{aligned}
\int (a^2 + x^2) \ln(a^2 + x^2)\, dx &= x \left( \frac{x^3}{3} + a^2 \right) \ln(a^2 + x^2) + \frac{4}{3} a^3 \tan^{-} 1 \left( \frac{x}{a} \right) \\
&\quad - \frac{4}{3} a^2 x - \frac{2}{9} x^3
\end{aligned}
\tag{5.40}
$$

we obtain the expression given in Ricoy and Volakis (1989). For our case, since $\beta = y = 0$, we obtain the simplified expression

$$
Z_{nn} = R_e + \frac{k_o \Delta x}{4} \left[ 1 - j\frac{1}{\pi} \left( \ln \gamma - \frac{5}{3} + \frac{8}{(k_o \Delta x)^2} + \ln \frac{k_o \Delta x}{4} \right) \right]. \tag{5.41}
$$

It is seen that as $\Delta x \to 0$ this expression reduces to

$$
Z_{nn}\big|_{\Delta x \to 0} = R_e = j\frac{2}{\pi (k_o \Delta x)} \tag{5.42}
$$

implying that $Z_{nn} \to \infty$ as $\Delta x \to 0$. This points to the limitations of the collocation method in solving the electric field integral equation (EFIE) for $H$-polarization incidence. Nevertheless, it has been found that this method is quite reliable if the matrix condition is maintained within acceptable guidelines (usually less than $10^6$) and provided, of course, the chosen interval is sufficiently small to accurately represent the current distribution. In contrast to $E$-polarization, for $H$-polarization the near-diagonal matrix elements do not diminish as rapidly, and this is attributed to the presence of $H_1^{(2)}$, which is more singular than $H_o^{(2)}$. Consequently, the condition of the $H$-polarization impedance matrix is generally much larger (compared with the $E$-polarization matrix) and may deteriorate further as $\Delta x \to 0$.

Given the current distribution, the far-zone scattered field can be computed from (2.133). We have

$$H_{zff}^s \sim \frac{e^{-jk_o\rho}}{\sqrt{\rho}} e^{j\frac{\pi}{4}} \frac{k_o}{4} \sqrt{\frac{2}{\pi k_o}} \sin\phi \int_{\frac{-w}{2}}^{\frac{w}{2}} J_x(x') e^{jk_o x' \cos\phi}\, dx' \tag{5.43}$$

and upon substitution of (5.32) we obtain

$$H_{zff}^s \sim \frac{e^{jk_o\rho}}{\sqrt{\rho}} e^{j\frac{\pi}{4}} \frac{k_o}{4} \sqrt{\frac{2}{\pi k_o}} \sin\phi \operatorname{sinc}\left(\frac{k_o \Delta x}{2}\cos\phi\right) \Delta x \sum_{n=0}^{N-1} J_n e^{jk_o x_n \cos\phi}. \tag{5.44}$$

The associated echowidth is given by

$$\sigma_H = 2\pi\rho \lim_{\rho\to\infty} \frac{|\mathbf{H}^s|^2}{|\mathbf{H}^i|^2} \tag{5.45}$$

and from (5.44) we obtain

$$\sigma_H = \frac{k_o}{4}(\Delta x)^2 \sin^2\left(\frac{k_o\Delta x}{2}\cos\phi\right) \sin^2\phi \left|\sum_{n=0}^{N-1} J_n e^{jk_o x_n \cos\phi}\right|^2. \tag{5.46}$$

Plots of the $H$-polarization (also referred to as transverse electric [TE] polarization) current are given in Figure 5-9 for four metallic strips of different widths. The corresponding currents excited on a $2\lambda$ resistive strip are shown in Figure 5-10 for $R_e = 0.5$, 2, 5, and 10. As required, the $H$-polarization current $J_x(x)$ vanishes at the strip edges. Also, in contrast to the $E$-polarization current, it does not precisely reduce to its physical optics value

$$J_{xPO} = \frac{2\sin\phi_o}{2R_e + \sin\phi_o} e^{jk_x \cos\phi_o}$$

even when $\phi_o = \frac{\pi}{2}$ (normal incidence). Instead, the numerically computed current (which can be considered as exact) has a strong oscillatory behavior about the PO value. To a great extent, the $H$-polarization strip current is similar to the thin-wire

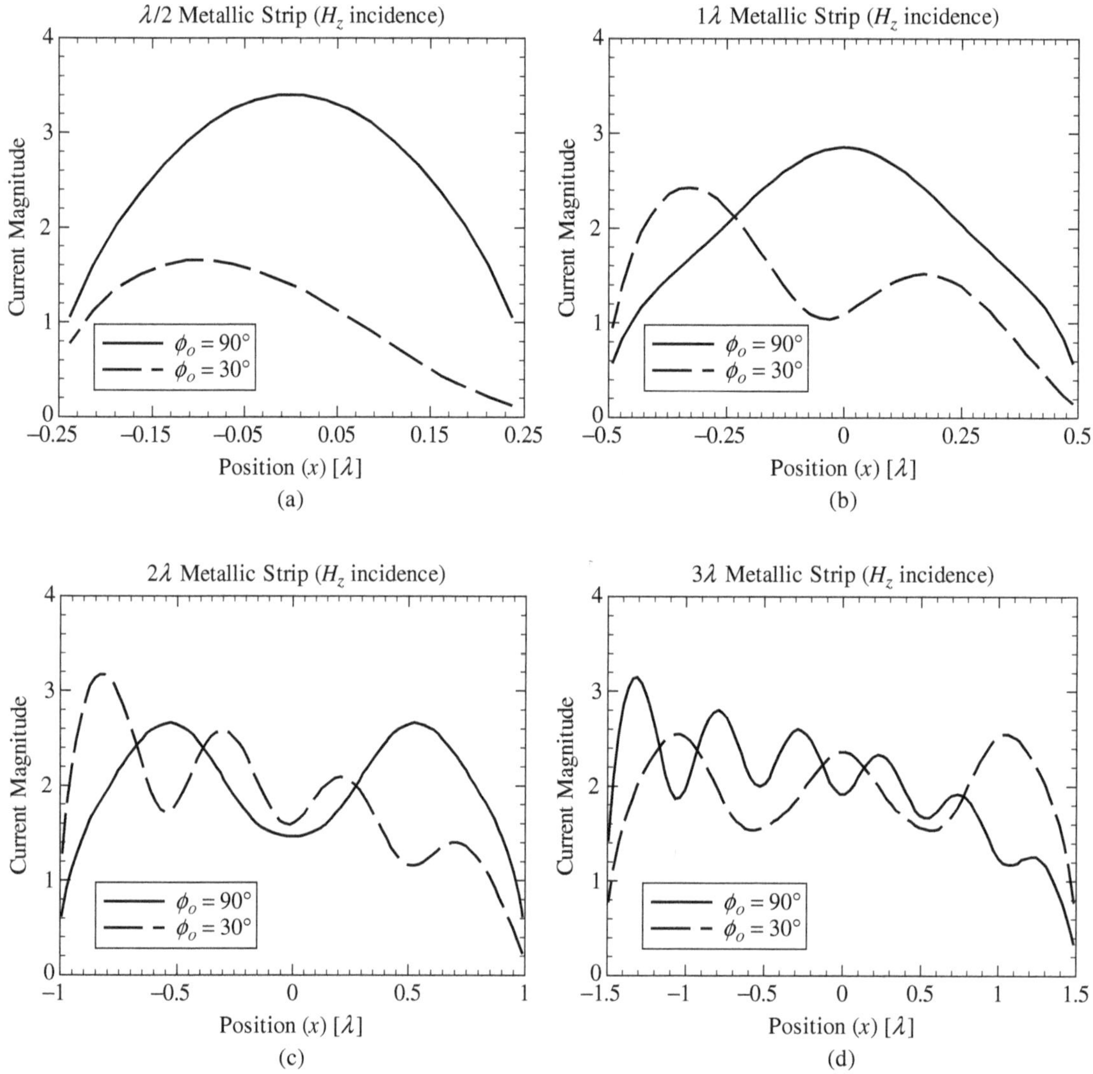


**FIGURE 5-9** ■ Currents on a metallic (flat) strip excited by an $H$-polarized plane wave incident at $\phi = 0°$, $30°$, and $90°$: (a) $w = \lambda/2$, (b) $w = \lambda$, (c) $w = 2\lambda$, and (d) $w = 3\lambda$ (Courtesy of K. Barkeshli).

current, and in fact its oscillatory behavior can be explained by again considering the traveling wave model discussed in Chapter 3. In this case, $J_{xPO} = J_{1x}$ is the current excited on an infinite width strip and can be referred to as the forced current component. When $J_{xPO}$ reaches one of the strip edges, it generates or reflects additional currents of the form $J_{2x}e^{-jk_o x}$ and $J_{3x}e^{+jk_o x}$, where $J_{2x}$ and $J_{3x}$ are some complex constants. Thus, the representation (4.62) can also be used to model the $H$-polarization strip current leading to a more efficient solution scheme. The oscillations observed in Figures 5-9 and 5-10 are simply due to the in-phase and out-of-phase additions of the aforementioned traveling current components.

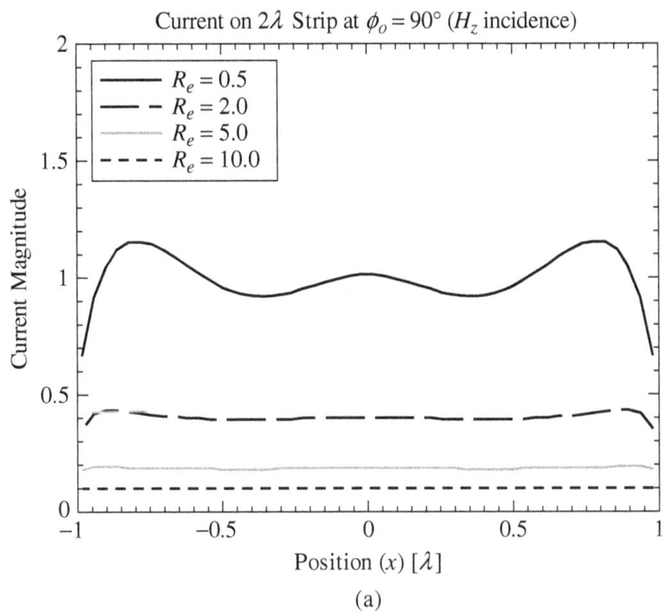


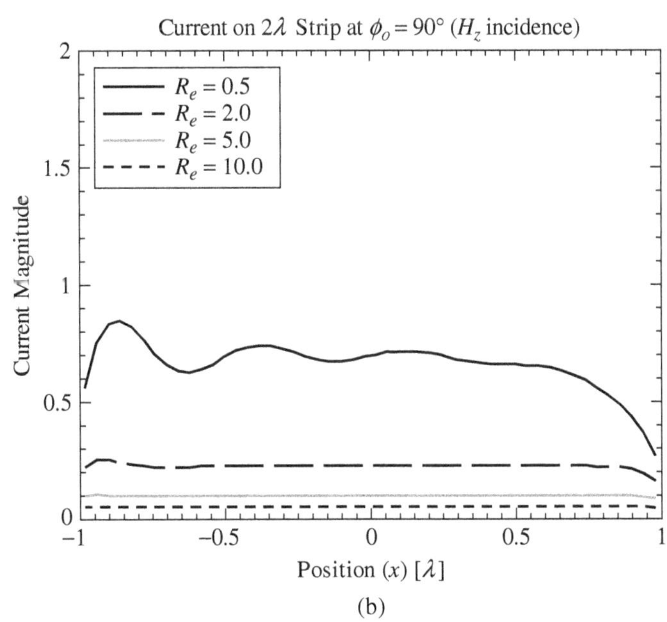


**FIGURE 5-10** $H$-polarization currents on a $2\lambda$ flat resistive strip at normal incidence for $R_e = 0.5$, 2, 5, and 10: (a) $\phi_o = 90°$, and (b) $\phi_o = 30°$. (Courtesy of K. Barkeshli)

Bistatic (fixed $\phi_o$) echowidth patterns for the flat resistive strip with $H$-polarization incidence are given in Figures 5-11 and 5-12. Figure 5-11 refers to metallic strips of different widths, whereas Figure 5-12 refers to the $2\lambda$ wide strip of different resistivities. Backscatter patterns are shown in Figure 5-13.

We remark that for $H$-polarization the presented bistatic and backscatter patterns display vanishing echowidths for observations near $\phi = 0$ or $\pi$ (i.e., for incidence in the plane of the strip). This is in contrast to the $E$-polarization patterns and is so

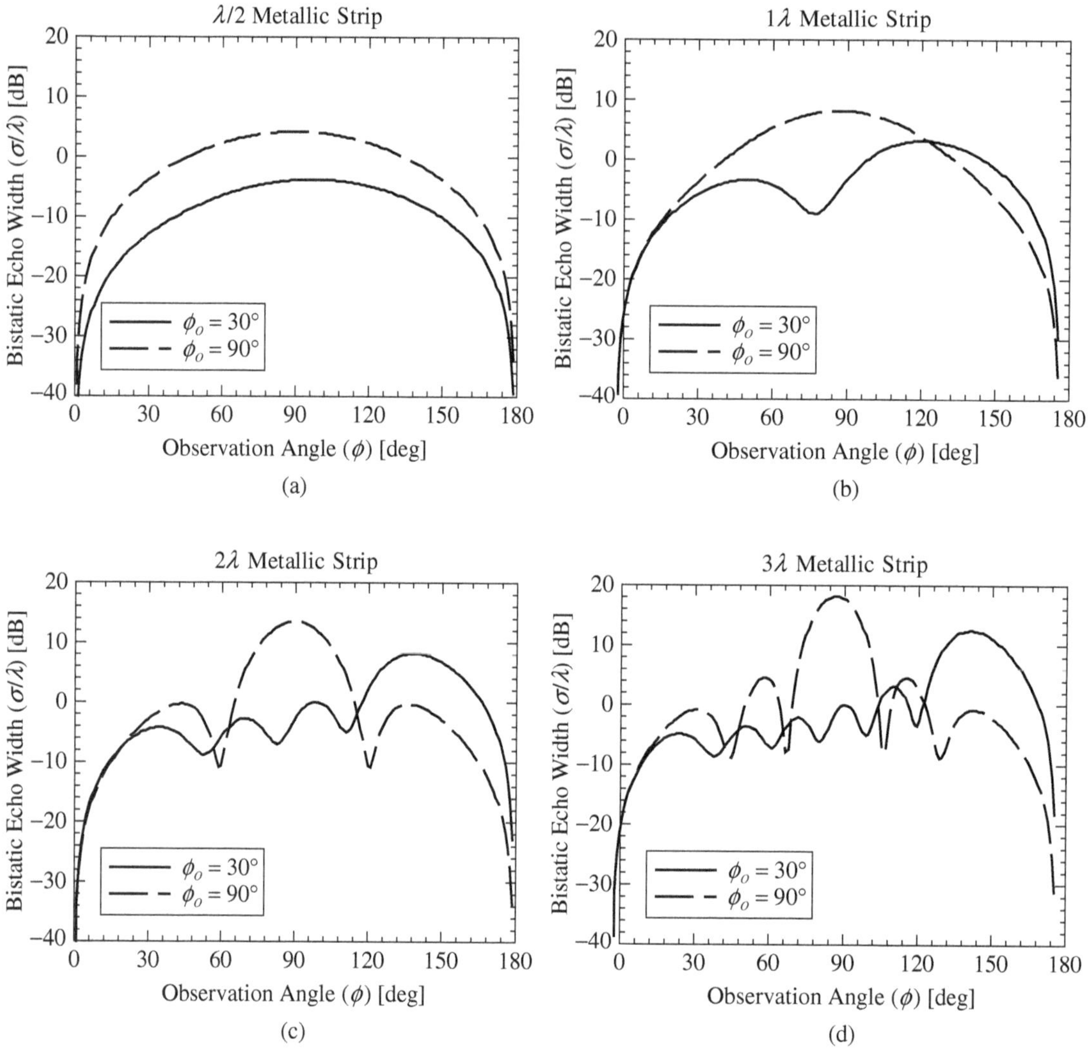


**FIGURE 5-11** ■ Bistatic echowidth patterns for a flat metallic strip excited by an $H$-polarized plane wave incident at $\phi_o = 30^\circ$ and $90^\circ$: (a) $w = \frac{\lambda}{2}$, (b) $w = \lambda$, (c) $w = 2\lambda$, and (d) $w = 3\lambda$. (Courtesy of K. Barkeshli)

because at grazing incidence the polarity of the incident field is normal to the strip's surface. It thus satisfies the boundary condition on the strip without the presence of a scattered field. However, for observations near $\phi_o = \phi = \frac{\pi}{2}$, both the $E$- and $H$-polarization patterns are identical and agree with those predicted via the physical optics approximation.

#### 5.1.2.2 Linear Basis-Galerkin's Solution

To avoid difficulties with the kernel singularity of the integral equation (5.30), the usual procedure is to employ a higher-order expansion for the current. This allows us to transfer one of the derivatives to the current by employing the divergence theorem

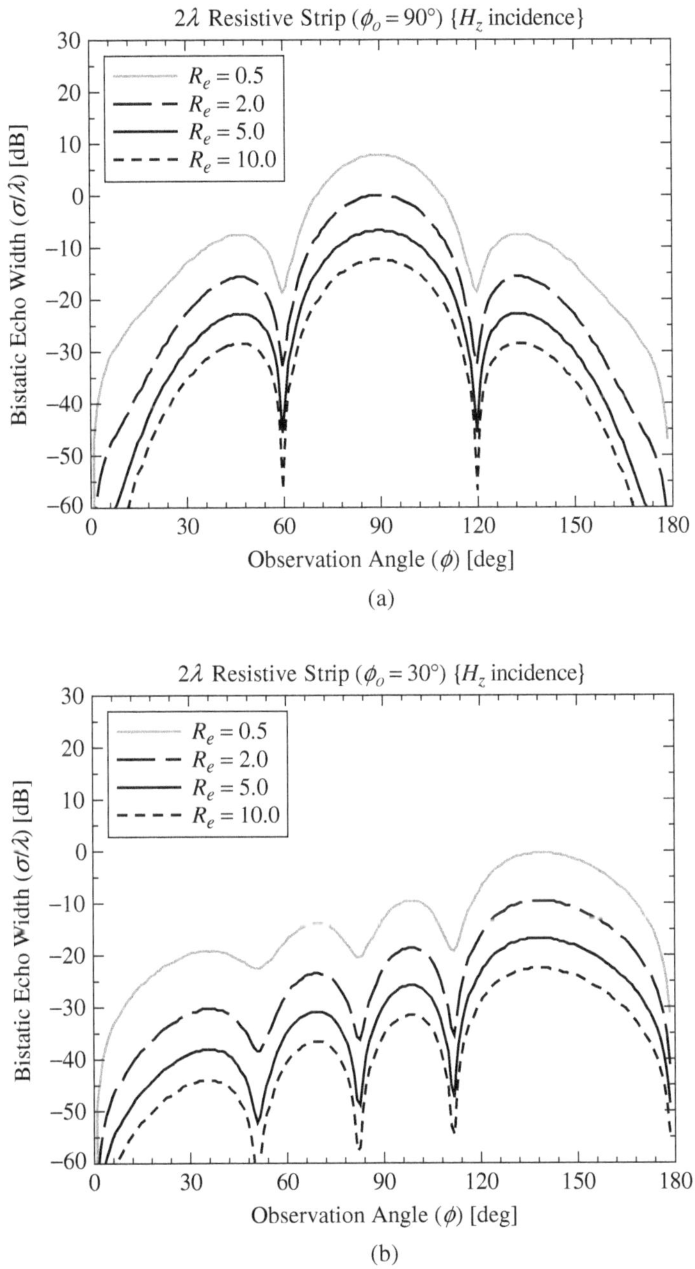


**FIGURE 5-12** *H*-polarization bistatic echowidth patterns for a 2λ resistive strip with $R_e = 0.5$, 2, 5, and 10: (a) $\phi_o = 90^\circ$, and (b) $\phi_o = 30^\circ$. (Courtesy of K. Barkeshli)

or integration by parts as done in (4.12). Noting that $J_x\left(\pm\frac{w}{2}\right) = 0$, we can rewrite (5.30) as

$$\sin\phi_o e^{+jk_o x\cos\phi} = R_e J_x(x) + \frac{k_o}{4}\int_{-w/2}^{w/2} J_x(x')H_o^{(2)}(k_o|x-x'|)\,dx' + \frac{1}{4k_o}\frac{\partial}{\partial x}\int \frac{dJ_x(x')}{dx'}H_o^{(2)}(k_o|x-x'|)\,dx' \tag{5.47}$$

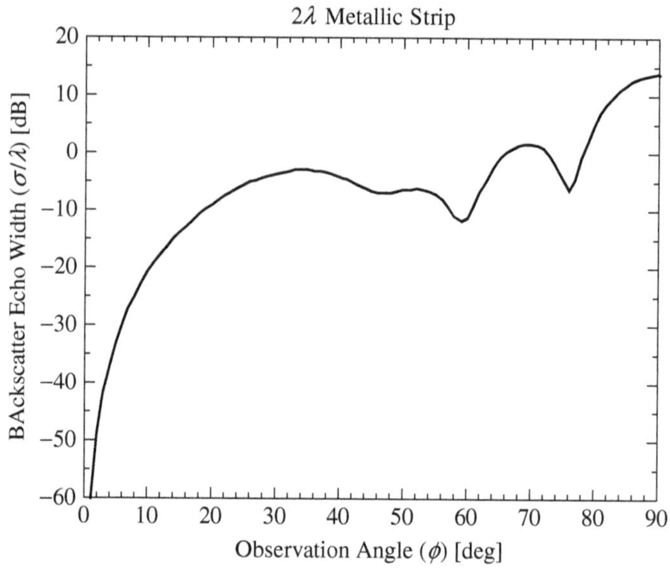


(a)

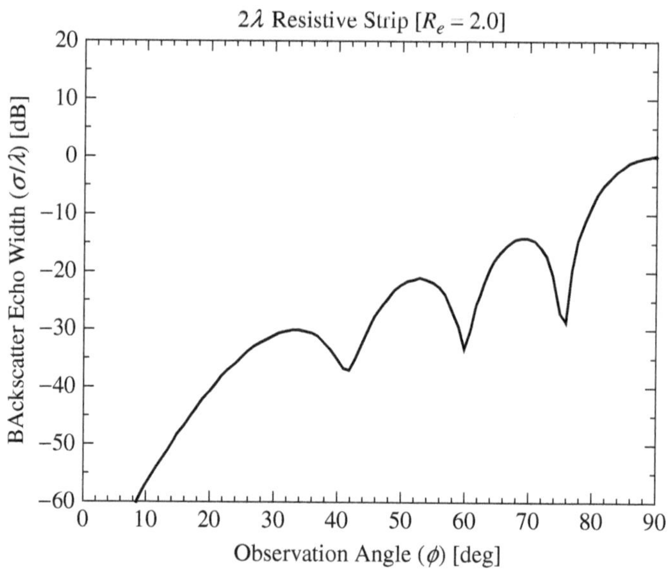


(b)

**FIGURE 5-13** *H*-polarization backscatter patterns for a resistive strip of different widths: (a) $R_e = 0$, and (b) $R_e = 2$. (Courtesy of K. Barkeshli)

which upon setting $R_e = 0$ is the same as the EFIE given in (3.99). To solve this integral equation, the current is expanded in terms of the triangle functions as

$$J_x(x) = \sum_{n=0}^{N-1} J_n T(x - x_n) \tag{5.48}$$

where now $x_n = -\frac{w}{2} + (n+1)\Delta x$. Choosing the weighting functions to be also equal to the expansion functions (Galerkin's solution) we obtain the set of equations

$$V_m = \sum_{n=0}^{N-1} J_n \zeta_{mn}^1 + \frac{k_o}{4} \sum_{n=0}^{N-1} J_n \zeta_{mn}^2 - \frac{1}{4k_o} \sum_{n=0}^{N-1} J_n \zeta_{mn}^3 \tag{5.49}$$

where

$$\begin{aligned} V_m &= \sin\phi_o \int_{x_m-\Delta x}^{x_m+\Delta x} \left(1 - \frac{|x - x_m|}{\Delta x}\right) e^{jk_o x \cos\phi_o}\, dx \\ &= -2 \sin\phi_o e^{jk_o x_m \cos\phi_o} \left[\frac{\cos(k_o \Delta x \cos\phi_o) - 1}{\Delta x k_o^2 \cos^2\phi_o}\right] \\ &\approx \Delta x \sin\phi_o e^{jk_o x_m \cos\phi_o} \end{aligned} \tag{5.50}$$

$$\zeta_{mn}^1 = \int_{x_m-\Delta x}^{x_m+\Delta x} R_e(x) \left(1 - \frac{|x - x_m|}{\Delta x}\right) \left(1 - \frac{|x - x_n|}{\Delta x}\right) dx \tag{5.51a}$$

$$\zeta_{mn}^2 = \int_{x_m-\Delta x}^{x_m+\Delta x} dx \left(1 - \frac{|x - x_m|}{\Delta x}\right) \int_{x_n-\Delta x}^{x_n+\Delta x} dx' \left(1 - \frac{|x - x_n|}{\Delta x}\right) H_o^{(2)}(k_o|x - x'|) \tag{5.51b}$$

$$\begin{aligned} \zeta_{mn}^3 = &\int_{x_m-\Delta x}^{x_m+\Delta x} H_o^{(2)}(k_o|x - x'|) \frac{d}{dx}\left(1 - \frac{|x - x_m|}{\Delta x}\right) \\ &\cdot \int_{x_n-\Delta x}^{x_n+\Delta x} \frac{d}{dx'}\left(1 - \frac{|x' - x_n|}{\Delta x}\right) dx'\, dx. \end{aligned} \tag{5.51c}$$

The last term $(\zeta_{mn}^3)$ in this expression was derived by again employing integration by parts to transfer the derivative from the Hankel function to the testing function $T_m(x)$ for the purpose of reducing even further the kernel's singularity. As usual, (5.49) can be written as

$$\{V_m\} = [Z_{mn}]\{J_n\} \tag{5.52}$$

in which the impedance matrix elements are identified to be

$$Z_{mn} = \zeta_{mn}^1 + \frac{k_o}{4}\zeta_{mn}^2 - \frac{1}{4k_o}\zeta_{mn}^3. \tag{5.53}$$

Given the resistivity profile $R_e(x)$, the integral $\zeta_{mn}^1$ can be readily evaluated analytically or numerically. In most cases, though, it is sufficient to replace $R_e(x)$ by its average value at the midpoint of each discretization segment of length $\Delta x$. Doing the latter permits an analytical evaluation of $\zeta_{mn}^1$. Specifically, we have

$$\zeta_{mn}^1 = \begin{cases} \frac{\Delta x}{3}\left[R_e\left(x = x_m - \frac{\Delta x}{2}\right) + R_e\left(x = x_m + \frac{\Delta x}{2}\right)\right] & n = m \\ \frac{\Delta x}{6} R_e\left(x = x_m + \frac{\Delta x}{2}\right) & n = m + 1 \\ \frac{\Delta x}{6} R_e\left(x = x_m - \frac{\Delta x}{2}\right) & n = m - 1 \\ 0 & \text{otherwise.} \end{cases} \tag{5.54}$$

To evaluate the integral $\zeta_{mn}^2$ we recognize that $T_m(x)$ and $T_n(x')$ are slowly varying functions in comparison with the Hankel function $H_o^{(2)}(k_o|x-x'|)$ particularly when $n=m$ or $n=m\pm 1$. Thus, without compromising accuracy, we can employ the five-point integration formula (since $\Delta x$ is very small) to rewrite it as

$$\zeta_{mn}^2 = \sum_{p=-2}^{2} \alpha_p \sum_{q=-2}^{2} \alpha_q I_{mn}^{pq} \tag{5.55}$$

where

$$I_{mn}^{pq} = \int_{\left(x_m+\frac{2p-1}{5}\right)\Delta x}^{\left(x_m+\frac{2p+1}{5}\right)\Delta x} \int_{\left(x_n+\frac{2q-1}{5}\right)\Delta x}^{\left(x_n+\frac{2q+1}{5}\right)\Delta x} H_o^{(2)}\left(k_o|x-x'|\right)\, dx'\, dx \tag{5.56}$$

and the coefficients

$$\alpha_p = \begin{cases} \frac{1}{5} & p=-2,2 \\ \frac{3}{5} & p=-1,1 \\ 1 & p=0 \end{cases} \tag{5.57}$$

are simply the values of $T(x_p - x_m)$ with $x_p = x_m + \frac{2p}{5}\Delta x$. The integral $I_{mn}^{pq}$ can be evaluated numerically using the midpoint integration formula provided $m \neq n$ or $p \neq q$. When $m=n$ and $p=q$, the integrand has a logarithmic singularity and must then be evaluated analytically. In this case the argument of the Hankel function is small, and the standard procedure is to replace it by its small argument expansion and proceed to perform the integration analytically. Using the two-term expansion (5.12), we obtain

$$\begin{aligned} I_{mm}^{pp} &= \int_{-\delta/2}^{\delta/2}\int_{-\delta/2}^{\delta/2}\left(1 - j\frac{2}{\pi}\ln\left|\frac{\gamma k_o(x-x')}{2}\right|\right) dx'\, dx \\ &= \delta^2 - j\frac{2}{\pi}\int_{-\delta/2}^{\delta/2}\left\{\int_{-\delta/2}^{x}\ln\frac{\gamma k_o(x-x')}{2}\,dx' + \int_{x}^{\delta/2}\ln\frac{\gamma k_o(x'-x)}{2}\,dx'\right\} dx \\ &= \delta^2 - j\frac{4}{\pi}\int_0^{\delta} t\ln\frac{\gamma k_o t}{2e}\,dt \end{aligned}$$

which can be integrated to give

$$I_{mm}^{pp}(\delta) = \delta^2\left[1 - j\frac{2}{\pi}\ln\left(\frac{\gamma k_o\delta}{2e^{3/2}}\right)\right] \tag{5.58}$$

where $\delta = \frac{2\Delta x}{5}$, $\gamma = 1.781$ and $\ln e = 1$. For $m \neq n$ or $p \neq q$ we may set

$$I_{mn}^{pq} \approx \left(\frac{2}{5}\Delta x\right)^2 H_o^{(2)}\left(k_o\left|\left[(m+1)+\frac{2p}{5}\right]\Delta x - \left[(n+1)+\frac{2q}{5}\right]\Delta x\right|\right).$$

However, when $\frac{2}{5}\Delta x$ is very small, it may be advisable to resort to an analytical evaluation of those adjacent elements for which $m = n \pm 1$ or $p = q \pm 1$. The procedure in this case is to employ a higher-order small argument expansion for the Hankel function. Up to $(x^4, x^4 \ln x)$ is sufficient, and the result of such an integration is given in Ricoy and Volakis (1989). Note that the dominant terms of this expression (with $\beta = 0$) should recover (5.58).

It remains to evaluate the integral $\zeta_{mn}^3$, which after carrying out the differentiations can be written as

$$\begin{aligned}\zeta_{mn}^3 &= \frac{1}{(\Delta x)^2}\left[\int_{m\Delta x}^{(m+1)\Delta x} - \int_{(m+1)\Delta x}^{(m+2)\Delta x}\right]\left[\int_{n\Delta x}^{(n+1)\Delta x} - \int_{(n+1)\Delta x}^{(n+2)\Delta x}\right] \\ &\quad \cdot H_o^{(2)}(k_o|x - x'|)\, dx'\, dx \\ &= \frac{1}{(\Delta x)^2}\int_0^{\Delta x}\int_0^{\Delta x} \{2H_o^{(2)}(k_o|(m-n)\Delta x + x - x'|) \\ &\quad - H_o^{(2)}(k_o|(m-n-1)\Delta x + x - x'|) \\ &\quad - H_o^{(2)}(k_o|m-n+1)\Delta x + x - x'|)\}\, dx\, dx'. \end{aligned} \tag{5.59}$$

It is clear that for $m = n$ the first of the previous integrands becomes singular over the integration range. The second and third integrals have the same behavior when $m = n + 1$ and $m = n - 1$, respectively. In those cases the result (5.58) should be employed to analytically evaluate the corresponding integrals. Specifically, by using (5.58) we have

$$\int_0^{\Delta x}\int_0^{\Delta x} H_o^{(2)}(k_o|x - x'|)\, dx\, dx' \approx (\Delta x)^2\left[1 - j\frac{2}{\pi}\ln\left(\frac{\gamma k_o \Delta x}{2e^{3/2}}\right)\right]. \tag{5.60}$$

Consequently, we may approximate $\zeta_{mn}^3$ as follows:

$$\zeta_{mn}^3 \approx \begin{cases} 2(\Delta x)^2\left[1 - j\frac{2}{\pi}\ln\left(\frac{\gamma k_o \Delta x}{2e^{3/2}}\right)\right] - 2H_o^{(2)}(k_o\Delta x) & m = n \\ 2H_o^{(2)}(k_o\Delta x) - (\Delta x)^2\left[1 - j\frac{2}{\pi}\ln\left(\frac{\gamma k_o \Delta x}{2e^{3/2}}\right)\right] - H_o^{(2)}(2k_o\Delta x) & m = n+1 \\ 2H_o^{(2)}(k_o\Delta x) - H_o^{(2)}(2k_o\Delta x) - (\Delta x)^2 \cdot \left[1 - j\frac{2}{\pi}\ln\left(\frac{\gamma k_o \Delta x}{2e^{3/2}}\right)\right] & m = n-1 \\ 2H_o^{(2)}(k_o|m-n|\Delta x) - H_o^{(2)}(k_o|m-n-1|\Delta x) - H_o^{(2)}(k_o|m-n+1|\Delta x) & \text{otherwise.} \end{cases} \tag{5.61}$$

Given the coefficients of the current expansion (5.48), the evaluation of the far zone fields and echowidth proceeds in the usual manner. Specifically, upon substitution of (5.48) into (5.43) we obtain

$$H_{zff}^s \sim -\frac{e^{-jk_o\rho}}{\sqrt{\rho}} e^{j\frac{\pi}{4}} \frac{k_o}{2}\sqrt{\frac{2}{\pi k_o}} \sin\phi \sum_{n=0}^{N-1} J_n e^{jk_o x_n \cos\phi_o}\left[\frac{\cos(k_o \Delta x \cos\phi_o) - 1}{\Delta x k_o^2 \cos^2\phi_o}\right] \tag{5.62}$$

from which we can compute the echowidth $\sigma_H$ as defined in (5.45). The resulting current distribution and echowidth is, of course, identical to the pulse-basis solution given in the previous section. This is so provided $N$ is sufficiently large, but, in general, fewer segments are required in connection with the Galerkin's linear expansion solution to reach convergence. The price one pays for the better convergence and accuracy of the Galerkin's linear expansion solution is a higher complexity in evaluating the impedance matrix elements, which in turn translates to a higher matrix fill time.

### 5.1.2.3 Narrow Strips

Similar to $E$-polarization, the $H$-polarization integral equation for the narrow perfectly conducting strip current can be solved analytically. To obtain a simplified integral equation for the narrow strip, we begin with (5.30) and replace the Hankel function with its two-term small argument expansion (5.12). Also, setting $R_e = 0$ and making the additional approximation $e^{jk_o x \sin\phi_o} \approx 1$ (since $k_o x \ll 1$) yields

$$\frac{k_o}{4}\int_{-w/2}^{w/2} J_x(x')\left(1+\frac{1}{k_o^2}\frac{d^2}{dx^2}\right)\left[1-j\frac{2}{\pi}\ln\left(\frac{k_o\gamma|x-x'|}{2}\right)\right]dx' = \sin\phi_o. \quad (5.63)$$

Upon introducing the change of variable (same as for $E$-polarization)

$$\xi = \frac{2x}{w}, \quad \xi' = \frac{2x'}{w} \quad (5.64)$$

this can be further simplified to yield

$$\frac{d^2}{d\xi^2}\int_{-1}^{1} J_x(\xi')\ln|\xi-\xi'|\,d\xi' = j\pi k_o w\sin\phi_o. \quad (5.65)$$

We now recall the identity (Tricomi, 1985)

$$\frac{d^2}{dx^2}\int_{-1}^{1}\sqrt{1-x'^2}\ln|x-x'|\,dx' = \pi \qquad x\in(-1,1) \quad (5.66)$$

and when this is compared with (5.60) we deduce that

$$J_x(\xi) = \chi_h\sqrt{1-\xi^2} = \chi_h\sqrt{1-\left(\frac{2x}{w}\right)^2} \quad (5.67)$$

where

$$\chi_h = jk_o w\sin\phi_o. \quad (5.68)$$

As expected, the current vanishes when $x = \frac{\pm w}{2}$, and furthermore this result displays that $J_x \sim \sqrt{d}$, where $d$ denotes the distance from the edge. This is the well-known current edge condition often invoked in diffraction theory and also observed in the results of Figure 5-11.

The corresponding echowidth is given by

$$\sigma_h = \frac{k_o}{4}\left|\sin\phi\int_{-w/2}^{w/2} J_x(x')\,dx'\right| \approx k_o\left|\frac{\pi w}{8}\chi_h\sin\phi\right|^2. \quad (5.69)$$

Plots of this echowidth formula as a function of $w$ are given in Figure 5-14.

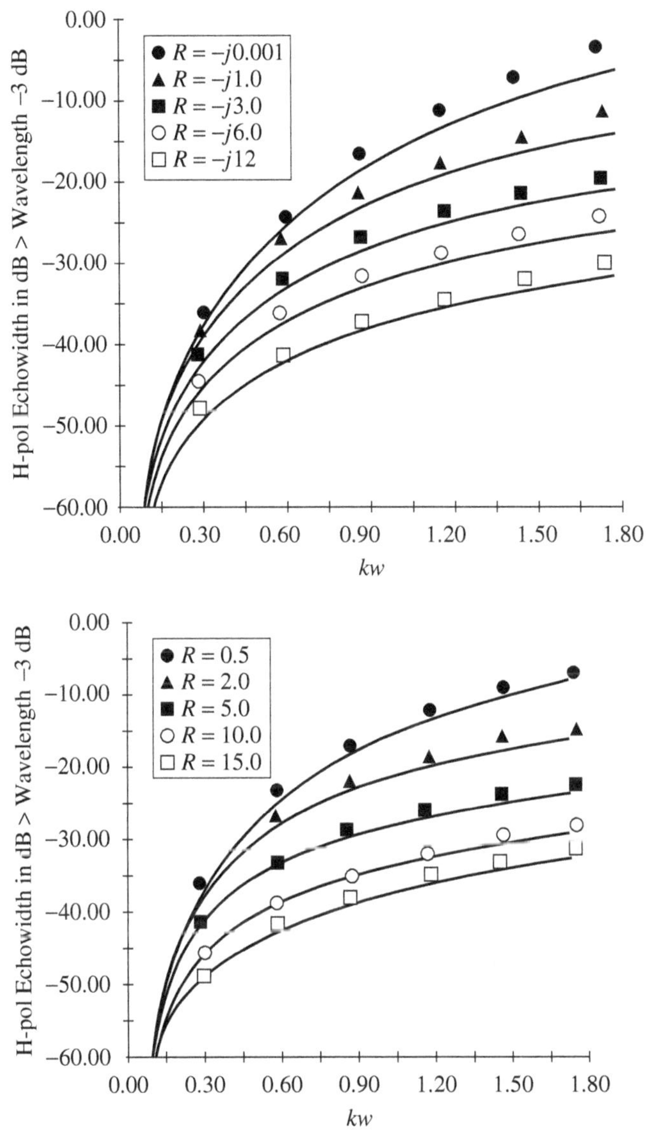


**FIGURE 5-14** *H*-polarization backscatter echowidth plots for a narrow resistive strip based on (5.69) and the formula stated in Senior and Volakis (1989). The method of moments data are displayed by the symbols. (After Senior and Volakis, 1989. With permission.)

## 5.2 METALLIC CYLINDERS

The geometry of the cylinder is shown in Figure 5-15 and we shall assume that the cylinder's boundary $C$ is of arbitrary shape. The pertinent integral equations for the equivalent surface currents are given by (3.98), and we note that the TM (or $E$-polarization) equation was already employed in Section 5.1.1. In that case, the TM integral equation was enforced on a flat strip to yield the surface current $J_z(x)$. Here we must discretize (3.98a) for an arbitrary curved surface where $\overline{\rho}$ and $\overline{\rho}'$ will now be functions of both $x$ and $y$. Thus, in performing the required integrations over the contour $C$, it will be necessary to employ some suitable parametric representation for the contour line. The same must, of course, be done for the TE integral equation (3.98b) whose implementation is more challenging because of the kernel's higher

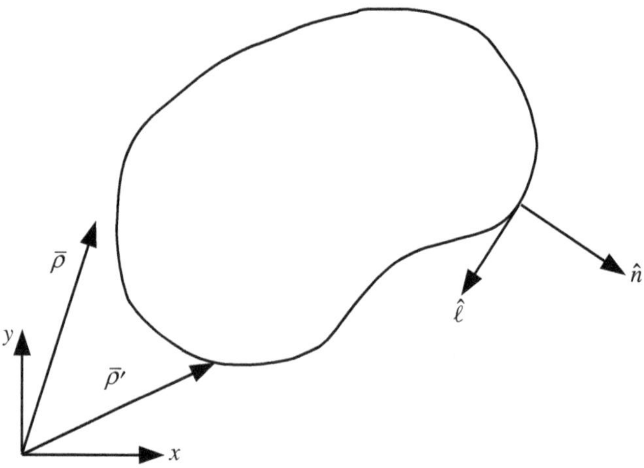


**FIGURE 5-15** ■ Illustration of the metallic cylinder.

singularity. As noted in Section 3.2.3, this integral equation was not employed to model the flat strip for $H$-polarization because it is valid only for closed surfaces, whereas the strip is an open surface.

For simplicity let us again begin with the solution of the TM integral equation.

### 5.2.1 *E*-Polarization

The field impinging upon the cylinder is assumed to be a plane wave of the same form as given in (5.3). Using this in (3.98a) yields the integral equation

$$\frac{k_o Z_o}{4} \int_C J_z(\overline{\rho}') H_o^{(2)}(k_o|\overline{\rho} - \overline{\rho}'|)\, dl' = e^{jk_o(x\cos\phi_o + y\sin\phi_o)}, \qquad \overline{\rho} \in C \qquad (5.70)$$

where $\overline{\rho} = x\hat{x} + y\hat{y}$ and $\overline{\rho}' = x'\hat{x} + y'\hat{y}$ are vectors that describe the integration and testing points on the contour C. To discretize this equation we must first divide it into $N$ small segments. For simplicity these can be chosen flat (linear) as shown in Figure 5-16. Although higher-order segments (e.g., quadratic, cubic) could be also used, flat segments are generally acceptable for most computational needs.

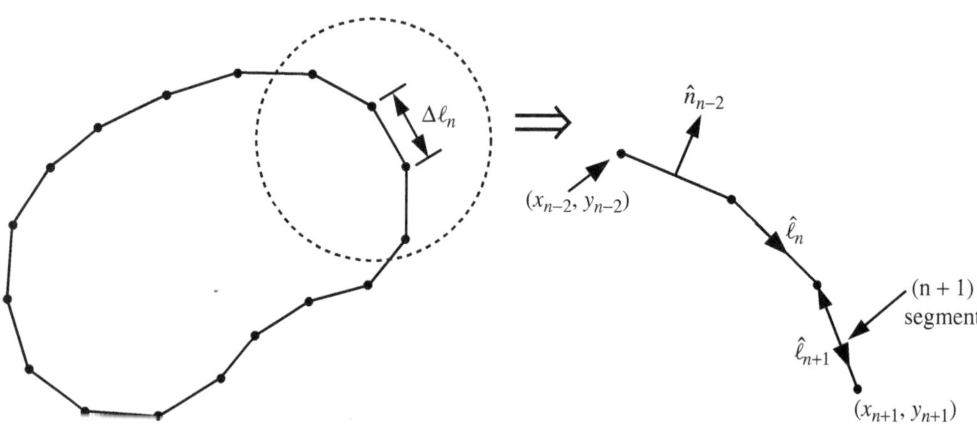


**FIGURE 5-16** ■ Segmentation of the cylinder's surface for numerical modeling.

We sequentially denote the points joining the flat segments by $(x_n, y_n)$, with $n = 0, 1, 2, \cdots, N$. Thus, the parametric representation for the points on the $(n+1)$th flat segment is

$$x = x_n + l\cos\theta_n \qquad y = y_n + l\sin\theta_n \tag{5.71}$$

in which

$$\theta_n = \tan^{-1}\left(\frac{y_n - y_{n-1}}{x_n - x_{n-1}}\right) \tag{5.72}$$

and $l$ is the distance measured along the $n$th segment starting from the point $(x_n, y_n)$. The current on each segment can be expressed by using one of the standard subsectional basis functions given in (4.19) through (4.21). Again, for the TM (or $E$-polarization) integral equation, pulse basis functions are suitable for this purpose. That is, we shall choose to express $J_z(l)$ on the $n$th segments as $J_n P_{\Delta l_n}(l - \frac{\Delta l_n}{2})$, where $l$ is measured from the point $(x_n, y_n)$, and we have assumed a segment of length $\Delta l_n$. Using this expansion and the parametric representation (5.71) in (5.70) permits a discretization of the TM integral equation. Then, by point matching at $x_{m+\frac{1}{2}} = x_m + \frac{\Delta l_m}{2}\cos\theta_m$, $y_{m+\frac{1}{2}} = y_m + \frac{\Delta l_m}{2}\sin\theta_m$ we obtain the linear equations

$$V_m = \frac{k_o Z_o}{4}\sum_{n=0}^{N-1} J_n \int_0^{\Delta l_n} H_o^{(2)}(k_o, R'_{mn})\,dl', \qquad m = 0, 1, \cdots, N-1 \tag{5.73}$$

where

$$R'_{mn} = \left[\left(x_m + \frac{\Delta l_m}{2}\cos\theta_m - x_n - l'\cos\theta_n\right)^2 + \left(y_m + \frac{\Delta l_m}{2}\sin\theta_m - y_n - l'\sin\theta_n\right)^2\right]^{\frac{1}{2}} \tag{5.74}$$

and

$$V_m = e^{jk_o\left[\left(x_m + \frac{\Delta l_m}{2}\cos\theta_m\right)\cos\phi_o + \left(y_m + \frac{\Delta l_m}{2}\sin\theta_m\right)\sin\phi_o\right]}. \tag{5.75}$$

For $m \neq n$, the integral in (5.73) can be readily evaluated numerically, but for $m = n$ it has the usual logarithmic singularity requiring analytical integration. As before, we may use the two-term expansion of the Hankel function (5.12) to obtain

$$\int_o^{\Delta l_m} H_o^{(2)}(k_o R'_{mm})\,dl' = \int_{\frac{-\Delta l_m}{2}}^{\frac{\Delta l_m}{2}} H_o^{(2)}(k_o|l'|)\,dl' \approx \Delta l_m\left[1 - \frac{j2}{\pi}\left(\ln\left(\frac{k_o\gamma\Delta l_m}{4}\right) - 1\right)\right]. \tag{5.76}$$

Consequently, the impedance matrix elements associated with the system (5.73) are explicitly given by

$$Z_{mn} = \frac{k_o Z_o}{4}\begin{cases}\Delta l_n H_o^{(2)}(k_o R_{mn}) & m \neq n \\ \Delta l_n\left[1 - \frac{j2}{\pi}\left(\ln\left(\frac{k_o\gamma\Delta l_n}{4}\right) - 1\right)\right] & m = n\end{cases} \tag{5.77}$$

in which

$$R_{mn} = \left[\left(x_m - x_n + \frac{\Delta l_m}{2}\cos\theta_m - \frac{\Delta l_n}{2}\cos\theta_n\right)^2 + \left(y_m - y_n + \frac{\Delta l_m}{2}\sin\theta_m - \frac{\Delta l_n}{2}\sin\theta_n\right)^2\right]^{\frac{1}{2}}. \tag{5.78}$$

The computation of the scattered field and echowidth proceeds in the usual manner. We have

$$\begin{aligned} E^s_{zff} &\sim \sqrt{\frac{k_o}{8\pi}}\frac{e^{-j(k_o\rho-\frac{\pi}{4})}}{\sqrt{\rho}}\int_C Z_o\hat{\rho}\times\hat{\rho}\times\hat{z}J_z(l')e^{jk_o\hat{\rho}\cdot\bar{\rho}'}dl' \\ &\sim -\frac{e^{-jk_o\rho}}{\sqrt{\rho}}e^{j\frac{\pi}{4}}\frac{k_oZ_o}{4}\sqrt{\frac{2}{\pi k_o}} \\ &\quad\cdot\sum_{n=0}^{N-1}J_n\int_0^{\Delta l_n}e^{jk_o[(x_n+l'\cos\theta_n)\cos\phi+(y_n+l'\sin\theta_n)\sin\phi]}dl' \\ &\sim -\frac{e^{jk_o\rho}}{\sqrt{\rho}}e^{j\frac{\pi}{4}}\frac{k_oZ_o}{4}\sqrt{\frac{2}{\pi k_o}}\sum J_ne^{jk_o(x_n\cos\phi+y_n\sin\phi)} \\ &\quad\cdot\Delta l_ne^{jk_o\frac{\Delta l_n}{2}(\cos\theta_n\cos\phi+\sin\theta_n\sin\phi)}\operatorname{sinc}\left[\frac{\Delta l_n}{2}(\cos\theta_n\cos\phi+\sin\theta_n\sin\phi)\right] \end{aligned} \tag{5.79}$$

from which we can compute the echowidth defined in (5.21). Patterns of the bistatic and backscatter echowidth for a circular and square metallic cylinder are given in Figures 5-17 and 5-18, respectively. The eigenfunction or series solution used for validating the method of moments results was obtained from

$$\sigma_E = \frac{2}{\pi}\left|\sum_{n=0}^{\infty}\frac{2}{1+\delta(n)}\frac{J_n(k_oa)}{H_n^{(2)}(k_oa)}\cos(n\phi)\right|^2$$

where $a$ denotes the cylinder's radius, and $\delta(n)$ is the Dirac delta function defined in (5.11).

As noted earlier, the transverse magnetic (TM) integral equation (5.70) is valid for open as well as closed metallic surfaces and can thus be used to study the effect of curvature on the scattering cross section (echowidth) of a metallic strip. To do so, we consider the five circular strips curved as illustrated in Figure 5-19a, but all have the same circumference of $2\lambda_o$. The curvature ($\kappa = 1/a$, where $a$ is the strip radius) of the strips varies from 0 to $\pi/\lambda_o$ and the one with the highest curvature is actually a circular cylinder of radius $\lambda_o/\pi$. The bistatic echowidths for each of the strips with $\phi = \phi_o = 90^\circ$ is given in Figure 5-19b. It is seen that the echowidth of each strip peaks at the backscatter and forward directions regardless of the strip's curvature. Also, at these locations, the echowidth is larger for the flatter strips and, in fact, at backscatter ($\phi = 90^\circ$) it is equal to $k_ow = 8\pi\lambda_o$ for the flat strip and closer to $\pi a = \pi/\kappa$ for the curved strips. The last result is based on a geometrical optics (GO) approximation of the scattered field. More specifically, on the basis of

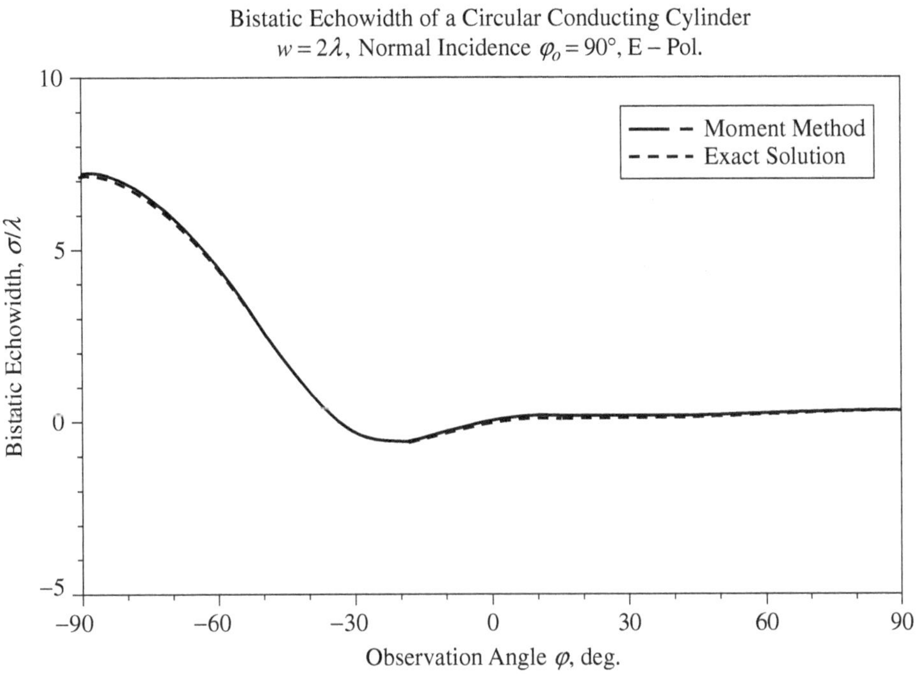


**FIGURE 5-17** Computed (bistatic) $E$-polarization echowidth of a circular metallic cylinder whose circumference is 2λ. Comparison of method of moments and eigenfunction/series solution (20 terms) (Barkeshli and Volakis, 2004).

the GO approximation, the scattered field by a circular cylinder along the direction of optical reflection is given by

$$E^s_{zff}\left(\text{or } H^s_{zff}\right) = \sqrt{\frac{1}{2\kappa}}\frac{e^{-jk_o\rho_r}}{\sqrt{\rho_r}}$$

where $\rho_r$ is the distance between the reflection and observation point. From (5.21) it follows that $\sigma_E = \pi/\kappa$ along the specular direction $\hat{\rho}_r$, such that $\hat{\rho}_r \cdot \hat{n} = -\hat{n} \cdot \hat{\rho}_o$, where $\hat{\rho}_o = \hat{x}\cos\phi_o + \hat{y}\sin\phi_o$ and $\hat{n}$ is the unit normal of the circular surface.

## 5.2.2 *H*-Polarization

The integral equation to be considered in this case is given by (3.98b) with the incident field as defined in (5.28). Specifically, we have

$$+\frac{1}{2}J_l(l) + \frac{j}{4}\oint_C J_l(l')\frac{\partial}{\partial n'}H_o^{(2)}(k_o|\overline{\rho} - \overline{\rho}'|)\,dl' = H_z^i(x, y) \tag{5.80}$$

where $J_l(l) = \hat{l} \cdot \mathbf{J}(l)$, $\hat{z} \times \hat{n} = \hat{l}$, and $\hat{n}$ denotes the unit normal to $C$. Using (3.92), this can be more explicitly written as

$$+\frac{1}{2}J_l(l) + \frac{jk_o}{4}\oint_C J_l(l')(\hat{n}' \cdot \hat{R})H_1^{(2)}(k_o|\overline{\rho} - \overline{\rho}'|)\,dl' = H_z^i(x, y) \tag{5.81}$$

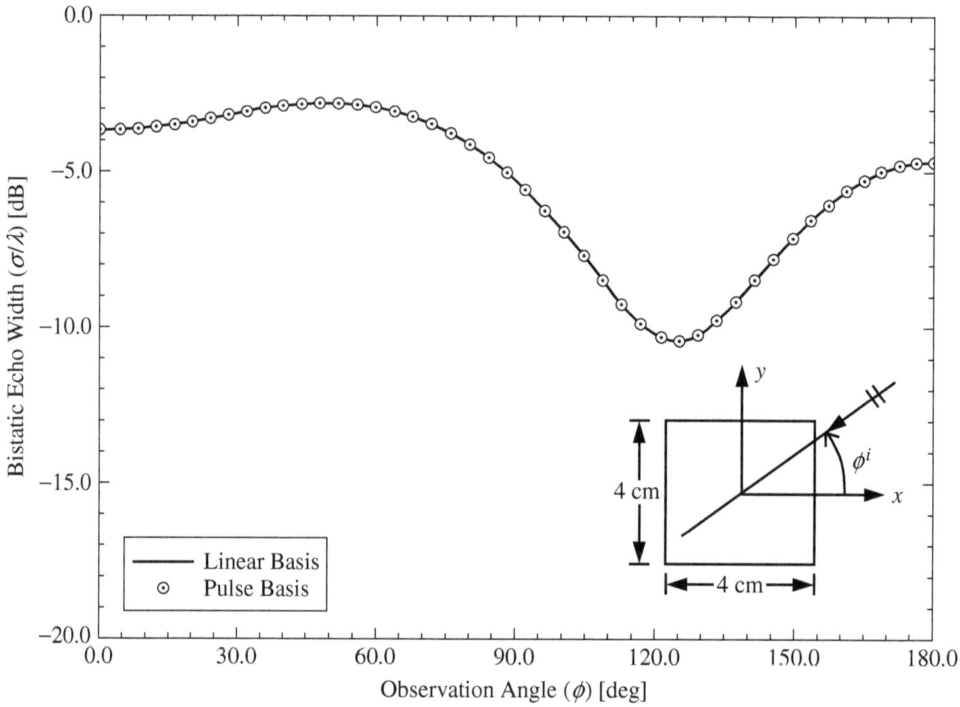


**FIGURE 5-18** *H*-polarization bistatic and backscatter echowidth patterns for a metallic rectangular cylinder at a frequency of $f = 2$ GHz (Barkeshli and Volakis, 2004).

with $\hat{R} = \frac{\bar{\rho}-\bar{\rho}'}{|\bar{\rho}-\bar{\rho}'|}$. We can now proceed with its discretization in the usual manner. Subdividing the contour $C$ into $N$ linear segments, making use of the parametric representations (5.71) and replacing the current with the pulse expansion

$$\mathbf{J} = \sum_{n=0}^{N-1} \hat{l}_n J_n P_{\Delta l_n}\left(l - \frac{\Delta l_n}{2}\right) \tag{5.82}$$

yields

$$H_z^i(x, y) = \sum_{n=0}^{N-1} J_n \left\{\frac{1}{2} + \frac{jk_o}{4} \int_0^{\Delta l_n} (\hat{n}' \cdot \hat{R}) H_1^{(2)}(k_o|\bar{\rho} - \bar{\rho}'|)\, dl'\right\}; (x, y) \in C. \tag{5.83}$$

Point matching this equation at the center of the segment, we obtain the usual matrix system (5.52). In this case

$$V_m = H_z^i(x_{m+1/2}, y_{m+1/2}) = e^{jk_o\left[(x_m + \frac{\Delta l_m}{2}\cos\theta_m)\cos\phi_o + (y_m + \frac{\Delta l_m}{2}\sin\theta_m)\sin\phi_o\right]} \tag{5.84}$$

and

$$Z_{mn} = \frac{1}{2}\delta(m - n) + \frac{jk_o}{4} \int_0^{\Delta l_n} (\hat{n}'_n \cdot \hat{R}'_{mn})\, H_1^{(2)}(k_o R'_{mn})\, dl' \tag{5.85}$$

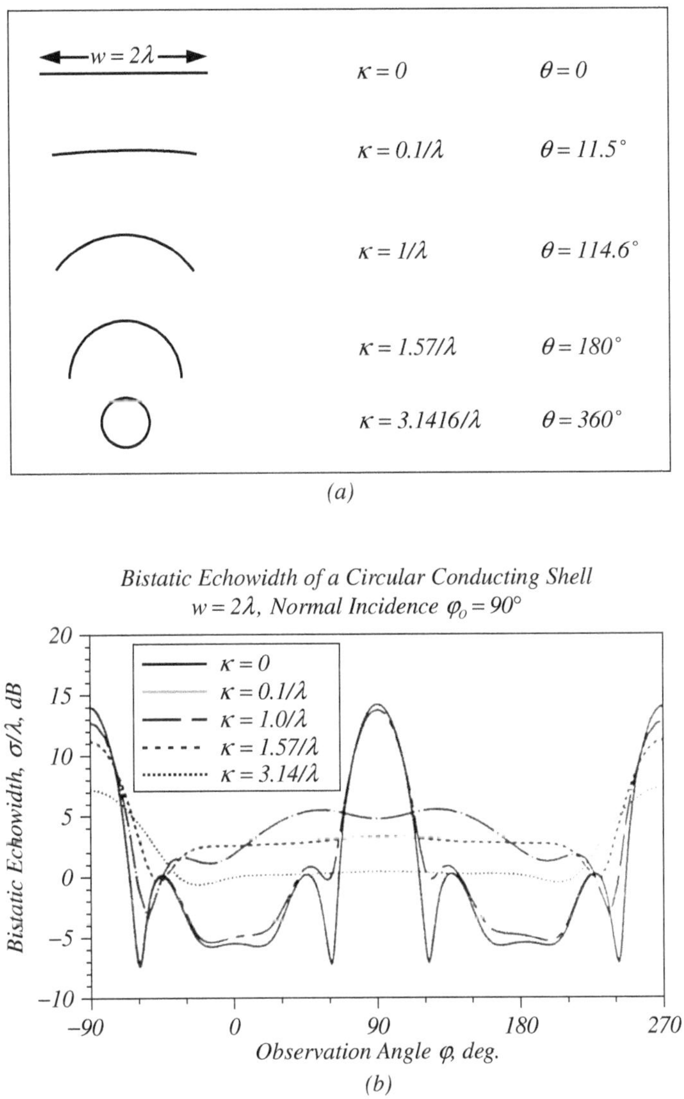


**FIGURE 5-19** Bistatic $E$-polarization echowidth patterns of five circular strips having different radii of curvature $a = \frac{1}{\kappa}$ but a constant circumference of $2\lambda$: (a) geometry, and (b) bistatic echowidth with $\phi = \phi_o = \frac{\pi}{2}$ (Barkeshli and Volakis, 2004).

in which $R'_{mn}$ is given by (5.74),

$$\hat{R}'_{mn} = \frac{\mathbf{R}'_{mn}}{|\mathbf{R}'_{mn}|} = \frac{\mathbf{R}'_{mn}}{R'_{mn}} \tag{5.86}$$

$$R'_{mn} = \hat{x}\left[(x_m - x_n) + \frac{1}{2}(\Delta l_m \cos\theta_m - 2l' \cos\theta_n)\right] + \hat{y}\left[(y_m - y_n) + \frac{1}{2}(\Delta l_m \sin\theta_m - 2l' \sin\theta_n)\right] \tag{5.87}$$

and

$$\hat{z} = \hat{l}_n \times \hat{n}_n \tag{5.88}$$

where $\hat{l}_n$ and $\hat{n}_n$ are the unit vectors tangent and normal to the $n$th segment, respectively, as illustrated in Figure 5-16.

When $m = n$, the integral in (5.85) vanishes because $\hat{R}'_{nn} \to \pm\hat{l}'_n$, implying $\hat{n}_n \cdot \hat{R}'_{nn} = 0$. For $m \neq n$, it may not be appropriate to employ a one-term midpoint integration for the elements adjacent to the matrix diagonal because of the singularity associated with $H_1^{(2)}$, and a higher-order numerical integration may be appropriate. Alternatively, since

$$\frac{\partial G_{2d}}{\partial l'} = \hat{l}' \cdot \nabla' G_{2d}(k_o R) = -\frac{jk_o}{4}(\hat{l}' \cdot \hat{R}) H_1^{(2)}(k_o R) \tag{5.89}$$

and $\hat{l}' \cdot \hat{R} \neq 0$ for $m \neq n$, we may rewrite $Z_{mn}$ as

$$\begin{aligned} Z_{mn} &\simeq \frac{j}{4}\int_0^{\Delta l_n} \frac{(\hat{n}_n \cdot \hat{R}'_{mn})}{(\hat{l}_n \cdot \hat{R}'_{mn})} \frac{d}{dl'} H_o^{(2)}(k_o R'_{mn})\, dl' \\ &\approx \frac{j}{4}\left(\frac{\hat{n}_n \cdot \hat{R}_{mn}}{\hat{l}_n \cdot \hat{R}_{mn}}\right)\left[H_o^{(2)}(k_o R_2) - H_o^{(2)}(k_o R_1)\right]; \qquad m \neq n. \end{aligned} \tag{5.90}$$

In this, $\hat{R}_{mn}$ is given by (5.87) upon setting $l' = \frac{\Delta l_n}{2}$, $R_1 = R'_{mn}|_{l'=0}$ and $R_2 = R'_{mn}|_{l'=\Delta l_n}$. We can thus summarize that

$$Z_{mn} = \begin{cases} \frac{1}{2} & m = n \\ \frac{j}{4}\frac{\hat{n}_n \cdot \hat{R}_{mn}}{\hat{l}_n \cdot \hat{R}_{mn}}\left[H_o^{(2)}(k_o R_1) - H_o^{(2)}(k_o R_2)\right] & m \neq n. \end{cases} \tag{5.91}$$

We note that another way to evaluate $Z_{nn}$ is to rewrite it as (see 5.80)

$$Z_{nn} = \frac{j}{4}\int_0^{\Delta l_n} \frac{\partial}{\partial n'} H_o^{(2)}(k_o|\overline{\rho} - \overline{\rho'}|)\, dl'. \tag{5.92}$$

However, since $Z_{nn}$ should be independent of the segment's orientation, we could instead consider a flat segment in the $x$-$y$ plane, and for this case the previous expression simplifies to

$$\begin{aligned} Z_{nn} &= \frac{j}{4}\int_{-\Delta l_n/2}^{\Delta l_n/2} \left[\frac{\partial}{\partial y} H_o^{(2)}\left(k_o\sqrt{x'^2 + y^2}\right)\right]_{y\to 0} dx' \\ &= \frac{j}{4}\left[\frac{\partial}{\partial y}\int_{-\Delta l_n/2}^{\Delta l_n/2} H_o^{(2)}\left(k_o\sqrt{(x_n - x')^2 + y^2}\right) dx'\right]_{y\to 0}. \end{aligned}$$

Next, we replace the Hankel function with its small argument expansion, and this gives (see (5.38))

$$\begin{aligned} Z_{nn} &= +\frac{j}{4}\frac{\partial}{\partial y}\left[\int_{-\Delta l_n/2}^{\Delta l_n/2}\left[c_1 - \frac{j2}{\pi}\ln\left(k_o\sqrt{x'^2 + y^2}\right)\right] dx'\right]_{y\to 0} \\ &= \frac{1}{\pi}\frac{\partial}{\partial y}\left[\frac{\Delta l_n}{2}\ln\left(\frac{k_o}{2}\sqrt{(\Delta l_n)^2 + y^2}\right) - \frac{\Delta l_n}{2} + y\tan^{-1}\left(\frac{\Delta l_n}{2y}\right)\right]_{y\to 0} \\ &= \frac{1}{\pi}\tan^{-1}\left(\frac{\Delta l_n}{2y}\right)\bigg|_{y\to 0} = \frac{1}{2} \end{aligned}$$

which is precisely the result in (5.91). This alternative procedure simply points out that the self-cell element can be evaluated without a need to invoke Cauchy's principal value theorem. Instead, by retaining the derivative outside the integral, the integration can be carried out in the standard manner. The appropriate differentiation can then be performed on the integrated quantity. In other words, the integral equation is enforced at a point just off the boundary surface, thus avoiding the improper integral altogether. The testing point is then moved onto the surface upon completion of the required integration and differentiations.

From (3.93) with $\psi(\overline{\rho}) = 0$ or from (2.133b), the far-zone scattered field is given by

$$
\begin{aligned}
H^s_{zff} &\sim \sqrt{\frac{k_o}{8\pi}} \frac{e^{-j(k_o\rho-\frac{\pi}{4})}}{\sqrt{\rho}} \int_C J_l(\overline{\rho}')(\hat{l}' \times \hat{\rho}) e^{jk_o(\hat{\rho}\cdot\overline{\rho}')}\, dl' \\
&\sim -\sqrt{\frac{k_o}{8\pi}} \frac{e^{-j(k_o\rho-\frac{\pi}{4})}}{\sqrt{\rho}} \int_C (\hat{n}' \cdot \hat{\rho}) J_l(\overline{\rho}') e^{jk_o(\hat{\rho}\cdot\overline{\rho}')}\, dl'.
\end{aligned}
\tag{5.93}
$$

When we substitute for $J_l(\overline{\rho}')$, as obtained from the method of moments solution, we have

$$
\begin{aligned}
H^{ff}_z &\sim -\sqrt{\frac{k_o}{8\pi}} e^{-j\left(k_o\rho-\frac{\pi}{4}\right)} \\
&\quad \cdot \sum_{n=0}^{N-1} J_n \int_0^{\Delta l_n} (\hat{n}_n \cdot \hat{\rho}) e^{jk_o[(x_n+l'\cos\theta_n)\cos\phi+(y_n+l'\sin\theta_n)sin\phi]}\, dl' \\
&\sim -\frac{e^{-jk_o\rho}}{\sqrt{\rho}} e^{j\frac{\pi}{4}} \frac{k_o}{4} \sqrt{\frac{2}{\pi k_o}} \sum_{n=0}^{N-1} J_n e^{jk_o(x_n\cos\phi+y_n\sin\phi)} \\
&\quad \cdot \Delta l_n e^{jk_o\frac{\Delta l_n}{2}(\cos\theta_n\cos\phi+\sin\theta_n\sin\phi)} \operatorname{sinc}\left[\frac{\Delta l_n}{2}(\cos\theta_n\cos\phi+\sin\theta_n\sin\phi)\right]
\end{aligned}
\tag{5.94}
$$

from which we can compute the echowidth defined in (5.45).

## 5.3 | *H*-POLARIZED (TE) SCATTERING BY CURVED RESISTIVE STRIPS

As discussed in Section 3.2.3, the magnetic field integral equation (MFIE) (5.80), by its derivation, is suitable only for modeling closed boundaries such as those of metallic cylinders. When modeling resistive strips, flat or curved (see Figure 5-20), it is necessary to employ the EFIE given in (3.99), which reduces to (5.30) when generalized to treat flat resistive strips. In the case of curved resistive strips, (3.99) takes the form

$$
\begin{aligned}
Y_o\hat{l}\cdot\mathbf{E}^i &= R_e J_l(\overline{\rho}') + \frac{k_o}{4}\int_C (\hat{l}\cdot\hat{l}') J_l(\overline{\rho}') H_o^{(2)}(k_o|\overline{\rho}-\overline{\rho}'|)\, dl' \\
&\quad + \frac{1}{4}\int_C J_l(\overline{\rho}') \frac{\partial}{\partial l'}\{(\hat{l}\cdot\hat{R}) H_1^{(2)}(k_o|\overline{\rho}-\overline{\rho}'|)\}\, dl'
\end{aligned}
\tag{5.95}
$$

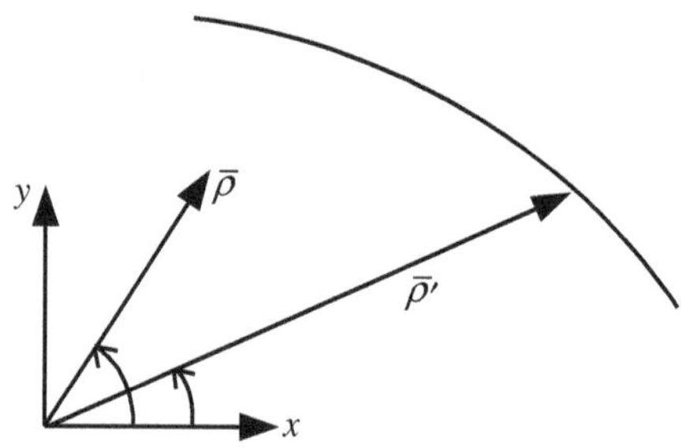


**FIGURE 5-20** Geometry of the curved strip.

where $R_e$ denotes, as usual, the normalized strip resistivity, $J_l(\overline{\rho})$ is the current density at the strip point $\overline{\rho}$, and $\hat{l}$ is the unit tangent to the strip at $\overline{\rho}$.

To solve (5.95) for $J_l(\overline{\rho})$, it is necessary to first discretize it by introducing an approximate expansion for the current density. The following sections present the numerical solution of (5.95) using the pulse-basis–point-matching and linear-basis–Galerkin's method.

### 5.3.1 Pulse-Basis–Point-Matching Solution

After introducing in (5.95) the pulse-basis expansion (5.82), upon point matching we obtain the usual system (5.52). For $n \neq m$ the impedance matrix elements are now given by

$$Z_{mn} = \frac{k_0}{4}(\hat{n}_m \cdot \hat{n}_n)\Delta l_n H_o^{(2)}(k_o R'_{mn}) + \frac{1}{4}\left[(\hat{l}_m \cdot \hat{R}_2)H_1^{(2)}(k_o R_2) - (\hat{l}_m \cdot \hat{R}_1)H_1^{(2)}(k_o R_1)\right]; \qquad n \neq m \tag{5.96}$$

in which $R'_{mn}$ is given by (5.74), $\hat{R}_1 = \hat{R}'_{mn}|_{l'=0}$, and $\hat{R}_2 = \hat{R}'_{mn}|_{l'=\Delta l_n}$. The distance $R'_{mn}$ and unit vector $\hat{R}'_{mn}$ were defined in (5.86) and (5.87) using the segment's parametric representation (5.71). The self-cell element of the impedance matrix is given by (5.41) since the curved strip was approximated using a collection of flat strip segments. Finally, for the curved strip, the excitation column elements are given by

$$V_m = \hat{l}_m \cdot (\hat{x}\sin\phi_o - \hat{y}\cos\phi_o)e^{jk_o\left[\left(x_m+\frac{\Delta l_m}{2}\cos\theta_m\right)\cos\phi_o+\left(y_m+\frac{\Delta l_m}{2}\sin\theta_m\right)\sin\phi_o\right]}. \tag{5.97}$$

The scattered field can be computed using (5.94).

An application of resistive sheets is the simulation of dielectric layers as discussed in Section 1.12. The idea of approximating a thin dielectric layer by a resistive sheet can be generalized to model thick dielectric cylinders. Specifically, the cylinder is first decomposed into stacks or an arrangement of dielectric layers as illustrated, for example, in Figure 5-21. Each dielectric layer or shell is then replaced by a sheet of resistivity $R_e Z_o = -jZ_o/[k_o\tau(\epsilon_r - 1)]$, where $\tau$ denotes the shell-layer thickness, and $\epsilon_r$ is the corresponding relative dielectric constant (which can be a function of $\overline{\rho}$). The backscatter echowidths of a rectangular $1\lambda_o \times 0.4\lambda_o$ dielectric cylinder and a thin circular dielectric shell using the proposed sheet simulation are given in

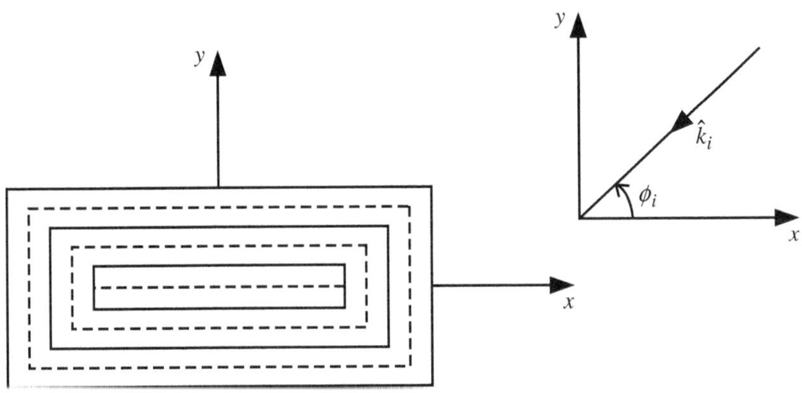


**FIGURE 5-21** Simulation of a dielectric cylinder with resistive sheets at the geometric center of each layer. (Peters, Volakis, Liepa, Senior and Ricoy, 1986)

Figure 5-22. It is seen that the sheet simulation yields quite accurate results when compared with a more rigorous formulation (see Richmond, 1965) which will be discussed later. However, one should recognize that such sheet simulations are not always as accurate, and furthermore the model's accuracy depends on the particular arrangement of the layers. For example, were we to model the rectangular dielectric block in Figure 5-21 using five planar dielectric layers (rather than rectangular shells), the computed echowidth on the basis of this sheet model will not be as accurate for small $\phi$ angles. This is because the sheet model composed of parallel flat layers neglects altogether the presence of the $y$-directed polarization currents, which are most important for incidences close to the $x$-axis. The importance of the vertical polarization currents is illustrated in Figure 5-23 where we consider the scattering by a $0.1\lambda_o$ thick dielectric layer.

Another example calculation that cannot be done with the MFIE is shown in Figure 5-24. This figure shows the backscatter echowidth of an ogive cylinder with a strip attached to its left. The total length of the cylinder is $10\lambda_o$, and about 110 segments were used for its numerical simulation. It is seen that the pulse-basis–point-matching solution is in good agreement with a solution based on the volume-surface integral equation (VSIE) given by (3.114) and whose solution is discussed later. As expected, the echowidth is higher around the broadside angular region and drops to low values near grazing.

## 5.3.2 Linear Basis-Galerkin's Solution

As was done for the flat strip, to reduce the kernel singularity of the integral equation (5.95) we may again employ a Galerkin's solution with linear basis. The implied discretization and expansion functions are illustrated in Figure 5-25. Each linear subdomain basis function now spans two adjacent segments of the discretized strip, and to avoid restrictions on the length of each of these strips we introduce the modified triangular subdomain bases

$$T_m(l) = \begin{cases} T_m^-(l) = \frac{l}{\Delta l_m^-} & 0 < l < \Delta l_m^- \text{ (in left segment)} \\ T_m^+(l) = \frac{\Delta l_m^+ - l}{\Delta l_m^+} & 0 < l < \Delta l_m^+ \text{ (in right segment).} \end{cases} \tag{5.98}$$

**FIGURE 5-22** Comparison of echowidth results based on the sheet and volume equivalent currents with H-incidence: (a) rectangular dielectric cylinder $1\lambda \times 0.4\lambda$, and (b) thin dielectric shell having an inner radius $R_{in} = 0.5\lambda$ and an outer radius of $R_{out} = 0.6\lambda$. (After Peters, Volakis and Liepa, 1986. With permission.)

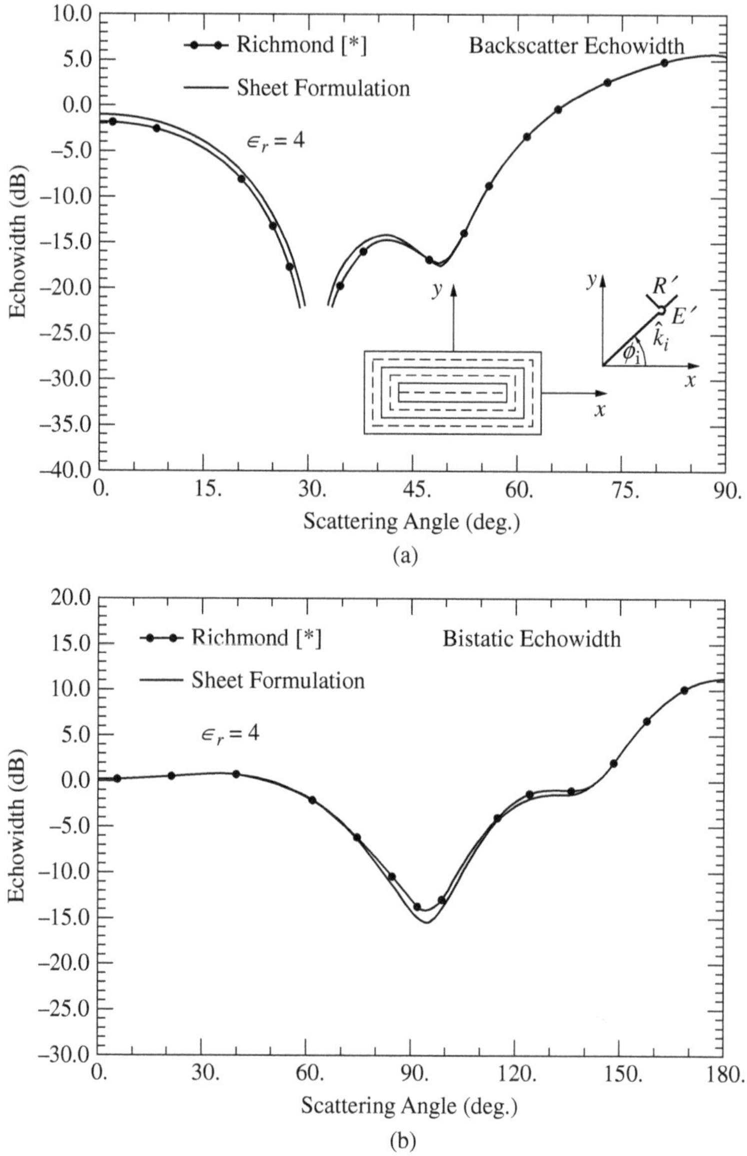


[*] Richmond, AP-T, Vol. 13, May 1965, pp. 334–341.

In this, $\Delta l_m^-$ and $\Delta l_m^+$ denote the lengths of the adjacent segments, and it is clear that for $\Delta l_m^- = \Delta l_m^+$ the $T_m(l)$ shown in (5.98) reduces to the simplified triangular basis given in Chapter 4 and employed in Section 5.1.2. The associated parametric equations for the (left) linear segment are

$$x = x_{m-1} + l\cos\theta_m^- \quad y = y_{m-1} + l\sin\theta_m^- \tag{5.99a}$$

whereas those for the right linear segment are

$$x = x_m + l\cos\theta_m^+ \quad y = y_m + l\sin\theta_m^+. \tag{5.99b}$$

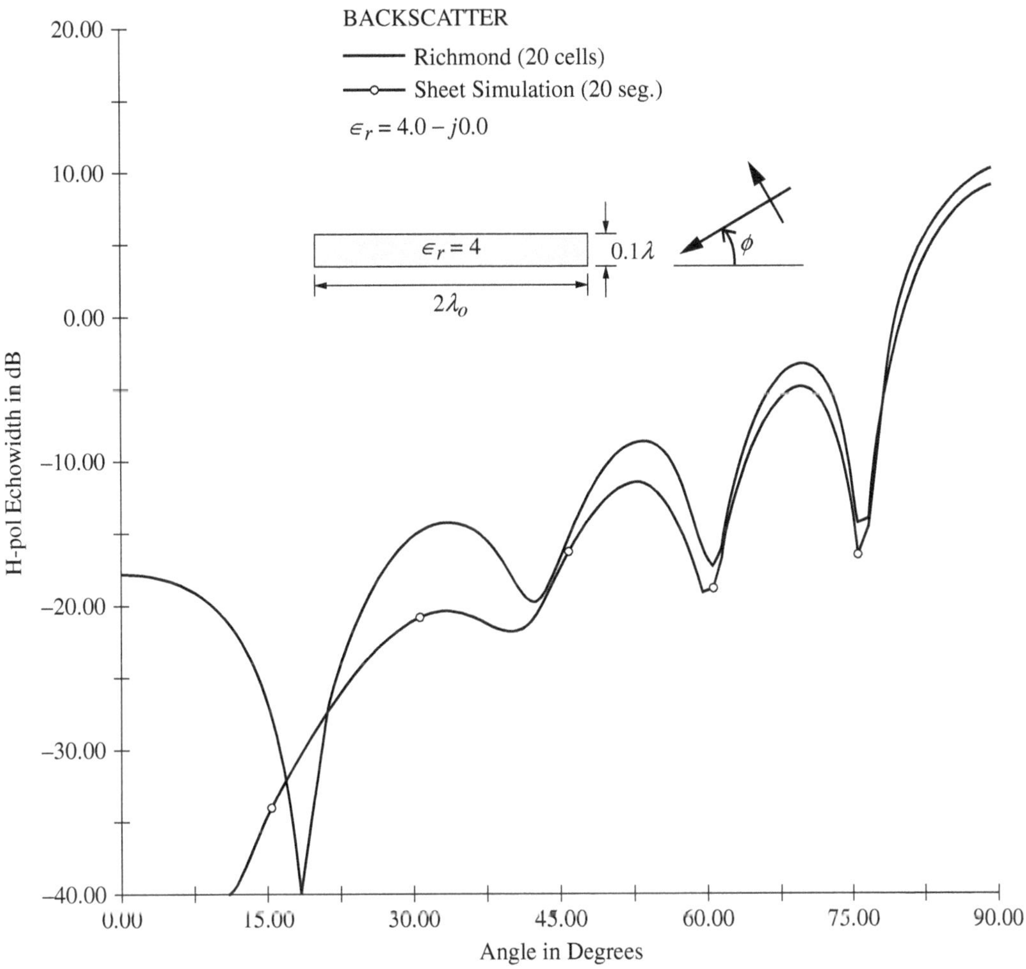


**FIGURE 5-23** Comparison of the computed backscatter echowidth by a thin $2.0\lambda_o \times 0.1\lambda_o$ dielectric slab using the rigorous volume integral equation (see Section 5.6) and the single sheet approximation. (Peters, Volakis, Liepa, Senior and Ricoy, 1986)

Similarly to (5.72), the definitions of $\theta_m^{\pm}$ are

$$\theta_m^- = \tan^{-1}\left(\frac{y_m - y_{m-1}}{x_m - x_{m-1}}\right) \tag{5.100a}$$

$$\theta_m^+ = \tan^{-1}\left(\frac{y_{m+1} - y_m}{x_{m+1} - x_m}\right) \tag{5.100b}$$

with the unit tangent vectors to each segment given by

$$\hat{l}_m^- = \hat{x}\cos\theta_m^- + \hat{y}\sin\theta_m^- \tag{5.101a}$$

$$\hat{l}_m^+ = \hat{x}\cos\theta_m^+ + \hat{y}\sin\theta_m^+. \tag{5.101b}$$

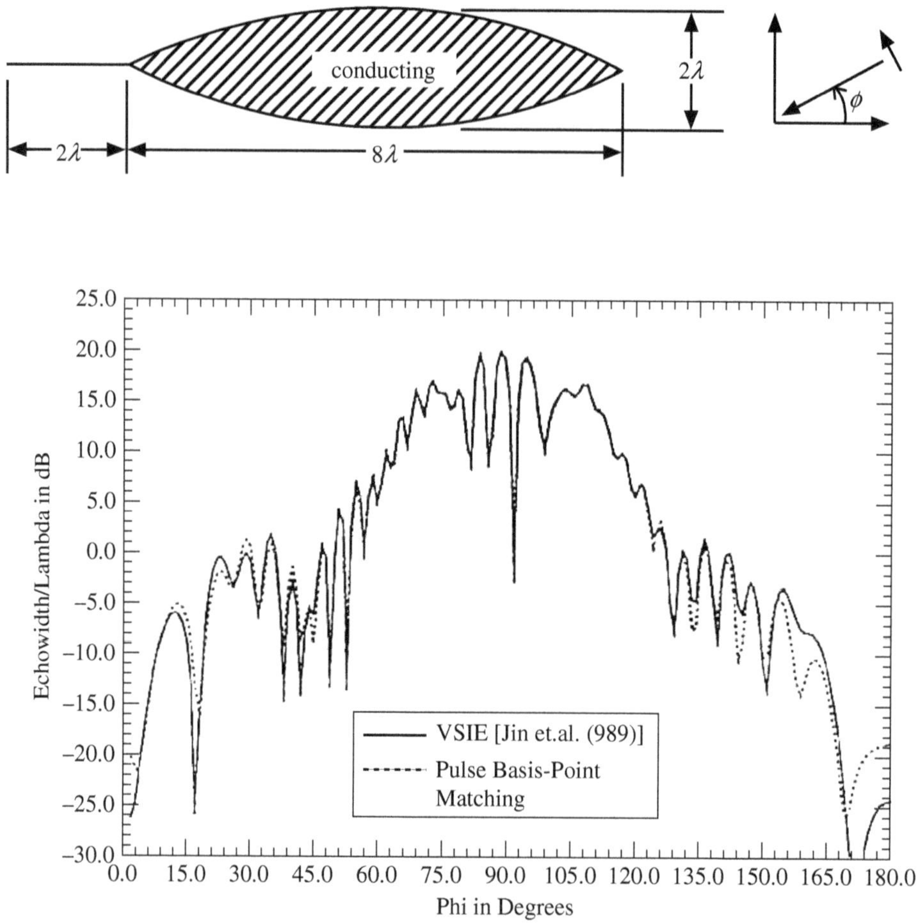


**FIGURE 5-24** ■ $H$-polarization (TE) backscatter echowidth of the illustrated $8\lambda_o$ long and $2\lambda_o$ wide metallic ogive with a $2\lambda_o$ metallic strip attached to its left tip. (After Jin, Volakis, and Liepa, 1989. With permission.)

We may now proceed with the discretization of (5.95) by first replacing the current $J_l(l)$ with the expansion

$$\mathbf{J}(l) = \sum_{n=0}^{N-2} J_n\left[\hat{l}_n^- T_n^-(l) - \hat{l}_n^+ T_n^+(l)\right] = \sum_{n=0}^{N-1} J_n \mathbf{T}_n(l) \tag{5.102}$$

where in this case $N$ is equal to the number of segments. Note also that $T_n^-(l)$ and $T_n^+(l)$ are nonzero only on the left and right segments of the $m$th node as defined in (5.98). Substituting (5.102) in (5.95) and employing Galerkin's procedure yields the usual system

$$[Z_{mn}]\{J_n\} = \{b_m\} \tag{5.103}$$

where the elements of the matrix $[Z_{mn}]$ and those of the excitation column can be defined in terms of the parameters introduced in (5.99) through (5.102). For the

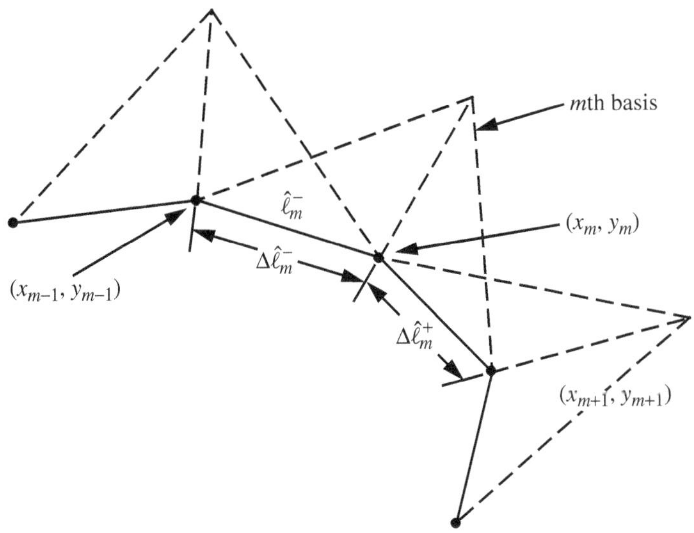


**FIGURE 5-25** Geometrical parameters for the adjacent segments simulating a curved strip.

excitation matrix elements we have

$$
\begin{aligned}
b_n =& \left(\sin\phi_o \cos\theta_n^- - \cos\phi_o \sin\theta_n^-\right) \\
&\cdot \int_0^{\Delta l_n^-} \frac{l}{\Delta l_n^-} e^{jk_o\left\{\left(x_{n-1}+l\cos\theta_n^-\right)\cos\phi_o+\left(y_{n-1}+l\sin\theta_n^-\right)\sin\phi_o\right\}}\, dl \\
&+ \left(\sin\phi_o \cos\theta_n^+ - \cos\phi_o \sin\theta_n^+\right) \\
&\cdot \int_0^{\Delta l_n^+} \frac{\Delta l_n^+ - l}{\Delta l_n^+} e^{jk_o\left\{\left(x_n+l\cos\theta_n^+\right)\cos\phi_o+\left(y_n+l\sin\theta_n^+\right)\sin\phi_o\right\}}\, dl
\end{aligned} \tag{5.104}
$$

and after integration we obtain

$$
\begin{aligned}
b_n =& \frac{1}{\Delta l_n^-}\left(\sin\phi_o \cos\theta_n^- - \cos\phi_o \sin\theta_n^-\right) e^{jk_o[x_{n-1}\cos\phi_o + y_{n-1}\sin\phi_o]} \\
&\cdot \left[\frac{e^{jk_o l g_n^-}}{-k_o^2\left(g_n^-\right)^2}\left(jk_o g_n^- - 1\right)\right]_0^{\Delta l_n^-} \\
&+ \frac{1}{\Delta l_n^+}(\sin\phi_o \cos\theta_n^+ - \cos\phi_o \sin\theta_n^+) e^{jk_o(x_n\cos\phi_o + y_n\sin\phi_o)} \\
&\cdot \left[\frac{\Delta l_n^+ e^{jk_o l g_n^+}}{jk_o g_n^+} + \frac{e^{jk_o l g_n^+}}{k_o^2\left(g_n^+\right)^2}\left(jk_o g_n^+ - 1\right)\right]_0^{\Delta l_n^+}
\end{aligned} \tag{5.105}
$$

in which

$$
g_n^{\pm} = \cos\theta_n^{\pm}\cos\phi_o + \sin\theta_n^{\pm}\sin\phi_o. \tag{5.106}
$$

The impedance matrix elements can be compactly expressed as

$$Z_{nm} = \zeta_{nm}^1 + \frac{k_o}{4}\zeta_{nm}^2 - \frac{1}{4k_o}\zeta_{nm}^3 \tag{5.107}$$

where

$$\zeta_{nm}^1 = \int_{C_m} R(l)\mathbf{T}_m(l)\cdot\mathbf{T}_n(l)\,dl \tag{5.108}$$

$$\zeta_{nm}^2 = \int_{C_m}\int_{C_n} \mathbf{T}_m(l)\cdot\mathbf{T}_n(l')H_o^{(2)}(k_oR)\,dl'\,dl \tag{5.109}$$

$$\zeta_{nm}^3 = \int_{C_m}\int_{C_n} \frac{dT_m(l)}{dl}\frac{dT_n(l')}{dl'}H_o^{(2)}(k_oR)\,dl'\,dl \tag{5.110}$$

in which $C_n$ denotes the contour segment encompassing the pair of strips to the left and right of $(x_n, y_n)$. The integrals in (5.108) can be readily integrated once the resistivity is specified over the $m$th segment. Approximating the resistivity by a constant within each segment, where the $m$th segment joins the $(m-1)$th and $m$th nodes, it follows that

$$\zeta_{nm}^1 = \begin{cases} R_n^e\left(\frac{\Delta l_n^-}{3}\right) + R_e^{n+1}\left(\frac{\Delta l_n^+}{3}\right) & n = m \\ \frac{R_e^n\Delta l_n^-}{6} & n = m+1 \\ \frac{R_e^m\Delta l_m^-}{6} & n = m-1 \\ 0 & \text{otherwise} \end{cases} \tag{5.111}$$

where $R_e^m$ denotes the average resistivity of the $m$th segment.

To evaluate $\zeta_{mn}^2$, we must proceed with the substitution of the expansion and weighting functions as given in (5.98). Doing so, we obtain

$$\begin{aligned} \zeta_{nm}^2 = {} & (\hat{l}_n^-\cdot\hat{l}_m^-)\int_0^{\Delta l_n^-}\frac{l}{\Delta l_n^-}\int_0^{\Delta l_m^-}\frac{l'}{\Delta l_m^-}H_o^{(2)}(k_oR_{11})\,dl'\,dl \\ & + (\hat{l}_n^-\cdot\hat{l}_m^-)\int_0^{\Delta l_n^-}\frac{l}{\Delta l_n^-}\int_0^{\Delta l_m^-}\left\{\frac{\Delta l_m^+ - l'}{\Delta l_m^+}\right\}H_o^{(2)}(k_oR_{12})\,dl'\,dl \\ & + (\hat{l}_n^+\cdot\hat{l}_m^-)\int_0^{\Delta l_n^+}\frac{\Delta l_n^+ - l}{\Delta l_n^+}\int_0^{\Delta l_m^-}\left\{\frac{l'}{\Delta l_m^-}\right\}H_o^{(2)}(k_oR_{21})\,dl'\,dl \\ & + (\hat{l}_n^+\cdot\hat{l}_m^+)\int_0^{\Delta l_n^+}\left\{\frac{\Delta l_n^+ - l}{\Delta l_n^+}\right\}\int_0^{\Delta l_m^+}\left\{\frac{\Delta l_m^+ - l'}{\Delta l_m^+}\right\}H_o^{(2)}(k_oR_{22})\,dl'\,dl \end{aligned} \tag{5.112}$$

where

$$\begin{aligned} R_{11} = \Big[ & \left(x_{n-1} + l\cos\theta_n^- - x_{m-1} - l'\cos\theta_m^-\right)^2 \\ & + \left(y_{n-1} + l\sin\theta_n^- - y_{m-1} - l'\sin\theta_m^-\right)^2\Big]^{\frac{1}{2}} \end{aligned} \tag{5.113}$$

$$R_{12} = \left[ \left(x_{n-1} + l\cos\theta_n^- - x_m - l'\cos\theta_m^+\right)^2 + \left(y_{n-1} + l\sin\theta_n^- - y_m - l'\sin\theta_m^+\right)^2 \right]^{\frac{1}{2}} \tag{5.114}$$

$$R_{21} = \left[ \left(x_n + l\cos\theta_n^+ - x_{m-1} - l'\cos\theta_m^-\right)^2 + \left(y_n + l\sin\theta_n^+ - y_{m-1} - l'\sin\theta_m^-\right)^2 \right]^{\frac{1}{2}} \tag{5.115}$$

$$R_{22} = \left[ \left(x_n + l\cos\theta_n^+ - x_m - l'\cos\theta_m^+\right)^2 + \left(y_n + l\sin\theta_n^+ - y_m - l'\sin\theta_m^+\right)^2 \right]^{\frac{1}{2}}. \tag{5.116}$$

All of the integrals in (5.112) can be evaluated numerically except when $n = m$ and $n = m \pm 1$. In this case, some of the integrands become singular, and although their singularity is integrable the associated integrals must, nevertheless, be evaluated with care. For the evaluation of these self-cell and near-diagonal matrix elements the reader is referred to Section 5.2.1. Specifically, the expressions given in (5.55) through (5.58), which refer to a flat strip, can still be used here with only minor modification. Such modifications pertain to the introduction of appropriate dot products $\hat{l}_m^\pm \cdot \hat{l}_n^\pm$ in front of the different integrals appearing (5.55).

The third term in the expression for the impedance matrix elements can be expanded as

$$\begin{aligned} \zeta_{nm}^3 = &+ \frac{1}{\Delta l_n^- \Delta l_m^-} \int_0^{\Delta l_n} \int_0^{\Delta l_m^-} H_o^{(2)}(k_o R_{11})\, dl'\, dl \\ &- \frac{1}{\Delta l_n^- \Delta l_m^+} \int_0^{\Delta l_n^-} \int_0^{\Delta l_m^+} H_o^{(2)}(k_o R_{12})\, dl'\, dl \\ &- \frac{1}{\Delta l_n^+ \Delta l_m^-} \int_0^{\Delta l_n^+} \int_0^{\Delta l_m^-} H_o^{(2)}(k_o R_{21})\, dl'\, dl \\ &+ \frac{1}{\Delta l_n^+ \Delta l_m^+} \int_0^{\Delta l_n^+} \int_0^{\Delta l_m^+} H_o^{(2)}(k_o R_{22})\, dl'\, dl. \end{aligned} \tag{5.117}$$

Again, the resulting double integrals can be readily evaluated numerically unless the argument of the Hankel function vanishes. In that case, the result in (5.60) should be employed whenever $\Delta l_n^\pm$ are chosen equal to one another. For arbitrary values of $\Delta l_n^\pm$, we must instead use the result

$$\int_0^{\Delta_1} \int_0^{\Delta_2} H_o^{(2)}(k_o|x - x'|)\, dx\, dx' = \Delta_1 \Delta_2 \left[ 1 - j\frac{2}{\pi} ln \left( \frac{\gamma k_o \sqrt{\Delta_1^2 + \Delta_2^2}}{2e} \right) \right] \tag{5.118}$$

given in Ricoy and Volakis (1989).

Upon solution of $\{J_n\}$ for (5.103), the computation of the scattered field can be carried out by substituting (5.102) into (5.93).

## 5.4 | PIECEWISE HOMOGENEOUS DIELECTRIC CYLINDERS

Boundary integral equations for modeling piecewise homogeneous cylinders were developed in Sections 3.2.2 through 3.2.4. In this section we present the discretization and solution of these integral equations. To begin, let us first consider the illumination of the homogeneous cylinder shown in Figure 3-10. The fields of the surface of the dielectric cylinder satisfy the coupled integral equation pair

$$\begin{aligned} &\frac{-j}{4}\oint_C \Big[\phi(\overline{\rho}')\frac{\partial}{\partial n'}H_o^{(2)}(k_1|\overline{\rho}-\overline{\rho}'|) \\ &\quad - \upsilon_1\psi(\overline{\rho}')H_o^{(2)}(k_1|\overline{\rho}-\overline{\rho}'|)\Big]dl' + \phi^{inc}(\overline{\rho}) = \frac{1}{2}\phi(\overline{\rho}) \\ &\frac{-j}{4}\oint_C \Big[\phi(\overline{\rho}')\frac{\partial}{\partial n'}H_o^{(2)}(k_2|\overline{\rho}-\overline{\rho}'|) \\ &\quad - \upsilon_2\psi(\overline{\rho}')H_o^{(2)}(k_2|\overline{\rho}-\overline{\rho}'|)\Big]dl' = -\frac{1}{2}\phi(\overline{\rho}). \end{aligned} \tag{5.119}$$

For TM illumination ($H_z = 0$)

$$\phi = E_{z1}, \qquad \psi = \frac{1}{\mu_{r1}}\frac{\partial E_{z1}}{\partial n}$$

$$\upsilon_1 = \mu_{r1}, \qquad \upsilon_2 = \mu_{r2}$$

and $\phi^{inc} = E_z^{inc}$, where $E_z^{inc}$ denotes the excitation field. Alternatively, for TE illumination ($E_z = 0$)

$$\phi = H_{z1}, \qquad \psi = \frac{1}{\epsilon_{r1}}\frac{\partial H_{z1}}{\partial n}$$

$$\upsilon_1 = \epsilon_{r1} \qquad \upsilon_2 = \epsilon_{r2}$$

and $\psi^{inc} = H_z^{inc}$ as dictated by duality.

To discretize (5.119) we subdivide the boundary $C$ into $N$ segments and introduce the expansions

$$\phi(\overline{\rho}) = \sum_{n=0}^{N-1}\phi_n L_n(\overline{\rho}) \qquad \psi(\overline{\rho}) = \sum_{n=0}^{N-1}\psi_n M_n(\overline{\rho}) \tag{5.120}$$

where $L_n(\overline{\rho})$ and $M_n(\overline{\rho})$ represent the chosen basis functions. Substituting (5.120) into (5.119), on employing the weighted residual method (see Section 4.7) we obtain the system

$$\begin{bmatrix} [A] & [B] \\ [C] & [D] \end{bmatrix} \begin{Bmatrix} \{\phi\} \\ \{\psi\} \end{Bmatrix} = \begin{Bmatrix} \{b\} \\ \{0\} \end{Bmatrix}. \tag{5.121}$$

The elements of the submatrices [A], [B], [C], and [D] are readily found to be given by

$$A_{mn} = \frac{1}{2}\int_{C_m} W_m(\overline{\rho})L_n(\overline{\rho}')\,dl + \frac{j}{4}\int_{C_m} W_m(\overline{\rho}) \fint_{C_n} L_n(\overline{\rho}')\frac{\partial H_o^{(2)}(k_1|\overline{\rho}-\overline{\rho}'|)}{\partial n'}\,dl'\,dl \tag{5.122a}$$

$$B_{mn} = -\frac{j}{4}\int_{C_m} W_m(\overline{\rho})\int_{C_n} \upsilon_1 M_n(\overline{\rho}')H_o^{(2)}(k_1|\overline{\rho}-\overline{\rho}'|)\,dl\,dl' \tag{5.122b}$$

$$C_{mn} = +\frac{1}{2}\int_{C_m} W_m(\overline{\rho})L_n(\overline{\rho}')\,dl - \frac{j}{4}\int_{C_m} W_m(\overline{\rho})\int_{C_n} -L_n(\overline{\rho}')\frac{\partial}{\partial n'}H_o^{(2)}(k_2|\overline{\rho}-\overline{\rho}'|)\,dl\,dl' \tag{5.122c}$$

$$D_{mn} = +\frac{j}{4}\int_{C_m} W_m(\overline{\rho})\int_{C_n} \upsilon_2 M_n(\overline{\rho}')H_o^{(2)}(k_2|\overline{\rho}-\overline{\rho}'|)\,dl\,dl'. \tag{5.122d}$$

In these expressions, $C_n$ denotes the integration interval over the $n$th segment or simply the segment of $C$ on which $L_n(\overline{\rho})$ or $M_n(\overline{\rho})$ are nonzero. Also, $W_m(\overline{\rho})$ represents the testing functions. For point matching, $W_m(\overline{\rho}) = \delta(\overline{\rho} - \overline{\rho}_m)$, with $\overline{\rho}_m$ denoting the testing point. In the case of Galerkin's method

$$W_m(\overline{\rho}) = \begin{cases} L_m(\overline{\rho}) \\ \text{or} \\ M_m(\overline{\rho}) \end{cases}$$

and to yield a symmetric system it is necessary that $L_m(\overline{\rho}) = M_m(\overline{\rho})$. Finally, the excitation column elements in (5.121) are given by

$$b_m = \int_{C_m} W_m(\overline{\rho})\phi^{inc}(\overline{\rho})\,dl. \tag{5.123}$$

The numerical computation of the integrals in (5.122) for arbitrary dielectric boundaries can be carried out by introducing parametric representations for each of the segments and following the procedure outlined in Section 5.2. When employing a pulse-basis–point-matching solution, we find that $A_{mn}$ is given by (5.91) upon setting $k_1 \to k_o$, and $B_{mn}$ is given by (5.77) multiplied by $\frac{+j\upsilon_1}{k_1 Z_1}$, again on setting $k_1 \to k_o$ and $Z_1 \to Z_o$. Similar expressions apply for the matrix elements $C_{mn}$ and $D_{mn}$

$$A_{mn} = C_{mn} \qquad n \neq m$$

and

$$\upsilon_2 B_{mn} = \upsilon_1 D_{mn}.$$

When employing the linear basis Galerkin's solution, the procedure outlined in Section 5.3 is appropriate for evaluating the matrix elements.

Upon evaluation of the boundary fields from (5.121), the far-zone scattered field is obtained from (3.93). To compute the fields within the dielectric cylinder, it

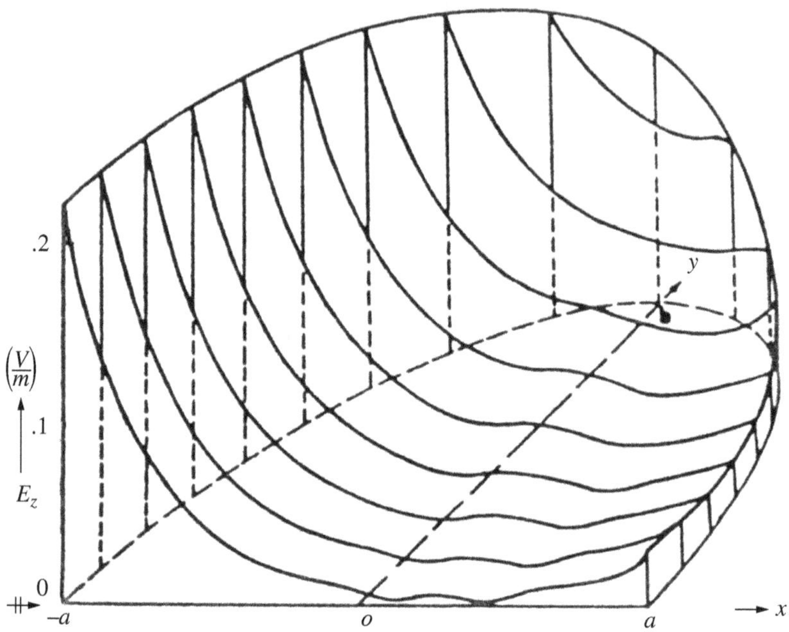


**FIGURE 5-26** E-field plot for an elliptical cylinder of muscle ($\epsilon_r = 60$, $\sigma = l'mho$ m, $a = 0.159$ m, $b = 0.239$ m, $f = 300$ MHz, TM incidence). (After Wu and Tsai, 1977. With permission.)

is necessary to resort to the exact expression (see (3.91))

$$\phi(\overline{\rho}) = +\frac{j}{4}\int_C \phi(\overline{\rho}')\Big[(\hat{n}\cdot\hat{R})k_o\sqrt{\mu_{r2}\epsilon_{r2}}\,H_1^{(2)}\left(k_o\sqrt{\epsilon_{r2}\mu_{r2}}\,|\overline{\rho}-\overline{\rho}'|\right) - \upsilon_2 H_o^{(2)}\left(k_o\sqrt{\epsilon_{r2}\mu_{r2}}\,R\right)\Big]\,dl'.$$

The fields interior to an elliptical lossy dielectric cylinder due to a TM plane wave incidence are displayed in Figure 5-26 (Wu and Tsai, 1977). This cylinder has a dielectric constant of $\epsilon_{r2} = 60(1 - j1)$ and is intended to simulate a human muscle at a frequency of $f = 300$ MHz. Such numerical results are useful in cancer treatments where of interest is the absorption of electromagnetic energy by biological tissues.

This solution procedure can be applied to the treatment of multilayered dielectric cylinders such as that shown in Figure 3-12. The pertinent coupled integral equations are given by (3.101a) through (3.101b). Discretizing these yields a matrix system involving nine submatrices compared with the four submatrices in the system (5.121).

Integral equations for piecewise homogeneous cylinders, such as that shown in Figure 5-27, can be carried out by reverting back to the generic equations (3.82). The specific integral equations are derived by considering each homogeneous section of

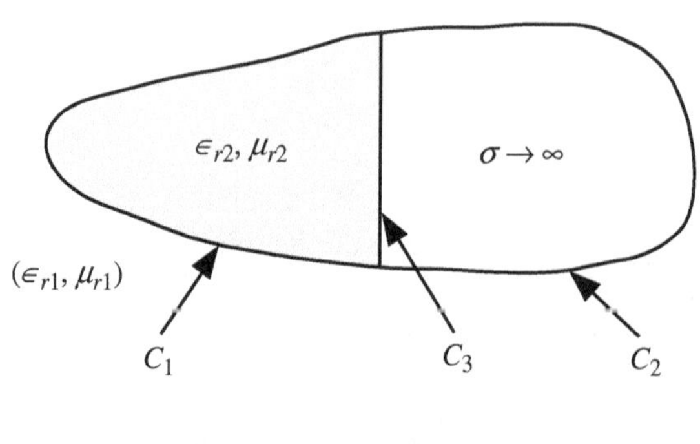


**FIGURE 5-27** Illustration of the contours for simulating a piecewise homogeneous cylinder.

the cylinder as a different scatterer together with the conditions that the **E** and **H** fields within the metal are zero and $E_z = \frac{\partial H_z}{\partial n} = 0$ on the metallic surfaces. For TM excitation it is found that

$$-\frac{j}{4} - \int_{C_1} \phi(\overline{\rho}') \frac{\partial}{\partial n'} H_o^{(2)}(k_1|\overline{\rho} - \overline{\rho}'|)\, dl' + \frac{j}{4} \int_{C_1+C_2} \mu_{r1} \psi(\overline{\rho}') H_o^{(2)}(k_1|\overline{\rho} - \overline{\rho}'|)\, dl'$$

$$+ \phi^{inc}(\overline{\rho}) = \begin{cases} \frac{1}{2}\phi(\overline{\rho}) & \rho \in C_1 \\ 0 & \rho \in C_2 \end{cases} \qquad (5.124)$$

$$-\frac{j}{4} - \int_{C_1} \phi(\overline{\rho}') \frac{\partial}{\partial n'} H_o^{(2)}(k_2|\overline{\rho} - \overline{\rho}'|)\, dl' + \frac{j}{4} \int_{C_1+C_2} \mu_{r2} \psi(\overline{\rho}') H_o^{(2)}(k_2|\overline{\rho} - \overline{\rho}'|)\, dl'$$

$$= \begin{cases} -\frac{1}{2}\phi(\overline{\rho}) & \overline{\rho} \in C_1 \\ 0 & \overline{\rho} \in C_3 \end{cases} \qquad (5.125)$$

where, as before, $\phi = E_{z1}$ and $\psi = \frac{1}{\mu_{r1}} \frac{\partial E_{z1}}{\partial \mu}$.

## 5.5 | ELIMINATION OF INTERIOR RESONANCES

As discussed in Section 3.1.5, the electric field integral equations (EFIE) and magnetic field integral equations (MFIE), whether applied to two-dimensional or three-dimensional problems, are prone to yielding erroneous results at certain frequencies. Failure of a certain integral equation solution to yield accurate results is manifested in the form of an ill-conditioned system (a matrix system with a condition number of $10^7$ or greater is usually considered as ill-conditioned). This occurs only when the boundary integral equations are enforced on closed-boundary contours as was the case of metallic and dielectric cylinders considered in Sections 5.2 and 5.4. The EFIE integral equations applied to planar and curved strips (not forming a closed surface) do not suffer from such failures, which can be identified to occur at and near the resonant frequencies of the cavity/waveguide formed by the closed boundary. To appreciate the severity of the problem, the reader is referred to Figure 5-28 (Collins, Jin, and Volakis, 1992) which displays the TM backscatter echowidth of the metallic circular cylinder as a function of frequency obtained from a solution of the MFIE given in Section 5.2.1. It is seen that the MFIE solution is replete with "spikes" that occur precisely at the resonant frequencies of the circular cylinder marked for reference by the vertical bars along the frequency axis. The spikes are obviously nonphysical and uncharacteristic of the exact series solution given in Section 5.2.1 and plotted for reference in Figure 5-28a. As expected, at the resonant frequencies the value of the computed scattered field on the surface of the circular cylinder via the MFIE is also in error when compared with the exact series solution (see Figure 5-28b).

In Section 3.1.5 we presented the combined field integral equation (CFIE) formulation as a means of overcoming the inherent failure of the EFIE and MFIE solutions. This technique is one of the most popular remedies to the problem and has been discussed extensively in the literature (Mautz and Harrington, 1978; Govind and Wilton, 1979; Peterson, 1990; Putnam and Medgyesi-Mitchang, 1991). Here we

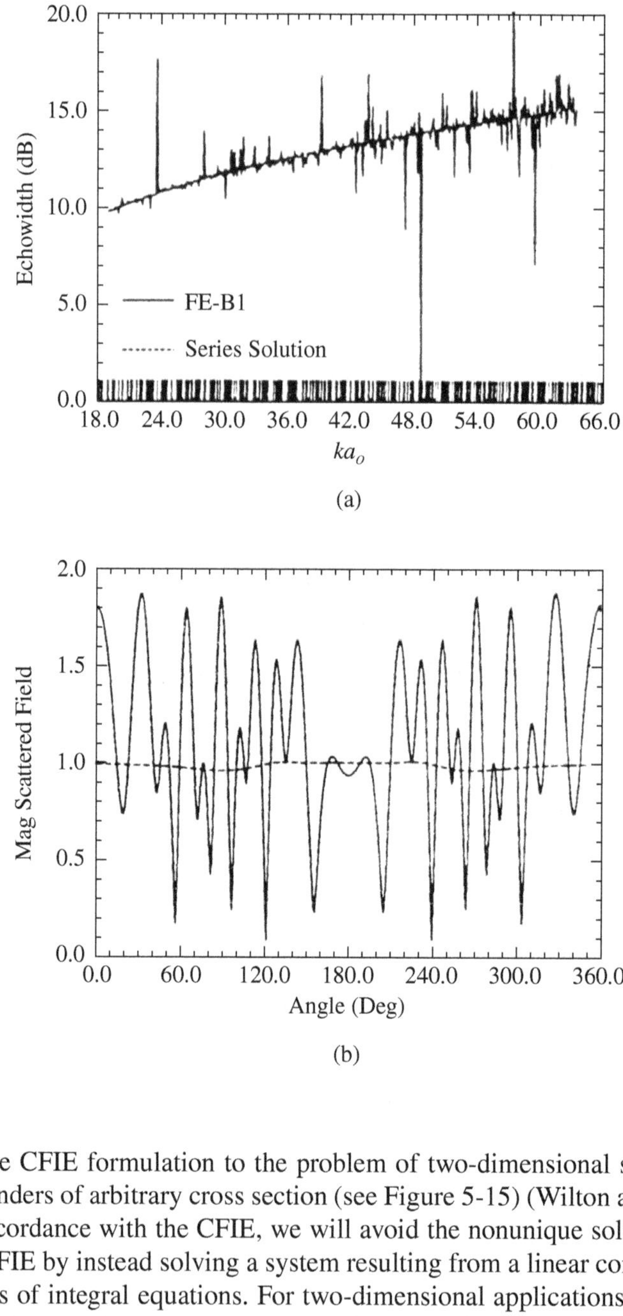


**FIGURE 5-28** Comparison of the far-zone and near-zone fields for TM plane wave incidence on a circular metallic cylinder as computed by the total field finite element-boundary integral (FE-BI) method and the eigenfunction series: (a) backscatter echo width vs. $ka_o$ (the lines over the horizontal axis correspond to the eigenvalues of a circular conducting waveguide), and (b) magnitude of the TM scattered field on the enclosure at the resonant frequency $ka_o = 23.586$. (After Collins, Jin, and Volakis, 1992. With permission.)

specialize the CFIE formulation to the problem of two-dimensional scattering by metallic cylinders of arbitrary cross section (see Figure 5-15) (Wilton and Wheeler, 1991). In accordance with the CFIE, we will avoid the nonunique solutions of the EFIE and MFIE by instead solving a system resulting from a linear combination of the two types of integral equations. For two-dimensional applications, it is appropriate to consider the integral equation resulting from the first or second of (3.45) since these decouple when specialized to two dimensions. In contrast, (3.44) remain coupled even for two-dimensional fields.

For TM incidence, the EFIE resulting from the first of (3.41a) is given by (5.70) and the MFIE as obtained from (3.41) reduces to (Wilton and Wheeler, 1991)

$$\frac{1}{2}J_z(\overline{\rho}) + \frac{jk_o}{4}\fint_C (\hat{n}\cdot\hat{R})J_z(\overline{\rho}')H_1^{(2)}(k_o|\overline{\rho}-\overline{\rho}'|)\,dl' = H_l^i(\overline{\rho}) \quad \overline{\rho}\in C \quad (5.126)$$

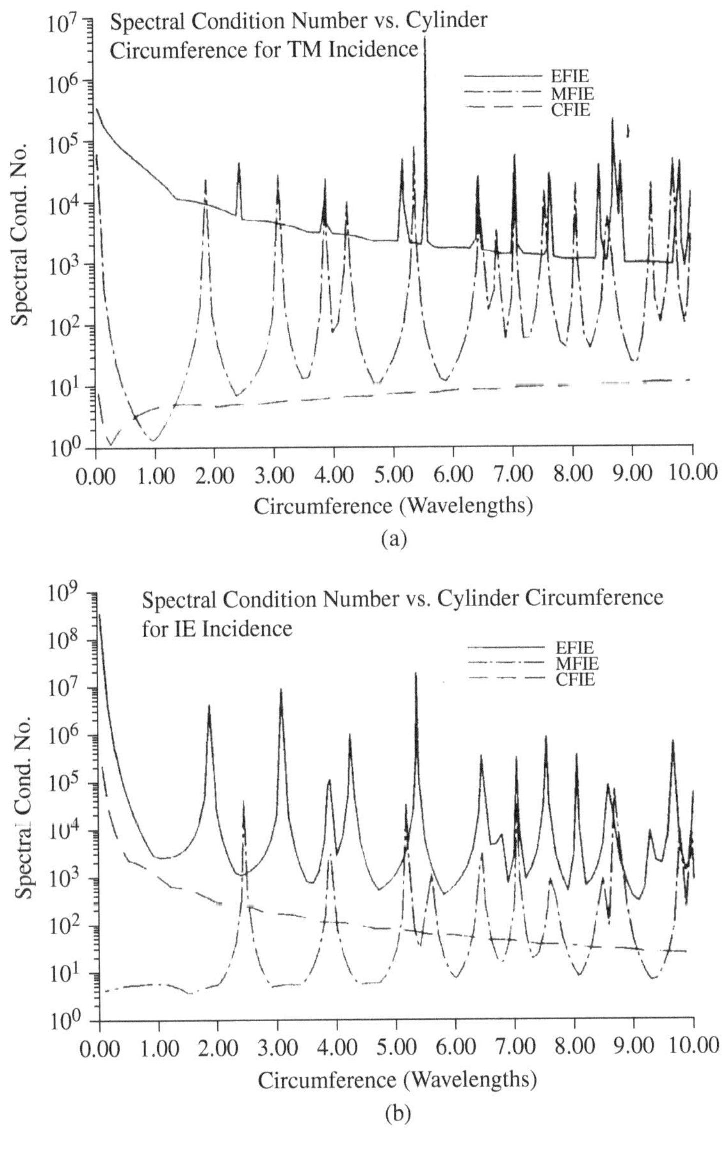


**FIGURE 5-29** Spectral condition number vs. circumference of a circular cylinder of the MFIE, EFIE, and CFIE matrix systems. For numerical implementation, the cylinder's geometry was approximated by 200 straight segments: (a) TM ($E$-polarization) incidence, and (b) TE ($H$-polarization) incidence. (After Wilton and Wheeler, 1991. With permission.)

where $H_l^i$ denotes the $\hat{l}$ components of the incident magnetic field. Combining (5.70) and (5.126), as dictated by (3.45), yields the integral equation

$$\frac{1}{2}J_z(\overline{\rho}) + \frac{jk_o}{4}\oint_C J_z(\overline{\rho}')\left[(\hat{n}\cdot\hat{R})H_1^{(2)}(k_o|\overline{\rho}-\overline{\rho}'|) - j\alpha H_o^{(2)}(k_o|\overline{\rho}-\overline{\rho}'|)\right]dl'$$

$$= \alpha Y_o E_z^i(\overline{\rho}) + H_l^i(\overline{\rho}) \qquad \overline{\rho}\in C \tag{5.127}$$

where $\alpha$ is a nonzero scalar, usually set to unity (experience shows that choosing $\alpha$ to be imaginary, between $\pm j0.5$ and $\pm j1$, also results in a good matrix condition). This CFIE can be readily discretized and uniquely solved for $J_z(\overline{\rho})$ using the procedures outlined in Sections 5.2 and 5.3. The far-zone scattered field is then computed from (5.79). Figure 5-29a clearly shows the improved condition of the CFIE system for a circular cylinder, even at the resonances.

The CFIE for TE incidence on a metallic cylinder is derived in a similar manner. Again, we consider a linear combination of the EFIE and MFIE resulting from the first of (3.41a) and (3.41b), respectively. These correspond to the integral equations given by (5.95) and (5.81), which when combined in accordance with (3.45) yield the CFIE

$$H_z^i(\overline{\rho}) + \alpha Y_o \hat{l} \cdot \mathbf{E}^i(\overline{\rho}) = \frac{J_l(\overline{\rho})}{2} + \frac{jk_o}{4} \oint_C J_l(\overline{\rho}') \big[ (\hat{n}' \cdot \hat{R}) H_1^{(2)}(k_o|\overline{\rho} - \overline{\rho}'|)$$
$$- j\alpha \hat{l} \cdot \hat{l}' H_o^{(2)}(k_o|\overline{\rho} - \overline{\rho}'|) - jk_o\alpha \frac{\partial}{\partial l'} (\hat{l} \cdot \hat{R}) H_1^{(2)}(k_o|\overline{\rho} - \overline{\rho}'|) \big] dl'. \tag{5.128}$$

This integral equation can again be discretized and solved as described in Sections 5.2 and 5.3. The improved condition of the $H$-polarization CFIE is demonstrated in Figure 5-29b.

We remark that the combined field formulation is only one method in overcoming the non-uniqueness of the EFIE and MFIE when applied to closed boundaries. Alternative approaches have been considered including the overspecification of the boundary conditions (Yaghjian, 1981; Mittra and Klein, 1975) and the combination of electric and magnetic currents (Mautz and Harrington, 1979), the latter being useful in radiation problems. Recently, a rather interesting technique based on the "complexification" of the wave number was introduced (Collins, Jin, and Volakis, 1992). In accordance with this method, no new integral equation needs to be derived. Instead, it requires only that the wave number in the MFIE or EFIE be replaced by $\alpha k_o$, where $\alpha$ is a complex constant with a very small imaginary part. It has been shown that the interior resonances of the EFIE, MFIE, or any other boundary integral equation are eliminated by simply set $\alpha$ equal to $1 - j0.001$ up to $1 - j0.005$. The success of this approach is indeed a consequence of the uniqueness theorem.

## 5.6 | SIMULATION OF INHOMOGENEOUS DIELECTRIC CYLINDERS

The aforementioned boundary integral formulations are suitable for homogeneous and piecewise homogeneous dielectric cylinders. Natural material (e.g., soil, vegetation, human tissue, or even man-made composites) are typically inhomogeneous. Consequently, we cannot break up the computational domain in separate homogeneous sections for analysis, as was illustrated in Figure 3-5. To solve for the fields scattered by or in the presence of an inhomogeneous structure, it is necessary to consider the volume and volume-surface integral equations discussed in Section 3.2.5. In this section we first present the solution of the traditional volume integral equation before proceeding with the solution of the more efficient volume-surface integral equation given in (3.114).

### 5.6.1 Volume Integral Equation

Consider the dielectric cylinder shown in Figure 5-30a. The cylinder is assumed to have varying permittivity and permeability given by $\epsilon = \epsilon_o \epsilon_r(x, y)$ and $\mu = \mu_o \mu_r(x, y)$, respectively. We are interested in computing the near, and far-zone

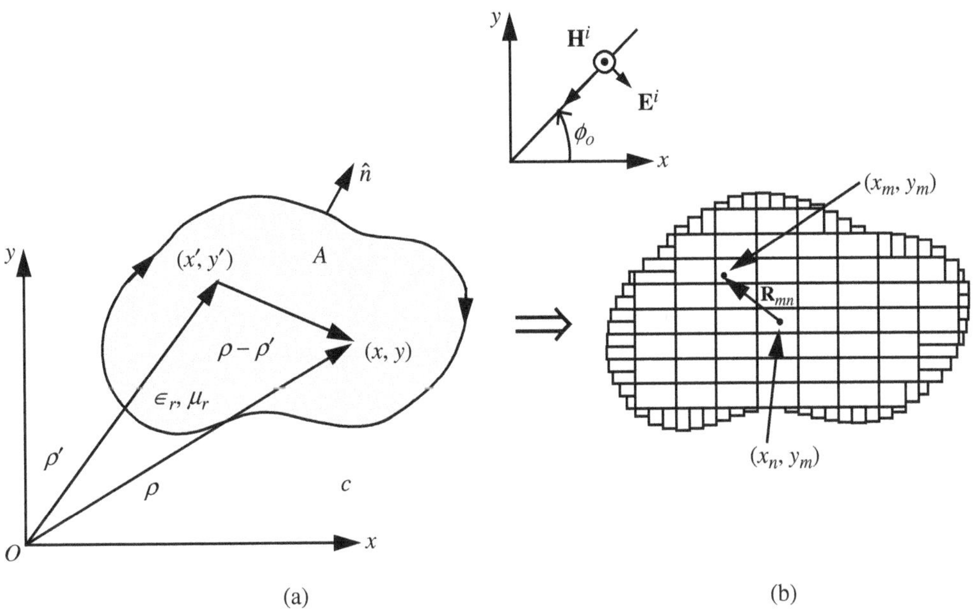


**FIGURE 5-30** ▪ Dicsretization of the inhomogeneous dielectric cylinder using rectangular elements: (a) original geometry, and (b) discretized geometry using rectangular elements/cells.

scattered fields from this cylinder due to a given excitation, e.g., TM incidence

$$\mathbf{E}^i = \hat{z}e^{jk_o(x\cos\phi_o + y\sin\phi_o)}.$$

As discussed in Section 3.2.5, a traditional approach in modeling the scattered fields involves the introduction of volume electric and magnetic polarization currents. These extend over only the domain of the dielectric region and are related to the total fields as illustrated in Figure 5-31. In accordance with the equivalence principle, the dielectric is removed and replaced by the polarization currents whose purpose is to recover the same scattered field as that generated by the original scatterer. Different equivalent currents can be selected for modeling the same scatterer as illustrated in Figure 3-7.

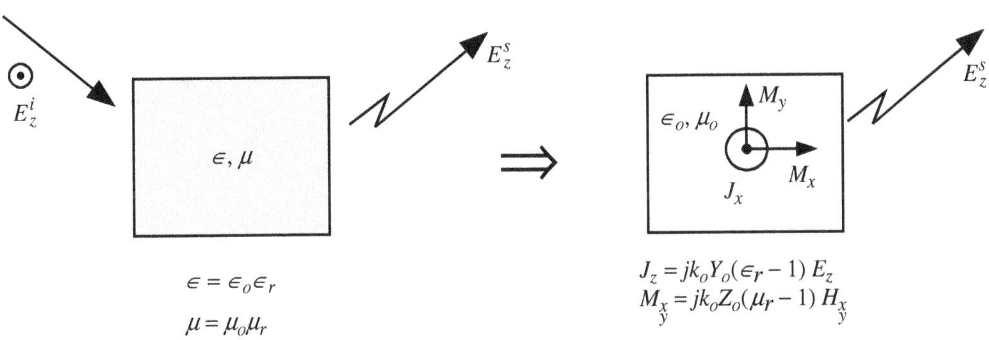


**FIGURE 5-31** ▪ Traditional volume currents for $E_z$ incidence.

To determine the volume-equivalent currents for a given inhomogeneous dielectric, we must solve the integral equations implied by (3.104) and (3.107), for the TM and TE incidence, respectively. The solution of each of these was first considered by Richmond (1965, 1966) for non-magnetic materials. This section generalizes Richmond's solution for cylinders having nontrivial $\epsilon_r$ and $\mu_r$, both of which can vary or be discontinuous across the cylinder. For TM incidence the three volume currents to be determined ($J_z$, $M_z$, and $M_y$) satisfy the integral equations

$$\frac{Z_o J_z(\overline{\rho})}{jk_o(\epsilon_r - 1)} + \frac{j}{4}\iint_A \left[M_x(\overline{\rho}')\frac{\partial}{\partial y} - M_y(\overline{\rho}')\frac{\partial}{\partial x}\right] H_o^{(2)}(k_o R)\, ds' + \frac{k_o Z_o}{4}\iint_A J_z(\overline{\rho}') H_o^{(2)}(k_o R)\, ds' = E_z^i(\overline{\rho}) \tag{5.129a}$$

$$\frac{M_x(\overline{\rho})}{jk_o(\mu_r - 1)} + \frac{jZ_o}{4}\iint_A J_z \frac{\partial}{\partial y} H_o^{(2)}(k_o R)\, ds' + \frac{1}{4k_o}\iint_A \left[M_x(\overline{\rho}')\left(k_o^2 + \frac{\partial^2}{\partial x^2}\right) + M_y(\overline{\rho}')\frac{\partial^2}{\partial x \partial y}\right] H_o^{(2)}(k_o R)\, ds' = Z_o H_x^i(\overline{\rho}) \tag{5.129b}$$

$$\frac{M_y(\overline{\rho})}{jk_o(\mu_r - 1)} - \frac{jZ_o}{4}\iint_A J_z \frac{\partial}{\partial x} H_o^{(2)}(k_o R)\, ds' + \frac{1}{4k_o}\iint_A \left[M_x(\overline{\rho}')\frac{\partial^2}{\partial y \partial x} + M_y(\overline{\rho}')\left(k_o^2 + \frac{\partial^2}{\partial y^2}\right)\right] H_o^{(2)}(k_o R)\, ds' = Z_o H_y^i(\overline{\rho}). \tag{5.129c}$$

On our way to numerically solving (5.129) for the equivalent currents $J_z$, $M_x$, and $M_y$, it is necessary to discretize the original dielectric cylinder and choose the basis functions for each of these currents. The simplest approach is to approximate the cylinder as a collection of small rectangular subsections (this subdivision is referred to as tessellation) and use the expansions

$$J_z = \sum_{n=0}^{N-1} J_{z_n} \mathrm{rect}(\Delta x_n, \Delta y_n) \tag{5.130a}$$

$$M_x = \sum_{n=0}^{N-1} M_{x_n} \mathrm{rect}(\Delta x_n, \Delta y_n) \tag{5.130b}$$

$$M_y = \sum_{n=0}^{N-1} M_{y_n} \mathrm{rect}(\Delta x_n, \Delta y_n) \tag{5.130c}$$

in which

$$\mathrm{rect}(\Delta x_n, \Delta y_n) = P_{\Delta x_n}(x - x_n) P_{\Delta y_n}(y - y_n)$$

with $P_{\Delta x_n}(x - x_n)$ being the usual pulse function centered at $x_n$ an of width $\Delta x_n$. The rect function is graphically illustrated in Figure 5-32, and clearly the expansions (5.130) will lead to unrealistic current discontinuities at the cell boundaries. This, of course, implies the presence of fictitious charges at those boundaries, thus compromising the accuracy of the numerical solution. Nevertheless, experience has shown that the simplistic expansion (5.130) leads to a solution of acceptable accuracy. The

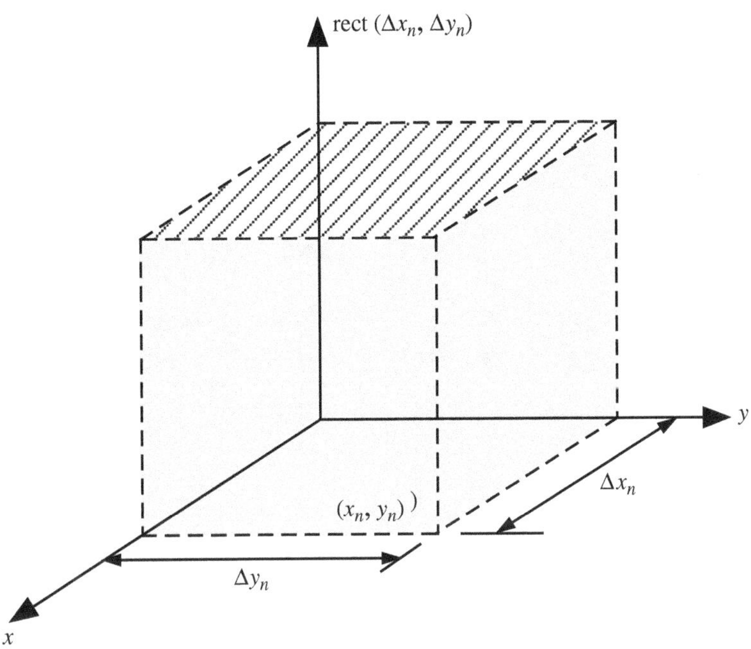


**FIGURE 5-32** Graphical illustration of the rect($\Delta x_n$, $\Delta y_n$) function.

lengths $\Delta x_n$ and $\Delta y_n$ denote the width and height of the $n$th rectangular element, $N$ of which are used for the total cylinder subdivision. For better numerical accuracy and stability, $\Delta x_n$ and $\Delta y_n$ should be chosen to be approximately equal. Also if $\Delta x_n$ and $\Delta y_n$ are all equal (i.e., $\Delta s_n = \Delta x$, $\Delta y_n = \Delta y$), it is then possible to employ the FFT in conjugation with an iterative solver to reduce the storage requirements of the solution method (as was discussed in Section 4.11).

Substituting the expansions (5.130) into the integral equations and point matching at the center of each rectangular cell, that is, at $\bar{\rho}_m = \hat{x}x_m + \hat{y}y_m$, yields the system

$$\begin{bmatrix} [A^{(1)}] & [A^{(2)}] & [A^{(3)}] \\ [B^{(1)}] & [B^{(2)}] & [B^{(3)}] \\ [C^{(1)}] & [C^{(2)}] & [C^{(3)}] \end{bmatrix} \begin{Bmatrix} \{J_{z_n}\} \\ \{M_{x_n}\} \\ \{M_{y_n}\} \end{Bmatrix} = \begin{Bmatrix} \{E^i_{z_m}\} \\ \{Z_o H^i_{x_m}\} \\ \{Z_o H^i_{y_m}\} \end{Bmatrix} \tag{5.131}$$

for the solution of current coefficients. In this

$$E^i_{z_m} = E^i_z(\overline{\rho}_m), \quad Z_o H^i_{x_m} = Z_o \hat{x} \cdot \mathbf{H}^i(\rho_m), \quad Z_o \mathbf{H}^i_{y_m} = Z_o \hat{y} \cdot \mathbf{H}^i(\rho_m)$$

and the nine square submatrices represent the interactions of the different currents with each other. For example, $[A^{(1)}]$ represents the interactions among the electric current elements, $[A^{(2)}]$ accounts for the interactions of the electric and $x$-directed magnetic currents and so on. Making use of the definitions (see Figure 5-33)

$$x_n^{\pm} = x_n \pm \frac{\Delta x_n}{2} \quad y_n^{\pm} = y_n \pm \frac{\Delta y_n}{2} \tag{5.132}$$

$$R_m = \sqrt{(x_m - x')^2 + (y_m - y')^2} \tag{5.133}$$

$$\epsilon_{r_n} = \epsilon_r(x_n, y_n), \quad \mu_{r_n} = \mu_r(x_n, y_n) \tag{5.134}$$

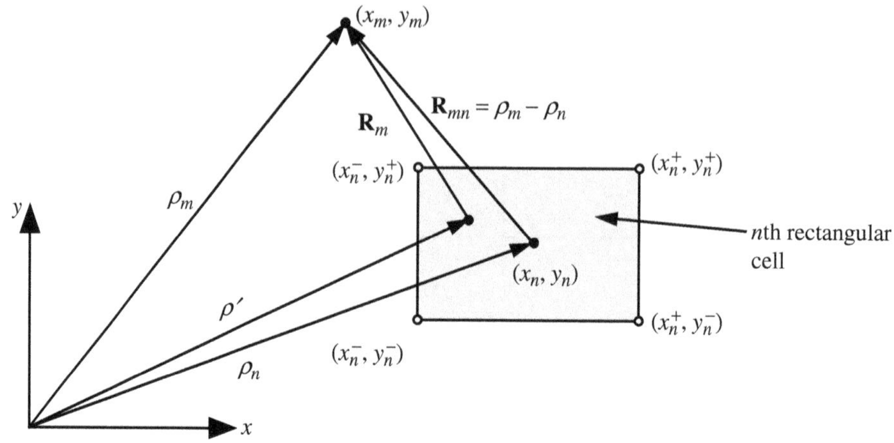


**FIGURE 5-33** Definition of parameters used in computing the matrix elements of the system (5.131).

the matrix elements may be formally defined as follows:

$$
\begin{aligned}
A_{mn}^{(1)} &= \frac{Z_o}{jk_o(\epsilon_{r_m}-1)}\delta(n-m) + \frac{k_o Z_o}{4}\int_{x_n^-}^{x_n^+}\int_{y_n^-}^{y_n^+} H_o^{(2)}(k_o R_m)\,dx'\,dy' \\
A_{mn}^{(2)} &= \frac{-j}{4}\int_{x_n^-}^{x_n^+}\left[H_o^{(2)}\left(k_o\sqrt{(x_m-x')^2+(y_m-y_n^+)^2}\right)\right. \\
&\quad \left. - H_o^{(2)}\left(k_o\sqrt{(x_m-x')^2+(y_m-y_n^-)^2}\right)\right] dx' \\
A_{mn}^{(3)} &= \frac{+j}{4}\int_{y_n^-}^{y_n^+}\left[H_o^{(2)}\left(k_o\sqrt{(x_m-x_n^+)^2+(y_m-y')^2}\right)\right. \\
&\quad \left. - H_o^{(2)}\left(k_o\sqrt{(x_m-x_n^-)^2+(y_m-y')^2}\right)\right] dx' \qquad (5.135) \\
B_{mn}^{(1)} &= +Z_o A_{mn}^{(2)} \\
B_{mn}^{(2)} &= \frac{-j}{k_o(\mu_{r_m}-1)}\delta(n-m) \\
&\quad + \frac{1}{4k_o}\frac{\partial}{\partial y}\int_{x_n^-}^{x_n^+}\left[H_o^{(2)}\left(k_o\sqrt{(x_m-x')^2+(y-y_n^+)^2}\right)\right. \\
&\quad \left. - H_o^{(2)}\left(k_o\sqrt{(x_m-x')^2+(y-y_n^-)^2}\right)\right]\Bigg|_{y=y_m} dx' \\
B_{mn}^{(3)} &= +\frac{1}{4k_o}\left[H_o^{(2)}\left(k_o\sqrt{(x_m-x_n^+)^2+(y_m-y_n^+)^2}\right)\right. \\
&\quad - H_o^{(2)}\left(k_o\sqrt{(x_m-x_n^+)^2+(y_m-y_n^-)^2}\right)
\end{aligned}
$$

$$- H_o^{(2)}\left(k_o\sqrt{(x_m - x_n^-)^2 + (y_m - y_n^+)^2}\right)$$
$$+ H_o^{(2)}\left(k_o\sqrt{(x_m - x_n^-)^2 + (y_m - y_n^-)^2}\right)\Bigg] \tag{5.136}$$

$$C_{mn}^{(1)} = Z_o A_{mn}^{(3)}$$
$$C_{mn}^{(2)} = B_{mn}^{(3)}$$
$$C_{mn}^{(3)} = \frac{-j}{k_o(\mu_r - 1)}\delta(m-n) + \frac{1}{4}\frac{\partial}{\partial x}\int_{y_n^-}^{y_n^+}\left[H_o^{(2)}\left(k_o\sqrt{(x-x')^2 + (y_m - y_n^+)^2}\right)\right.$$
$$\left. - H_o^{(2)}\left(k_o\sqrt{(x-x')^2 + (y_m - y_n^-)^2}\right)\right]\Bigg|_{x=x_m} dx'. \tag{5.137}$$

All integrals in the aforementioned expressions can be evaluated using the techniques discussed in the earlier part of this chapter. Specifically, the reader is referred to equations (5.13) through (5.16), (5.56) through (5.58), (5.90) through (5.92), (5.118)(see also Ricoy and Volakis (1989)).

In his original paper, Richmond (1965, 1966) chose to evaluate the matrix elements $A_{mn}^{(i)}$, $B_{mn}^{(i)}$ and $C_{mn}^{(i)}$ by instead integrating over a circular cell centered at $(x_n, y_n)$ and having the same cross sectional area as the original rectangular cell. Richmond's definition of the equivalent cylinder is illustrated in Figure 5-34, and he was motivated in introducing this approximation because the resulting integrals over the circular domain can be evaluated exactly. For example, it can be shown that

$$\int_{x_n^-}^{x_n^+}\int_{y_n^-}^{y_n^+} H_o^{(2)}(k_o R_m)\,dx'\,dy' \approx \int_0^{2\pi}\int_0^{a_n} H_o^{(2)}(k_o|\bar{\rho}_m - \bar{\rho}'|)\rho'\,d\rho'\,d\phi'$$
$$= \begin{cases} \frac{2}{k_o^2}\left[\pi k_o a_n H_1^{(2)}(k_o a) - 2j\right] & m = n \\ 2\pi a_n J_1(k_o a) H_o^{(2)}(k_o R_{mn}) & m \neq n \end{cases} \tag{5.138}$$

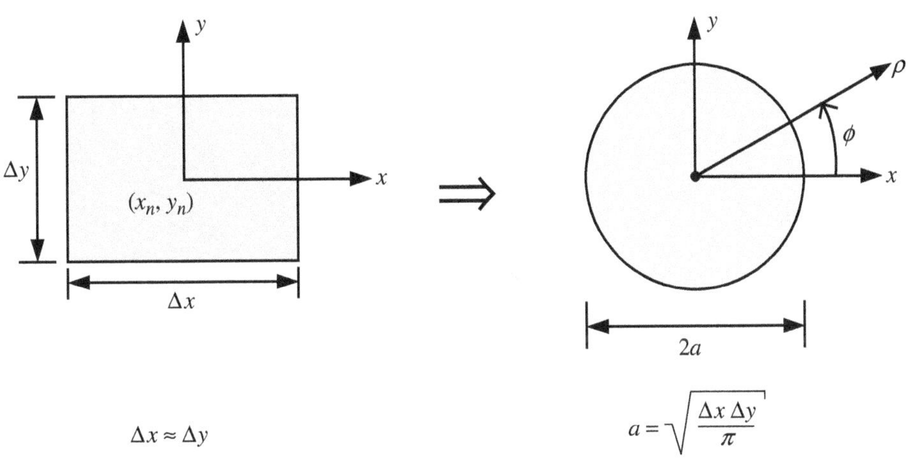


**FIGURE 5-34** Richmond's approximation of a rectangular dielectric cell by an "equivalent" circular cell.

in which $R_{mn} = \sqrt{(x_m - x_n)^2 + (y_m - y_n)^2}$ and $a_n = \sqrt{\frac{\Delta x_n \Delta y_n}{\pi}}$ assuming that $\Delta x_n \approx \Delta y_n$. The evaluation of the self-cell in (5.138) is rather trivial upon noting that

$$|\overline{\rho}_m - \overline{\rho}'| = \rho'$$

when $m = n$ and the identity

$$\frac{d}{d\rho'}\left[\rho' H_1^{(2)}(k_o\rho')\right] = k_o H_o^{(2)}(k_o\rho')\rho'.$$

From (5.138), it follows that we can approximate $A_{mn}^{(1)}$ as

$$A_{mn}^{(1)} = \begin{cases} \dfrac{Z_o}{jk_o(\epsilon_{r_m} - 1)} + \dfrac{Z_o}{2k_o}\left[\pi k_o a_m H_1^{(2)}(k_o a_m) - 2j\right] & n = m \\ Z_o \dfrac{\pi a_n}{2} J_1(k_o a_n) H_o^{(2)}(k_o R_{mn}) & n \neq m. \end{cases} \tag{5.139a}$$

Also by using the cylindrical cell approximations, it can be shown that (Richmond, 1966)

$$A_{mn}^{(2)} = -\frac{jk_o}{4}\int_0^{2\pi}\int_0^{a_n} \hat{y}\cdot\hat{R}_m H_1^{(2)}(k_o R_m)\rho'\, d\rho'\, d\phi' \tag{5.139b}$$

$$A_{mn}^{(3)} = +\frac{jk_o}{4}\int_0^{2\pi}\int_0^{a_n} \hat{x}\cdot\hat{R}_m H_1^{(2)}(k_o R_m)\rho'\, d\rho'\, d\phi' \tag{5.139c}$$

$$B_{mn}^{(1)} = Z_o A_{mn}^{(2)} \tag{5.139d}$$

$$B_{mn}^{(2)} = \begin{cases} \dfrac{1}{jk_o(\mu_r - 1)} + \dfrac{1}{jk_o}\left[\dfrac{j\pi k_o a_m}{4} H_1^{(2)}(k_o a_m) + 1\right], & m = n \\ \dfrac{\pi a_n J_1(k_o a_n)}{2k_o^2 R_{mn}^3}\Big[k_o R_{mn}(y_m - y_n)^2 H_o^{(2)}(k_o R_{mn}) \\ \quad +((x_m - x_n)^2 - (y_m - y_n)^2) H_1(k_o R_{mn})\Big], & m \neq n \end{cases} \tag{5.139e}$$

$$B_{mn}^{(3)} = \frac{\pi a_n J_1(k_o a_n)}{2k_o R_{mn}^3}(x_m - x_n)(y_m - y_n)\left[2H_1^{(2)}(k_o R_{mn}) - k_o R_{mn} H_o^{(2)}(k_o R_{mn})\right] \tag{5.139f}$$

$$C_{mn}^{(1)} = Z_o A_{mn}^{(3)} \tag{5.139g}$$

$$C_{mn}^{(2)} = B_{mn}^{(3)} \tag{5.139h}$$

and

$$C_{mn}^{(3)} = \begin{cases} B_{mm}^{(2)}, & m = n \\ \dfrac{\pi a_n J_1(k_o a_n)}{2k_o R_{mn}^3}\Big[k_o R_{mn}(x_m - x_n)^2 H_o^{(2)}(k_o R_{mn}) \\ +((y_m - y_n)^2 - (x_m - x_n)^2) H_1^{(2)}(k_o R_{mn})\Big], & m \neq n \end{cases} \tag{5.139i}$$

in which $\hat{R}_m = [(x_m - x')\hat{x} + (y_m - y')\hat{y}]/R_m$. We note that since $\hat{R}_m = -(\hat{x}\cos\phi' + \hat{y}\sin\phi')$ when $n = m$, it follows that $\hat{y}\cdot\hat{R}_m = -\sin\phi'$ and $H_1^{(2)}(k_o R_{mn}) = H_1^{(2)}(k_o\rho')$ in (5.139b) and (5.139c). Also,

$$H_1^{(2)}(k_o\rho')\rho'\Big|_{\rho'\to 0} = \frac{-j2}{\pi} \tag{5.140}$$

implying that $A_{mm}^{(2)} = A_{mm}^{(3)} = 0$. For $m \neq n$, midpoint integration can be employed to evaluate $A_{mn}^{(2)}$ and $A_{mn}^{(3)}$. The circular cell approximation is based on the equal area concept and therefore offers a way of modeling the contribution of small non-rectangular or other shape elements that remain after the subdivision of the cylinder. No special treatment of these elements is required other than the introduction of a circular cylinder of radius $a = \sqrt{\Delta A/\pi}$, where $\Delta A$ is the area of the original non-rectangular element.

In solving the system (5.131) it is advisable to normalize each of the rows by the largest entry of that row to improve the matrix condition number before inversion. Also, to avoid large matrix element values when $\epsilon_r$ or $\mu_R$ are near unity, we can instead solve for the field quantities $E_{z_n} = \frac{J_{zn}}{jk_oY_o(\epsilon_r-1)}$, $H_{x_n} = \frac{M_{xn}}{jk_oZ_o(\mu_r-1)}$ and $H_{y_n} = \frac{M_{yn}}{jk_oZ_o(\mu_r-1)}$, which, of course, remain bounded throughout the dielectric.

Once the system (5.131) is solved, the far-zone scattered field can be evaluated from (2.133). It should be noted, however, that for better accuracy and consistency the cell integrations should be over the same cross sectional area used for the evaluation of the matrix elements. That is, if circular elements were used, the far-zone integral should also be evaluated by integrating over the sum of the circular domains defining those elements. Needless to say, since the previously given solution applies to dielectric cylinders with nontrivial permeability and permittivity, duality can be invoked for the solution of the fields under TE excitation (see (5.28)). Some illustrative scattering patterns using Richmond's formulation were shown in Figure 5-22.

### 5.6.2 Volume-Surface Integral Equation

The previously presented solution and in particular that employing circular elements, has been found quite accurate for modeling cylinders having nominal values of relative dielectric constants (e.g., up to 10). However, for large values of $\epsilon_r$ or $\mu_r$, it is necessary that the discretized geometry maintains as much as possible the shape and detail of the original geometry. Given that $\Delta x$ and $\Delta y$ must be generally selected to be less than $0.1\lambda_d$, where $\lambda_d$ is the wavelength in the dielectric, it becomes rather difficult to maintain geometrical fidelity for arbitrary cross section cylinders when modeling them as a collection of rectangular and circular elements. Moreover, as discussed in Section 3.2.4, the previous solution is not as efficient when both $\epsilon_r$ and $\mu_r$ are greater than unity.

To improve upon the geometrical modeling, one could choose to tessellate the cylinder's cross section using triangles or second-order quadrilaterals. The latter are four-sided elements whose sides, instead of being flat, are second-order curves. Further, to reduce the number of unknown current coefficients, we could instead choose to represent the scattered fields directly in terms of the total electric or magnetic fields as given in (3.114). Next we consider the solution of the VSIE (3.114) using isoparametric quadrilaterals (four-sided elements) to tessellate the cylinder's cross section. The word *isoparametric* ("iso" is the Greek word for equal) implies elements that have the same order of approximation in representing the fields and the geometry/contour of the dielectric. It is, of course, logical to choose isoparametric elements because, as can be expected, a quadratic field expansion cannot improve the solution accuracy when we insist on using square rectangular elements for discretizing a dielectric scatterer having circular or other nonlinear contours. The second-order (meaning the sides are quadratic curves) quadrilateral, to be employed in the subsequent integral equation solution, is illustrated in Figure 5-35. The contour of the cylinder will be modeled using quadratic line segments that are depicted in Figure 5-36. An example of such a tessellation applied to a circular cylinder is illustrated in Figure 5-37.

A characteristic of each geometric element (line, area, or volume) is its shape function. These shape functions allow, for example, a curved line segment to be mapped to a straight segment or an arbitrary quadrilateral to be mapped to a square as shown in Figures 5-35 and 5-36. These mappings are necessary to perform the integrations required for evaluating the matrix elements.

The transformation that maps a quadratic line segment in the $x$-$y$ plane into a straight line in the $\zeta - \eta$ plane is given by

$$x = \sum_{i=1}^{3} L_i(\zeta)x_i \qquad y = \sum_{i=1}^{3} L_i(\zeta)y_i \tag{5.141}$$

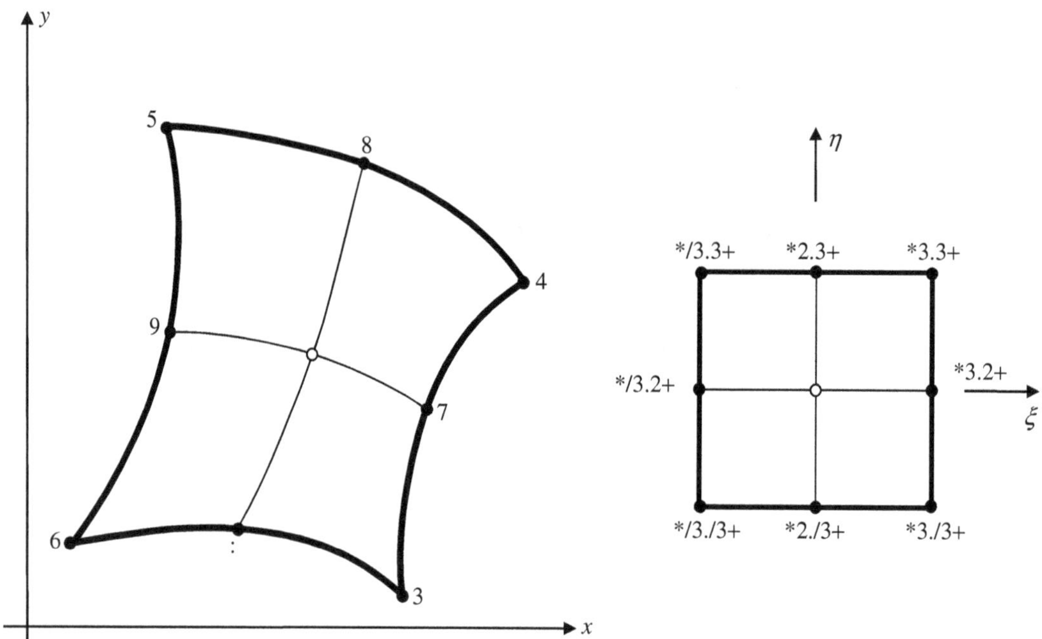


**FIGURE 5-35** ■ Illustration of a second-order quadrilateral in its original shape, and after transformation to a square onto the $\zeta\eta$ plane. The numerical circles denote the node locations, and there is a one-to-one correspondence between the nodes on the two planes (adapted from Jin, Volakis, and Liepa, 1989).

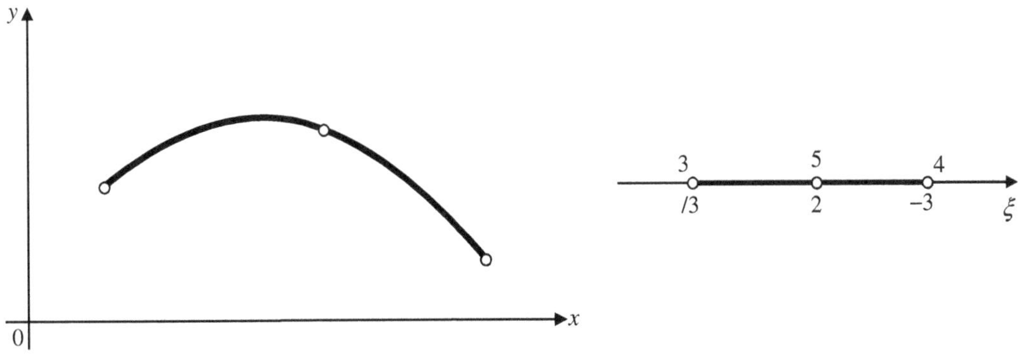


**FIGURE 5-36** Illustration of the second-order curved-line segment in its original form and after its transformation to a straight segment onto the $\zeta$-axis. There is a one-to-one correspondence between the nodes and segment points on the two planes (adapted from Jin, Volakis, and Liepa, 1989).

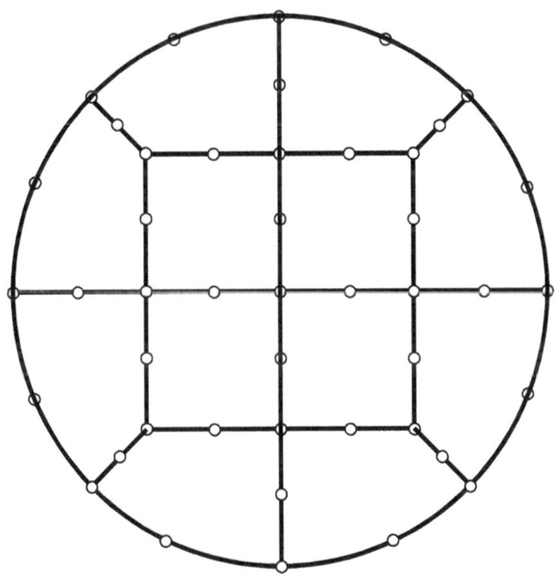

**FIGURE 5-37** Illustration of the discretization of a circular cylinder using second-order quadrilaterals and line segments (adapted from Jin, Volakis, and Liepa, 1989).

where Cartesian coordinates $(x_i, y_i)$ denote the location of the $i$th node and

$$L_1 = -\frac{1}{2}(1-\zeta)\zeta, \quad L_2 = \frac{1}{2}(1+\zeta)\zeta, \quad L_3 = 1-\zeta^2 \qquad (5.142)$$

are the pertinent shape functions. They can be readily determined by simply introducing the quadratic expansion (for quadratic elements)

$$x = c_1 + c_2\zeta + c_3\zeta^2$$

and then finding the constants $c_i$ such that $x = x_1$ at $\zeta = -1$, $x = x_2$ at $\zeta = 1$ and $x = x_3$ at $\zeta = 0$. It is important to observe that $L_j(\zeta)$ vanishes at all nodes except the $j$th and is equal to unity at the $j$th node. Also, note that the shape functions are independent of the element's location in the $x$-$y$ plane.

The transformation that maps a quadrilateral in the $x$-$y$ plane onto a square in the $\zeta - \eta$ plane is given by

$$x = \sum_{i=1}^{8} N_i(\zeta, \eta)x_i \qquad y = \sum_{i=1}^{8} N_i(\zeta, \eta)y_i \tag{5.143}$$

where the quadratic shape functions now take the form

$$\begin{aligned} N_1 &= -\tfrac{1}{4}(1-\zeta)(1-\eta)(\zeta+\eta+1), & N_2 &= \tfrac{1}{4}(1+\zeta)(1-\eta)(\zeta-\eta-1) \\ N_3 &= \tfrac{1}{4}(1+\zeta)(1+\eta)(\zeta+\eta-1), & N_4 &= \tfrac{1}{4}(1-\zeta)(1+\eta)(-\zeta+\eta-1) \\ N_5 &= \tfrac{1}{2}(1-\zeta^2)(1-\eta), & N_6 &= \tfrac{1}{2}(1+\zeta)(1-\eta)^2 \\ N_7 &= \tfrac{1}{2}(1-\zeta^2)(1+\eta), & N_8 &= \tfrac{1}{2}(1-\zeta)(1-\eta^2). \end{aligned} \tag{5.144}$$

These shape functions can again be determined beginning with the expansion

$$x = c_1 + c_2\zeta + c_3\eta + c_4\eta\zeta + c_5\eta^2 + c_6\zeta^2$$

and then proceeding to find the coefficients, $c_i$, by demanding that $x$ takes the specified values at the nodes. We observe, as before in (5.141), that $N_j(\zeta, \eta)$ vanishes at all nodes except the $j$th and is equal to unity when evaluated at the $j$th node.

Having introduced the new geometric elements let us now proceed with the discretization of the integral equation (3.114), which can be rewritten as (Jin and Volakis, 1991)

$$F^{inc}(\overline{\rho}) + \iint_A A(\overline{\rho}, \overline{\rho}')F(\overline{\rho}')\,ds' + \oint B(\overline{\rho}, \overline{\rho}')F(\overline{\rho}')\,ds'$$
$$= \begin{cases} uF(\overline{\rho}), & \overline{\rho} \in A, \quad \overline{\rho} \notin C \\ \frac{1}{2}[u(\overline{\rho}^+) + u(\overline{\rho}^-)]F(\overline{\rho}), & \overline{\rho} \in C \end{cases} \tag{5.145a}$$

where

$$A(\overline{\rho}, \overline{\rho}') = \frac{-jk_o^2}{4}[\upsilon(\overline{\rho}') - u(\overline{\rho}')]H_o^{(2)}(k_o|\overline{\rho} - \overline{\rho}'|)$$
$$\frac{-j}{4}\nabla' u(\overline{\rho}) \cdot \nabla' H_o^{(2)}(k_o|\overline{\rho} - \overline{\rho}'|) \tag{5.145b}$$

$$B(\overline{\rho}, \overline{\rho}') = \frac{-j}{4}\left[u(\overline{\rho}^+) - u(\overline{\rho}'^-)\right](\hat{n}' \cdot \hat{R})k_o H_1^{(2)}(k_o|\overline{\rho} - \overline{\rho}'|) \tag{5.145c}$$

$\hat{n}'$ denotes the outward normal along the integration segment and $\overline{\rho}'^{+/-}$ indicates the integration point just outside/inside the contour $C$. Note that $C$ encompasses the outer contour enclosing the cylinder and all boundaries associated with a material discontinuity. For TM incidence

$$F(\overline{\rho}) = E_z(\overline{\rho}), \quad u(\overline{\rho}) = \frac{1}{\mu_r(\overline{\rho})}, \quad \upsilon(\overline{\rho}) = \epsilon_r(\overline{\rho})$$

whereas for TE incidence

$$F(\overline{\rho}) = H_z(\overline{\rho}), \quad u(\overline{\rho}) = \frac{1}{\epsilon(\overline{\rho})}, \quad \upsilon(\overline{\rho}) = \mu_r(\overline{\rho}).$$

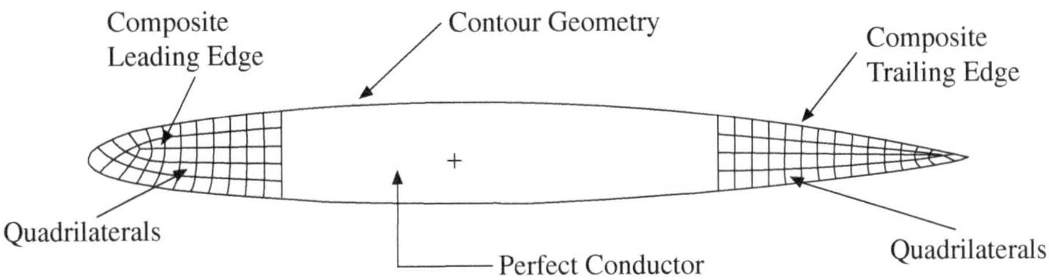


**FIGURE 5-38** Example discretization of an aircraft wing cross section composed of metallic and non-metallic components.

To discretize (5.145) we first tessellate the cylinder's volume into $M$ quadrilaterals and the contour $C$ into $N$ second-order curvilinear segments as illustrated in Figure 5-38. If certain sections of the cylinder are metallic, it is necessary only to discretize along the contour enclosing the metal. Further, for TM incidence it is not even required to consider the metallic contours because $F = E_z$ vanishes on these contours. The modeling of exposed metallic boundaries may, in this case, be done by considering a thin dielectric layer having $\epsilon_r = 1 - j10^6$ and a minimal thickness of $0.01\lambda$ or so.

Based on the previous geometrical discretization we introduce the expansions

$$F(\overline{\rho}) = \sum_{e=1}^{M} F^e(\overline{\rho}) = \sum_{e=1}^{M}\sum_{i=1}^{8} N_i(\overline{\rho})\phi_i^e, \qquad \rho \in A \tag{5.146a}$$

$$F(\overline{\rho}) = \sum_{l=1}^{N} F^l(\overline{\rho}) = \sum_{l=1}^{N}\sum_{i=1}^{3} L_i(\overline{\rho})\phi_i^l, \qquad \rho \in C \tag{5.146b}$$

for the fields appearing in the first and second integral of (5.145), respectively. The shape functions $L_i(\overline{\rho})$ and $N_i(\overline{\rho})$ are the same as those defined in (5.142) and (5.144). Consequently, the coefficients $\phi_i^l$ and $\phi_i^e$ are simply equal to the fields at the nodes of the curvilinear contour segment or the quadrilateral nodes. To determine the nodal fields we substitute (5.146) into (5.145) to obtain the system of discrete equations

$$F^{inc}(\rho_m^-) + \sum_{e=1}^{M}\sum_{i=1}^{8} \tilde{A}_{im}^e \phi_i^e + \sum_{l=1}^{N}\sum_{i=1}^{3} \tilde{B}_{im}^l \phi_i^l$$
$$= \begin{cases} u(\overline{\rho}_m)F(\overline{\rho}_m) & \rho \in A \\ \frac{1}{2}\left[u(\overline{\rho}_m^+) + u(\overline{\rho}_m^-)\right]F(\overline{\rho}_m) & \rho \in C \end{cases} \tag{5.147}$$

in which $\overline{\rho}_m$ denotes the testing point.

For point matching, the testing points are chosen to be the field nodes. In this manner, a total of $M' + N$ equations are constructed, where $M' < M$ excludes those nodes of the quadrilaterals coinciding with the contour $C$. The matrix elements $\widetilde{B}_{im}^l$ are given by

$$\widetilde{B}_{im}^l = \int_{C^l} B(\overline{\rho}_m, \overline{\rho}')L_i(\overline{\rho}')\,ds' = \int_{-1}^{+1} B(\overline{\rho}_m, \overline{\rho}')L_i(\xi')|J^l(\xi')|\,d\xi' \tag{5.148}$$

in which $C^l$ denotes the $l$th curvilinear contour segment and

$$|J^l(\xi)| = \sqrt{\left(\frac{\partial x}{\partial \xi}\right)^2 + \left(\frac{\partial y}{\partial \xi}\right)^2}. \tag{5.149}$$

They can be evaluated by employing a four-point Gaussian integration formula. We have

$$\widetilde{B}^l_{im} = \sum_{p=1}^{4} w_p B(\overline{\rho}_m, \overline{\rho}'_p) L^l_i(\xi_p)|J^l(\xi_p)| \tag{5.150}$$

where $\xi_1 = -0.8611363116$, $\xi_2 = -0.3399810436$, $\xi_3 = 0.3399810436$, $\xi_4 = 0.8611363116$, $w_1 = w_4 = 0.3478548451$, $w_2 = w_3 = 0.6521451549$ and $\overline{\rho}_p = x(\xi_p)\hat{x} + y(\xi_p)\hat{y}$. The unit normal appearing in (5.145) can be evaluated from the expression

$$\hat{n} = \left(\hat{x}\frac{\partial y}{\partial \xi} - \hat{y}\frac{\partial x}{\partial \xi}\right) |J^l(\xi)|^{-1} \tag{5.151}$$

with $x(\xi)$ and $y(\xi)$ as given in (5.141). It should be noted that (5.150) is valid even for the self-cell because the integral discontinuity defined in (3.81) is not picked up when performing a numerical evaluation.

In a similar manner, we can express $\widetilde{A}^e_{im}$ as

$$\begin{aligned}\widetilde{A}^e_{im} &= \iint_{A^e} A(\overline{\rho}_m, \overline{\rho}') N_i(\overline{\rho}')\, ds' \\ &= \int_{-1}^{+1}\int_{-1}^{+1} A(\overline{\rho}_m, \overline{\rho}') N_i(\xi', \eta')|J^e(\xi', \eta')|\, d\xi' d\eta' \end{aligned} \tag{5.152}$$

where $A^e$ denotes the $e^{\text{th}}$ quadrilateral and

$$J(\xi, \eta) = \frac{\partial x}{\partial \xi}\frac{\partial y}{\partial \eta} - \frac{\partial x}{\partial \eta}\frac{\partial y}{\partial \xi} \tag{5.153}$$

is the determinant of the Jacobian transformation matrix. They can be evaluated by employing a nine-point Gaussian integration. We have

$$\widetilde{A}^e_{im} = \sum_{p=1}^{3}\sum_{q=1}^{3} w_p w_q A(\overline{\rho}_m, \overline{\rho}'_{pq}) N_i(\xi_p, \eta_q)|J^e(\xi_p, \eta_q)| \tag{5.154}$$

where $\xi_1 = \eta_1 = -0.7745966692$, $\xi_2 = \eta_2 = 0.0$, $\xi_3 = \eta_3 = 0.7745966692$, $w_1 = w_3 = 0.5555555556$, $w_2 = 0.8888888889$, and $\overline{\rho}'_{pq} = x(\xi_p, \eta_q)\hat{x} + y(\xi_p, \eta_q)\hat{y}$ with $x(\xi, \eta)$ and $y(\xi, \eta)$ as defined in (5.143). Again, we remark that (5.154) remains valid even when $\rho_{pq} = \rho_m$.

Results depicting the accuracy of the VSIE are illustrated in Figure 5-39. Figure 5-39 shows the fields computed via the VSIE in comparison to the exact eigenfunction solution for a circular dielectric cylinder have $\epsilon_r = 72 - j162$ and a diameter of $0.1\lambda_o$. This value of $\epsilon_r$ corresponds to the relative permittivity of a muscle tissue at about 100 MHz. Backscatter patterns for a nonuniform dielectric cylinder with nominal values of $\epsilon_r$ and $\mu_r$ are shown in Figure 5-40 (Ricoy, Kempel, and Volakis, 1990).

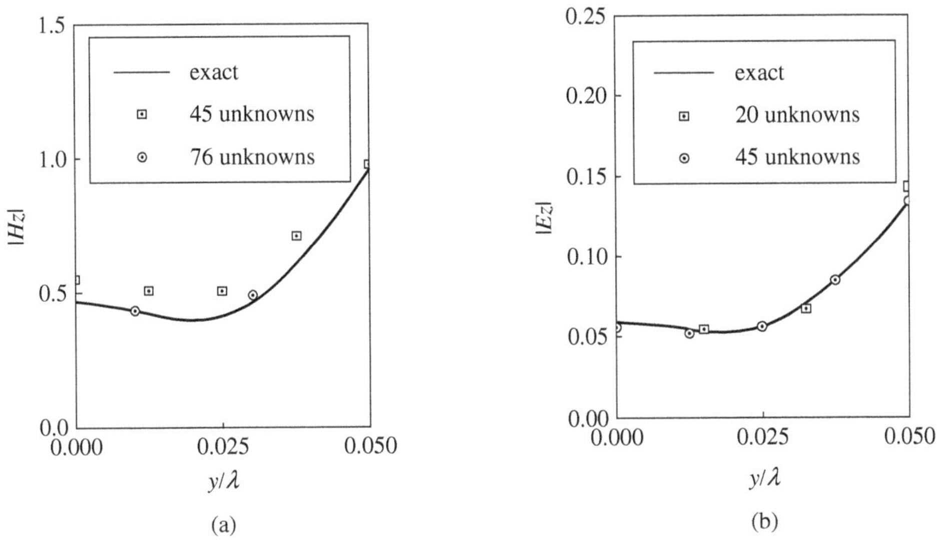


**FIGURE 5-39** Fields in a circular dielectric cylinder of radius $0.05\lambda_o$ as computed via the VSIE and the eigenfunction series (exact). The cylinder has $\epsilon_r = 70 - j162$, and the plane wave illumination is along the $x$-direction: (a) TE incidence and (b) TM incidence (adapted from Jin, Volakis, and Liepa, 1989).

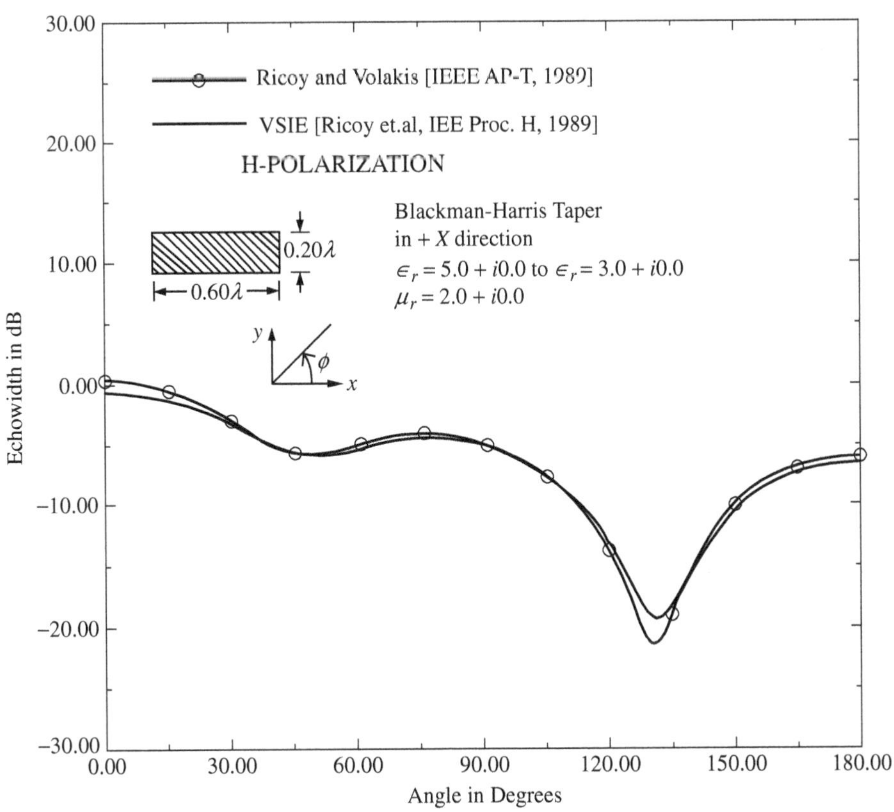


**FIGURE 5-40** Backscatter patterns of the illustrated non-uniform square dielectric cylinder (Ricoy *et al.*, 1990).

## BIBLIOGRAPHY

Barkeshli, K. and Volakis, J. L. (2004). Electromagnetic scattering from thin stripspart ii: Numerical solution for strips of arbitrary size. *IEEE Trans. on Education*, *47*(1), 107–113.

Collins, J. D., Jin, J.-M., and Volakis, J. L. (1992). Eliminating interior resonances in finite element-boundary integral methods for scattering. *IEEE Trans. Antennas Propagat.*, *40*(12), 1583–1585.

Govind, S. and Wilton, D. R. (1979). Numerical computation of electromagnetic scattering by inhomogeneous penetrable bodies. *Rome Air Dev. Center, U.S. Air Force, report RADC-TR-79-172.*

Jin, J. J. and Volakis, J. L. (1990). Electromagnetic scattering by a perfectly conducting patch array on a dielectric slab. *IEEE Trans. Antennas Propagat.*, *38*(4), 556–563.

Jin, J.-M. and Volakis, J. L. (1991). A finite element-boundary integral formulation for scattering by three-dimensional cavity-backed apertures. *IEEE Trans. Antennas and Propagat.*, *39*, 97–104.

Jin, J.-M., Volakis, J. L., and Liepa, V. V. (1989). A moment method solution of a volume-surface integral equation using isoparametric elements and point matching TE scattering. *IEEE Trans. Microwave Theory Tech.*, *37*(10), 1641–1645.

Mautz, J. R. and Harrington, R. F. (1978). H-field, E-field and combined field solutions for conducting bodies of revolution. *Arch. Elekron. Ubertragungstech (AEU)*, *32*, 159–164.

Mautz, J. R. and Harrington, R. F. (1979). Electromagnetic scattering from a homogeneous material body of revolution. *Arch. Elek. Ubertragung.*, *33*, 71–80.

Mei, K. and Bladel, J. V. (1963). Scattering by perfectly-conducting rectangular cylinders. *IEEE Trans. Antennas Propagat.*, *11*(2), 185–192.

Mittra, R. and Klein, C. A. (1975). *"Stability and convergence of moment method solutions"* (chap. in *Numerical and Asymptotic Techniques in Electromagnetics*, R. Mittra, ed.), New York: Springer-Verlag.

Peters, T. T., Volakis, J. L., and Liepa, V. V. (1986). Simulation of 2d dielectric/metallic structures. Report 388967-4-T, University of Michigan, Radiation Laboratory.

Peters, T. T., Volakis, J. L., Liepa, V. V., Senior, T. B. A., and Ricoy, M. A. (1986). Simulation of two-dimensional dielectric structures with resistive sheets. Report 389055-1-F, University of Michigan, Radiation Laboratory.

Peterson, A. F. (1990). The interior resonance problem associated with surface integral equations of electromagnetics: Numerical consequences and a survey of remedies. *Electromagnetics*, *10*, 293–312.

Putnam, J. M. and Medgyesi-Mitchang, L. N. (1991). Combined field formulation for inhomogeneous two- and three-dimensional bodies: The junction problem. *IEEE Trans. Antennas Propagat.*, *39*, 667–672.

Richmond, J. H. (1965). Scattering by a dielectric cylinder of arbitrary cross section shape. *IEEE Trans. Antennas Propagat.*, *13*(3), 334–341.

Richmond, J. H. (1966). TE-wave scattering by a dielectric cylinder of arbitrary cross-section shape. *IEEE Trans. Antennas Propagat.*, *AP-14*, 460–464.

Ricoy, M., Kempel, L., and Volakis, J. (1990). Simple integral equations for two dimensional scattering with further reduction in unknowns ans scattering code manual. *University of Michigan RadLab report*, *RL-857*.

Ricoy, M. A. and Volakis, J. L. (1989). Integral equations with reduced unknowns for the simulation of two-dimensional composite structures. *IEEE Trans. Antennas Propagat.*, *37*(3), 362–372.

Senior, T. B. A. and Volakis, J. L. (1989). Scattering by gaps and cracks. *IEEE Trans. Antennas Propagat.*, *37*(6), 744–750.

Tricomi, F. G. (1985). *Integral Equations*. Dover.

Wilton, D. R. and Wheeler, J. E. I. (1991). *"Comparison of the convergence rates of the conjugate gradient method applied to various integral equation formulations"* (chap. in *PIER 5,* J.A. Kong and T.K. Sarkar, eds.), New York: Elsevier, chap. 5.

Wu, T.-K. and Tsai, L. L. (1977). Electromagnetic fields induced inside arbitrary cylinders of biological tissue. *IEEE Trans. Microwave Theory and Techn.*, *25*, 61–65.

Yaghjian, A. D. (1981). Augmented electric and magnetic field integral equations. *Radio Sci.*, *16*, 987–1001.

CHAPTER 6

# Three-Dimensional Scattering

**Chapter Outline**



The previous chapters provided the framework for solving integral equations. But so far, we have yet to consider scattering or radiation from general three-dimensional (3-D) structures. The subject of scattering by 3-D structures using full-wave methods was a focus during the 1980s; Rao, Wilton and Glisson's (1982) classic paper was among the first on the subject. Previously, Livesay and Chen (1974) had considered volumetric modeling of biological tissues via the method of moments (MoM) solution of volume integral equations using simple cubic elements. Surface integral equation formulations for homogeneous material bodies were developed by Putnam and Medgyesi-Mitchang (1991). Later, parametric elements were introduced for both curvilinear volumetric modeling (Graglia, 1988) and for accurate modeling of curved surfaces (Song and Chew, 1995). Better geometry modeling of conformal elements were complemented by higher-order basis functions (BFs) (Graglia, Wilton, and Peterson, 1997) for much improved MoM accuracy and reduction in computational cost. Additionally, surface integral equations were hybridized by the finite element methods (Jin and Volakis, 1991; Antilla and Alexopoulos, 1994), resulting in state-of-the-art simulation tools that can efficiently model arbitrary material domains as well as conducting regions in real-life problems.

In this chapter, we summarize the method of moments for surface and volume integral equations. In the first half, surface and volume modeling techniques are outlined along with the definitions of frequently used flat and curvilinear elements. Basis functions and differential operators (gradient, divergence, and curl) are given in curvilinear coordinates, and details of the MoM implementation are presented. Construction of the MoM matrix, numerical integrations, and careful handling of singular integrals are given in detail, as these constitute the main steps in the implementation. We conclude by providing volumetric modeling for material regions and several numerical examples.

## 6.1 SCATTERING BY METALLIC BODIES

The unique radar return or signature of an airborne or ground vehicle, as seen by the radar, is called the radar cross section (RCS). Consider a representative fighter aircraft, as in Figure 6-1, illuminated by a transmitting radar antenna placed at a substantial distance from the target. The electromagnetic wave that impinges on the target can be approximated by a plane wave. To keep our analysis simple, we will also assume that the aircraft is a metallic surface denoted by $S$. Thus, for our radar scattering analysis, we may proceed to use the electric field integral equation (EFIE), magnetic field integral equation (MFIE), or the combined field integral equation (CFIE) developed in Chapter 3. This section considers these integral equations as applied to perfectly electrically conducting (PEC) (i.e., metallic) scatterers.

### 6.1.1 Electric, Magnetic, and Combined Field Integral Equations

Let us consider the problem of plane wave scattering by an airframe structure as in Figure 6-1. For simplicity, we will assume that this scatterer is PEC. Therefore, our first step in computing the associated RCS is to evaluate the current density $\mathbf{J}$ on the surface $S$ of the airframe. This can be done by invoking the EFIE given in (3.42a). For PEC structures, the EFIE is generated by enforcing the boundary condition $\hat{t}\cdot(\mathbf{E}^i+\mathbf{E}^s)=0$, where $\hat{t}$ is the unit tangent vector to the body's surface $S$ (see Figure 6-1). Doing so yields

$$\hat{t}\cdot\int_S d\mathbf{r}'\ \overline{\mathbf{\Gamma}}(\mathbf{r},\mathbf{r}')\cdot\mathbf{J}(\mathbf{r}')=\frac{-j}{kZ_0}\hat{t}\cdot\mathbf{E}^i(\mathbf{r}) \tag{6.1}$$

where $\mathbf{E}^i(\mathbf{r})$ is the incident electric field, and $Z_0=\sqrt{\mu_0/\epsilon_0}$ is the background impedance.

As usual, $\overline{\mathbf{\Gamma}}(\mathbf{r},\mathbf{r}')$ is the dyadic Green's function given in (2.95)

$$\overline{\mathbf{\Gamma}}(\mathbf{r},\mathbf{r}')=\left[\overline{\mathbf{I}}+\frac{1}{k^2}\nabla\nabla\right]G(\mathbf{r},\mathbf{r}'). \tag{6.2}$$

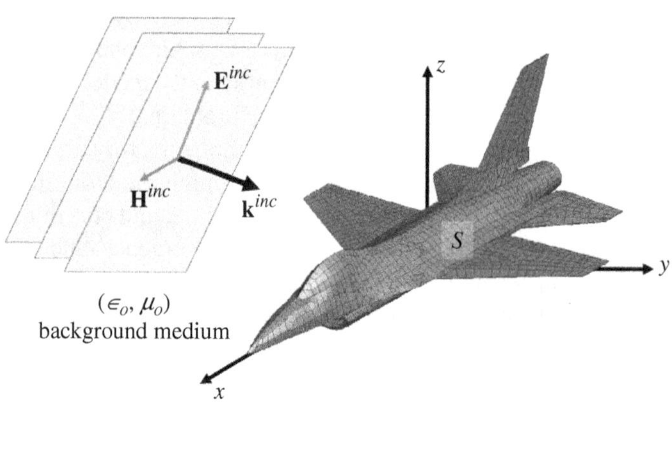


**FIGURE 6-1** Illustration of a plane wave impinging upon an airframe structure.

Correspondingly, using the condition $\mathbf{J} = \hat{n} \times (\mathbf{H}^i + \mathbf{H}^s)$[1], we can instead employ the MFIE valid for closed scatterers given by (see (3.42b))

$$-T\hat{n} \times \mathbf{J}(\mathbf{r}) + \fint_S d\mathbf{r}' \, \mathbf{J}(\mathbf{r}') \times \nabla G(\mathbf{r}, \mathbf{r}') = \hat{n} \times \mathbf{H}^i(\mathbf{r}). \tag{6.3}$$

In this, $T = 1 - \Omega/4\pi$, and $\Omega$ is the solid angle subtended by the observation point $\mathbf{r}$. The bar through the integral sign represents the principal value, excluding the observation point $\mathbf{r}$. For smooth surfaces, $\Omega = 2\pi$, and hence $T = 1/2$. The derivation of the MFIE is given in (3.42b), and as noted there $\hat{n}$ is the outward normal. Next, we proceed with the discretization and solution of the previous EFIE and MFIE equations.

### 6.1.2 Triangular Element Mesh Representations

A first step in the numerical solution of (6.1) and (6.3) is the discretization of the scatterer surface $S$. This is done by breaking up $S$ as a collection of triangles or quadrilaterals (see Figures 6-1 and 6-2), a process referred to as *gridding* or *meshing* of the surface. It constitutes the first approximation in the numerical solution process. A next step is to define a suitable current density representation on each surface element.

Let's for the moment focus on triangular elements to grid (or mesh) the surface $S$. This is a popular approach for modeling arbitrary structures and is widely used due to the availability of gridding software tools for triangular meshes. As depicted in Figure 6-2, the geometry is approximated by a collection of flat triangular elements, or patches. In electromagnetics, the use of flat triangulations was sparked by the

**FIGURE 6-2** Approximation of the sphere's surface by a mesh of flat triangles.

[1]The enforced boundary condition is the usual $\mathbf{J} = \hat{n} \times (\mathbf{H}^+ - \mathbf{H}^-)$, where $\mathbf{H}^+ = \mathbf{H}^i + \mathbf{H}^s$ is the exterior field, and $\mathbf{H}^- = 0$ within the PEC body; $\mathbf{H}^i$ refers to the incident magnetic field.

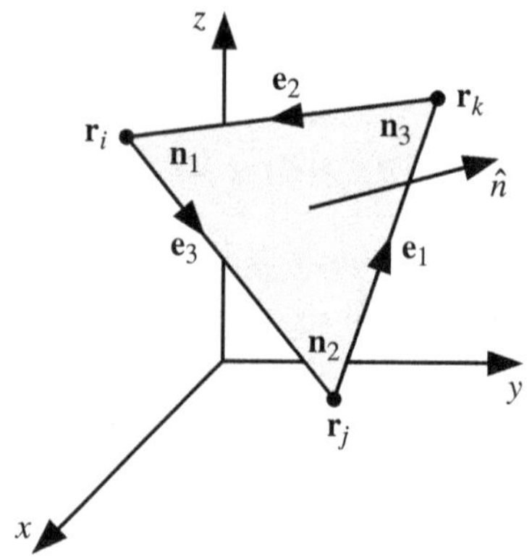


**FIGURE 6-3** Local and global numbering assignments for representing each triangular element in the surface grid representing the antenna or scatterer. Note that the local numbering scheme is counterclockwise, and this choice must be maintained for all triangles forming the surface.

paper by Rao, Wilton and Glisson (1982) and associated development of Rao–Wilton–Glisson (RWG) basis or expansion functions (Rao, Wilton and Glisson, 1982; Rao and Wilton, 1991; Horng, McKinzie, and Alexopoulos, 1992).

To represent the geometry and induced surface currents we begin by defining each triangular element in terms of its three vertices ($\mathbf{r}_i, \mathbf{r}_j, \mathbf{r}_k$) shown in Figure 6-3. These vertices are associated with local and global numbering schemes. The local numbers (for the triangle's nodes and edges) are conveniently chosen as (1, 2, 3). Correspondingly, the global numbers for the nodes will be denoted as $(i, j, k)$ and are unique in the entire surface grid. As a rule, we choose to denote local edge $e_p$ to have the same local number as the node $n_p$ located across the edge $e_p$. With this numbering scheme, the local edge 3 of the triangular element connects local nodes 1 and 2; local edge 1 connects local nodes 2 and 3, and local edge 2 connects local nodes 3 and 1. This numbering convention also defines the unit vector normal to the triangular element in the right-hand rule sense.

The local and global numbering of the nodes and elements provide a convenient way to define the mesh of the scattering surface as an ordered list. For example, consider the section of a surface mesh depicted in Figure 6-4. The depicted mesh around triangle 34($t_{34}$) is defined by the vectors $\mathbf{r}_{10}$, $\mathbf{r}_{55}$, and $\mathbf{r}_{38}$. Likewise, triangle 22 is defined by the three nodes $\mathbf{r}_{29}$, $\mathbf{r}_{55}$, and $\mathbf{r}_{10}$. That is, the local edge 3 of triangle 34 coincides with local edge 1 of triangle 22, implying that the same edge in a mesh can be identified in multiple ways. Thus, an interconnectivity relationship (or table) that uniquely identifies all edges and nodes comprising the grid is an essential aspect of integral equation solutions. As part of this process, a node connectivity table and a coordinates listing table for each node are generated, as illustrated in Tables 6-1 and 6-2.

Given the three nodes defining each triangle, any surface location within the triangle can be obtained via the simplex or barycentric coordinates ($\zeta_1, \zeta_2, \zeta_3$) (Volakis, Chatterjee, and Kempel, 1998). Specifically, the position vector at any point within the triangle takes the form

$$\mathbf{r}(\zeta_1, \zeta_2, \zeta_3) = \mathbf{r}_1\zeta_1 + \mathbf{r}_2\zeta_2 + \mathbf{r}_3\zeta_3 \tag{6.4}$$

where, $\mathbf{r}_{1,2,3}$ denote the position vectors defining the local nodes (1, 2, 3).

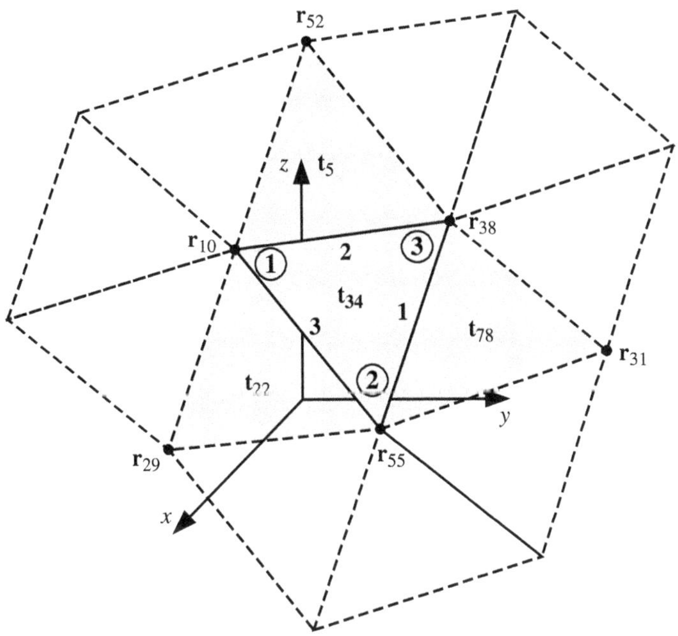


**FIGURE 6-4** ■ A representative triangular patch (34, or $t_{34}$) in a typical mesh surrounded by neighboring triangles carrying random global numbering to define their vertices (see also Volakis, 1992).

**TABLE 6-1** ■ Example data structure for the 3D coordinates of the mesh nodes.

| Node # | Node Coordinates $(x, y, z)$ |
|---|---|
| $n_1$ | $(x_1, y_1, z_1)$ |
| $n_2$ | $(x_2, y_2, z_2)$ |
| $n_3$ | $(x_3, y_3, z_3)$ |
| ... | ... |
| 22 | (0.217, 0.822, 630.2) |
| ... | ... |

**TABLE 6-2** ■ Example data structure for defining the triangular elements using three corner nodes.

| Triangle # | Global Triangle Nodes $(n, p, q)$ |
|---|---|
| $t_1$ | $(n_1, p_1, q_1)$ |
| $t_2$ | $(n_2, p_2, q_2)$ |
| $t_3$ | $(n_3, p_3, q_3)$ |
| ... | ... |
| 34 | (10, 55, 38) |
| ... | ... |

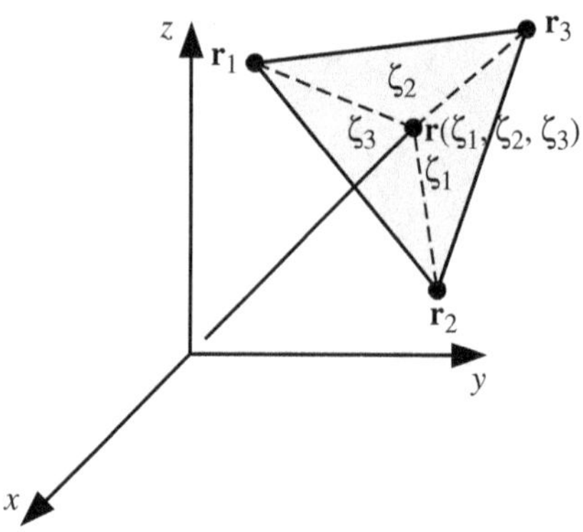


**FIGURE 6-5** Definition of the barycentric coordinates used to define the position vector **r** within a flat triangle.

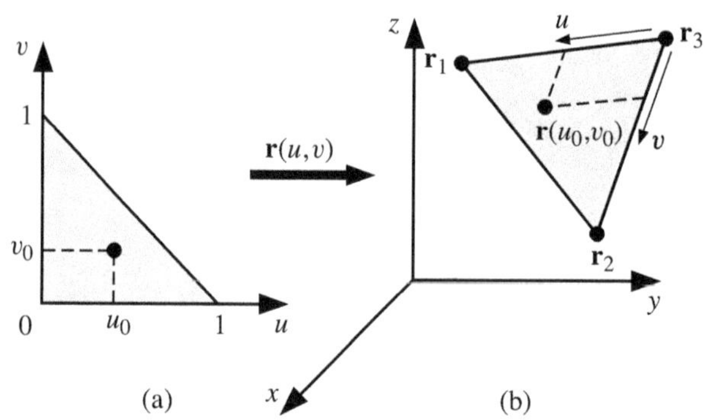


**FIGURE 6-6** Parametric mapping of (a) a (*u*, *v*) parametrized right triangle onto (b) a flat triangular element having the arbitrary node positions $\mathbf{r}_1$, $\mathbf{r}_2$, and $\mathbf{r}_3$.

Referring to Figure 6-5, the barycentric coordinates $(\zeta_1, \zeta_2, \zeta_3)$ are given by

$$\zeta_1 = \frac{Area(\mathbf{r}, \mathbf{r}_2, \mathbf{r}_3)}{Area(\mathbf{r}_1, \mathbf{r}_2, \mathbf{r}_3)}, \quad \zeta_2 = \frac{Area(\mathbf{r}_1, \mathbf{r}, \mathbf{r}_3)}{Area(\mathbf{r}_1, \mathbf{r}_2, \mathbf{r}_3)} \quad \zeta_3 = \frac{Area(\mathbf{r}_1, \mathbf{r}_2, \mathbf{r})}{Area(\mathbf{r}_1, \mathbf{r}_2, \mathbf{r}_3)} \tag{6.5}$$

where $Area(.)$ denotes the surface area of the sub-triangle formed by the three nodes in its argument. It is readily observed that $\zeta_i = 1$ when $\mathbf{r} = \mathbf{r}_i$. Also, $\zeta_1 + \zeta_2 + \zeta_3 = 1$.

Alternatively, the surface can be defined using a $(u, v)$ parametrization. As depicted in Figure 6-6, every point $(u, v)$, with $u + v \leq 1$ on the parametric triangle (Figure 6-6a) can be mapped onto the triangular element of the mesh (Figure 6-6b). Using this mapping, it is straightforward to write

$$\mathbf{r}(u, v) = \mathbf{r}_{13}u + \mathbf{r}_{23}v + \mathbf{r}_3 \tag{6.6}$$

where $\mathbf{r}_{13} = \mathbf{r}_1 - \mathbf{r}_3$ and $\mathbf{r}_{23} = \mathbf{r}_2 - \mathbf{r}_3$. This parametric definition of the triangular element is particulary useful when we consider curvilinear elements to represent the surface of the structure. Details on the definition of curvilinear triangles are given in Section 6.2.

Using the chosen triangular surface discretization, we now proceed to rewrite the integral equation (6.1) over each element of the mesh as

$$\hat{t} \cdot \sum_{i=1}^{N_{tri}} \int_{S_i} d\mathbf{r}' \; \overline{\Gamma}(\mathbf{r}, \mathbf{r}') \cdot \mathbf{J}(\mathbf{r}') = \frac{-j}{k_0 Z_0} \hat{t} \cdot \mathbf{E}^i(\mathbf{r}) \tag{6.7}$$

where $\hat{t}$ denotes the tangential unit vector at the observation point $\mathbf{r}$ on the triangular surface mesh. Also, $S_i$ refers to the surface of the $i$th triangle, and $N_{tri}$ is the number of triangles used to discretize the surface.

To solve for the surface current density $\mathbf{J}(\mathbf{r})$, we proceed to express it as a summation of the chosen basis functions $\mathbf{j}_n(\mathbf{r})$. Doing so, we have

$$\mathbf{J}(\mathbf{r}) = \sum_{n=1}^{N} x_n \, \mathbf{j}_n(\mathbf{r}) \tag{6.8}$$

where $N$ denotes the dimension of the expansion, and $x_n$ are the unknown coefficients to be determined (by solving the EFIE). As can be realized, the basis functions $\mathbf{j}_n(\mathbf{r})$ are intrinsically related to the geometrical discretization of the surface. That is, they should have a support over one or more of the triangles defining $S$.

Using (6.8), (6.7) becomes

$$\hat{t} \cdot \int_{S_n} d\mathbf{r}' \, \overline{\Gamma}(\mathbf{r}, \mathbf{r}') \cdot \sum_{n=1}^{N} x_n \, \mathbf{j}_n(\mathbf{r}') = \frac{-j}{k_0 Z_0} \hat{t} \cdot \mathbf{E}^i(\mathbf{r}) \tag{6.9}$$

or equivalently

$$\sum_{n=1}^{N} x_i \hat{t} \cdot \int_{S_n} d\mathbf{r}' \, \overline{\Gamma}(\mathbf{r}, \mathbf{r}') \cdot \mathbf{j}_n(\mathbf{r}') = \frac{-j}{k_0 Z_0} \hat{t} \cdot \mathbf{E}^i(\mathbf{r}). \tag{6.10}$$

To determine the coefficients $\{x_n\}$, a linear system is formed by testing (6.10) $N$ times using suitable weighting or testing functions (see Chapter 4). Denoting the testing functions as $\mathbf{t}_m(\mathbf{r})$ and adopting Galerkin's testing, viz. $\mathbf{t}_m(\mathbf{r}) = \mathbf{j}_m(\mathbf{r})$ we get

$$\sum_{n=1}^{N} x_n \int_{S_m} d\mathbf{r} \, \mathbf{j}_m(\mathbf{r}) \cdot \int_{S_n} d\mathbf{r}' \, \overline{\Gamma}(\mathbf{r}, \mathbf{r}') \cdot \mathbf{j}_n(\mathbf{r}') = \frac{-j}{k_0 Z_0} \int_{S_m} d\mathbf{r} \, \mathbf{j}_m(\mathbf{r}) \cdot \mathbf{E}^i(\mathbf{r}). \tag{6.11}$$

By setting $m = 1, 2, 3, \ldots, N_{tri}$ we can generate $N$ linear equations. These can be readily cast into a matrix of the form

$$[Z]\{x\} = \{b\} \tag{6.12}$$

where

$$Z_{mn} = \int_{S_m} d\mathbf{r} \, \mathbf{j}_m(\mathbf{r}) \cdot \left[ \int_{S_n} d\mathbf{r}' \, G(\mathbf{r}, \mathbf{r}') \mathbf{j}_n(\mathbf{r}') + \frac{1}{k_0^2} \int_{S_n} d\mathbf{r}' \, \nabla\nabla G(\mathbf{r}, \mathbf{r}') \cdot \mathbf{j}_n(\mathbf{r}') \right] \tag{6.13}$$

and

$$b_m = \frac{-j}{k_0 Z_0} \int_{S_m} d\mathbf{r} \, \mathbf{j}_m(\mathbf{r}) \cdot \mathbf{E}^i(\mathbf{r}). \tag{6.14}$$

Over the next few sections, we will provide various ways to define the expansion currents $\mathbf{j}_n(\mathbf{r})$ over the elements. Subsequently, we will consider the evaluation of the integrals in (6.13). As the integrands entail singularities of the Green's function when $m = n$, their accurate numerical evaluation is a major challenge.

### 6.1.3 Rao–Wilton–Glisson Basis Functions

The triangular mesh proposed in Figure 6-4 leads to a restricted set of basis functions to be used in defining $\mathbf{j}_n(\mathbf{r})$ appearing in (6.8). The RWG basis functions (Rao, Wilton and Glisson, 1982) are such that the unknowns $\{x_n\}$ represent the current flowing perpendicular to the edge joining two triangles. Two such triangles are depicted in Figure 6-7 and share the edge defined by the location vectors $\mathbf{r}_j$ and $\mathbf{r}_k$. A key aspect of the RWG basis is that they taper linearly to zero at the nodes opposite to the edge (see Figure 6-7). The mathematical representation of the RWG basis is rather simple and given by

$$\mathbf{j}_n(\mathbf{r}) = \begin{cases} \dfrac{l_c}{2A^+}\mathbf{r}^+ & \mathbf{r} \in S^+ \\ \dfrac{l_c}{2A^-}\mathbf{r}^- & \mathbf{r} \in S^- \\ 0 & \text{otherwise.} \end{cases} \tag{6.15}$$

Here, $S^+$ and $S^-$ refer to the surfaces of the triangles forming the $n$th edge of length $l_c$ as depicted in Figure 6-7 ($S^+$ and $S^-$ do, of course, belong to the $S_{i,j}$ surface set). They each have areas $A^+$ and $A^-$, respectively. As seen, the vectors $\mathbf{r}^{+/-}$ have their origin at the nodes of the $S^{+/-}$ triangle opposite to the common edge, respectively. Clearly, $\mathbf{j}_n(\mathbf{r})$ vanishes to zero at the nodes opposite to the edge as $\mathbf{r}^{+/-}$ have their origin there. Also, its amplitude at the edge is maximum and equal to unity.

The RWG basis functions belong to the class of *subdomain* functions as each is nonzero only on a single pair of triangles. As such, the current density on each triangle is a superposition of 3 weighted RWG basis (contributed by the three edges forming the triangle).

With this in mind, the matrix entries in (6.13) take the more explicit form

$$Z_{mn} = \int_{S_m^+ + S_m^-} d\mathbf{r}\, \mathbf{j}_m(\mathbf{r}) \cdot \int_{S_n^+ + S_n^-} d\mathbf{r}' \left[ G(\mathbf{r},\mathbf{r}')\, \mathbf{j}_n(\mathbf{r}') + \frac{1}{k_0^2} \nabla\nabla G(\mathbf{r},\mathbf{r}') \cdot \mathbf{j}_n(\mathbf{r}') \right]. \tag{6.16}$$

**FIGURE 6-7** Vector construction of Rao–Wilton–Glisson basis functions on a pair of flat triangular elements.

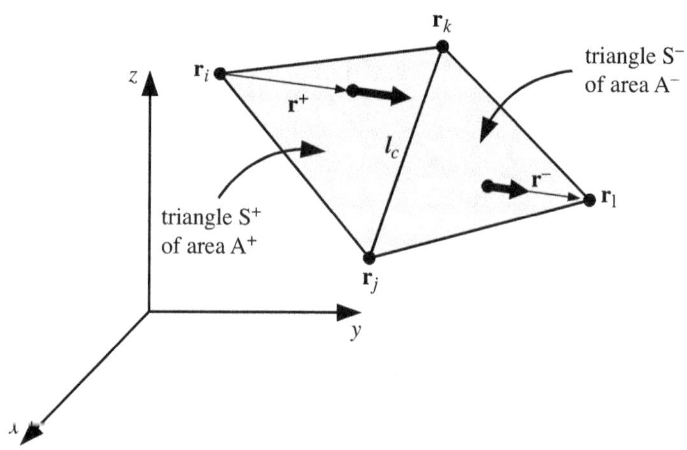

Also, the right-hand side elements are

$$b_m = \frac{-j}{k_0 Z_0} \int_{S_m^+ + S_m^-} d\mathbf{r}\, \mathbf{j}_m(\mathbf{r}) \cdot \mathbf{E}^i(\mathbf{r}). \tag{6.17}$$

In these, $n$th basis function refers to the expansion on the triangles $S_n^+$ and $S_n^-$. Likewise, the $m$th testing function is defined over the $S_m^+$ and $S_m^-$ triangular surfaces that share the $m$th edge.

### 6.1.4 Method of Moments Matrix Assembly

Typically, the number of edges are 1.5 times the number of triangles comprising the structures. However, for open surfaces, the unpaired edges (at the periphery of the surface) must have their normal currents set to zero a priori, thus, reducing the unknowns to be solved. Assuming that a given edge is away from the surface's periphery, the matrix element $Z_{mn}$ in (6.16) involves the evaluation of four double surface integrals. Specifically, $Z_{mn} = Z_{mn}^{++} + Z_{mn}^{+-} + Z_{mn}^{-+} + Z_{mn}^{--}$, where

$$Z_{mn}^{++} = \int_{S_m^+} d\mathbf{r}\, \mathbf{t}_m(\mathbf{r}) \cdot \int_{S_n^+} d\mathbf{r}' \left[ G(\mathbf{r}, \mathbf{r}')\, \mathbf{j}_n(\mathbf{r}') + \frac{1}{k^2} \nabla\nabla G(\mathbf{r}, \mathbf{r}') \cdot \mathbf{j}_n(\mathbf{r}') \right] \tag{6.18}$$

$$Z_{mn}^{+-} = \int_{S_m^+} d\mathbf{r}\, \mathbf{t}_m(\mathbf{r}) \cdot \int_{S_n^-} d\mathbf{r}' \left[ G(\mathbf{r}, \mathbf{r}')\, \mathbf{j}_n(\mathbf{r}') + \frac{1}{k^2} \nabla\nabla G(\mathbf{r}, \mathbf{r}') \cdot \mathbf{j}_n(\mathbf{r}') \right] \tag{6.19}$$

$$Z_{mn}^{-+} = \int_{S_m^-} d\mathbf{r}\, \mathbf{t}_m(\mathbf{r}) \cdot \int_{S_n^+} d\mathbf{r}' \left[ G(\mathbf{r}, \mathbf{r}')\, \mathbf{j}_n(\mathbf{r}') + \frac{1}{k^2} \nabla\nabla G(\mathbf{r}, \mathbf{r}') \cdot \mathbf{j}_n(\mathbf{r}') \right] \tag{6.20}$$

$$Z_{mn}^{--} = \int_{S_m^-} d\mathbf{r}\, \mathbf{t}_m(\mathbf{r}) \cdot \int_{S_n^-} d\mathbf{r}' \left[ G(\mathbf{r}, \mathbf{r}')\, \mathbf{j}_n(\mathbf{r}') + \frac{1}{k^2} \nabla\nabla G(\mathbf{r}, \mathbf{r}') \cdot \mathbf{j}_n(\mathbf{r}') \right]. \tag{6.21}$$

For $m \neq n$, the evaluation of (6.18) to (6.21) can be done using a suitable numerical quadrature. We may, for example, use a seven-point quadrature as defined in Figure 6-8 with the corresponding samples and weights given in Table 6-3. This seven-point rule provides $O(h^6)$ accuracy, where $h$ is the radius of the circumscribing circle (Abramowitz and Stegun, 1972) with its tips lying on the circle: the $(x_i, y_i)$ points in Figure 6-8 are referred to the origin of that circle. Also, when $m = n$, integrand singularities need to be carefully addressed. This will be discussed later in more detail.

Direct and iterative solvers discussed in Section 6.3.8 and in Chapter 4 can be employed to solve the matrix system (6.12).

The fidelity of the geometrical representation has direct impact on the solution of $\mathbf{J}$ via (6.12). At a minimum, it is important to make triangles (forming the mesh) small enough so that the rendered geometrical and electrical fidelity is maintained. Typically, the edge length of the triangles need be less than $\lambda_0/10$, where $\lambda_0$ is the free-space wavelength at the solution frequency. The uniformity of the RWG (determined by the uniformity of the mesh) is important in generating a well-conditioned matrix system. However, the density and size of the triangular elements may be constrained by the geometrical detail of the structure. In such cases, the mesh may contain very small elements in localized regions, often leading to ill-conditioned matrices. This is in part due to the so-called low-frequency breakdown of the EFIE

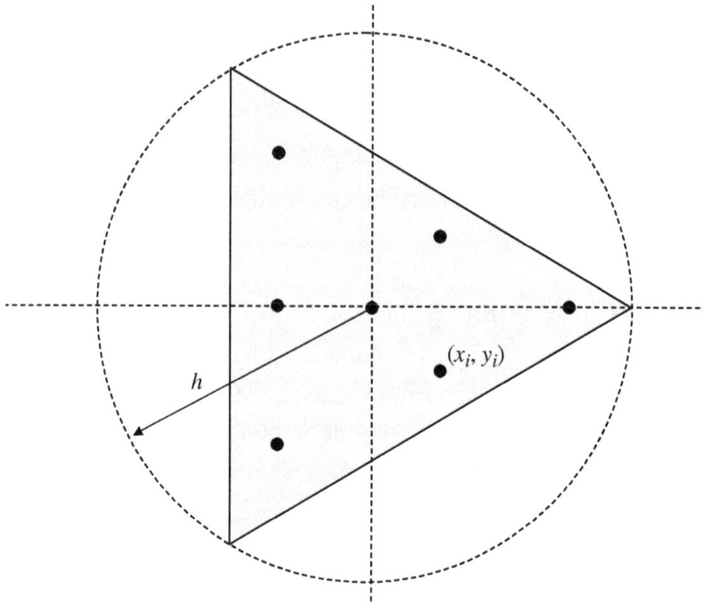


**FIGURE 6-8** Numerical quadrature points defined on an equilateral triangle.

**TABLE 6-3** Coordinates $(x_i, y_i)$ and the associated weights $w_i$ for the numerical quadrature on the unit triangle shown in Figure 6-8.

| $(x_i, y_i)$ | $w_i$ |
|---|---|
| (0,0) | 270/1200 |
| $\left(\left(\frac{\sqrt{15}+1}{7}\right)h, 0\right)$ | $\frac{155-\sqrt{15}}{1200}$ |
| $\left(\left(\frac{-\sqrt{15}+1}{14}\right)h, \pm\left(\frac{\sqrt{15}+1}{14}\right)\sqrt{3}h\right)$ | $\frac{155-\sqrt{15}}{1200}$ |
| $\left(\left(-\frac{\sqrt{15}-1}{7}\right)h, 0\right)$ | $\frac{155+\sqrt{15}}{1200}$ |
| $\left(\left(\frac{\sqrt{15}-1}{14}\right)h, \pm\left(\frac{\sqrt{15}-1}{14}\right)\sqrt{3}h\right)$ | $\frac{155+\sqrt{15}}{1200}$ |

operator (Peterson, 1991; Lee, Lee, and Burkholder, 2003; Bluck and Walker, 2007). Thus, abrupt changes to the triangular mesh size must be avoided to alleviate some of the ill-conditioning issues. Several approaches to improve ill-conditioning have been considered, and the subject continues to be a research issue. For example, several authors (ibid.) considered preconditioner basis functions. Others considered transformations based on the Calderon identities (Andriulli, Cools, Bagci, Olyslager, Buffa, Christiansen, and Michielssen, 2008; Stephanson and Lee, 2009). We discuss Calderon-based preconditioners in more detail in Section 6.3.8.1.

In the above, we developed the main steps for the MoM solution of surface integral equations using triangular elements with RWG basis functions. Although triangular elements are often sufficient for modeling 3-D geometries, geometrical errors are introduced, particularly for curved surfaces. As can be realized, surface representations using flat facets introduce sharp wedges at their joints that are not in the original geometry (see Figure 6-2). To minimize such geometrical and solution errors, higher-order curvilinear elements were introduced (particularly in modeling airframe fine details). These higher-order elements also lead to fewer unknowns as there is no need to overdiscretize the geometry. The following section examines popular curvilinear elements and their geometrical framework using curvilinear coordinates.

## 6.2 | CURVED TRIANGULAR AND QUADRILATERAL ELEMENTS

Due to their mathematical simplicity and availability of related (commercial) mesh generation software, flat surface triangulations is a first choice in computational electromagnetic modeling (CEM). However, flat triangles have several shortcomings in modeling highly curved structures. Coarse meshes fail to capture geometry features, leading to inaccurate MoM solutions. An immediate remedy is to increase the density of the triangulation around the curved regions; however, this leads to an unnecessarily large number of unknowns in the MoM system since the number of basis functions is proportional to the total number of triangular elements in the mesh. More importantly, matrix conditioning deteriorates significantly as the triangulation is highly nonuniform, consisting on regions with very large and very small elements in proximity.

An alternative to using facets to represent curved surfaces is to use curvilinear elements (triangular or quadrilateral). In this section, we develop the geometrical representation of several curvilinear elements. Conformal basis functions are also defined on these curved elements as tangential vector fields. Using curvilinear elements, we expect that high-fidelity geometry details of real-life structures will be mathematically represented without the need to increase mesh density (or the MoM matrix size). We first proceed with parametric representations of curves, surfaces, and volumes, providing concise definitions of line, surface, and volume elements as well as vector fields defined over these elements.

### 6.2.1 Parametric Representations

A space curve in 3-D can be represented as a mapping that transforms a unit element (e.g., a line of unit length) in the parameter space onto an arbitrary position vector in the $(x, y, z)$. Such a representation is simply written as

$$\mathbf{r}(x, y, z) = \mathbf{r}(u) \tag{6.22}$$

with the condition $u \in [0, 1]$. Several such parametric representations are shown in Figure 6-9. Typically, surface and volume mappings are constructed by a Cartesian product of one-dimensional transformations. For example, a curvilinear quadrilateral element in the $(x, y, z)$ space can be constructed as the image of a unit square in the $(u, v)$ space through the relation

$$\mathbf{r} = \mathbf{r}(u, v) \tag{6.23}$$

where again $u \in [0, 1]$ and $v \in [0, 1]$. Several typical curvilinear elements are shown in Figure 6-10. As in the case of facets, the geometrical surfaces can now be represented as a mesh of curvilinear elements (see Figure 6-11).

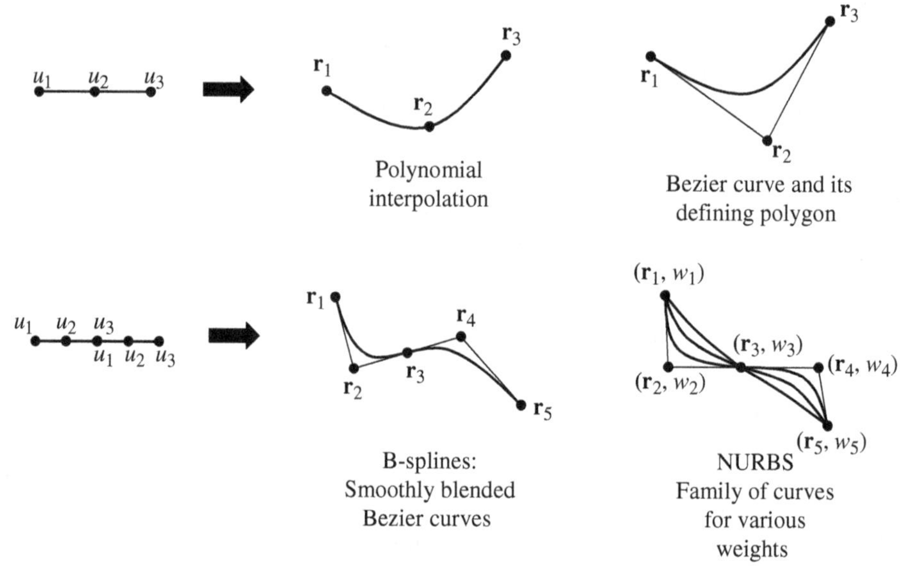


**FIGURE 6-9** Examples of parametric mappings.

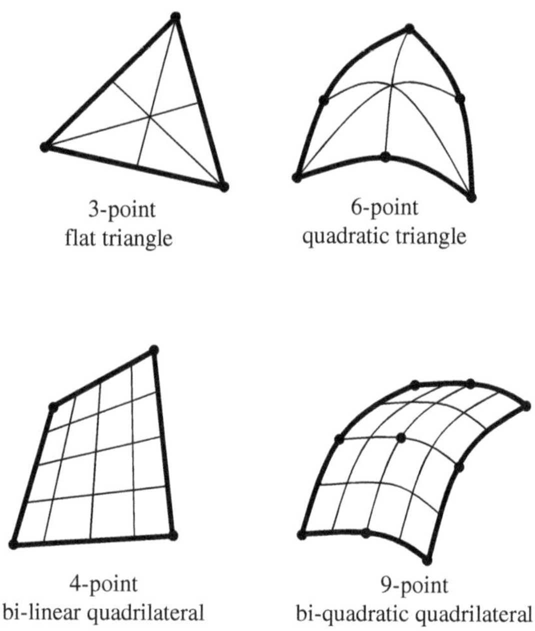


**FIGURE 6-10** Examples of parametric surface elements.

Before we proceed further, the parametric mappings in (6.22) and (6.23) need be explicitly defined. Among many alternatives, interpolation-based parametric representations (see Figure 6-9) lend themselves to general tools with a flexibility to model a broad range of geometries with mathematical ease.

### 6.2.2 Polynomial Interpolations

One of the most widely used representations in computer-aided modeling is that of polynomial curve fitting (also called interpolation). This provides an approximation to the actual geometry since the interpolation points are the only ones required to

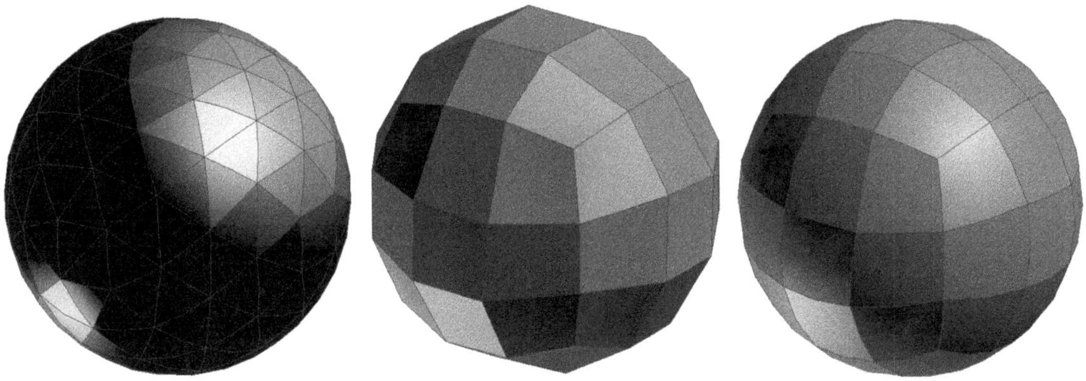

**FIGURE 6-11** ■ Approximations of a sphere using (a) a mesh of six-point quadratic triangles, (b) a mesh of four-point bilinear quadrilaterals, and (c) a mesh of nine-point biquadratic quadrilaterals.

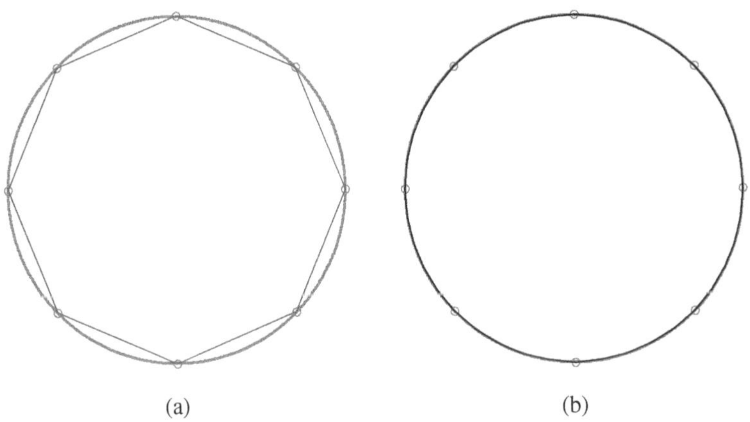


**FIGURE 6-12** ■ Approximations of a circle: (a) using eight flat lines, and (b) using four quadratic elements.

coincide with those on the original geometry. Depending on the degree of polynomial (employed for interpolation) or the number of interpolation points, the geometrical accuracy of the representation varies. Figure 6-12a depicts an approximation of a circle using piecewise linear sections. Likewise, Figure 6-12b gives an approximation to the same circle using second-order polynomials. The superior accuracy and fidelity of the curvilinear elements are clearly depicted in this representation. More importantly, the curvilinear elements provide for a systematic representation of higher-order curves (the straight ones being first-order elements).

We have yet to provide a mathematical basis for the polynomial representations and their accuracy. To do so, we begin by defining the points of interpolation as a set of location vectors $\{\mathbf{r}_1, \mathbf{r}_2, \ldots, \mathbf{r}_k\}$, where $k$ is the total number of interpolation points. These interpolation points may be generated from an available mathematical model or by scanning the surface of an actual structure of interest via ultrasonic or optical scanning hardware.

Having $k$ discrete points on the curve, we proceed to generate a polynomial curve of degree $n$ in three-dimensional space using an ordered set of $(n + 1)$ nodes. These nodes will be the known points (see, e.g., Figure 6-9) on the curves denoted by $\{\mathbf{r}_i, \mathbf{r}_{i+1}, \ldots, \mathbf{r}_{i+n}\}$. Using Lagrange interpolation polynomials, a continuous curve

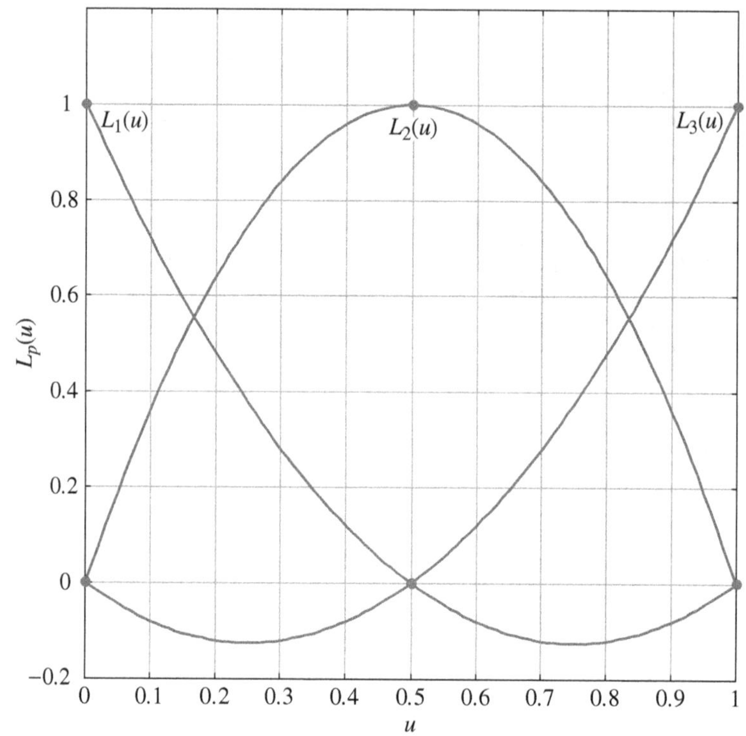


**FIGURE 6-13** Second-order Lagrange interpolators.

that traverses the known points can be represented as

$$\mathbf{r}(u) = \sum_{j=i}^{j+n} \mathbf{r}_j L_j^{(n)}(u) \tag{6.24}$$

where again $u \in [0, 1]$. Also, $L_j^{(n)}(u)$ are the Lagrange interpolation polynomial of degree $n$ multiplying the interpolation point $\mathbf{r}_j$. Thus, three Lagrange polynomials are needed for a second-order representation (see Figure 6-13). Note that $L_j^{(n)}(u)$ vanish at all nodes except the $j$th.

An extension of the Lagrange interpolation (6.24) for surface representation is a straightforward process. It simply amounts to using a product of interpolation polynomials, one for $u$ and another for $v$ direction. Using the set of surface points $\mathbf{r}_{ii'}$ to be interpolated, we have

$$\mathbf{r}(u, v) = \sum_{j=i}^{j+n} \sum_{j'=i'}^{j'+n'} \mathbf{r}_{jj'} L_{jj'}^{(nn')}(u, v) \tag{6.25}$$

where

$$L_{jj'}^{(nn')}(u, v) = L_j^{(n)}(u) \times L_{j'}^{(n')}(v). \tag{6.26}$$

Following the same approach, a third parameter $w$ can be introduced to represent volumetric domains.

As we mentioned already, the successful use of interpolatory representations depends on the polynomial degree and associated interpolation points. However, for complicated real-life geometries, blindly applying higher-order interpolatory polynomials to represent arbitrary geometries can lead to restrictive geometrical representations, particularly in regions where the geometry changes rapidly. Thus, in practice, polynomial interpolations are rarely used with higher-order polynomials.

Alternatively, free-form representations, such as Bézier, B-spline, and nonuniform rational B-splines (NURBS), are more widely used in manufacturing industry due to their flexibility and their unique representation properties; the following section provides a discussion.

### 6.2.3 Free-Form Representations

As already noted, interpolation meshes are constructed by constraining the surface elements to pass through *predefined* data points on the original surfac; that is, they are *surface-fitting* techniques. These methods are suitable surface approximations when a set of sampled data on the surface is available. However, when the surface body design also depends on functional and aesthetic requirements (e.g., the body of a custom designed automobile), one has to resort to a combination of computational and heuristic methods to mathematically represent the geometry. Bézier and B-spline curves and their rational counterparts (see, e.g., Farin, 1997) were developed with this goal in mind and are the backbone of mathematical models in computer-aided design (CAD) tools. In essence, these representations allow us to define both the points traversed by the curve and also the derivatives of the curve at that point. As such, continuity can be enforced at the interpolation points. The following sections discuss these tools, as they are becoming more widely used in CEM geometry representations as well.

#### 6.2.3.1 Bézier and B-Spline Curves and Surfaces

A Bézier curve is specified by a set of points associated with a *defining polygon*. As such, the shape of the actual curve closely follows the shape of the defining polygon. For example, Figure 6-14 shows a generic third-order Bézier curve and its defining polygon.

The Bézier curves have the following properties that make them particularly popular for generating and representing artistic designs:

- The degree of the polynomial defining the curve segment is one less than the number of defining polygon points (this also holds for interpolation polynomials).

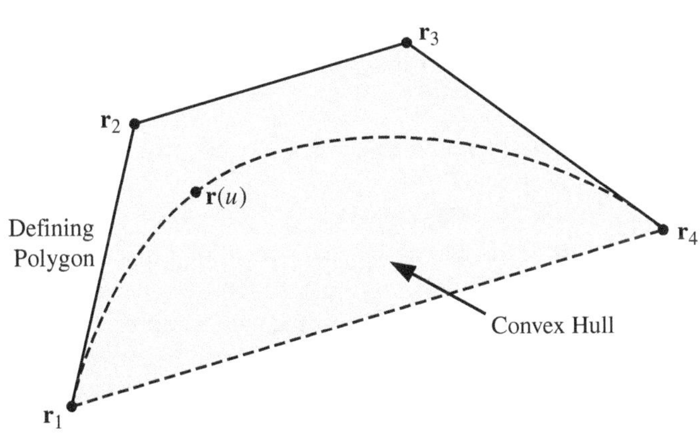


**FIGURE 6-14** A generic Bézier curve and its defining polygon.

- The curve follows the shape of the defining polygon.
- The first and last points on the curve are coincident with the first and last points of the defining polygon.
- The tangent vectors at the ends of the curve have the same direction as the first and last polygon sides, respectively. These tangent vectors can be controlled by changing the polygon points $\mathbf{r}_2$ and $\mathbf{r}_3$.
- The curve is contained within the convex hull of the defining polygon, that is, within the largest convex polygon obtainable with the defining polygon vertices.
- The curve exhibits the variation-diminishing property. This means that the curve does not oscillate about a straight line more than the defining polygon.
- The curve is invariant under an affine transformation (a combination of linear transformations such as translation and rotation).

Similarly to any interpolation-based representation, a Bézier curve is defined by the expression

$$\mathbf{r}(u) = \sum_{i=0}^{n} \mathbf{r}_i B_i^n(u) \qquad 0 \le u \le 1 \tag{6.27}$$

where $B_i^n(u)$ are the Bézier or Bernstein blending functions given by

$$B_i^n(u) = \binom{n}{i} u^i (1-u)^{n-i} \tag{6.28}$$

with

$$\binom{n}{i} = \frac{n!}{i!(n-i)!} \tag{6.29}$$

and $\mathbf{r}_i$ define the polygon vertices as depicted in Figure 6-14.

B-spline curves (Rogers and Adams, 1990; Farin, 1997) are formed by blending several Bézier curves together. An $n$th degree B-spline curve is formed by connecting $n$th degree Bézier curves and imposing $(n-1)$th derivative continuity at the junction points. The local parameter of each Bézier curve runs from 0 to 1. A *knot vector* that joins each Bézier polygon must also be prespecified. If this knot vector is nonuniform then the resulting curve is called a nonuniform B-spline.

Alternatively, B-splines are also written using an expansion similar to (6.27)

$$\mathbf{r}(u) = \sum_{i=0}^{n} \mathbf{r}_i N_i^n(u) \qquad 0 \le u \le 1 \tag{6.30}$$

where $N_i^n(u)$ are B-spline blending functions (Rogers and Adams, 1990) and are also functions of the knot vector.

Unlike Bézier curves, B-splines have the desired property of local control; that is, when one of its vertices is moved to a new location only the part of the curve around that vertex changes. This is not the case for Bézier curves since the basis functions $B_i^n(u)$ are global (i.e., nonzero over the whole interval $0 \le u \le 1$). Hence, a change in the position of a single vertex affects the entire curve.

Rational extensions of Bézier and B-spline curves are appropriately named as rational Bézier and rational B-spline curves. They allow one to assign additional weights to each polygon vertex, giving these curves one more degree of freedom. This is accomplished by projecting the four-dimensional Bézier and B-spline curves onto a three-dimensional real space. A rational Bézier curve can be expressed as

$$\mathbf{r}(u) = \frac{\sum_{i=0}^{n} \mathbf{r}_i \omega_i B_i^n(u)}{\sum_{i=0}^{n} \omega_i B_i^n(u)} \qquad 0 \le u \le 1 \tag{6.31}$$

where $\omega_i$ is the additional weight of the $i$th vertex of the defining polygon.

Blending rational Bézier curves with a nonuniform knot vector leads to the industry standard NURBS representation. This powerful parametric representation is used in most of the available CAD tools.

#### 6.2.3.2 Bézier Patches

Making use of the previously defined Bézier and B-spline curves, we can proceed to define surface representations (Rogers and Adams, 1990; Farin, 1997). For surfaces, we introduce the parametric representation

$$\mathbf{r}(u, v) = \sum_{i=0}^{m} \sum_{j=0}^{n} \mathbf{r}_{ij} B_i^m(u) B_j^n(v) \tag{6.32}$$

where, as before, $u \in [0, 1]$ and $v \in [0, 1]$. This definition can also be provided in matrix form as

$$\mathbf{r}(u, v) = [U][N][R][M]^T[V] \tag{6.33}$$

where

$$[U] = \begin{bmatrix} u^n & u^{n-1} & \cdots & 1 \end{bmatrix} \tag{6.34}$$

$$[V] = \begin{bmatrix} v^m & v^{m-1} & \cdots & 1 \end{bmatrix}^T \tag{6.35}$$

$$[R] = \begin{bmatrix} \mathbf{r}_{00} & \cdots & \mathbf{r}_{0m} \\ \vdots & \ddots & \vdots \\ \mathbf{r}_{n0} & \cdots & \mathbf{r}_{nm} \end{bmatrix}. \tag{6.36}$$

Also, $[N]$ and $[M]$ are given by

$$[N] = \begin{bmatrix} k_{00}^{(n)} & \cdots & k_{0n}^{(n)} \\ \vdots & \ddots & \vdots \\ k_{n0}^{(n)} & \cdots & k_{nn}^{(n)} \end{bmatrix}, \qquad [M] = \begin{bmatrix} k_{00}^{(m)} & \cdots & k_{0m}^{(m)} \\ \vdots & \ddots & \vdots \\ k_{m0}^{(m)} & \cdots & k_{mm}^{(m)} \end{bmatrix} \tag{6.37}$$

with

$$k_{ij}^l = (-1)^{j-i} \binom{l}{j} \binom{j}{i} \quad \text{and} \quad \binom{l}{j} = \frac{l!}{i!(l-i)!}. \tag{6.38}$$

For quadrilateral surface patches, the defining polygon net must be topologically rectangular; that is, the net must have the same number of vertices in each "row." Figure 6-15 shows a generic quadratic Bézier patch and its defining polygon mesh. Evidently, each boundary of a Bézier surface is a Bézier curve.

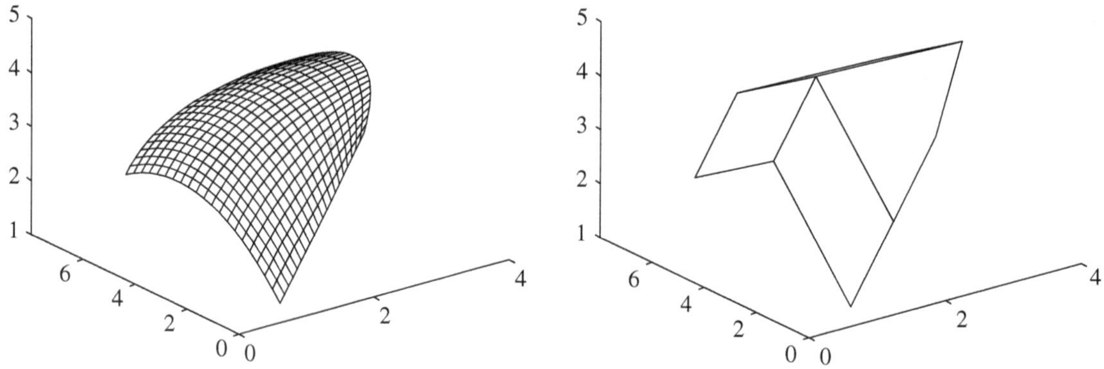

FIGURE 6-15 ■ Tensor product Bézier surface and its defining polygon net.

As already noted, more complex surfaces can be represented by multiple Bézier surface patches smoothly joined to form B-spline surfaces and NURBS as discussed next.

#### 6.2.3.3 B-Spline Surfaces

Cartesian-product B-spline surfaces are the natural extensions of Cartesian-product Bézier surfaces, defined by

$$\mathbf{r}(u, v) = \sum_{i=0}^{m}\sum_{j=0}^{n} \mathbf{r}_{ij} N_i^m(u) N_j^n(v) \tag{6.39}$$

where $N_i^m(u)$ and $N_j^n(v)$ are the B-spline basis functions in the biparametric $u$ and $v$ directions, respectively. These are blended Bézier surfaces used to transform a B-spline surface to a set of connected Bézier surfaces (Farin, 1997). As with B-spline curves, the shape and character of a B-spline surface is significantly influenced by the knot vectors. Open, periodic, and nonuniform knot vectors are frequently used in practice. For example, it is possible to use an open knot vector for one parametric direction and a periodic knot vector for the other. This leads to a cylindrical surface of varying cross sectional area. The local control properties of the B-spline curves also carry over to B-spline surfaces.

#### 6.2.3.4 NURBS Surfaces

Bézier and B-spline surfaces can be generalized to their rational counterparts by defining a projection of a four-dimensional tensor product Bézier or B-spline surfaces. Specifically, a rational Bézier patch takes the form

$$\mathbf{r}(u, v) = \frac{\sum_{i=0}^{m}\sum_{j=0}^{n} \omega_{ij}\mathbf{r}_{ij} B_i^m(u) B_j^n(v)}{\sum_{i=0}^{m}\sum_{j=0}^{n} \omega_{ij} B_i^m(u) B_j^n(v)} \tag{6.40}$$

and a rational B-spline surface is written as

$$\mathbf{r}(u, v) = \frac{\sum_{i=0}^{m}\sum_{j=0}^{n} \omega_{ij}\mathbf{r}_{ij} N_i^m(u) N_j^n(v)}{\sum_{i=0}^{m}\sum_{j=0}^{n} \omega_{ij} N_i^m(u) N_j^n(v)}. \tag{6.41}$$

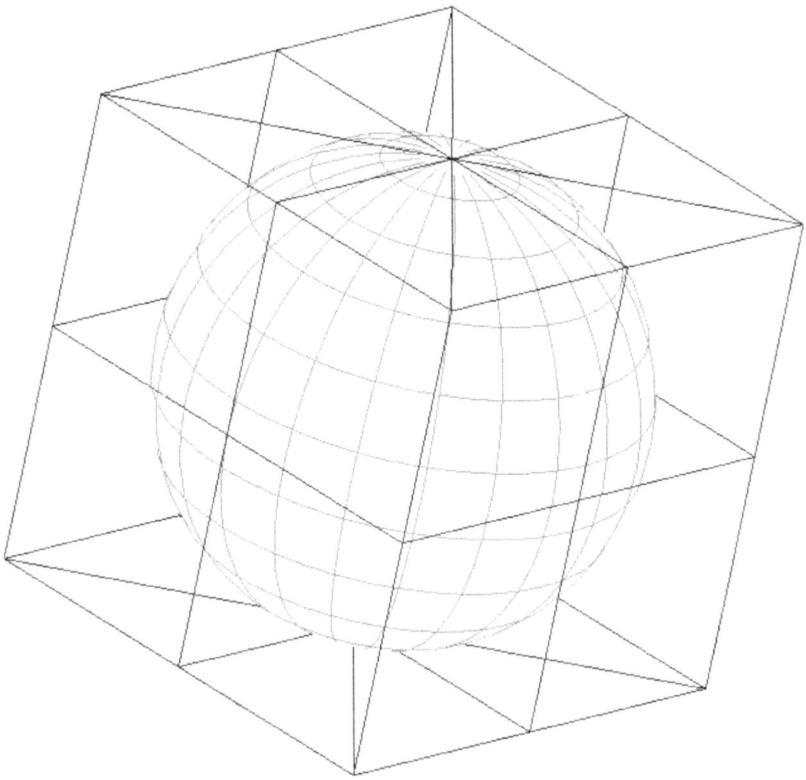

**FIGURE 6-16** Surface of a sphere using nonuniform rational B-splines and their defining polygon.

It should be noted that these surfaces are not tensor product surfaces themselves. Like their nonrational counterparts, open uniform, periodic uniform, and nonuniform knot vectors can be used to represent various rational Bézier and B-spline surfaces.

In closing our discussion of free-form surface representations, we should note that a strong attraction of rational B-spline surfaces is their ability to represent quadric surfaces in 3-D. The general expression for quadric surfaces is given by

$$Ax^2 + By^2 + Cz^2 + Dxy + Eyz + Fxz + Gx + Hy + Jz + K = 0. \quad (6.42)$$

NURBS surfaces can represent all such quadrics and blend them into higher-order sculptured surfaces. For example, one can represent a sphere exactly using a single rational B-spline surface as a collection of smoothly blended rational Bézier patches. The sphere and defining polygon *net* are shown in Figure 6-16.

### 6.2.4 Curvilinear Coordinates

Once the geometry is mathematically defined with a mesh of interconnected elements, the next step is to define basis functions (current density expansions) conformal to these elements. This allows us to convert the surface integral equations (SIEs) or volume integral equations (VIEs) into a matrix system through the MoM procedure outlined in Section 6.1. The construction of vector basis functions conformal to the curvilinear elements requires a local coordinate frame attached to the surface (or volume) as shown in Figures 6-17 and 6-18. This local frame can be easily

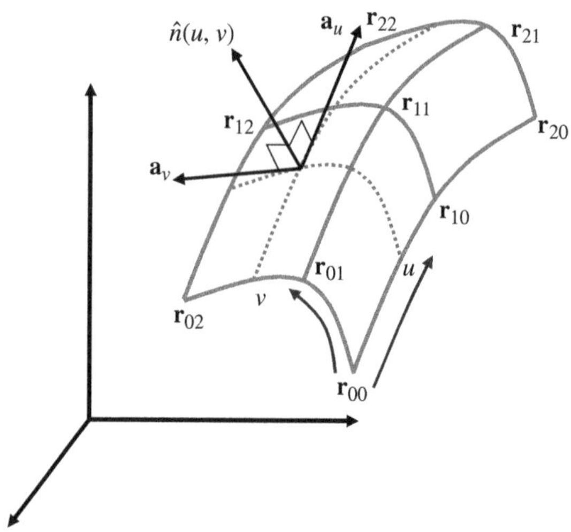


FIGURE 6-17 ■ Basis vectors ($\mathbf{a}_u$, $\mathbf{a}_v$), and normal $\hat{n}(u, v)$.

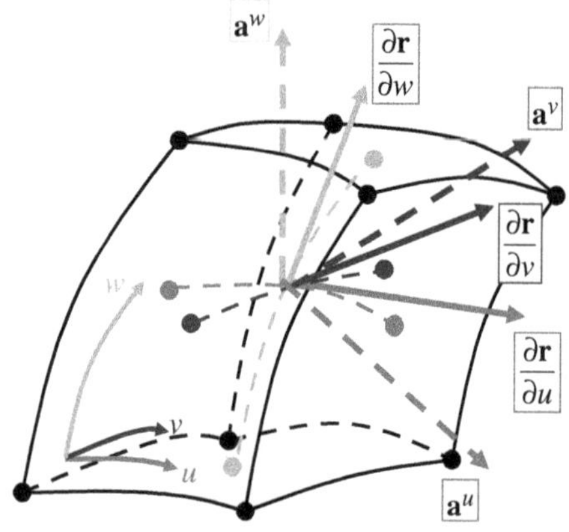


FIGURE 6-18 ■ Illustration of covariant and contravariant unitary vectors defined in a curvilinear hexahedral elements.

constructed using directional derivatives with respect to the normalized $(u, v, w)$ parameters. Likewise, various differential operations, such as the divergence and curl can be evaluated using these parametric variables. Next, we outline the mathematical tools needed to carry out calculations in terms of the $(u, v, w)$ variables.

An important quantity in any parametric representation is the tangent to the curve given by

$$\mathbf{a}_u = \frac{\partial \mathbf{r}(u)}{\partial u}. \tag{6.43}$$

Here, we intentionally used the partial derivatives to indicate that the representation may have additional control variables. As shown in Figure 6-17, for surfaces we can define two basis vectors along the two parametric directions given by

$$\mathbf{a}_u = \frac{\partial \mathbf{r}(u, v)}{\partial u}, \qquad \mathbf{a}_v = \frac{\partial \mathbf{r}(u, v)}{\partial v}. \tag{6.44}$$

Although tangent to the surface, $\mathbf{a}_u$ and $\mathbf{a}_v$ are not expected to be perpendicular. Thus, in using them to define the unit normal to each surface point, we must adapt the operation

$$\hat{n} = \frac{\mathbf{a}_u \times \mathbf{a}_v}{|\mathbf{a}_u \times \mathbf{a}_v|}. \tag{6.45}$$

We remark that the triplet of vectors $(\mathbf{a}_u, \mathbf{a}_v, \hat{n})$ forms a local conformal reference frame (at each surface point). It is assumed that no two such vectors are colinear (the surface is called "degenerate" if $\mathbf{a}_u$ and $\mathbf{a}_v$ are colinear).

Typically, the vector functions $\mathbf{F}(\mathbf{r}) = \mathbf{F}(u, v)$ defined on a parametric surface $\mathbf{r}(u, v)$ can be represented in this coordinate frame as

$$\mathbf{F}(u, v) = F^{(u)}(u, v)\mathbf{a}_u(u, v) + F^{(v)}(u, v)\mathbf{a}_v(u, v) \tag{6.46}$$

where $F^{(u)}(u, v)$ and $F^{(v)}(u, v)$ are the contravariant components of the vector $\mathbf{F}(u, v)$. The basis vectors $\mathbf{a}_u(u, v)$ and $\mathbf{a}_v(u, v)$ are correspondingly called covariant vectors and form one of the two complementary basis in expressing the vector function $\mathbf{F}(u, v)$ on $\mathbf{r}(u, v)$. This representation can be extended to vector fields as outlined in Section 6.3 and are particularly useful in formulating and implementing conformal MoM solutions. Most subsequent implementations in this book are based on such parametric representations.

A typical mesh of a geometry constructed using curvilinear elements (quadrilaterals) is shown in Figure 6-19. As indicated, conformal elements result in higher fidelity in geometrical representation including finer details that would otherwise require a much denser grid. As a result, curvilinear elements reduce the overall unknown count, leading to less memory and central processing unit (CPU) time. Indeed, the modeling fidelity, accuracy, and efficiency of MoM solutions using these elements have been reported in the literature from the 1990s on (Jin, Volakis, and

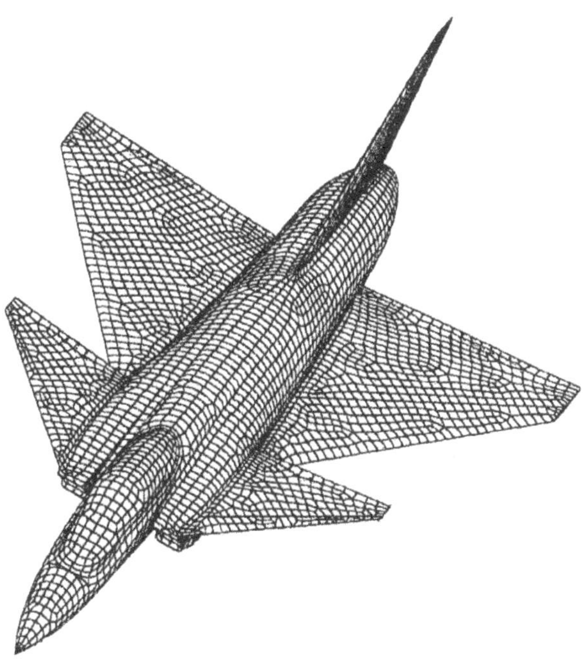

**FIGURE 6-19** Generic VFY218 aircraft modeled by quadrilateral patches.

Liepa, 1989; Antilla and Alexopoulos, 1994; Song and Chew, 1995; Sertel and Gürel, 1997; Graglia, Wilton, and Peterson, 1997).

### 6.2.4.1 Parametric Representation of Volume Elements

Generalizing the curvilinear surface elements in (6.25) leads to hexahedral finite elements. For example, given a set of 27 points in space $\{\mathbf{r}_{ijk}, i, j, k = 0, 1, 2\}$ on a topologically cubic grid, a curvilinear triquadratic volume element (see Figure 6-20) takes the form

$$\mathbf{r}(u, v, w) = \sum_{i=0}^{2}\sum_{j=0}^{2}\sum_{k=0}^{2} L_{ijk}(u, v, w)\mathbf{r}_{ijk} \tag{6.47}$$

where $L_{ijk}(u, v, w)$ are the Cartesian products of the usual Lagrange interpolation functions, and $0 \le u \le 1, 0 \le v \le 1, 0 \le w \le 1$.

For such a volume (cubic) element, three basis vectors used for the parametric derivatives are

$$\begin{aligned}
\mathbf{a}_u(u, v, w) &= \frac{\partial \mathbf{r}(u, v, w)}{\partial u} \\
\mathbf{a}_v(u, v, w) &= \frac{\partial \mathbf{r}(u, v, w)}{\partial v} \\
\mathbf{a}_w(u, v, w) &= \frac{\partial \mathbf{r}(u, v, w)}{\partial w}.
\end{aligned} \tag{6.48}$$

In proper parametric spaces (without warps and singularities), the three vectors form a suitable local coordinate frame to be used in representing the electric fields and currents as well as derivative and integral operators. Formally, the basis vectors in (6.48) are called the covariant unitary vectors and are not necessarily orthogonal to each other. This degree of freedom makes definition (6.48) a powerful tool in representing conformal entities when carrying out various mathematical and engineering problems.

Although the relaxed orthogonality for the triplet $(\mathbf{a}_u, \mathbf{a}_v, \mathbf{a}_w)$ allows for more flexibility, the covariant unitary vectors draw their attractiveness from an alternative

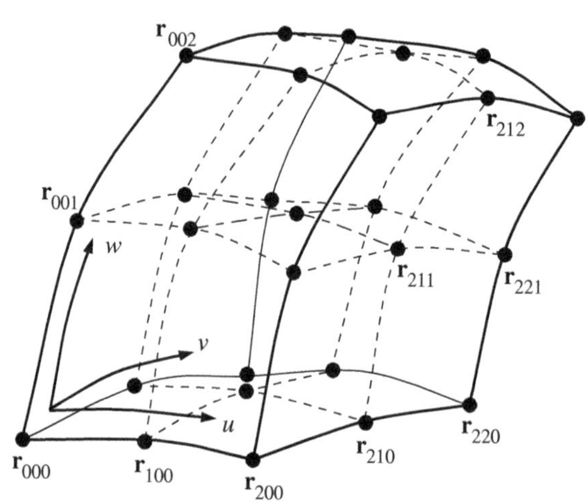


**FIGURE 6-20** Quadratic hexahedral finite element.

set of basis vectors, namely, the contravariant basis vectors. The contravariant basis vectors are derived from the covariant ones via the following mutual orthonormality rules:

$$\mathbf{a}^u(u, v, w) = \frac{1}{V+}\mathbf{a}^v(u, v, w) \times \mathbf{a}^w(u, v, w)$$

$$\mathbf{a}^v(u, v, w) = \frac{1}{V+}\mathbf{a}^w(u, v, w) \times \mathbf{a}^u(u, v, w)$$

$$\mathbf{a}^w(u, v, w) = \frac{1}{V+}\mathbf{a}^u(u, v, w) \times \mathbf{a}^v(u, v, w). \tag{6.49}$$

In these expressions, the scalar $V^+$ is chosen so that the two triplets are mutually orthonormal to each other and satisfy the condition $\mathbf{a}_\eta \cdot \mathbf{a}^\xi(u, v, w) = \delta_{\eta\xi}$, where $\delta_{\eta\xi}$ is the Kronecker delta function.

The Jacobian of the parametric transformation from $(x, y, z)$ to $(u, v, w)$ coordinates is given by

$$[J] = \begin{bmatrix} \frac{\partial x}{\partial u} & \frac{\partial y}{\partial u} & \frac{\partial z}{\partial u} \\ \frac{\partial x}{\partial v} & \frac{\partial y}{\partial v} & \frac{\partial z}{\partial v} \\ \frac{\partial x}{\partial w} & \frac{\partial y}{\partial w} & \frac{\partial z}{\partial w} \end{bmatrix} \tag{6.50}$$

and the metric tensor is given by

$$[G] = \begin{bmatrix} g_{uu} & g_{uv} & g_{uw} \\ g_{vu} & g_{vv} & g_{vw} \\ g_{wu} & g_{wv} & g_{ww} \end{bmatrix} \tag{6.51}$$

where $g_{ij} = \mathbf{a}_i \cdot \mathbf{a}_j$. This metric completely characterizes the geometrical properties of the 3-D parametric space.

When working with the $(u, v, w)$ variables, it is important to note that the differential volume and the area (for a constant $w$-surface) elements $dv = \sqrt{|G|}\, du\, dv\, dw$ and $ds = (g_{uu}g_{vv} - g_{uv}g_{vu})\, du\, dv$ are used. Here, $|G|$ is the determinant of the metric tensor given in (6.51).

An arbitrary vector $\mathbf{F}$ can now be represented in either the covariant projection form using $(\mathbf{a}_u, \mathbf{a}_v, \mathbf{a}_w)$

$$\mathbf{F} = (\mathbf{F} \cdot \mathbf{a}_u)\mathbf{a}^u + (\mathbf{F} \cdot \mathbf{a}_v)\mathbf{a}^v + (\mathbf{F} \cdot \mathbf{a}_w)\mathbf{a}^w \tag{6.52}$$

or contravariant projection form using $(\mathbf{a}^u, \mathbf{a}^v, \mathbf{a}^w)$

$$\mathbf{F} = (\mathbf{F} \cdot \mathbf{a}^u)\mathbf{a}_u + (\mathbf{F} \cdot \mathbf{a}^v)\mathbf{a}_v + (\mathbf{F} \cdot \mathbf{a}^w)\mathbf{a}_w. \tag{6.53}$$

Obviously, these follow from the orthonormality properties of the covariant and contravariant vectors.

Using the expressions (6.52) and (6.53), the differential operations in the curvilinear coordinate system can be summarized as follows.

Gradient of a scalar function $\phi(u, v, w)$:

$$\nabla\phi = \frac{\partial \phi}{\partial u}\mathbf{a}^u + \frac{\partial \phi}{\partial v}\mathbf{a}^v + \frac{\partial \phi}{\partial w}\mathbf{a}^w. \tag{6.54}$$

Divergence of a vector function $\mathbf{F}(u, v, w)$:

$$\nabla \cdot \mathbf{F} = \frac{1}{\sqrt{|G|}} \left\{ \frac{\partial}{\partial u} \left[ (\mathbf{F} \cdot \mathbf{a}^u) \sqrt{|G|} \right] + \frac{\partial}{\partial v} \left[ (\mathbf{F} \cdot \mathbf{a}^v) \sqrt{|G|} \right] + \frac{\partial}{\partial w} \left[ (\mathbf{F} \cdot \mathbf{a}^w) \sqrt{|G|} \right] \right\}. \tag{6.55}$$

Curl of a vector function $\mathbf{F}(u, v, w)$:

$$\nabla \times \mathbf{F} = \frac{1}{\sqrt{|G|}} \left\{ \left[ \frac{\partial (\mathbf{F} \cdot \mathbf{a}_w)}{\partial v} - \frac{\partial (\mathbf{F} \cdot \mathbf{a}_v)}{\partial w} \right] \mathbf{a}_u + \left[ \frac{\partial (\mathbf{F} \cdot \mathbf{a}_u)}{\partial w} - \frac{\partial (\mathbf{F} \cdot \mathbf{a}_w)}{\partial u} \right] \mathbf{a}_v + \left[ \frac{\partial (\mathbf{F} \cdot \mathbf{a}_v)}{\partial u} - \frac{\partial (\mathbf{F} \cdot \mathbf{a}_u)}{\partial v} \right] \mathbf{a}_w \right\}. \tag{6.56}$$

The corresponding operators acting on surfaces in three dimensions can be given as follows:

Surface gradient of a scalar function $\phi(u, v)$:

$$\nabla_s \phi(u, v) = g^{uu} \frac{\partial \phi}{\partial u} \mathbf{a}_u + g^{uv} \frac{\partial \phi}{\partial u} \mathbf{a}_v + g^{vu} \frac{\partial \phi}{\partial v} \mathbf{a}_u + g^{vv} \frac{\partial \phi}{\partial v} \mathbf{a}_v \tag{6.57}$$

where $g^{ij}, i = (u, v)$ and $j = (u, v)$ are the elements of the inverse metric tensor.

Surface divergence of a vector function $\mathbf{f}(u, v) = f_u \mathbf{a}_u + f_v \mathbf{a}_v$:

$$\begin{aligned} \nabla_s \cdot \mathbf{f}(u, v) &= g^{uu} \frac{\partial \mathbf{f}}{\partial u} \cdot \mathbf{a}_u + g^{uv} \frac{\partial \mathbf{f}}{\partial u} \cdot \mathbf{a}_v + g^{vu} \frac{\partial \mathbf{f}}{\partial v} \cdot \mathbf{a}_u + g^{vv} \frac{\partial \mathbf{f}}{\partial v} \cdot \mathbf{a}_v \\ &= \frac{1}{\sqrt{G_s}} \left( \frac{\partial (f_u \sqrt{G_s})}{\partial u} + \frac{\partial (f_v \sqrt{G_s})}{\partial v} \right). \end{aligned} \tag{6.58}$$

## 6.2.5 Parametric Representations of Surface and Volume Elements

### 6.2.5.1 Parametric Representation of Volume Elements

Tetrahedral and prismatic elements are alternatives to hexahedral volume elements. These can be considered as straightforward extensions of flat triangular elements and are traditionally expressed in terms of barycentric coordinates. That is, any point within a tetrahedron (defined by four distinct and non-colinear vertices $\mathbf{r}_1, \mathbf{r}_2, \mathbf{r}_3$, and $\mathbf{r}_4$) can be obtained using its barycentric coordinates $(\zeta_1, \zeta_2, \zeta_3, \zeta_4)$ as

$$\mathbf{r}(\zeta) = \sum_{i=1}^{4} \zeta_i \mathbf{r}_i. \tag{6.59}$$

Similarly to flat triangles, the barycentric coordinates $\zeta_i$ are defined as the normalized volumes of the subtetrahedral domains constructed by connecting the point $\mathbf{r}(\zeta)$ to the vertices as shown in Figure 6-21a. That is, $\zeta_i = V_{jk}/V$. Using a similar logic, it is possible to define curvilinear extensions of these tetrahedral elements, as shown in Figure 6-21b. We direct the reader to Graglia (1987, 1988) and Graglia, Wilton, and Peterson (1997) for a more detailed discussion on curvilinear tetrahedra.

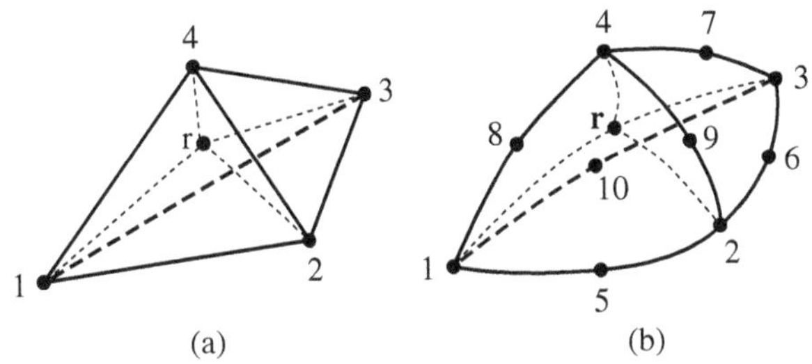


**FIGURE 6-21** Tetrahedral finite element: (a) conventional flat tetrahedra and (b) second-order curvilinear tetrahedra.

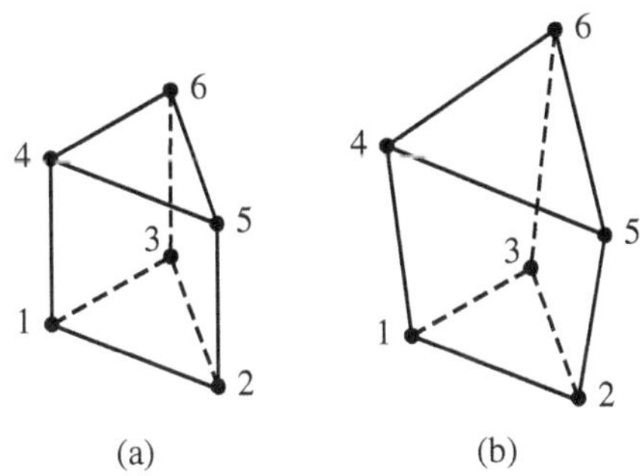


**FIGURE 6-22** Triangular prism elements: (a) right-angle prism and (b) distorted prism.

Triangular prism elements, shown in Figure 6-22, are also useful in modeling layered structures with arbitrary metallizations (as encountered in microwave circuit designs). We refer the reader to Özdemir and Volakis (1997); Volakis, Chatterjee, and Kempel (1998); Graglia, Wilton, Peterson, and Gheorma (1998) for an extended discussion of prism elements in the context of finite element CEM modeling.

### 6.2.6 Example Representations of Surface and Volume Basis Functions

In this section, we provide the reader with explicit definition of the conformal basis functions used in MoM implementations of SIEs and VIEs. We start by introducing rooftop basis functions on flat quadrilaterals and provide an alternative expression for the (RWG) BFs on flat triangular elements. Subsequently, we introduce their *curved* counterparts defined on curved parametric surfaces described above.

#### 6.2.6.1 Rooftop Basis Functions: Flat and Conformal Representations

Rooftop BFs are defined on pairs of flat rectangular elements (see Figure 6-23). They approximate the unknown surface current through the edge joining the elements in a

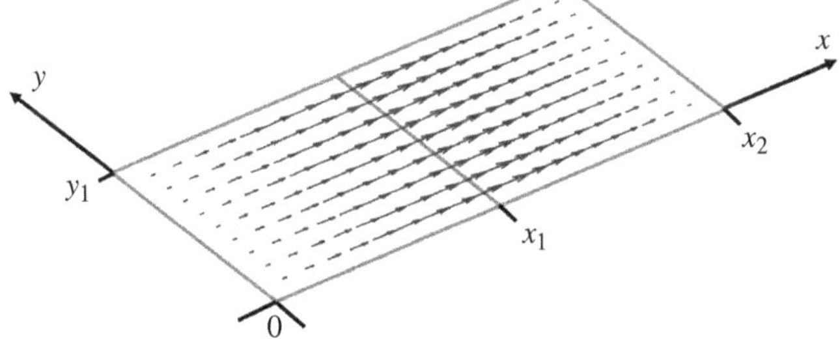


**FIGURE 6-23** Rooftop basis function defined as a pair of flat rectangular regions.

piecewise linear manner (much like the wire basis functions described in Chapter 4). In contrast to the RWG functions, the rooftop basis functions have constant value in the transverse direction. Referring to Figure 6-23, they can be mathematically expressed as

$$\mathbf{b}(x, y) = \begin{cases} \dfrac{x}{x_1} P(y/y_1)\hat{x} & 0 \le x \le x_1 \\ \dfrac{(x_2 - x)}{(x_2 - x_1)} P(y/y_1)\hat{x} & x_1 \le x \le x_2 \end{cases} \tag{6.60}$$

where $P(y/y_1)$ defines the pulse function that is nonzero for $y$ values between 0 and $y_1$

$$P(y/y_1) = \begin{cases} 1 & 0 \le y \le y_1 \\ 0 & \text{otherwise.} \end{cases} \tag{6.61}$$

Unfortunately, (6.60) is suitable only for flat rectangular elements. For curved surfaces, this expression must be generalized to curved rooftop BFs. As described in Section 6.2.1, a suitable coordinate frame for representing the curved BFs can be constructed by the covariant vectors

$$\mathbf{a}_u = \frac{\partial \mathbf{r}}{\partial u}, \qquad \mathbf{a}_v = \frac{\partial \mathbf{r}}{\partial v}. \tag{6.62}$$

Now we can use the mapping $\mathbf{r}(u, v)$ that transforms the unit square of the $(u, v)$ domain into a curvilinear quadrilateral element on the geometry surface to define conformal rooftop BFs (see Figure 6-24). The curved rootfop BF that flows along isoparametric directions can simply be defined on the $(u, v)$ domain in terms of local $u$ and $v$ parameters as

$$\mathbf{j}(u, v) = \frac{1}{\sqrt{g(u, v)}} u \frac{\partial \mathbf{r}(u, v)}{\partial u} \tag{6.63}$$

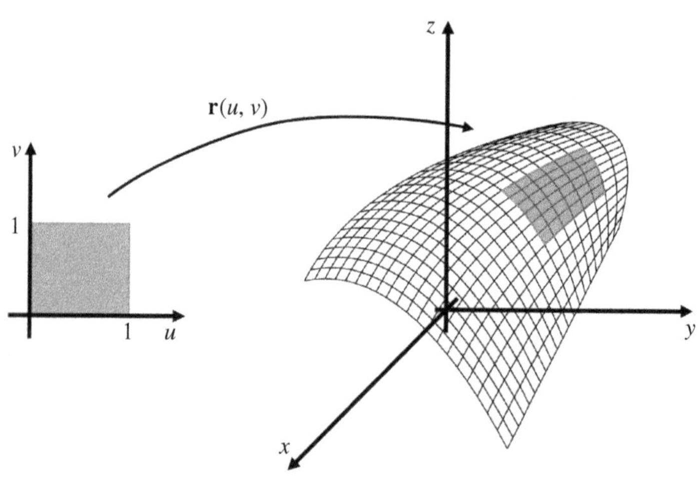


**FIGURE 6-24** Mapping of the parametric unit square in the $(u, v)$ domain onto a conformal quadrilateral element on a curved surface.

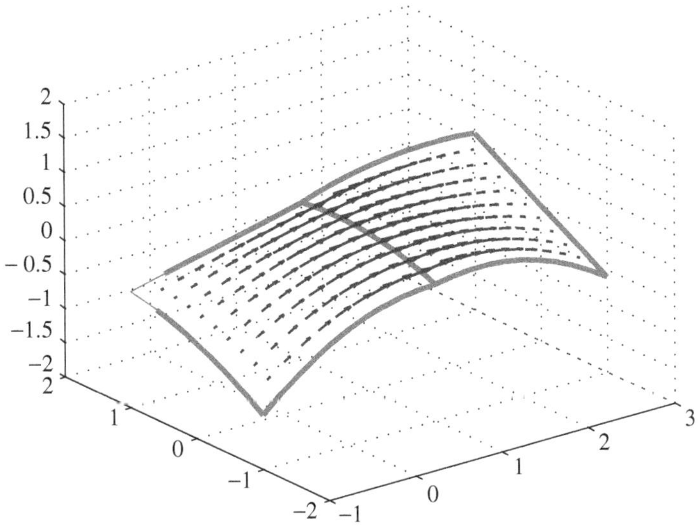

**FIGURE 6-25** Curvilinear rooftop basis function defined on the parametric space (arrows show the current flow and vector length represents current strength).

where $g(u, v)$ is the determinant of the metric tensor, given by

$$g = \begin{vmatrix} g_{11} & g_{12} \\ g_{21} & g_{22} \end{vmatrix} \tag{6.64}$$

where

$$g_{11} = \frac{\partial \mathbf{r}}{\partial u} \cdot \frac{\partial \mathbf{r}}{\partial u}, \qquad g_{12} = \frac{\partial \mathbf{r}}{\partial u} \cdot \frac{\partial \mathbf{r}}{\partial v}, \qquad g_{21} = \frac{\partial \mathbf{r}}{\partial v} \cdot \frac{\partial \mathbf{r}}{\partial u}, \qquad g_{22} = \frac{\partial \mathbf{r}}{\partial v} \cdot \frac{\partial \mathbf{r}}{\partial v}. \tag{6.65}$$

Equation (6.63) defines only one half of the curved rooftop (see Figure 6-25). Using the corresponding vertices of the adjacent element, one can properly pair the two halves to form the basis function depicted in Figure 6-25.

In addition to the basis function, the EFIE MoM formulation requires evaluations of the basis function's divergence. Using the previously provided definitions and the formula (6.58), the surface divergence of the conformal rooftop BF can be found to be

$$\sigma(\mathbf{r}) = \nabla_s \cdot \mathbf{j}(\mathbf{r}) = \frac{1}{\sqrt{g(u, v)}}. \tag{6.66}$$

Likewise, the surface charge density in each quadrilateral element is given by

$$\sigma\, ds = \frac{-j}{\omega} \nabla_s \cdot \mathbf{j} \sqrt{g(u, v)}\, du\, dv. \tag{6.67}$$

It is important to note that the surface charge density of the curved rooftop BF is constant in the $(u, v)$ parametric space, viz. $\sigma\, ds/du\, dv = -j/\omega$ and thus is the case for flat rectangular elements. We also show next that the current density across the common edge per unit parameter is independent of the geometry. Thus, the rooftop BF flow is continuous across a common edge between the two elements supporting the BF. That is, there is no line-charge accumulation on the common edge. This can be shown by evaluating the current flowing normal to the common edge. To do so, we proceed to evaluate the unit normal $\hat{t}_\perp$ as depicted in Figure 6-25. We have

$$\hat{t}_\perp = \hat{t}_\parallel \times \hat{n} = \frac{1}{\sqrt{gg_{22}}}\left[g_{22}\frac{\partial \mathbf{r}}{\partial u} - g_{12}\frac{\partial \mathbf{r}}{\partial v}\right] \tag{6.68}$$

where

$$\hat{t}_\parallel = \frac{1}{\sqrt{g_{22}}}\frac{\partial \mathbf{r}}{\partial v}. \tag{6.69}$$

Using (6.63) and (6.68), the normal component of the current BF flowing across the edge is found to be

$$\hat{t}_\perp \cdot \mathbf{j} = \frac{1}{\sqrt{g_{22}(u = 1, v)}}. \tag{6.70}$$

As the denominator is simply the differential length along the common edge, it follows that the current density across the edge per unit parameter is independent of the rest of the geometry. Therefore, the normal components corresponding to the two halves of the rooftop BFs (at the common edge) are equal, implying continuous current flow.

In closing, we remark that the differentiation $\frac{\partial \mathbf{r}}{\partial u}$ in (6.23) may not appear as straightforward when B-spline representations are used. However, the recursive relation

$$\frac{\partial}{\partial s}B_i^n(s) = n\left[B_{i-1}^{n-1}(s) - B_i^{n-1}(s)\right] \tag{6.71}$$

for Bernstein polynomials allows substantial simplification. As $B_{i-1}^{n-1}(s)$ and $B_i^{n-1}(s)$ are already available in the calculation of $\mathbf{r}(u, v)$, no extra effort is needed in evaluating the surface tangents, used in the definition of the BFs.

### 6.2.6.2 RWG Basis Functions: Flat and Conformal Representations

Due to the flexibility of flat triangulations in modeling arbitrary surfaces, RWG basis functions have been used widely in electromagnetic scattering and radiation analysis. As described in Section 6.1.3, RWG BFs are defined over pairs of flat triangules in (6.15). For 3-D triangular surface meshes, the RWG BFs are depicted in Figure 6-26 and are linearly varying vector fields.

When curved triangular subdomain functions are considered, the conformal RWG BF can be defined with the aid of the transformation $\mathbf{r}(u, v)$ as in the case of rooftop BFs (see Figure 6-27).

The conformal RWG BFs can be defined on the $(u, v)$ domain (Wandzura, 1992) using the representation

$$\mathbf{j}(\mathbf{r}) = \frac{1}{\sqrt{g(u, v)}}\left(u\frac{\partial \mathbf{r}}{\partial u} + v\frac{\partial \mathbf{r}}{\partial v}\right) \tag{6.72}$$

whose divergence is simply

$$\nabla_s \cdot \mathbf{b}(\mathbf{r}) = \frac{2}{\sqrt{g(u, v)}}. \tag{6.73}$$

The actual current flow on a curvilinear triangle pair is given in Figure 6-28. We note that (6.72) defines only one half of the curved RWG BF pair.

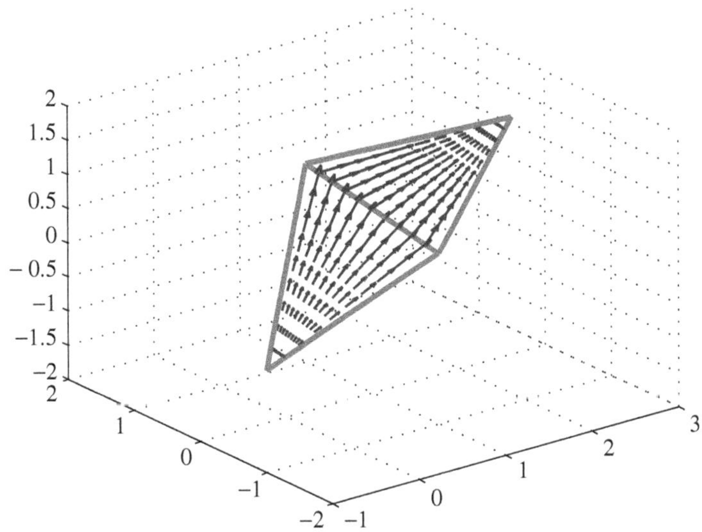


**FIGURE 6-26** Rao–Wilton–Glisson basis function on a pair of flat triangular regions.

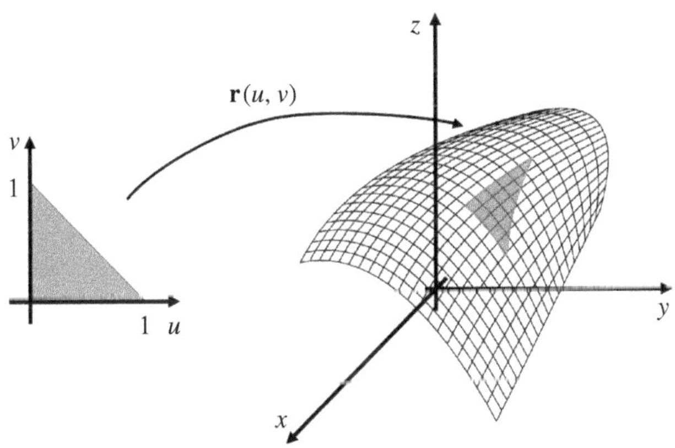


**FIGURE 6-27** Mapping of the parametric unit triangle on the (u, v) domain onto a conformal triangular element on a curved surface.

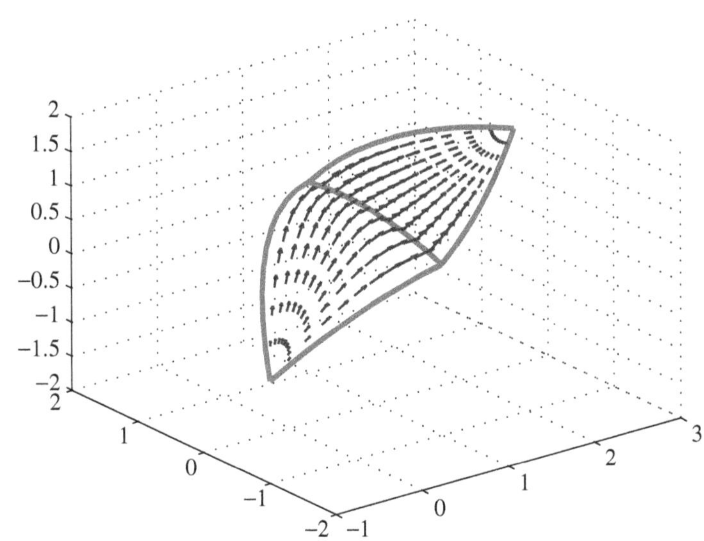


**FIGURE 6-28** CRWG BF defined on the unit triangle in the parametric space.

Similar to the curved rooftop basis functions, the surface charge density for the curvilinear RWG basis is also constant when expressed in the parametric $(u, v)$ domain, viz.

$$\sigma\, ds = \frac{-j}{\omega}\nabla_s \cdot \mathbf{b}\,\sqrt{g(u, v)}\, du\, dv. \tag{6.74}$$

Likewise, the current density across the common edge is independent of the triangle's geometry. Thus, as desired, there is no line-charge accumulation on the common edge. To show this, let us once again consider the unit vector parallel to the common edge shared by the triangles

$$\hat{t}_{\parallel} = \frac{1}{\sqrt{g_{11} + g_{22} - 2g_{12}}}\left(\frac{\partial \mathbf{r}}{\partial u} - \frac{\partial \mathbf{r}}{\partial v}\right). \tag{6.75}$$

Also, let us consider the unit tangent vector perpendicular to the common edge. It is given by

$$\hat{t}_{\perp} = \hat{t}_{\parallel} \times \hat{n} = \frac{1}{\sqrt{g(g_{11} + g_{22} - 2g_{12})}}\left[(g_{22} - g_{12})\frac{\partial \mathbf{r}}{\partial u} + (g_{11} - g_{12})\frac{\partial \mathbf{r}}{\partial v}\right] \tag{6.76}$$

where $\hat{n}$ is the normal to the triangle. Thus, from (6.72)

$$\hat{t}_{\perp} \cdot \mathbf{b} = \frac{1}{\sqrt{g_{11} + g_{22} - 2g_{12}}}. \tag{6.77}$$

Clearly, this is independent of the triangles' geometry since the right-hand side is just the differential length along the common edge. Further, as the result in (6.77) is the same regardless of which triangle in the pair is considered, we conclude that no charge accumulation exists at the common edge.

We note that in addition to the curved RWGs, other higher-order basis on triangular and rectangular subdomains have been employed for MoM solutions (Wandzura, 1992; Popović and Kolundžija, 1994; Notaros and Popovic, 1996; Notaros, Popovic, Weem, Brown, and Popovic, 2001). Next, we focus on curvilinear rooftop and curvilinear RWG basis functions for scattering solutions involving arbitrary scatterers.

## 6.3 | EVALUATION OF MOM MATRIX ENTRIES

To illustrate the implementation of a solution, we must first populate the MoM matrix entries for a particular mesh and chosen basis functions. If we specifically choose biquadratic quadrilateral elements, a typical conformal basis function can be defined on a pair of neighboring quadrilateral elements as displayed in Figure 6-29. Each biquadratic surface element can potentially support four half-basis functions (defined in terms of the parametric surface derivatives). These half-basis functions must, of course, be associated with one of the quadrilateral's four edges. The four half-basis

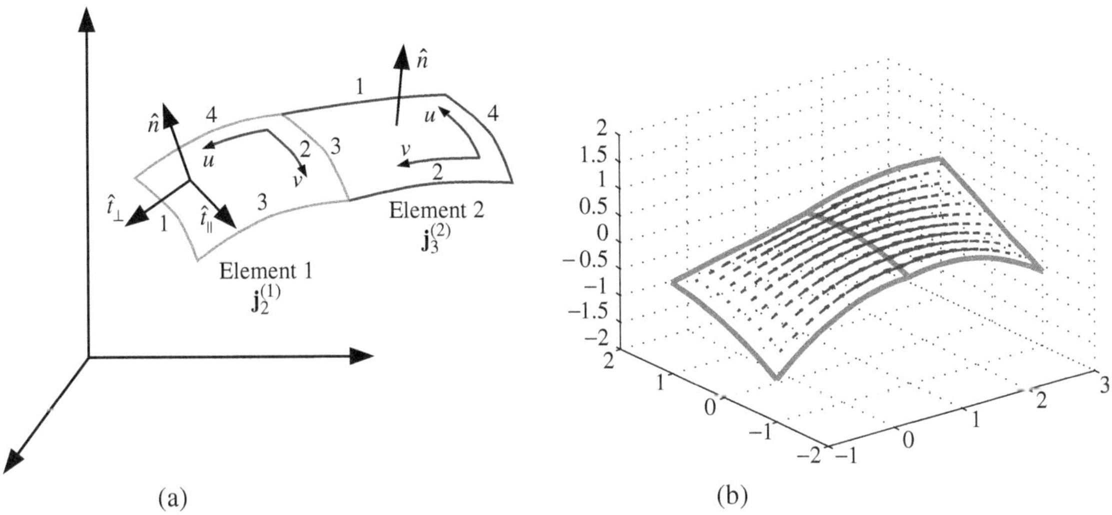


**FIGURE 6-29** Pair of quadrilaterals for defining basis functions: (a) pair of patches forming the support of basis function associated with the common edge and (b) quiver plot of the current density defined at the common edge shared by element pair.

functions for the left element 1 (see Figure 6-29a) are given by (see Section 6.2.6)

$$\mathbf{j}_1^{(1)} = \frac{u}{\sqrt{G_s}}\mathbf{a}_u, \qquad \mathbf{j}_2^{(1)} = \frac{1}{\sqrt{G_s}}(1-u)\mathbf{a}_u \tag{6.78}$$

$$\mathbf{j}_3^{(1)} = \frac{v}{\sqrt{G_s}}\mathbf{a}_v \qquad \mathbf{j}_4^{(1)} = \frac{1}{\sqrt{G_s}}(1-v)\mathbf{a}_v.$$

As before, $\mathbf{a}_u = \partial\mathbf{r}/\partial u$, $\mathbf{a}_v = \partial\mathbf{r}/\partial v$, and $G_s$ is the determinant of the metric tensor for the transformation (6.22) given in (6.51). The actual current flow across edge 2 of element 1 (or edge 3 of element 2) is depicted in Figure 6-29b. Using $\mathbf{a}_u$ and $\mathbf{a}_v$, the surface normal to the quadrilateral is determined. The triad $(\mathbf{a}_u, \mathbf{a}_v, \hat{n})$ form a right-handed system, but $\mathbf{a}_u$, $\mathbf{a}_v$, $\hat{n}$ are not necessarily orthonormal.

The vector field representing the basis function associated with the common edge of the quadrilateral pair is shown in Figure 6-29b. This BF consists of two half BFs (half BF 2 on element 1 and half BF 3 on element 2). Properly pairing these half-rooftop BFs on neighboring patches allows for a continuous current representation on the entire body. As required, the normal component of the current across neighboring patches is continuous; that is, no artificial charge accumulation exists on the common edge joining the pair of elements. More general higher-order basis functions can, of course, be constructed if the linear variations in (6.78) are replaced by higher-order functions. However, the accuracy of such representations should be correlated to the accuracy of the actual geometry model. That is, one should not pursue a higher-order representation of the current density aiming to improve solution accuracy if the geometry representation is not equally accurate. We do remark, however, that higher-order have been reported to generate smaller MoM systems. Hence, they can be advantageous as they may require less computational resources for the same accuracy solutions (Jorgensen, Volakis, Meincke, and Breinbjerg, 2004).

For the element pair depicted in Figure 6-29, the half-basis function defined in element 2 is $\mathbf{j}_3^{(2)} = -\frac{v}{\sqrt{G_s}}\mathbf{a}_v$. This is, of course, identical to $\mathbf{j}_3^{(1)}$ except for a minus sign introduced to correct for the direction of current flow. We also note that the normal component of $\mathbf{j}_2^{(1)}$ across the common edge can be evaluated using the tangential and normal unit vectors to that edge given by

$$\hat{t}_{\parallel} = \frac{1}{\sqrt{g_{vv}}}\mathbf{a}_v, \quad \hat{t}_{\perp} = \hat{t}_{\parallel} \times \hat{n} = \frac{1}{\sqrt{G_s g_{vv}}}\left[g_{vv}\mathbf{a}_u - g_{uv}\mathbf{a}_v\right]. \tag{6.79}$$

Using these, $\mathbf{j}_2^{(1)}$ and $\mathbf{j}_3^{(2)}$ at the common edge can be written as

$$\hat{t}_{\perp} \cdot \mathbf{j}_2^{(1)} = \frac{1}{\sqrt{g_{vv}(u=1, v)}} \tag{6.80}$$

$$\hat{t}_{\perp} \cdot \mathbf{j}_3^{(2)} = \frac{1}{\sqrt{g_{uu}(u, v=0)}}. \tag{6.81}$$

As $g_{vv}(u = 1, v) = g_{uu}(u, v = 0)$, it follows that continuity is ensured across the common edge. We can also verify that the surface divergence for both basis functions is constant. Specifically, $\nabla_s \cdot \mathbf{j}_2^{(1)} = 1/\sqrt{G_s}$, and $\nabla_s \cdot \mathbf{j}_3^{(2)} = -1/\sqrt{G_s}$, implying constant surface charge $\sigma = \nabla_s \cdot \mathbf{j}$ in the parametric space (i.e., $\sigma\, ds/du\, dv = \pm 1$)[2].

Having defined the BFs, we can now define the surface current density as $\mathbf{J}(\mathbf{r}) = \sum_{i=1}^{N} x_i \mathbf{j}_i(\mathbf{r})$, where $\{x_i\}$ are the unknown coefficients of the expansion. We now proceed to evaluate the matrix entries $Z_{mn}$ so that $\{x_i\}$ can be solved.

### 6.3.1 Element Matrices and Assembly Process

The double surface integrals in the matrix element expression (6.13) need to be numerically evaluated to construct the MoM matrix using a suitable quadrature rule. Since each basis and testing function is defined on a pair of elements, it is advantageous to compute the contributions to the matrix entries from pairs of basis and testing domains (i.e., compute element submatrices).

These contributions can be subsequently added into the actual matrix using the mesh connectivity, as also done in standard finite element methods. Specifically, the MoM matrix entries can be assembled as a combination of the $4 \times 4$ matrix elements noted in Figure 6-30. Each of these takes the form

$$\left(\widetilde{Z}_{mn}\right)_{kl} = \int_S d\mathbf{r}\, \mathbf{j}_m^k(\mathbf{r}) \cdot \int_{S'} d\mathbf{r}'\, \overline{\boldsymbol{\Gamma}}(\mathbf{r}, \mathbf{r}') \cdot \mathbf{j}_n^l(\mathbf{r}'). \tag{6.82}$$

Here, $\mathbf{j}_m^k$ $m = 1, 2, 3, 4$ denote the four half-basis functions on $k$th element, and $\mathbf{j}_n^l$ $n = 1, 2, 3, 4$ refers to the four half-basis function on element $l$. Their geometry is depicted in Figure 6-30. Each entry of the $\widetilde{Z}_{mn}$ submatrix will contribute to a specific global matrix element as dictated by mesh connectivity tables (see, e.g., Tables 6-1 and 6-2). We can surmise that each global MoM matrix entry is the sum of four contributions.

The numerical evaluation of (6.82) is straightforward when the source and testing elements are geometrically well separated. However, when the source and testing

[2] $ds = \sqrt{G_s}\, du\, dv$.

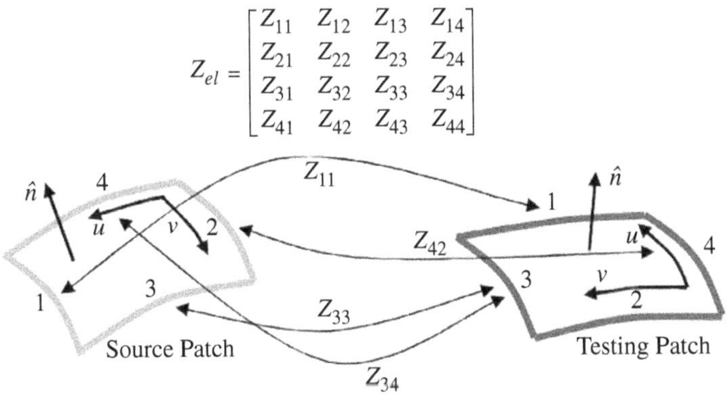


**FIGURE 6-30** Geometry of the interacting quadrilateral pair and their associate matrix elements.

elements overlap (self-cells in the MoM matrix), the singular behavior of the Green's function when $\mathbf{r} = \mathbf{r}'$ precludes use of a standard numerical evaluation. The subject of self-cell evaluation has been a focus in many journal papers. Indeed, several integral evaluation techniques based on singularity extraction of flat elements were presented in Wilton, Rao, Glisson, Schaubert, Al-Bundak, and Butler (1984) and Eibert and Hansen (1996). However, these are not readily applicable to curvilinear elements. The following section explores singularity annihilation techniques for curvilinear elements. These are sufficiently general to be equally applicable for flat patches as well.

### 6.3.2 Evaluation of Integrals with Singular Kernels

To evaluate the integrals when $k = l$ or even when the integrand is nearly singular (when the $k$th element is adjacent to the $l$th), we will first pursue a number of steps to reduce the singularity. We begin by first rewriting (6.82) as

$$Z_{mn} = \int_S \sqrt{G_s}\, du\, dv\, \mathbf{j}'_m(\mathbf{r}) \cdot \int_{S'} \sqrt{G'_s}\, du'\, dv'\, \overline{\Gamma}(\mathbf{r}, \mathbf{r}') \cdot \mathbf{j}'_n(\mathbf{r}') \tag{6.83}$$

on introducing the Jacobians $d\mathbf{r} = \sqrt{G_s}\, du\, dv$ and $d\mathbf{r}' = \sqrt{G'_s}\, du'\, dv'$. From (6.78), we see that the normalization with respect to $1/\sqrt{g}$ amounts to canceling the Jacobian. As usual, the primed variables refer to the integration over $l$th element and the unprimed to the $k$th element.

A key step in numerical evaluation of $Z_{mn}$ is the use of the divergence theorem (see Chapter 3) to relax the singularity of (6.83). Rewriting (6.83) more explicitly by substituting for the Green's function, we have

$$Z_{mn} = \int_S d\mathbf{r}\, \mathbf{j}_m(\mathbf{r}) \cdot \left[ \int_{S'} d\mathbf{r}'\, G(\mathbf{r}, \mathbf{r}')\, \mathbf{j}_n(\mathbf{r}') + \frac{1}{k^2} \int_{S'} d\mathbf{r}'\, \nabla\nabla G(\mathbf{r}, \mathbf{r}') \cdot \mathbf{j}_n(\mathbf{r}') \right] \tag{6.84}$$

or,

$$Z_{mn} = \int_S d\mathbf{r}\, \mathbf{j}_m(\mathbf{r}) \cdot \left[ \int_{S'} d\mathbf{r}'\, G(\mathbf{r}, \mathbf{r}')\, \mathbf{j}_n(\mathbf{r}') - \frac{1}{k^2} \nabla \int_{S'} d\mathbf{r}'\, \nabla' G(\mathbf{r}, \mathbf{r}') \cdot \mathbf{j}_n(\mathbf{r}') \right]. \tag{6.85}$$

The hypersingularity of the kernel in the second integral is certainly due to the $\nabla'$ operator. To reduce this singularity, we will employ the identity $\nabla' \cdot (G\mathbf{j}_i) = \nabla' G \cdot \mathbf{j}_n + G\nabla' \cdot \mathbf{j}_n$ and divergence theorem (see Chapters 2 and 3). Doing so, we have

$$\begin{aligned} Z_{mn} &= \int_S d\mathbf{r}\, \mathbf{j}_m(\mathbf{r}) \cdot \left[ \int_{S'} d\mathbf{r}'\, G(\mathbf{r}, \mathbf{r}')\, \mathbf{j}_n(\mathbf{r}') \right. \\ &\quad \left. - \frac{1}{k^2} \nabla \int_{S'} d\mathbf{r}' \left\{ \nabla' \cdot \left[ G(\mathbf{r}, \mathbf{r}')\, \mathbf{j}_n(\mathbf{r}') \right] - G(\mathbf{r}, \mathbf{r}') \nabla' \cdot \mathbf{j}_n(\mathbf{r}') \right\} \right] \\ &= \int_S d\mathbf{r}\, \mathbf{j}_m(\mathbf{r}) \cdot \left[ \int_{S'} d\mathbf{r}'\, G(\mathbf{r}, \mathbf{r}')\, \mathbf{j}_n(\mathbf{r}') + \frac{1}{k^2} \nabla \int_{S'} d\mathbf{r}'\, G(\mathbf{r}, \mathbf{r}') \nabla' \cdot \mathbf{j}_n(\mathbf{r}') \right] \\ &\quad - \int_S d\mathbf{r}\, \mathbf{j}_m(\mathbf{r}) \cdot \left[ \frac{1}{k^2} \nabla \oint_{C'} d\hat{n}_{c'} \cdot G(\mathbf{r}, \mathbf{r}')\, \mathbf{j}_n(\mathbf{r}') \right] \end{aligned} \tag{6.86}$$

where the latter closed line integral is over the boundary of the surface $S$ (testing surface or the $k$th element) (see Figure 6-31). As noted in Chapter 3, $d\hat{n}_{C'}$ is perpendicular to the contour $C$ (enclosing $s'$). Thus, since $d\hat{n}_{C'} \cdot \mathbf{j}_n(\mathbf{r}') = 0$, the contour integral vanishes, allowing us to reduce (6.86) to

$$Z_{mn} = \int_S d\mathbf{r}\, \mathbf{j}_m(\mathbf{r}) \cdot \left[ \int_{S'} d\mathbf{r}' G(\mathbf{r}, \mathbf{r}') \mathbf{j}_n(\mathbf{r}') + \frac{1}{k^2} \nabla \int_{S'} d\mathbf{r}' G(\mathbf{r}, \mathbf{r}') \nabla' \cdot \mathbf{j}_n(\mathbf{r}') \right]. \tag{6.87}$$

We can again use the divergence theorem $\nabla \cdot (\phi \mathbf{A}) = \nabla\phi \cdot \mathbf{A} + \phi \nabla \cdot \mathbf{A}$ on the second term of (6.87) with $\mathbf{A} = \mathbf{j}_m(\mathbf{r})$ and $\phi = \int_s d\mathbf{r}' G(\mathbf{r}, \mathbf{r}') \nabla' \cdot \mathbf{j}_n(\mathbf{r}')$. Doing so leads to

$$\begin{aligned} Z_{mn} &= \int_S d\mathbf{r}\, \mathbf{j}_m(\mathbf{r}) \cdot \int_{S'} d\mathbf{r}' G(\mathbf{r}, \mathbf{r}') \mathbf{j}_n(\mathbf{r}') \\ &\quad + \frac{1}{k^2} \oint_C d\hat{n}_{C'} \cdot \left[ \mathbf{j}_m(\mathbf{r}) \int_{S'} d\mathbf{r}' G(\mathbf{r}, \mathbf{r}') \nabla' \cdot \mathbf{j}_n(\mathbf{r}') \right] \\ &\quad - \frac{1}{k^2} \int_S d\mathbf{r} \nabla \cdot \mathbf{j}_m(\mathbf{r}) \int_{S'} d\mathbf{r}' G(\mathbf{r}, \mathbf{r}') \nabla' \cdot \mathbf{j}_n(\mathbf{r}') \end{aligned} \tag{6.88}$$

where $d\hat{n}_{C'}$ is again normal to the boundary $C$ of the testing domain $S$ (as depicted in Figure 6-31). Then, since $d\hat{n}_{c'}$ and $\mathbf{j}_m(\mathbf{r})$ are always perpendicular, it follows that

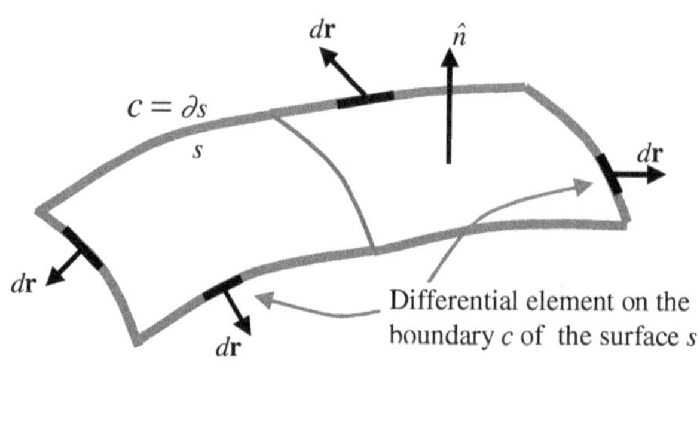


**FIGURE 6-31** Parameters for the application of the divergence theorem on the testing function.

the line integral in (6.88) vanishes, giving

$$Z_{mn} = \int_S d\mathbf{r}\,\mathbf{j}_m(\mathbf{r}) \cdot \int_{S'} d\mathbf{r}' G(\mathbf{r}, \mathbf{r}')\mathbf{j}_n(\mathbf{r}') - \frac{1}{k^2}\int_S d\mathbf{r}\nabla \cdot \mathbf{j}_m(\mathbf{r}) \int_{S'} d\mathbf{r}' G(\mathbf{r}, \mathbf{r}')\nabla' \cdot \mathbf{j}_n(\mathbf{r}'). \tag{6.89}$$

We note that both integrals in (6.89) now have $1/R$ singularity (where $R$ is the distance between the testing and observation points in the $m$th and $n$th surface element, respectively). This is an integrable singularity, but nevertheless it must be carefully evaluated numerically. In the next section, we present several annihilation techniques for integrating the singular integrals in (6.89) for curved elements. This is particularly important when $n = m$ (i.e., for the MoM matrix self-terms). After these annihilation transformations, standard Gaussian quadrature (see Figure 6-8 and Table 6-3) can be used to numerically evaluate $Z_{mn}$.

The MoM matrix elements appearing in the formulations as double surface integrals on testing and basis subdomains are evaluated numerically differently depending on the distance between the testing and the basis functions. Generally, fixed-order Gaussian quadrature (Abramowitz and Stegun, 1972) is adapted for the evaluation of all MoM matrix integrals that do not involve overlapping of the integration and testing elements. For the latter, the integrands are singular and require more careful consideration, as discussed next.

It is advantageous to evaluate the matrix elements (6.89) by integrating in the $(u, v)$ space. To do so, we first replace the basis and testing functions by their expression in the $(u, v)$ space. Specifically, we introduce Figure 6-28 to get

$$\mathbf{j}(u, v) = \begin{pmatrix} \mathbf{f}_{CRT}(u, v) \\ \mathbf{f}_{CRWG}(u, v) \end{pmatrix} = \frac{1}{\sqrt{g}}\left[\begin{pmatrix} u \\ u \end{pmatrix}\frac{\partial \mathbf{r}}{\partial u} + \begin{pmatrix} 0 \\ v \end{pmatrix}\frac{\partial \mathbf{r}}{\partial v}\right] \tag{6.90}$$

for the curved rectangular (*CRT*) and curved triangular (*CRWG*) subdomains. Also noting that

$$\nabla \cdot \mathbf{j}(u, v) = \frac{1}{\sqrt{g}}\begin{pmatrix} 1 \\ 2 \end{pmatrix} \tag{6.91}$$

$Z_{mn}$ reduces to

$$Z_{mn} = \int_S ds \int_{S'} ds \left[\mathbf{t}_m(\mathbf{r}) \cdot \mathbf{j}_n(\mathbf{r}') - \frac{1}{k^2}\nabla \cdot \mathbf{t}_m(\mathbf{r})\nabla' \cdot \mathbf{j}_n(\mathbf{r}')\right]\frac{e^{-jkR}}{4\pi R}. \tag{6.92}$$

From (6.72) and (6.78), we can cancel the presence of $\sqrt{(g)}$ in the numerator of (6.92) (viz. $ds = \sqrt{g}\,du\,dv$). To do so, we introduce the normalized basis functions

$$\tilde{\mathbf{j}}(u, v) = \left[\begin{pmatrix} u \\ u \end{pmatrix}\frac{\partial \mathbf{r}}{\partial u} + \begin{pmatrix} 0 \\ v \end{pmatrix}\frac{\partial \mathbf{r}}{\partial v}\right] \tag{6.93}$$

and divergence

$$\nabla \cdot \tilde{\mathbf{j}}(u, v) = \begin{pmatrix} 1 \\ 2 \end{pmatrix}. \tag{6.94}$$

Using these in (6.92) gives

$$Z_{mn} = \int_S du\,dv \int_{S'} du'\,dv' \left[\tilde{\mathbf{j}}_m \cdot \tilde{\mathbf{j}}_n - \frac{1}{k^2}\nabla \cdot \tilde{\mathbf{j}}_m \nabla' \cdot \tilde{\mathbf{j}}_n\right] \frac{e^{-jkR}}{4\pi R} \tag{6.95}$$

$$Z_{mn} = \int_S du\,dv \int_{S'} du'\,dv' \left[\tilde{\mathbf{j}}_m \cdot \tilde{\mathbf{j}}_n - \frac{1}{k^2}\left(\frac{1}{4}\right)\right] \frac{e^{-jkR}}{4\pi R}. \tag{6.96}$$

Thus, for rooftop basis functions, the surface integrals are simply over the square domain bounded by $u = [0, 1]$ and $v = [0, 1]$. Also, for RWGs, the domains must, of course, be bounded by the lines $u = 0, \quad v = 0$ and $u + v = 1$. As this triangular domain can also be mapped to the same square domain in an auxiliary $(\eta, \zeta)$ parametric space, the same integration routines designed for square integration domains domains can be adapted for RWGs as well.

With all of this in mind, we may rewrite (6.96) as

$$Z_{mn} = \iint_{\square,\triangle} \iint_{\square,\triangle} du\,dv\,du'\,dv' \frac{g(u, v, u', v')}{R(u, v, u', v')} \tag{6.97}$$

$$= \iint_{\square,\triangle} \iint_{\square,\triangle} du\,dv\,du'\,dv' h(u, v, u', v') \tag{6.98}$$

where $\square$ implies rectangular elements, and $\triangle$ refers to triangular ones. The outer integrals over the testing subdomain, and are numerically straightforward to evaluate using low-order Gaussian quadrature. For each quadrature point of the outer integral, the inner surface integral (which contains an integrable singularity) must be carefully evaluated. As noted, this is due to the singularity of order $1/R$ in (6.97) when $u = u'$ and $v = v'$. Although this singularity is integrable, "blindfold" application of standard numerical integration routines will lead to inaccuracies when $u \approx u'$ and $v \approx v'$. Of course, for the case when $(u, v)$ and $(u', v')$ are not close to each other, standard integration routines are applicable. That is, when testing and basis elements do not overlap or are adjacent to each other, standard numerical quadrature can be employed.

Thus, in the following we will focus on the evaluation of the inner integral

$$I(u_0, v_0) = \iint_{\square,\triangle} du'dv' \frac{g(u', v', u_0, v_0)}{R(u', v', u_0, v_0)} \tag{6.99}$$

where $R(u', v')$ vanishes at the $(u_0, v_0)$ point. The techniques presented in the next subsection are intended to recast the inner integral in (6.97) so that it can be evaluated using standard numerical means.

### 6.3.3 Singularity Annihilation Techniques

To evaluate singular integrals, a commonly used approach is to add and subtract a term that can be analytically evaluated but has the same order of singularity (Wilton, Rao, Glisson, Schaubert, Al-Bundak, and Butler, 1984; Barkeshli and Volakis, 1990; Eibert and Hansen, 1996). For our case, this approach works well for flat (faceted) discretizations of the geometry and relies on the addition and subtraction of the $1/R$

singularity, viz.

$$I(u_0, v_0) = \iint_{\square,\triangle} du\,dv \left[\frac{g(u,v)}{R(u,v,u_0,v_0)} - \frac{g_0}{R(u,v,u_0,v_0)}\right] + \iint_{\square,\triangle} du\,dv \frac{g_0}{R(u,v,u_0,v_0)} \tag{6.100}$$

where $g_0 = g(u_0, v_0)$. The last integral can be evaluated analytically (ibid.). Thus, although singular, it is accurate. The first integral no longer contains a singularity and can therefore be evaluated using standard methods. In particular, for triangular elements and RWG basis functions, the following formula can be readily used to evaluate the extracted singular term in (6.100) (Makarov, 2002), viz.

$$\begin{aligned}\frac{1}{4A^2}\int_S\int_S \frac{1}{|\mathbf{r}-\mathbf{r}'|}dS\,dS' &= \frac{1}{6\sqrt{a}}\ln\left[\frac{(a-b+\sqrt{ad})(b+\sqrt{ac})}{(-a+b+\sqrt{ad})(-b+\sqrt{ac})}\right] \\ &+ \frac{1}{6\sqrt{c}}\ln\left[\frac{(-b+c+\sqrt{cd})(b+\sqrt{ac})}{(b-c+\sqrt{cd})(-b+\sqrt{ac})}\right] \\ &+ \frac{1}{6\sqrt{d}}\ln\left[\frac{(a-b+\sqrt{ad})(-b+c+\sqrt{cd})}{(b-c+\sqrt{cd})(-a+b+\sqrt{ad})}\right]\end{aligned} \tag{6.101}$$

where $A$ refers to the area of the triangular element, $d = a - 2b + c$, $a = (\mathbf{r}_3 - \mathbf{r}_1)\cdot(\mathbf{r}_3 - \mathbf{r}_1)$, $b = (\mathbf{r}_3 - \mathbf{r}_1)\cdot(\mathbf{r}_3 - \mathbf{r}_2)$, and $c = (\mathbf{r}_3 - \mathbf{r}_2)\cdot(\mathbf{r}_3 - \mathbf{r}_2)$, in which the $\mathbf{r}_i$, $i = 1, 2, 3$ refer to the three vertices of the triangle.

However, for curved surfaces, the addition and subtraction of the $1/R$ (Song and Chew, 1995) term may not lead to a well-behaved first term in (6.100). This is true even for a Taylor series approximation of the curved subdomain about the singular point $(u_0, v_0)$.

Alternatively, it is possible to annihilate the singularity (Graglia, 1993) using suitable parametric transformations on $u$ and $v$ parameters in (6.100). Specifically, let us introduce the transformation from $(u, v)$ to $(\eta, \zeta)$ coordinates

$$(u, v) \rightarrow (\eta, \zeta) \tag{6.102}$$

leading to the integral

$$\iint du\,dv\, f(u,v) = \iint d\eta\, d\zeta\; f(\eta,\zeta)\; J(\eta,\zeta) \tag{6.103}$$

where $f(u, v) = g(u, v)/R(u, v)$ and

$$J(\eta,\zeta) = \begin{vmatrix} \frac{\partial u}{\partial \eta} & \frac{\partial u}{\partial \zeta} \\ \frac{\partial v}{\partial \eta} & \frac{\partial v}{\partial \zeta} \end{vmatrix} \tag{6.104}$$

is the *Jacobian* of the transformation. It is clear from (6.104) that if a transformation leads to a Jacobian that has a zero at the same location where $f(\eta, \zeta)$ is singular, the integrand $f(\eta, \zeta)J(\eta, \zeta)$ becomes nonsingular in the integration domain. Thus, the associated integrals can be evaluated in a standard manner. Next, we discuss such simple transformations to regularize the integral in (6.100).

### 6.3.4 Regularization for Triangular Subdomains

In this subsection, we consider the integration of (6.100) over a curvilinear triangle as in Figure 6-32. The integration will be done in the $(u, v)$ space, and it is assumed that the integrand is singular at $\mathbf{r}(x, y, z) = \mathbf{r}_0$ or at $(u, v) = (u_0, v_0)$ in the $(u, v)$ space. Two methods are next presented to annihilate the singularity at $(u, v) = (u_0, v_0)$. As such, the integral (6.100) can then be integrated using standard numerical techniques.

#### 6.3.4.1 Annihilation Method I for Triangular Subdomains

To evaluate (6.103) using Method I, the unit triangle to the left of Figure 6-32 is subdivided into three subtriangles that have a common vertex at $(u_0, v_0)$, as shown in Figure 6-33. Each subtriangle is transformed to another unit triangle via the $(u, v) \to (\eta, \zeta)$ mapping

$$\begin{bmatrix} u \\ v \end{bmatrix} = \eta \begin{bmatrix} u_{13} \\ v_{13} \end{bmatrix} + \zeta \begin{bmatrix} u_{23} \\ v_{23} \end{bmatrix} + \begin{bmatrix} u_3 \\ v_3 \end{bmatrix}. \tag{6.105}$$

Here, $u_{ij} = u_i - u_j$, $v_{ij} = v_i - v_j$ in which $i$, $j$ denote the vertex numbers of the subtriangles (see Figures 6-34 through 6-36). As expected, the transformation is

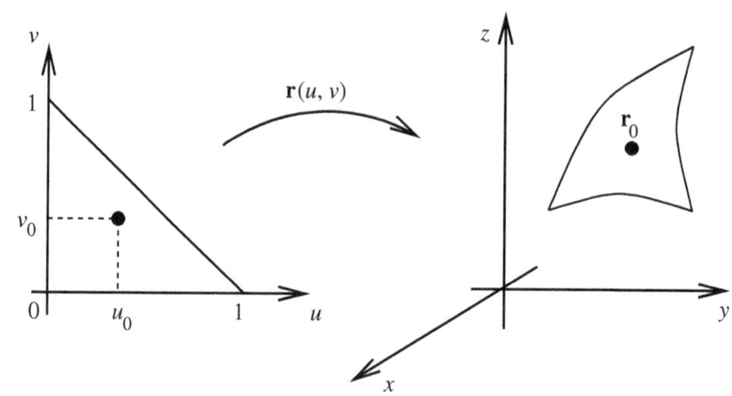


**FIGURE 6-32** Parametric mapping of a curvilinear triangle to a unit triangle in the $(u, v)$ space.

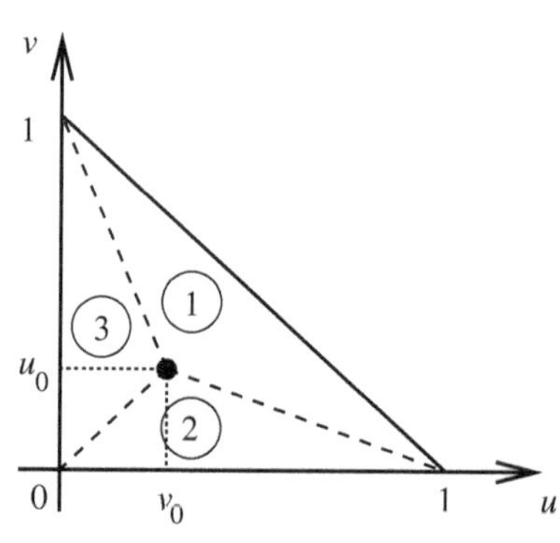


**FIGURE 6-33** Subdivision of the parametric unit triangle to three subtriangles for singularity annihilation.

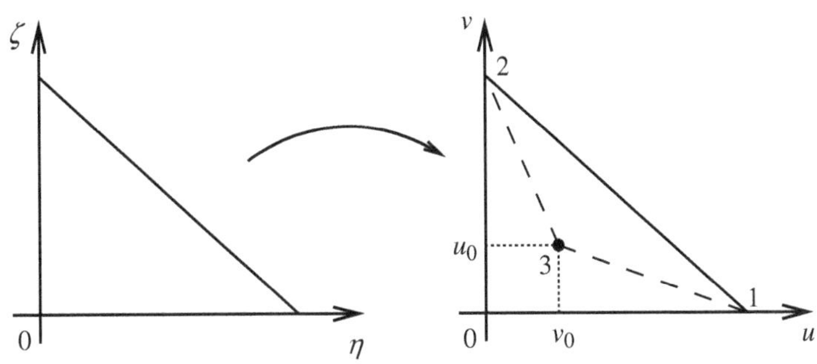


**FIGURE 6-34** Mapping of subtriangle 1 in Figure 6-33 to the $(\eta, \zeta)$ space.

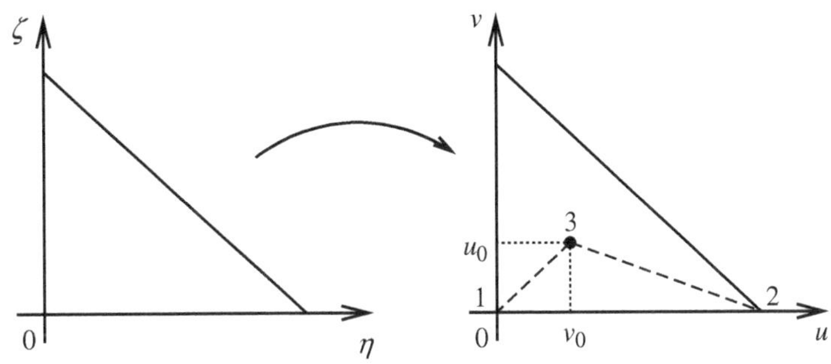


**FIGURE 6-35** Mapping of subtriangle 2 in Figure 6-33 to the $(\eta, \zeta)$ space.

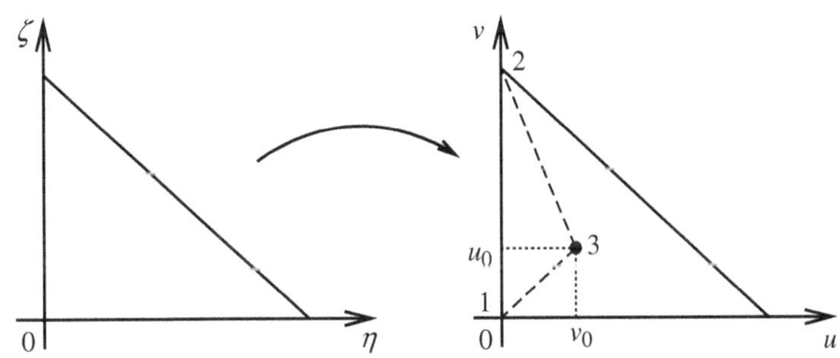


**FIGURE 6-36** Mapping of subtriangle 3 in Figure 6-33 to the $(\eta, \zeta)$ space.

different for each subtriangle since the vertices at $(u_i, v_i)$ are different. It is essential that vertex 3 of each subtriangle be associated with the singular point $(u_0, v_0)$. As such, the singularity is now shifted to the origin of each triangle in the $(\eta, \zeta)$ space. The corresponding integral (6.100) or (6.103) can now be written as

$$\int_{\triangle} du\,dv\; f(u, v) = \int_{\triangle_1} du\,dv f(u, v) + \int_{\triangle_2} du\,dv\; f(u, v) + \int_{\triangle_3} du\,dv\; f(u, v). \tag{6.106}$$

Introducing the transformations just described, we have

$$\int_{\triangle_i} du\,dv\; f(u, v) = \int_{\triangle} d\eta\, d\zeta\; f_i(\eta, \zeta) J_i(\eta, \zeta) \tag{6.107}$$

in which

$$J_i(\eta, \zeta) = (u_{13}v_{23} - u_{23}v_{13}). \tag{6.108}$$

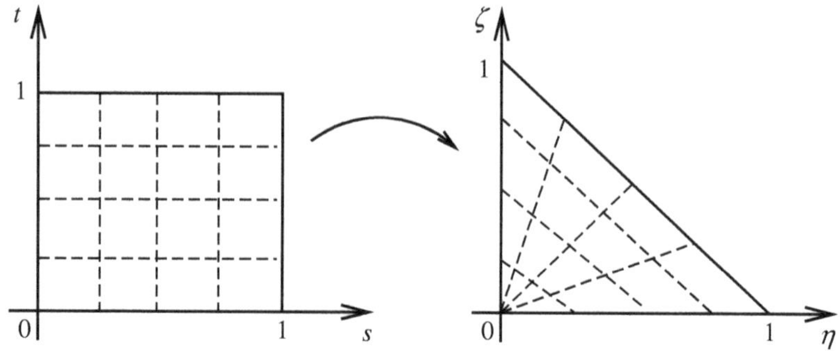


**FIGURE 6-37** Mapping from a triangle to a rectangular domain to annihilate the singularity at the origin.

Thus, of interest is the evaluation of the integral

$$\int_{\triangle} du\, dv\, f(u, v) = \int_{\triangle} d\eta\, d\zeta\, g(\eta, \zeta). \tag{6.109}$$

Here $g(\eta, \zeta) = f_i(\eta, \zeta) J_i(\eta, \zeta)$ is still singular in the $(\eta, \zeta)$ parametric space, but the singularity is now at the origin regardless of $(u_0, v_0)$ point location.

To completely annihilate the singularity in (6.109), we proceed to introduce the transformation in Figure 6-37. Specifically, the triangle in $(\eta, \zeta)$ space is transformed to a unit square in the $(s, t)$ domain. The specific transformation for doing so is

$$\begin{aligned} \eta &= st \\ \zeta &= s(1-t) \end{aligned} \tag{6.110}$$

with the corresponding Jacobian given by

$$J(s, t) = \begin{vmatrix} t & s \\ (1-t) & -s \end{vmatrix} = -st - s(1-t) = -s \tag{6.111}$$

Since the origin of the $(\eta, \zeta)$ domain is the map of $s = 0$ line in the $(s, t)$ domain, this Jacobian has a zero at the origin of the $(\eta, \zeta)$ domain to cancel the singularity. Hence, the integral in the $(s, t)$ domain becomes

$$\int_{\triangle} du\, dv\, f(u, v) = \int_{\square} ds\, dt\, G(s, t)\, J(s, t). \tag{6.112}$$

Because the right hand side integral is nonsingular, it can be evaluated using standard numerical techniques including Gaussian quadrature methods.

#### 6.3.4.2 Annihilation Method II for Triangular Subdomains

An alternative regularization technique is to map the $(u, v)$ triangle directly onto a unit square using the $(u, v) \to (\eta, \zeta)$ transformation.

$$u = \eta\zeta \tag{6.113}$$

$$v = \eta(1-\zeta). \tag{6.114}$$

The Jacobian is then

$$J(\eta, \zeta) = -\eta \tag{6.115}$$

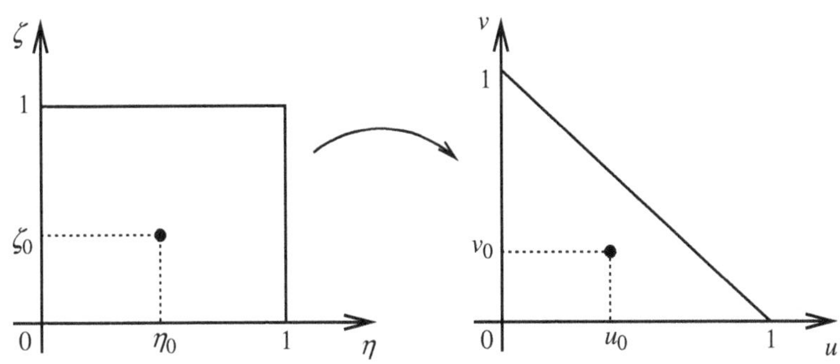


**FIGURE 6-38** Mapping using the transformation in (6.113) and (6.114).

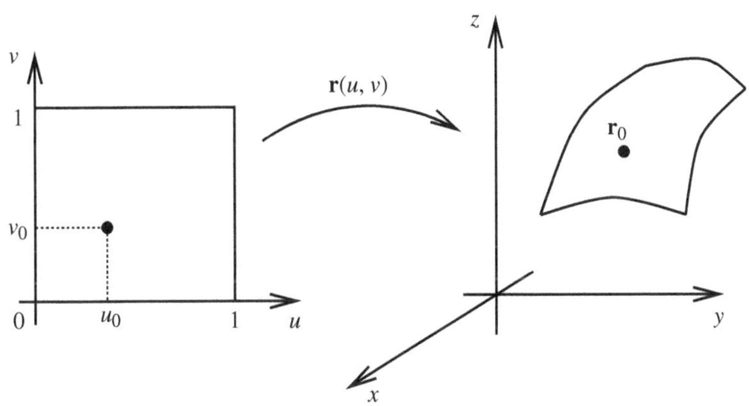


**FIGURE 6-39** Mapping of a curved rectangular patch to a parametric unit square.

and the singular integral then becomes

$$\int_{\triangle} du\, dv\, f(u, v) = \int_{\square} d\eta\, d\zeta\, f(\eta, \zeta)\, J(\eta, \zeta) \tag{6.116}$$

$$= \int_{\square} d\eta\, d\zeta\, g(\eta, \zeta). \tag{6.117}$$

Of course, $g(\eta, \zeta)$ is still singular at $(\eta_0, \zeta_0)$ in the $(\eta, \zeta)$ (see Figure 6-38). However, given that the integration is over a rectangular domain, the annihilation of the singularity is easier, as explored in the next section.

### 6.3.5 Annihilation Transforms for Square Subdomains

Referring to Figure 6-17, we recall that any curvilinear quadrilateral can be transformed to a unit square. This is again depicted in Figure 6-39 for convenience. We also recall that the observation point $\mathbf{r}_0$ where the kernel singularity occurs is not mapped to $(u_0, v_0)$. In the following, we present four methods to annihilate this singularity.

#### 6.3.5.1 Annihilation Method I for Quadrilateral Subdomains

As for the triangles, this approach amounts to subdividing the unit square into four smaller subtriangles. This is depicted in Figure 6-40, where each subtriangle is shown to have a common vertex at $(u_0, v_0)$.

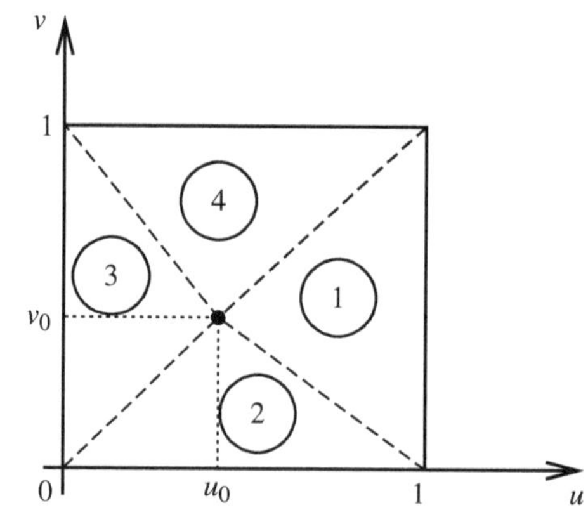


**FIGURE 6-40** Subdivision of the unit square into four subtriangles for numerical evaluation.

For each subtriangle we now proceed to use the same parameter transformations used in (6.105) and repeated here

$$\begin{bmatrix} u \\ v \end{bmatrix} = \eta \begin{bmatrix} u_{13} \\ v_{13} \end{bmatrix} + \zeta \begin{bmatrix} u_{23} \\ v_{23} \end{bmatrix} + \begin{bmatrix} u_3 \\ v_3 \end{bmatrix} \tag{6.118}$$

with

$$J(u, v) = u_{13}v_{23} - u_{23}v_{13}. \tag{6.119}$$

The singular integral (6.103) is thus transformed to

$$\int_{\square} du\, dv\, f(u, v) = \int_{\triangle} d\eta\, d\zeta\, G(\eta, \zeta) \tag{6.120}$$

where

$$g(\eta, \zeta) = \sum_{i=1}^{4} f_i(\eta, \zeta) J_i(\eta, \zeta) \tag{6.121}$$

in which $J_i$ is given as in (6.119). Method I for triangular subdomains in Section 6.3.4 can then be applied to annihilate the singularity at the origin of the $(\eta, \zeta)$ parametric space.

#### 6.3.5.2 Annihilation Method II for Quadrilateral Subdomains

For this method, we propose the $(u, v) \to (s, t)$ transformation having the Jacobian

$$J(s, t) = \frac{\partial u}{\partial s}\frac{\partial v}{\partial t}. \tag{6.122}$$

To annihilate the singularity at $(u_0, v_0)$, the Jacobian must have a zero at $(u_0(s_0), v_0(t_0))$. This can be achieved by separately setting

$$\left.\frac{\partial u}{\partial s}\right|_{s=s_0} = 0 \quad \text{and} \quad \left.\frac{\partial v}{\partial t}\right|_{t=t_0} = 0 \tag{6.123}$$

where $(s_0, t_0)$ refers to the point $(u_0, v_0)$ in the $(u, v)$ space. An additional constraint on the transformation is that the unit square in the $(s, t)$ domain should be mapped to the unit square in the $(u, v)$ domain.

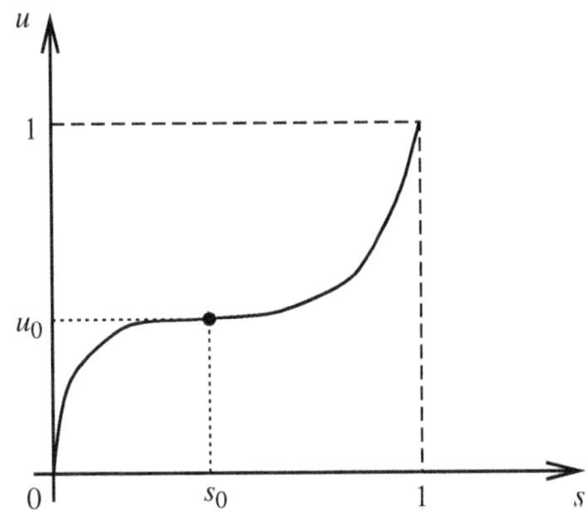


**FIGURE 6-41** Transformation applied on each parameter.

The previous constraints, namely, $\frac{\partial u}{\partial s}\Big|_{s=s_0} = 0$ along with $u(s = 0) = 0$ and $u(s = 1) = 1$, can be satisfied by a function $u(s)$ that has a generic form as depicted in Figure 6-41. The lowest possible order polynomial satisfying the previous criteria is of the form

$$u(s) = \frac{(s - s_0)^3 + s_0^3}{(1 - s_0)^3 + s_0^3} \tag{6.124}$$

giving

$$\frac{\partial u}{\partial s} = \frac{3(s - s_0)^2}{(1 - s_0)^3 + s_0^3}. \tag{6.125}$$

Then from $\partial u/\partial s|_{s=s_0} = 0$, it follows that

$$u_0 = \frac{s_0^3}{(1 - s_0)^3 + s_0^3} \Rightarrow s_0 = \frac{1}{1 + \sqrt[3]{\dfrac{1}{u_0} - 1}}. \tag{6.126}$$

Similarly, for $v$, we have

$$v(t) = \frac{(t - t_0)^3 + t_0^3}{(1 - t_0)^3 + t_0^3} \tag{6.127}$$

$$\frac{\partial v}{\partial t} = \frac{3(t - t_0)^2}{(1 - t_0)^3 + t_0^3}. \tag{6.128}$$

On enforcing $\partial v/\partial t|_{t=t_0} = 0$, it follows that

$$v_0 = \frac{s_0^3}{(1 - t_0)^3 + t_0^3} \Rightarrow t_0 = \frac{1}{1 + \sqrt[3]{\dfrac{1}{v_0} - 1}}. \tag{6.129}$$

Thus, the Jacobian of the transformation becomes

$$J(s,t) = J_s(s)J_t(t) = \frac{9(s-s_0)^2(t-t_0)^2}{\left[(1-s_0)^3+s_0^3\right]\left[(1-t_0)^3+t_0^3\right]} \tag{6.130}$$

and the associated singular integral is transformed into the nonsingular integral

$$\int_{\square} du\, dv\, f(u,v) = \int_{\square} ds\, dt\, f(s,t)\, J(s,t). \tag{6.131}$$

We remark that due to the Jacobian the integrand $f(s,t)J(s,t)$ is zero on the $t = t_0$ and $s = s_0$ lines. As such, the Jacobian cancels the singularity of $f(s,t)$ at $(s_0, t_0)$ corresponding to $(u_0, v_0)$ in the original $(u, v)$ domain.

### 6.3.5.3 Annihilation Method III for Quadrilateral Subdomains

This method is a mixture of Methods I and II. The integration domain (unit square) in the $(u, v)$ parameter space is first divided into four rectangular subdomains having a common node $u = u_0$, $v = v_0$ as depicted in Figure 6-42. Each subdomain is then mapped as a unit square in the $(\eta, \zeta)$ parametric space. Figure 6-43 depicts such a mapping. The associated transformation and the Jacobian are

$$\begin{aligned} u &= u_0(1-\eta) \\ v &= v_0(1-\zeta) \\ J_1(\eta,\zeta) &= u_0 v_0. \end{aligned} \tag{6.132}$$

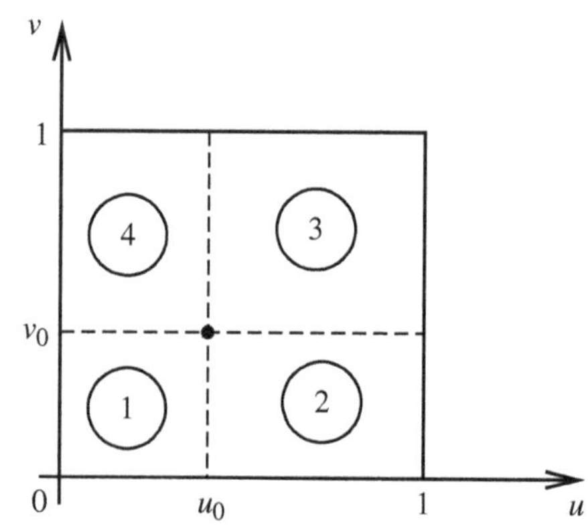


**FIGURE 6-42** Subdivision of the unit square for annihilation Method III.

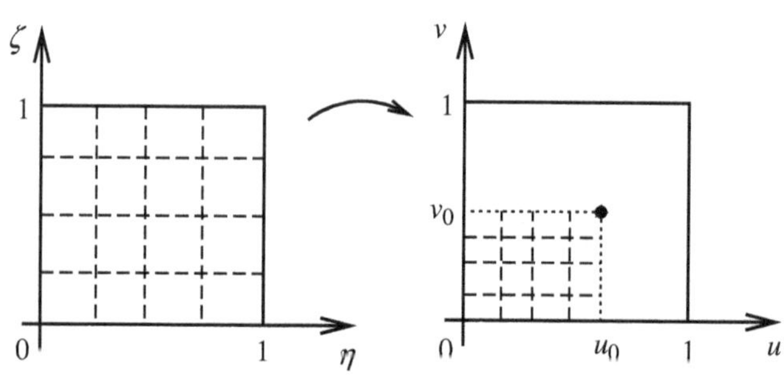


**FIGURE 6-43** Transformation for the first subdomain.

Similarly, the other subdomain the transformations are

$$
\begin{aligned}
u &= (1 - u_0)\eta \\
v &= v_0(1 - \zeta) \\
J_2(\eta, \zeta) &= (1 - u_0)v_0
\end{aligned}
\tag{6.133}
$$

$$
\begin{aligned}
u &= (1 - u_0)\eta \\
v &= (1 - v_0)\zeta \\
J_3(\eta, \zeta) &= (1 - u_0)(1 - v_0)
\end{aligned}
\tag{6.134}
$$

$$
\begin{aligned}
u &= u_0(1 - \eta) \\
v &= (1 - v_0)\zeta \\
J_4(\eta, \zeta) &= u_0(1 - v_0).
\end{aligned}
\tag{6.135}
$$

As the singularity is at the origin of the $(\eta, \zeta)$ space, we can again employ the transformation given by (6.124) and (6.127) with $(s, t) = (0, 0)$. Doing so we have

$$\eta(s) = s^3, \qquad \zeta(t) = t^3 \tag{6.136}$$

and

$$J(s) = 3s^2, \qquad J(t) = 3t^2 \tag{6.137}$$

for the Jacobian.

However, the transformations

$$\eta(s) = s^2, \qquad \zeta(t) = t^2 \tag{6.138}$$

with the corresponding Jacobians

$$J(s) = 2s, \qquad J(t) = 2t \tag{6.139}$$

work equally well in a general, nonsingular integral.

#### 6.3.5.4 Annihilation Method IV for Quadrilateral Subdomains

This method is a variation of Method III. Specifically, the first transformation is exactly as in Method III. That is, the unit $(u, v)$ square is subdivided again into four subdomains as in Figure 6-42. However, as opposed to Method III, the square domain in the $(\eta, \zeta)$ space is divided into two triangular subdomains separated by the diagonal $\eta = \zeta$ line (see Figure 6-44). Each triangular subdomain can then be

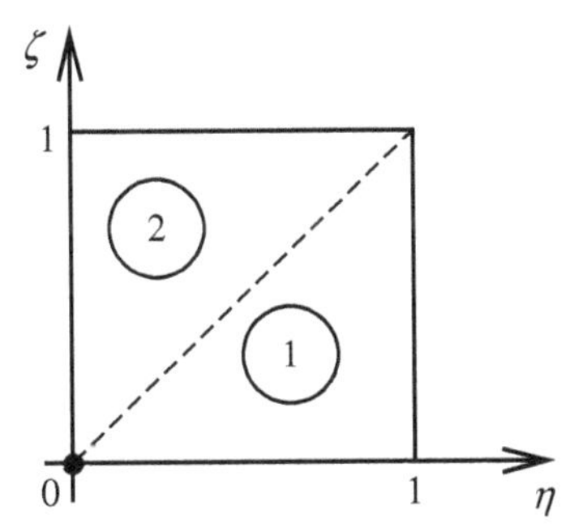


**FIGURE 6-44** Subdivision of the unit square for application of annihilation Method IV.

considered as the map of a unit triangle in the $(\alpha, \beta)$ parametric domain. The latter $(\eta, \zeta) \to (\alpha, \beta)$ transformation is given by

$$\begin{bmatrix} \eta \\ \zeta \end{bmatrix} = \alpha \begin{bmatrix} \eta_{13} \\ \zeta_{13} \end{bmatrix} + \beta \begin{bmatrix} \eta_{23} \\ \zeta_{23} \end{bmatrix} + \begin{bmatrix} \eta_3 \\ \zeta_3 \end{bmatrix} \tag{6.140}$$

having the Jacobian

$$J(\alpha, \beta) = \eta_{13}\zeta_{23} - \eta_{23}\zeta_{13}. \tag{6.141}$$

The singular integral whose singularity is at $(\eta, \zeta) = (0, 0)$ becomes

$$\int_{\square} d\eta\, d\zeta\, G(\eta, \zeta) = \int_{\triangle} d\alpha\, d\beta\, [g_1(\alpha, \beta) J_1(\alpha, \beta) + g_2(\alpha, \beta) J_2(\alpha, \beta)]\,. \tag{6.142}$$

To evaluate (6.142) numerically, we proceed as in Method I for triangular subdomains. To do so, we must again transform the singular integrand $\sum_{i=1}^{2} g_i(\alpha, \beta) J_i(\alpha, \beta)$ in the $(\alpha, \beta)$ space to a nonsingular integrand in $(\mu, \nu)$ parameter space. The transformation is simply given by

$$\alpha = \mu\nu \tag{6.143}$$

$$\beta = \mu(1 - \nu) \tag{6.144}$$

with the Jacobian given by $J(\alpha, \beta) = -\mu$.

### 6.3.6 Numerical Integration

Other variations of the aforementioned techniques to annihilate the singularity of the Green's function on curved subdomains are, of course, possible (Duffy, 1982; Graglia, 1987; Khayat and Wilton, 2005). From experience, it can be stated that for triangular subdomains Method I performs best. Method I for quadrilateral subdomains also performs best as it requires the least amount of computational resources.

Upon reaching a nonsingular integral using any of the methods presented in Subsections 6.3.3–6.3.5, a fixed point Gaussian quadrature integration is acceptable to evaluate the resulting integrals. Of course, greater accuracy can be achieved by increasing the integration order.

Once the matrix elements are computed, the resulting matrix system must be solved to determine the unknown column $\{x_i\}$ for $i = 1, 2, \ldots, N$. A direct solution method, such as Gaussian elimination or lower-upper (LU) decomposition can always be used to solve for $\{x_i\}$. However, the computational resources required by a direct solver are very high ($O(N^3)$) as was noted in Chapter 4. As iterative algorithms (Saad, 1996; Topsakal, Kindt, Sertel, and Volakis, 2001) have $O(N^2)$ complexity (per iteration), they are much more attractive for matrix systems where $N$ can reach several millions of unknowns. The hope with iterative methods is to achieve convergence in few iterations. Several preconditioning methods can be employed to achieve quick convergence (Sertel and Volakis, 2000). In addition, the fast multipole method (FMM) (Coifman, Rokhlin, and Wandzura, 1993) and the adaptive integral method (AIM) (Bleszynski, Bleszynski, and Jaroszewicz, 1996) are ways to speed up the matrix-vector products executed in any iterative solver. These are examined in Chapter 7.

### 6.3.7 Source Modeling and Antenna Applications

Two types of excitation will be used in the examples considered next. Some source or excitation models were given in Chapter 4 (Section 4.4) and the reader is also referred to that chapter as it relates to wire-type antennas and scattering.

#### 6.3.7.1 Plane Wave Incidence for RCS Calculations (Monostatic and Bistatic RCS)

For radar scattering applications, the excitation is typically a plane wave illumination (see Section 4.4.3 and Figure 6-1) of the form

$$\begin{aligned}\mathbf{E}^i &= E_0^i \hat{e} e^{-jk_0 \hat{k}^i \cdot \mathbf{r}} \\ \mathbf{H}^i &= Y_0 \hat{k}^i \times \mathbf{E}^i = -Y_0 \hat{r}^i \times \mathbf{E}^i\end{aligned} \tag{6.145}$$

where $\hat{e} = \hat{x}e_x + \hat{y}e_y + \hat{z}e_z$ (see Section 1.3),

$$\begin{aligned}\hat{k}^i &= -(\hat{x}\cos\phi_i \sin\theta_i + \hat{y}\sin\phi_i \sin\theta_i + \hat{z}\cos\theta_i) \\ &= -\hat{r}^i\end{aligned} \tag{6.146}$$

is the incidence field direction, and $(\theta_i, \phi_i)$ are the spherical angles as in Figure 2-2. In general, the incident field polarization can be arbitrary, viz.

$$\begin{bmatrix} \hat{E}_0^i e_x \\ \hat{E}_0^i e_y \\ \hat{E}_0^i e_z \end{bmatrix} = E_\theta^i \begin{bmatrix} \cos\phi_i \cos\theta_i \\ \sin\phi_i \cos\theta_i \\ -\sin\theta_i \end{bmatrix} + E_\phi^i \begin{bmatrix} -\sin\phi_i \\ \cos\phi_i \\ 0 \end{bmatrix}. \tag{6.147}$$

Once the current density due to the excitation (6.145) is computed via the matrix system solution, the RCS is given by (4.61) and repeated here:

$$\sigma^{3D}(\theta, \phi) = \lim_{r\to\infty} 4\pi r^2 \frac{|\mathbf{E}^s(\theta, \phi)|^2}{|\mathbf{E}^i|^2} = \lim_{r\to\infty} 4\pi r^2 \frac{|\mathbf{H}^s(\theta, \phi)|^2}{|\mathbf{H}^i|^2}. \tag{6.148}$$

By definition, $\mathbf{E}^s(\theta, \phi)$ refers to the scattered field given in (2.77) or more concisely as

$$\mathbf{E}^{FF}(\mathbf{J}) = \frac{jk_0}{4\pi} \frac{e^{-jk_0 r}}{r} \hat{r} \times \int_{s'} \mathbf{J}(\mathbf{r}') e^{jk_0 \mathbf{r}' \cdot \hat{r}}\, dr' \tag{6.149}$$

$$\mathbf{H}^{FF}(\mathbf{J}) = -\frac{jk_0}{4\pi} \frac{e^{-jk_0 r}}{r} \hat{r} \times \hat{r} \times \int_{s'} \mathbf{J}(\mathbf{r}') e^{jk_0 \mathbf{r}' \cdot \hat{r}}\, dr' \tag{6.150}$$

where $\hat{r} = \hat{x}\cos\phi \sin\theta + \hat{y}\sin\phi \sin\theta + \hat{z}\cos\theta$ is the observation direction, and $\mathbf{r}' = x'\hat{x} + y'\hat{y} + z'\hat{z}$.

Typically, one is interested in the RCS for both the $\theta$ (vertical, $E_\phi^i = 0$) and $\phi$ (horizontal, $E_\theta^i = 0$) polarized incident waves, including the associated cross-polarization terms. Thus, the components for $\sigma^{3D}$ can be more completely written in a matrix form as

$$\overline{\sigma}^{3D} = \begin{bmatrix} \sigma_{\theta\theta} & \sigma_{\theta\phi} \\ \sigma_{\phi\theta} & \sigma_{\phi\phi} \end{bmatrix} \tag{6.151}$$

where

$$\sigma_{\gamma,\delta}(\theta,\phi) = \lim_{r\to\infty} 4\pi r^2 \frac{\left|\mathbf{E}_\gamma^s(\theta,\phi)\right|^2}{\left|\mathbf{E}_\delta^i\right|^2} \qquad \text{for } \gamma,\delta \in \{\theta,\phi\}. \tag{6.152}$$

### 6.3.7.2 Delta Gap and Current Excitation

For antenna analysis, the excitations can take the form of a delta gap or magnetic frill model. These were discussed in Chapter 4 and can be used here as well for general 3-D antenna structures.

Consider the simple model of a wire antenna modeled by a quadrilateral mesh, shown in Figure 6-45. We may proceed with the MoM procedure as outlined in Section 6.3 and generate the MoM matrix for all basis functions supported by the wire strip. A way to introduce the delta gap excitation in this geometry, we force the $p$th basis function of the strip (see Figure 6-45) to have a constant value of $x_p = 1$.

Specifying the $p$th unknown eliminates the $p$th matrix evaluation and makes the excitation of the form

$$b_j = Z_{jp} \qquad \text{for } j = 1,\ldots,p-1,p+1,\ldots,N. \tag{6.153}$$

The remaining matrix is then of the form

$$\begin{bmatrix} Z_{11} & \cdots & Z_{1(p-1)} & Z_{1(p+1)} & \cdots & Z_{1N} \\ \vdots & \ddots & \vdots & \vdots & \ddots & \vdots \\ Z_{(p-1)1} & \cdots & Z_{(p-1)(p-1)} & Z_{(p-1)(p+1)} & \cdots & Z_{(p-1)N} \\ Z_{(p+1)1} & \cdots & Z_{(p+1)(p-1)} & Z_{(p+1)(p+1)} & \cdots & Z_{(p+1)N} \\ \vdots & \ddots & \vdots & \vdots & \ddots & \vdots \\ Z_{N1} & \cdots & Z_{N(p-1)} & Z_{N(p+1)} & \cdots & Z_{NN} \end{bmatrix} \cdot \begin{bmatrix} x_1 \\ \vdots \\ x_{(p-1)} \\ x_{(p+1)} \\ \vdots \\ x_N \end{bmatrix} = \begin{bmatrix} Z_{1p} \\ \vdots \\ Z_{(p-1)p} \\ Z_{(p+1)p} \\ \vdots \\ Z_{Np} \end{bmatrix} \tag{6.154}$$

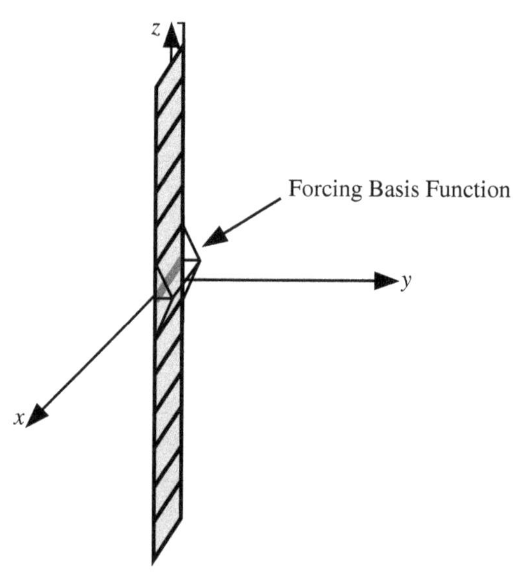


FIGURE 6-45 ■ PEC strip approximation of a wire antenna.

with $N - 1$ unknowns and $N - 1$ equations. In antenna analysis, it is desired to compute the voltage across the excitation terminals (the $p$th element). This voltage can simply be found by the path integration. Alternatively, we may need the total electric field integrated over the $p$th testing function to obtain

$$V_p = \sum_{i=1}^{N} Z_{pi} x_i \tag{6.155}$$

with $x_p = 1$.

Having the voltage across the feed, the input impedance is obtained from $Z_{in} = V_p/x_p$ as similarly done in (4.58). The same approach may also be adapted in calculating mutual impedance between neighboring antenna elements. However, accurate feed models typically require fine meshes for more accurate results.

#### 6.3.7.3 Aperture Excitations

For problems associated with waveguide feeds or open aperture excitations (where the incident field may be expressed by a known function or mode over the aperture), it is necessary to introduce the known aperture field representation in generating the right-hand excitation vector. For coaxial feeds attached to aperture or patch antennas, approximate equations must be devised (see Figure 6-46). As done in Chapter 4 for a coaxial feed, we introduce the magnetic current excitation of the form

$$\mathbf{M}^{coax} = \mathbf{E}^{coax} \times \hat{n} \qquad \text{with } \mathbf{E}^{coax} = \frac{V_i}{2}\hat{\rho}\ln(b/a) \tag{6.156}$$

valid over $a < \rho < b$. As in Section 4.4.2, this magnetic current is much like the magnetic frill generator used for wire antennas. The electric and magnetic fields

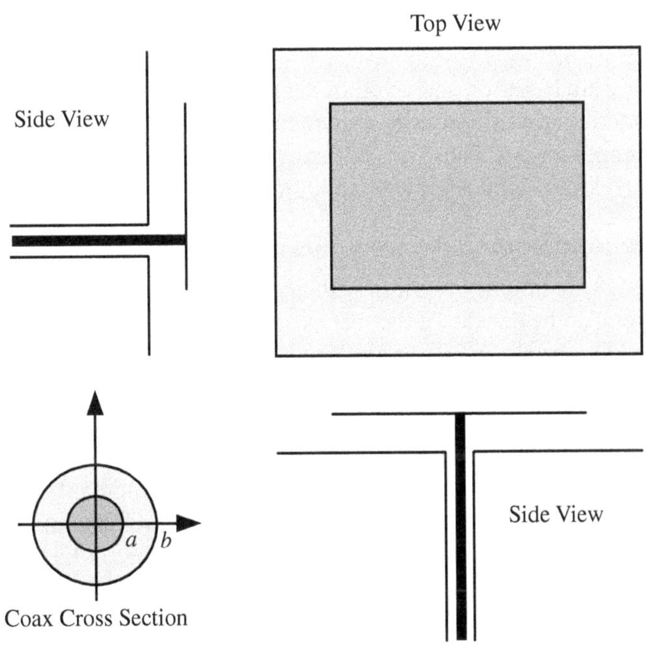


**FIGURE 6-46** Coaxial feed model for a simple patch antenna.

generated by this magnetic current

$$\mathbf{E}^i = \int_{s'} \nabla' G(\mathbf{r}, \mathbf{r}') \times \mathbf{M}^{coax}(\mathbf{r}')\, dr' \tag{6.157}$$

$$\mathbf{H}^i = jk_0 \int_{s'} G(\mathbf{r}, \mathbf{r}')\mathbf{M}^{coax}(\mathbf{r}')\, dr' + \frac{j}{k_0}\nabla \int_{s'} G(\mathbf{r}, \mathbf{r}')\nabla' \cdot \mathbf{M}^{coax}(\mathbf{r}')\, dr'. \tag{6.158}$$

These can then be used as incident field excitations identifying the right-hand side of the matrix equations.

## 6.3.8 Matrix Solution Methods

So far, we have discussed the MoM procedure to convert integral representations into a linear system of equations. The governing integro-differential equations were discretized concurrently representing the solution space as a discrete mesh of connected finite elements with the unknowns as a summation of weighted basis functions. This is the most common method for solving problems involving arbitrary geometries. Other MoM discretizations using entire domain basis functions, have also been employed (Popović and Kolundžija, 1994; Notaros and Popovic, 1996; Notaros, Popovic, Weem, Brown, and Popovic, 2001). However, these require some level of a priori knowledge of the field behavior. Although simpler specialized discretizations, such as mode matching or entire domain bases, result in much smaller systems, the applicability of such MoM implementations are rather limited.

Typically, when the size of the matrix system is less that a few thousand, the system is solved via a direct method, such as LU decomposition as discussed in Chapter 4 (Section 4.3). Alternatively, for larger problems iterative solvers provide a much faster solution that direct solvers (see Section 4.11). For large problems, fast and efficient methods that minimize storage of the matrix system are necessary as well. The fast methods will be discussed in Chapter 7 and they rely on iterative solvers as they are based on accelerating the matrix-vector products.

Direct and iterative algorithms were mentioned in Chapter 4. Next, we discuss possible preconditioners. These are critically needed in the iterative solvers. They are particularly essential for not well-conditioned matrix systems.

### 6.3.8.1 Preconditioning Approaches

For large-scale simulations (possibly with geometrical surface details, e.g., antenna arrays on aircraft), the density of the surface mesh is likely to be highly nonuniform. Nonuniform meshes are known to produce ill-conditioned MoM matrix equations. Such meshes are rather typical in most realistic antenna and scattering problems. Therefore, use of preconditioners to lower the matrix condition number is essential.

The *diagonal* preconditioner is simple and leads to significant convergence improvements. However, it does so only for diagonally dominant matrices. On the other hand, *block diagonal* preconditioners are more robust but require renumbering of the mesh or matrix rearranging so that the dominant matrix terms are clustered around the diagonal. This can be done more readily for 2-D problems but is quite difficult in 3-D. Alternatively, with iterative solvers, sparse matrix preconditioners can be used instead of block preconditioners.

A simple way to generate an effective preconditioning matrix is to execute an LU decomposition on the near-field (near-diagonal band) matrix. Depending on the sparsity of the near-field matrix, this LU process may require a significant CPU time. Thus, for large matrix systems this is not an attractive approach. Next, we mention some popular preconditioning approaches. To avoid the large CPU times in generating the preconditioner, we can instead employ an incomplete LU (ILU) factorization of the matrix (Saad, 1996). The ILU algorithm is identical to a direct LU algorithm, but it avoids fill-ins of elements in the decomposed LU matrices. Thus, much less CPU time is required. The ILU (and its variants) can be found in (Saad, 1996). A pseudo code is repeated herein for completeness:

```
for i = 2,...,n, do:
  for k = 1,...,i-1 and for (i,k) in NZ(Z) do:
    compute z_ik = z_ik/z_kk
    for j = k+1,...,n and for (i,j) in NZ(Z) do:
      compute z_ij = z_ij - z_ik z_kj
    end do
  end do
end do
```

Here NZ(Z) is the sparsity pattern of the near-field matrix [Z]. As the conventional LU decomposition algorithm is applied only to the nonzero entries of the matrix, the memory is not affected, and the sparsity pattern of the stored ILU matrix is the same as the original one.

Recently, approximate inverse preconditioners based on direct inversion of subsections of the MoM matrix were proposed. Their implementation details and performance evaluations can be found in (Carr, Bleszynski, and Volakis, 2004).

#### 6.3.8.2 Multiplicative Calderon Preconditioner

More recently, a very effective and computationally cheap multiplicative preconditioner based on the self-regularizing properties of the EFIE was introduced (Christiansen and Nedelec, 2003; Adams, 2004; Andriulli, Cools, Bagci, Olyslager, Buffa, Christiansen, and Michielssen, 2008; Stephanson and Lee, 2009). This preconditioner relies on the *Calderon projector* (Hsiao and Kleinman, 1997) which translates from (2.102), repeated here for convenience:

$$\mathbf{E}(\mathbf{r}) = \int_V \left\{\left[\nabla \times \overline{\boldsymbol{\Gamma}}(\mathbf{r},\mathbf{r}')\right] \cdot \mathbf{M}(\mathbf{r}') + jkZ\overline{\boldsymbol{\Gamma}}(\mathbf{r},\mathbf{r}') \cdot \mathbf{J}(\mathbf{r}')\right\} dv' \quad (6.159a)$$

$$\mathbf{H}(\mathbf{r}) = \int_V \left\{jkY\overline{\boldsymbol{\Gamma}}(\mathbf{r},\mathbf{r}') \cdot \mathbf{M}(\mathbf{r}') - \left[\nabla \times \overline{\boldsymbol{\Gamma}}(\mathbf{r},\mathbf{r}')\right] \cdot \mathbf{J}(\mathbf{r}')\right] dv'. \quad (6.159b)$$

Taking the "$\hat{n}\times$" of both sides and using $\mathbf{E} \times \hat{n} = \mathbf{M}$ and $\hat{n} \times \mathbf{H} = \mathbf{J}$ leads to

$$\begin{aligned} \hat{n} \times \mathbf{E}(\mathbf{r}) = \ & \hat{n} \times \int_V \left\{\left[\nabla \times \overline{\boldsymbol{\Gamma}}(\mathbf{r},\mathbf{r}')\right] \cdot (-\hat{n} \times \mathbf{E}(\mathbf{r}'))\right. \\ & \left. + \ jk\overline{\boldsymbol{\Gamma}}(\mathbf{r},\mathbf{r}') \cdot (Z\hat{n} \times \mathbf{H}(\mathbf{r}'))\right\} dv' \end{aligned} \quad (6.160a)$$

$$\begin{aligned} Z\hat{n} \times \mathbf{H}(\mathbf{r}) = \ & \hat{n} \times \int_V \left\{jk\overline{\boldsymbol{\Gamma}}(\mathbf{r},\mathbf{r}') \cdot (-\hat{n} \times \mathbf{E}(\mathbf{r}'))\right. \\ & - \left[\nabla \times \overline{\boldsymbol{\Gamma}}(\mathbf{r},\mathbf{r}')\right] \cdot (Z\hat{n} \times \mathbf{H}(\mathbf{r}'))\right] dv'. \end{aligned} \quad (6.160b)$$

We can introduce the following operator notations using an arbitrary vector field $\mathbf{X}(\mathbf{r}')$

$$\mathcal{T}\{\mathbf{X}(\mathbf{r})\} = jk\hat{n} \times \int_V \overline{\mathbf{\Gamma}}(\mathbf{r}, \mathbf{r}') \cdot \mathbf{X}(\mathbf{r}')\, dv' \tag{6.161}$$

and

$$\frac{\mathbf{X}(\mathbf{r})}{2} - \mathcal{K}\{\mathbf{X}(\mathbf{r})\} = \hat{n} \times \int_V \left[\nabla \times \overline{\mathbf{\Gamma}}(\mathbf{r}, \mathbf{r}')\right] \cdot \mathbf{X}(\mathbf{r}')\, dv' \tag{6.162}$$

where

$$\mathcal{K}\{\mathbf{X}(\mathbf{r})\} = \hat{n} \times \int_V \nabla G(\mathbf{r}, \mathbf{r}') \times \mathbf{X}(\mathbf{r}')\, dv' \tag{6.163}$$

in which $G(\mathbf{r}, \mathbf{r}')$ is the scalar Green's function as defined in Chapter 2. Now, we can write (6.160) in a compact, matrix form

$$\begin{pmatrix} \hat{n} \times \mathbf{E} \\ Z\hat{n} \times \mathbf{H} \end{pmatrix} = \begin{pmatrix} \frac{\mathcal{I}}{2} - \mathcal{K} & \mathcal{T} \\ -\mathcal{T} & \frac{\mathcal{I}}{2} - \mathcal{K} \end{pmatrix} \begin{pmatrix} \hat{n} \times \mathbf{E} \\ Z\hat{n} \times \mathbf{H} \end{pmatrix} \tag{6.164}$$

where $\mathcal{I}$ is the identity operator. The $2 \times 2$ matrix operator in (6.164) is a *projection* operator mapping the tangential field quantities $\hat{n} \times \mathbf{E}$ and $\hat{n} \times \mathbf{H}$ onto themselves. The two Calderon identities can be obtained by multiplying both sides of (6.164) by the same matrix operator, resulting in

$$\begin{pmatrix} \frac{\mathcal{I}}{2} - \mathcal{K} & \mathcal{T} \\ -\mathcal{T} & \frac{\mathcal{I}}{2} - \mathcal{K} \end{pmatrix} \begin{pmatrix} \hat{n} \times \mathbf{E} \\ Z\hat{n} \times \mathbf{H} \end{pmatrix} =$$
$$\begin{pmatrix} \left(\frac{\mathcal{I}}{2} - \mathcal{K}\right)^2 - \mathcal{T}^2 & \left(\frac{\mathcal{I}}{2} - \mathcal{K}\right)\mathcal{T} + \mathcal{T}\left(\frac{\mathcal{I}}{2} - \mathcal{K}\right) \\ -\mathcal{T}\left(\frac{\mathcal{I}}{2} - \mathcal{K}\right) - \left(\frac{\mathcal{I}}{2} - \mathcal{K}\right)\mathcal{T} & \left(\frac{\mathcal{I}}{2} - \mathcal{K}\right)^2 - \mathcal{T}^2 \end{pmatrix} \begin{pmatrix} \hat{n} \times \mathbf{E} \\ Z\hat{n} \times \mathbf{H} \end{pmatrix}. \tag{6.165}$$

Equating the diagonal matrix entries on both sides gives

$$\left(\frac{\mathcal{I}}{2} - \mathcal{K}\right)^2 - \mathcal{T}^2 = \frac{\mathcal{I}}{2} - \mathcal{K} \tag{6.166}$$

or

$$\mathcal{T}^2 = \frac{\mathcal{I}}{4} - \mathcal{K}^2 \tag{6.167}$$

which is the first Calderon identity. The off diagonal entries in (6.165) gives the second Calderen identity, viz.

$$\left(\frac{\mathcal{I}}{2} - \mathcal{K}\right)\mathcal{T} + \mathcal{T}\left(\frac{\mathcal{I}}{2} - \mathcal{K}\right) = \mathcal{T} \tag{6.168}$$

or

$$\mathcal{T}\mathcal{K} = -\mathcal{K}\mathcal{T}. \tag{6.169}$$

Since $\mathcal{K}$ is a compact operator (on smooth surface), the conditioning of (6.167) is primarily governed by the identity operator, implying that squared EFIE (6.167) is much better conditioned that the original as discussed in detail in Adams (2004) and Hsiao and Kleinman (1997). However, care must be taken in the discretization of (6.167) since separate sets of basis functions need be used in discretizing the two successive $\mathcal{T}$ operators. A rather intuitive set of basis functions, *dual* to the conventional RWG basis functions and defined on a barycentric refinement of the original surface triangulation (that results in a Voronoi-cell-like dual mesh), was presented in Buffa and Christiansen (BC; 2007). Since the construction of the Calderon preconditioner does not involve matrix inversions or decompositions, it can be readily computed during the MoM matrix computation phase of the solver. However, the generation of the dual complex and the associated BC basis functions requires careful bookkeeping of the various indices involved. Nonetheless, the computational savings of the Calderon preconditioner, in particular for low-frequency problems, were reported in Andriulli, Cools, Bagci, Olyslager, Buffa, Christiansen, and Michielssen (2008). Furthermore, a loop-star basis function version is given in Stephanson and Lee (2009) showing both improved convergence and immunity to low-frequency breakdown.

### 6.3.9 Performance of Preconditioned Conjugate Gradient Squared Solver

For example, to evaluate the performance of the popular ILU preconditioner, we considered the scattering by the PEC ogive depicted in Figure 6-47. The ILU preconditioner was implemented using all three formulations (EFIE, MFIE, and CFIE). As we already noted, the CFIE converged much faster due to its better conditioning. Thus, preconditioning of the CFIE may not have much of an effect in solution time. The size of the matrix system was $N = 480$ and refers to a 10″ long ×2″ diameter ogive (with its long axis coincident with the $x$-axis). All calculations were carried out at 5.91 GHz.

Figure 6-48 shows the residual error as a function of iteration number for the EFIE matrix. It is seen that due to the irregular mesh around the sharp tips, the conjugate gradient squared (CGS) solver does not show tendency for convergence even after 50 iterations. However, once the ILU preconditioner is employed, convergence is dramatically improved, requiring only $N/50$ iterations ($N$ = number of unknowns) to achieve a residual error of $10^{-5}$ (Section 4.11).

Nonpreconditioned and preconditioned solution data for the MFIE matrix are given in Figure 6-49. Since the MFIE formulation always leads to better conditioned systems (this is apart from its formulation accuracy), residual error shows some tendency for convergence. Regardless, convergence without preconditioning is very

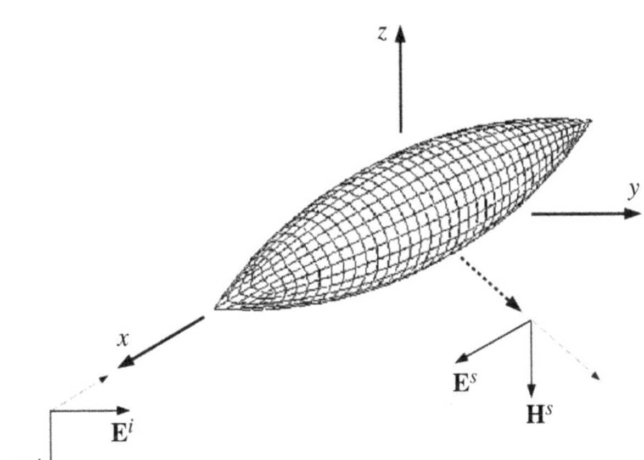


**FIGURE 6-47** Ogive geometry and coordinate system; excitation in a plane wave impinging along the $x$-axis.

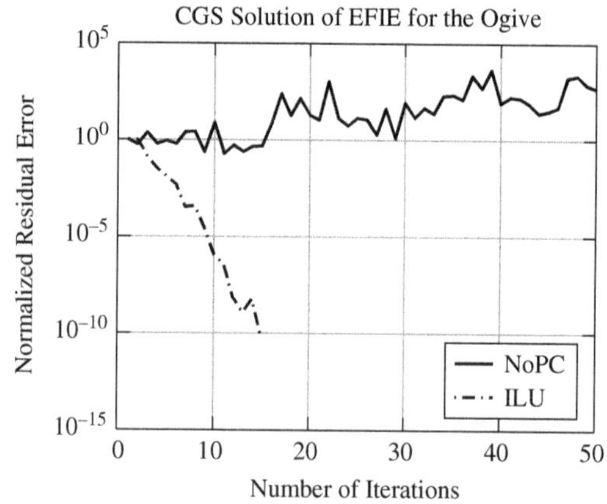


**FIGURE 6-48** ILU performance applied to an EFIE matrix system for the ogive in Figure 6-47

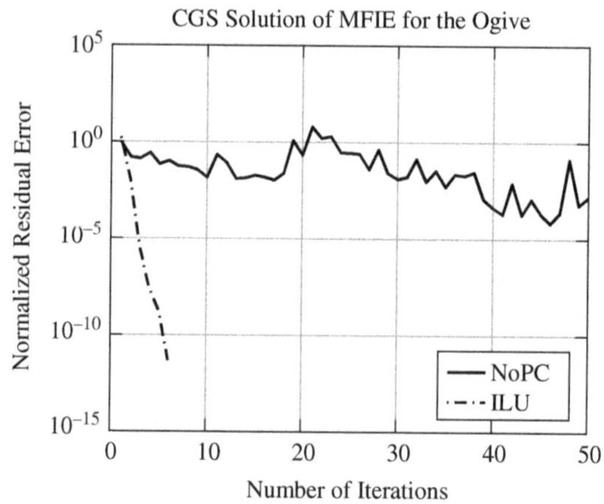


**FIGURE 6-49** ILU performance applied to the MFIE matrix system for the ogive in Figure 6-47.

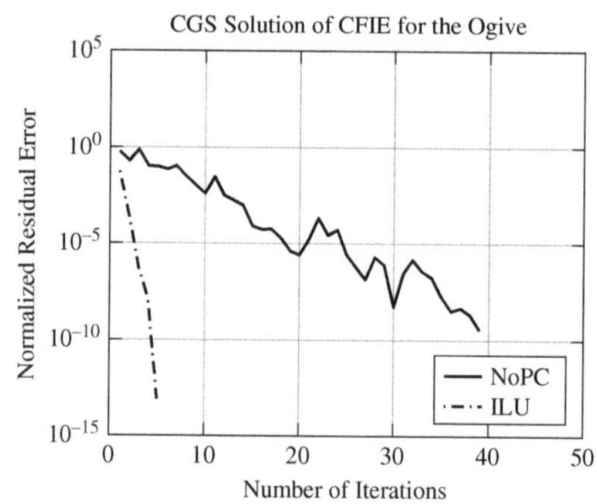


**FIGURE 6-50** ILU performance applied to the CFIE matrix system for the ogive in Figure 6-47.

slow. When the ILU preconditioner is included, convergence is reached down to $10^{-5}$ in about $N/150$ iterations. Figure 6-50 gives the corresponding convergence curves for the CFIE matrix. Since the CFIE system is better conditioned, convergence is now achieved even without preconditioning. Use of ILU simply improves convergence from $N/25$ down to $N/150$ to reach a residual error of $10^{-5}$.

Table 6-4 summarizes the performance of the ILU preconditioner for a larger problem using the CFIE. The scatterer in this simulation has sharp edges and tips

TABLE 6-4 ■ ILU performance for the ogive scattering target shown in Figure 6-47.

| Number of Unknowns | Matrix Fill Time(min) (8 proc.) | Precond. LU Time(min) (1 proc.) | Number of Iter. | Residual Error | Time per Solution(min) (8 proc.)[3] |
|---|---|---|---|---|---|
| 53000 | 77 | 81 | 5 | 0.001 | 5 |

as well as smooth sections. Also, the mesh is quite distorted and non-uniform around the edges. Nevertheless, the performance of the ILU preconditioner is quite impressive. Specifically, ILU improved the CFIE convergence ($\alpha = 0.5$) matrix down to $N/10{,}000$. As a result, the solution time is only five minutes for a 53,000 unknown system on an eight-processor SGI Origin 2000.

## 6.4 VOLUMETRIC MODELING

Volume integral equation formulations were discussed in Chapter 3 at some length (Peterson, 1991; Arvas, Rahhal-Arabi, and Sadigh, 1991; Volakis, 1992). Here, we revisit these formulations and present MoM solutions.

Indeed, with the recent success of fast integral methods, such as the multilevel FMM (MLFMM) (Coifman, Rokhlin, and Wandzura, 1993; Chew, Jin, Lu, Michielssen, and Song, 1997), it is important to reexamine VIE methods refering to Section 3.1.7 in modeling inhomogeneous volumetric structures (Lu, 2001). In this chapter, we develop a VIE solution method using curvilinear hexahedral elements for penetrable structures. We introduce the MLFMM formulation for the pertinent VIEs in Chapter 7, Section 7.5. It is demonstrated that even piecewise constant basis functions provide excellent accuracy at a nominal sampling rates when conformal elements are used.

### 6.4.1 Volume Integral Equation Formulation

To derive the volume integral equation to be discretized, we parallel the steps in Chapter 3. Here, we generalize that formulation to permittivity and permeability tensors. As in Chapter 3, we start with the wave equation

$$\nabla \times \left[\overline{\mu}^{-1}(\mathbf{r}) \cdot \nabla \times \mathbf{E}(\mathbf{r})\right] - \omega^2 \overline{\epsilon}(\mathbf{r}) \cdot \mathbf{E}(\mathbf{r}) = -j\omega \mathbf{J}(\mathbf{r}) \tag{6.170}$$

where $\overline{\epsilon}(\mathbf{r})$ and $\overline{\mu}(\mathbf{r})$ are the dyadic material parameters describing the electromagnetic behavior of the solution domain. The specific problem of interest is depicted in Figure 6-51. Our goal in the next few steps is to generate a wave equation for the fields in an otherwise uniform medium. The scattering from the volume is to be represented by equivalent currents to be found subject to boundary conditions.

To arrive at such a wave equation, we subtract $\nabla \times [\overline{\mu}_0^{-1}(\mathbf{r}) \cdot \nabla \times \mathbf{E}(\mathbf{r})] - \omega^2 \overline{\epsilon}_0(\mathbf{r}) \cdot \mathbf{E}(\mathbf{r})$ from both sides of (6.170). Upon rearranging, we then get

$$\begin{aligned} \nabla \times \left[\overline{\mu}_0^{-1}(\mathbf{r}) \cdot \nabla \times \mathbf{E}(\mathbf{r})\right] - \omega^2 \overline{\epsilon}_0(\mathbf{r}) \cdot \mathbf{E}(\mathbf{r}) = -j\omega \mathbf{J}(\mathbf{r}) + \omega^2 \left[\overline{\epsilon}(\mathbf{r}) - \overline{\epsilon}_0(\mathbf{r})\right] \\ \cdot \mathbf{E}(\mathbf{r}) - \nabla \times \{[\overline{\mu}^{-1}(\mathbf{r}) - \overline{\mu}_0^{-1}(\mathbf{r})] \\ \cdot \nabla \times \mathbf{E}(\mathbf{r})\} \end{aligned} \tag{6.171}$$

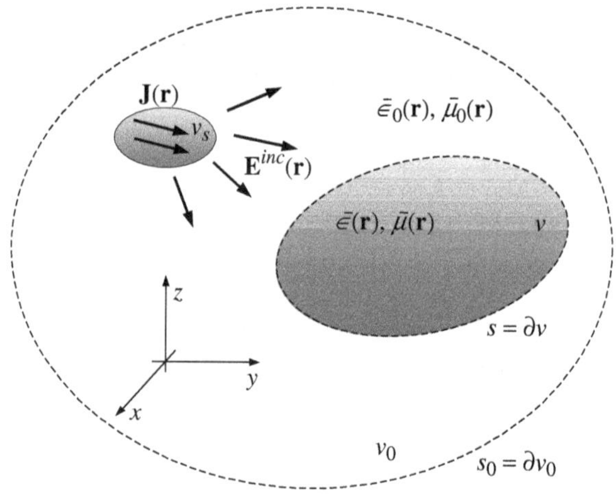


**FIGURE 6-51** Geometrical setup for the VIE derivations.

where $\overline{\epsilon}_0(\mathbf{r}) = \epsilon_0\overline{\mathbf{I}}$ and $\overline{\mu}_0(\mathbf{r}) = \mu_0\overline{\mathbf{I}}$ refer to the permittivity and permeability of the background medium (free space here), respectively, and $\overline{\mathbf{I}}$ is the identity dyad. The right-hand side of (6.171) can now be identified as a source term and accounts for the presence of the inhomogeneous medium. The solution for $\mathbf{E}(\mathbf{r})$ can now be carried out using the dyadic Green's function, satisfying

$$\nabla \times \left[\overline{\mu}_0^{-1}(\mathbf{r}) \cdot \nabla \times \overline{\mathbf{\Gamma}}(\mathbf{r}, \mathbf{r}')\right] - \omega^2 \overline{\epsilon}_0(\mathbf{r}) \cdot \overline{\mathbf{\Gamma}}(\mathbf{r}, \mathbf{r}') = \overline{\mu}_0^{-1}(\mathbf{r})\delta(\mathbf{r} - \mathbf{r}'). \quad (6.172)$$

The Green's function takes the well-known form

$$\overline{\mathbf{\Gamma}}(\mathbf{r}, \mathbf{r}') = \left(\overline{\mathbf{I}} + \frac{1}{k_0^2}\nabla\nabla\right) G(\mathbf{r}, \mathbf{r}'), \quad G(\mathbf{r}, \mathbf{r}') = \frac{e^{-jk_0|\mathbf{r}-\mathbf{r}'|}}{4\pi|\mathbf{r} - \mathbf{r}'|} \quad (6.173)$$

where $k_0 = \omega\sqrt{\epsilon_0\mu_0}$. Using the right-hand side of (6.171) as the source, we readily obtain (Chew, 1995)

$$\begin{aligned} \mathbf{E}(\mathbf{r}) = &-j\omega\mu_0 \int_{V+V_0} d\mathbf{r}'\overline{\mathbf{\Gamma}}(\mathbf{r}, \mathbf{r}') \cdot \mathbf{J}(\mathbf{r}') \\ &+k_0^2 \int_{V+V_0} d\mathbf{r}'\overline{\mathbf{\Gamma}}(\mathbf{r}, \mathbf{r}') \cdot (\overline{\epsilon}_r(\mathbf{r}') - \overline{\mathbf{I}}) \cdot \mathbf{E}(\mathbf{r}') \\ &- \int_{V+V_0} d\mathbf{r}'\overline{\mathbf{\Gamma}}(\mathbf{r}, \mathbf{r}') \cdot \nabla' \times \left[(\overline{\mu}_r^{-1}(\mathbf{r}') - \overline{\mathbf{I}}) \cdot \nabla' \times \mathbf{E}(\mathbf{r}')\right] \end{aligned} \quad (6.174)$$

where $\overline{\epsilon}_r = \overline{\epsilon}/\epsilon_0$, and $\overline{\mu}_r = \overline{\mu}/\mu_0$. The first term in (6.174) is identified as the excitation field, viz.

$$\mathbf{E}^i(\mathbf{r}) = -j\omega\mu_0 \int_{V_s} d\mathbf{r}'\overline{\mathbf{\Gamma}}(\mathbf{r}, \mathbf{r}') \cdot \mathbf{J}(\mathbf{r}') \quad (6.175)$$

where the integration domain $(V + V_0)$ is conveniently replaced by $V_s$ since this represents the only volume over which $\mathbf{J}(\mathbf{r})$ is nonzero. The other two terms vanish when $\overline{\epsilon}(\mathbf{r}) = \epsilon_0\overline{\mathbf{I}}$, $\overline{\mu}(\mathbf{r}) = \mu_0\overline{\mathbf{I}}$ and consequently represent the scattering due to material volume $V$.

To relax the differentiability requirement on $(\overline{\mu}_r^{-1}(\mathbf{r}) - \overline{\mathbf{I}}) \cdot \nabla' \times \mathbf{E}(\mathbf{r}')$ appearing in the third term of (6.174), we make use of the divergence theorem ($\hat{n}$ is outward from $V$), viz.

$$\int_V d\mathbf{r}\nabla\cdot[\overline{\mathbf{\Gamma}}\times\mathbf{y}] = \oint_S d\mathbf{r}\hat{n}\cdot[\overline{\mathbf{\Gamma}}\times\mathbf{y}] = \oint_S d\mathbf{r}\overline{\mathbf{\Gamma}}\cdot(\hat{n}\times\mathbf{y}) \tag{6.176}$$

and the dyadic identity

$$\nabla\cdot[\overline{\mathbf{\Gamma}}\times\mathbf{y}] = \overline{\mathbf{\Gamma}}\cdot\nabla\times\mathbf{y} - \nabla\times\overline{\mathbf{\Gamma}}\cdot\mathbf{y} \tag{6.177}$$

where $\mathbf{y}$ is an arbitrary vector chosen as $\mathbf{y} = \left(\overline{\mu}_r^{-1}(\mathbf{r}) - \overline{\mathbf{I}}\right)\cdot\nabla'\times\mathbf{E}(\mathbf{r}')$. This manipulation leads to

$$\begin{aligned}\mathbf{E}(\mathbf{r}) = \mathbf{E}^i(\mathbf{r}) &+ k_0^2\int_{V+V_0} d\mathbf{r}'\overline{\mathbf{\Gamma}}(\mathbf{r},\mathbf{r}')\cdot(\overline{\epsilon}_r(\mathbf{r}') - \overline{\mathbf{I}})\cdot\mathbf{E}(\mathbf{r}')\\ &- \int_{V+V_0} d\mathbf{r}'\nabla'\times\overline{\mathbf{\Gamma}}(\mathbf{r},\mathbf{r}')\cdot\left[\left(\overline{\mu}_r^{-1}(\mathbf{r}') - \overline{\mathbf{I}}\right)\cdot\nabla'\times\mathbf{E}(\mathbf{r}')\right]\\ &+ \oint_{S_0} d\mathbf{r}'\overline{\mathbf{\Gamma}}(\mathbf{r},\mathbf{r}')\cdot\hat{n}'\times\left[\left(\overline{\mu}_r^{-1}(\mathbf{r}') - \overline{\mathbf{I}}\right)\cdot\nabla'\times\mathbf{E}(\mathbf{r}')\right].\end{aligned} \tag{6.178}$$

In comparison with (6.174), this does not include any differentiation of $\left(\overline{\mu}_r^{-1}(\mathbf{r}') - \overline{\mathbf{I}}\right)\cdot\nabla'\times\mathbf{E}(\mathbf{r}')$. However, a surface integral is introduced enclosing $V + V_0$. Moreover, as $S_0 \to \infty$, the radiation condition implies that this surface integral vanishes, giving

$$\begin{aligned}\mathbf{E}(\mathbf{r}) = \mathbf{E}^i(\mathbf{r}) &+ k_0^2\int_{V+V_0} d\mathbf{r}'\overline{\mathbf{\Gamma}}(\mathbf{r},\mathbf{r}')\cdot(\overline{\epsilon}_r(\mathbf{r}') - \overline{\mathbf{I}})\cdot\mathbf{E}(\mathbf{r}')\\ &- \int_{V+V_0} d\mathbf{r}'\nabla'\times\overline{\mathbf{\Gamma}}(\mathbf{r},\mathbf{r}')\cdot\left[\left(\mu_r^{-1}(\mathbf{r}') - \overline{\mathbf{I}}\right)\cdot\nabla'\times\mathbf{E}(\mathbf{r}')\right].\end{aligned} \tag{6.179}$$

An alternative way to derive (6.179) is to use the standard equivalent electric and magnetic currents. In accordance with Harrington (1968), we could introduce the volumetric electric and magnetic current density sources (see (3.62))

$$\mathbf{J}_{eq}(\mathbf{r}) = j\omega\epsilon_0[\overline{\epsilon}_r(\mathbf{r}) - \overline{\mathbf{I}}]\cdot\mathbf{E}(\mathbf{r}),\quad \mathbf{M}_{eq}(\mathbf{r}) = -[\overline{\mu}_r(\mathbf{r}) - \overline{\mathbf{I}}]\cdot\mathbf{H}(\mathbf{r}). \tag{6.180}$$

These sources radiate in free space and are responsible for the scattered fields, viz.

$$\mathbf{E}^s(\mathbf{r}) = -j\omega\mu_0\int_v d\mathbf{r}'\left[\overline{\mathbf{\Gamma}}(\mathbf{r},\mathbf{r}')\cdot\mathbf{J}_{eq}(\mathbf{r}') + \nabla'\times\overline{\mathbf{\Gamma}}(\mathbf{r},\mathbf{r}')\cdot\mathbf{M}_{eq}(\mathbf{r}')\right] \tag{6.181}$$

with

$$\mathbf{E}(\mathbf{r}) = \mathbf{E}^i(\mathbf{r}) + \mathbf{E}^s(\mathbf{r}). \tag{6.182}$$

The dual expressions are also valid for $\mathbf{H}^s(\mathbf{r})$. Equation (6.182) and its dual yield a pair of integral equations for the solution of the unknown volume equivalent currents $\mathbf{J}_{eq}(\mathbf{r})$ and $\mathbf{M}_{eq}(\mathbf{r})$ (or $\mathbf{H}(\mathbf{r})$ and $\mathbf{E}(\mathbf{r})$, respectively).

The previous formulation involving equivalent volume currents was used by Richmond (1965) for two-dimensional applications, as discussed in Chapter 5. It resulted in six scalar unknowns per volume location for problems involving nontrivial permittivity and permeability. In contrast, (6.179) involves only a single vector field unknown (three scalar unknowns) per volume location. However, (6.179) requires differentiation (curl) of the unknown field $\mathbf{E}(\mathbf{r})$ in much the same way as the finite element method (Volakis, Chatterjee, and Kempel, 1998). One could choose to remove the differentiation from $\mathbf{E}(\mathbf{r})$ but would then obtain an integrand involving differentiation of the permeability across the volume. This was done in Volakis (1992) and although the final expression avoids the curl of the unknown $\mathbf{E}(\mathbf{r})$, the need to differentiate $\overline{\mu}_r(\mathbf{r})$ is undesirable for numerical implementation as $\overline{\mu}_r(\mathbf{r})$ is not available in analytic form. Clearly, among the various derived representations, the one given in (6.179) is most attractive for numerical implementation.

### 6.4.2 VIE Formulation for Dielectrics

For the specific case when $\mu_r = 1$, (6.179) simplifies to

$$\mathbf{E}(\mathbf{r}) = \mathbf{E}^i(\mathbf{r}) + k_0^2 \int_v d\mathbf{r}' \overline{\mathbf{\Gamma}}(\mathbf{r}, \mathbf{r}') \cdot (\overline{\epsilon}_r(\mathbf{r}') - \overline{\mathbf{I}}) \cdot \mathbf{E}(\mathbf{r}'). \tag{6.183}$$

To solve for $\mathbf{E}$ in (6.183) using the standard MoM, we introduce the expansion

$$\mathbf{E}(\mathbf{r}) = \sum_{i=1}^{N} x_i \mathbf{e}_i(\mathbf{r}) \tag{6.184}$$

for the unknown fields. As usual, $\mathbf{e}_i(\mathbf{r})$ refer to the basis functions, and $\{x_i\}$ are the unknowns of the expansion. Upon carrying out Galerkin's testing, we arrive at a matrix equation $[Z]\{x\} = \{b\}$ with

$$Z_{ji} = \langle \mathbf{e}_j(\mathbf{r}), \mathbf{e}_i(\mathbf{r}) \rangle - k_0^2 \langle \mathbf{e}_j(\mathbf{r}), \int_v d\mathbf{r}' \overline{\mathbf{\Gamma}}(\mathbf{r}, \mathbf{r}') \cdot (\overline{\epsilon}_r - \overline{\mathbf{I}}) \cdot \mathbf{e}_i(\mathbf{r}') \rangle. \tag{6.185}$$

The right-hand side of the matrix system takes the form $b_j = \langle \mathbf{e}_j(\mathbf{r}), \mathbf{E}^i(\mathbf{r}) \rangle$.

Similar to the SIE implementation in the previous chapter sections, the hypersingularity in (6.185) is relaxed using the divergence theorem. This gives

$$\begin{aligned} Z_{ji} = & \int_v d\mathbf{r} \mathbf{e}_j(\mathbf{r}) \cdot \mathbf{e}_i(\mathbf{r}) \\ & - k_0^2 \int_v d\mathbf{r} \mathbf{e}_j(\mathbf{r}) \cdot \int_v d\mathbf{r}' G(\mathbf{r}, \mathbf{r}') (\overline{\epsilon}_r - \overline{\mathbf{I}}) \cdot \mathbf{e}_i(\mathbf{r}') \\ & - \int_v d\mathbf{r} \nabla \cdot \mathbf{e}_j(\mathbf{r}) \int_v d\mathbf{r}' G(\mathbf{r}, \mathbf{r}') \nabla' \cdot [(\overline{\epsilon}_r - \overline{\mathbf{I}}) \cdot \mathbf{e}_i(\mathbf{r}')] \\ & + \oint_s d\mathbf{r} \mathbf{e}_j(\mathbf{r}) \cdot \int_v d\mathbf{r}' \nabla G(\mathbf{r}, \mathbf{r}') \nabla' \cdot [(\overline{\epsilon}_r - \overline{\mathbf{I}}) \cdot \mathbf{e}_i(\mathbf{r}')]. \end{aligned} \tag{6.186}$$

To further simplify (6.186), we assume that the permittivity is constant within each element; hence, we can rewrite (6.186) over a pair of source and testing hexahedra to yield

$$\begin{aligned} Z_{ji} = & \int_v d\mathbf{r} \mathbf{e}_j(\mathbf{r}) \cdot \mathbf{e}_i(\mathbf{r}) \\ & - k_0^2 \int_v d\mathbf{r} \mathbf{e}_j(\mathbf{r}) \cdot (\overline{\epsilon}_r - \overline{\mathbf{I}}) \int_{v'} d\mathbf{r}' G(\mathbf{r}, \mathbf{r}') \cdot \mathbf{e}_i(\mathbf{r}') \end{aligned}$$

$$- \int_v d\mathbf{r} \nabla \cdot \mathbf{e}_j(\mathbf{r})(\bar{\epsilon}_r - \bar{\mathbf{I}}) \int_{v'} d\mathbf{r}' G(\mathbf{r}, \mathbf{r}') \nabla' \cdot \mathbf{e}_i(\mathbf{r}')$$

$$+ \oint_s d\mathbf{r} \mathbf{e}_j(\mathbf{r}) \cdot (\bar{\epsilon}_r - \bar{\mathbf{I}}) \cdot \int_{v'} d\mathbf{r}' \nabla G(\mathbf{r}, \mathbf{r}') \nabla' \cdot \mathbf{e}_i(\mathbf{r}'). \qquad (6.187)$$

Numerical evaluation of these matrix elements can again be carried out using a suitable-order Gaussian quadrature. However, the singular integral encountered in the self-cell interactions is treated separately using an annihilation technique similar to that used for SIEs (Sertel and Volakis, 2002).

In the next section, we summarize several different orders of volumetric basis functions to be used in (6.184). In particular, we describe three different orders of basis functions, each having specific qualities attractive for volumetric modeling. For purely dielectric bodies, all three basis functions are observed to generate accurate results. However, when the target geometry has nontrivial permittivity and permeability, as discussed in the next section, second-order basis functions must be due to the need to carry out the differentiation in $\nabla \times \mathbf{E}$ appearing in (6.179).

### 6.4.3 Zeroth-Order Volumetric Basis Functions

For each curvilinear hexahedral volume element shown in Figure 6-52, three basis functions associated with the three parametric directions can be defined as

$$\mathbf{e}_1 = \frac{1}{\sqrt{g}} \mathbf{a}_u, \qquad \mathbf{e}_2 = \frac{1}{\sqrt{g}} \mathbf{a}_u, \qquad \mathbf{e}_3 = \frac{1}{\sqrt{g}} \mathbf{a}_u. \qquad (6.188)$$

It is easy to show that the divergence of the previous basis function (inside the element) is identically zero. This is a property shared by the electric field inside a uniform dielectric region. Thus, it is an important property to maintain. Another advantage of this expansion is that the basis functions are defined on separate elements. Hence, this expansion allows for a natural decomposition of the problem into different domains. It is important to note that the normal components of the basis functions are allowed to be discontinuous. We remark that these basis functions are

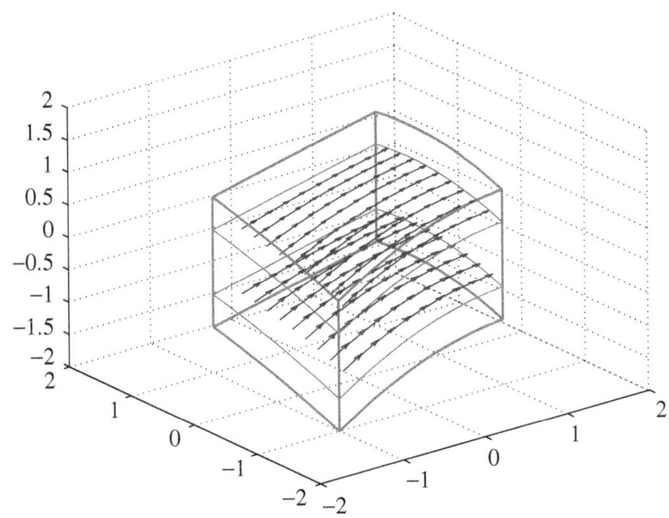


**FIGURE 6-52** Illustration of the zeroth-order electric field basis functions.

conformal generalizations of those presented in Livesay and Chen (1974) for rectangular domains. Figure 6-52 illustrates the behavior of the vector field inside the curvilinear hexahedron at two different isoparametric cuts.

### 6.4.4 First-Order Volumetric Basis Functions

Divergence conforming basis functions for hexahedral finite elements can be defined as (six basis functions associated with the six faces of the element)

$$\mathbf{e}_1 = \frac{1}{\sqrt{g}}(u)\mathbf{a}_u,\ \mathbf{e}_2 = \frac{1}{\sqrt{g}}(1-u)\mathbf{a}_u,\ \mathbf{e}_3 = \frac{1}{\sqrt{g}}(v)\mathbf{a}_v$$

$$\mathbf{e}_4 = \frac{1}{\sqrt{g}}(1-v)\mathbf{a}_v,\quad \mathbf{e}_5 = \frac{1}{\sqrt{g}}(w)\mathbf{a}_w,\quad \mathbf{e}_6 = \frac{1}{\sqrt{g}}(1-w)\mathbf{a}_w. \tag{6.189}$$

For this choice of basis functions, the divergence inside each element is found to be a "constant" (in the parametric space), viz. $\nabla \cdot \mathbf{e}_i = 1/\sqrt{g}$. The advantage of this expansion is that the normal components of the basis functions, defined across pairs of elements sharing a common face, are continuous. Figure 6-53 illustrates the behavior of the vector field inside a pair of curvilinear hexahedra (sharing a common face) at two different isoparametric cuts. We remark that these basis functions are analogous to the rooftop basis functions for SIEs.

### 6.4.5 Second-Order Volumetric Basis Functions

For each volume element, the second order basis functions correspond to 12 functions. These are associated with the 12 edges of the element defined as

$$\mathbf{e}_1 = \frac{1}{\sqrt{g}}(v)(w)\mathbf{a}_u, \qquad \mathbf{e}_2 = \frac{1}{\sqrt{g}}(1-v)(w)\mathbf{a}_u,$$

$$\mathbf{e}_3 = \frac{1}{\sqrt{g}}(v)(1-w)\mathbf{a}_u, \qquad \mathbf{e}_4 = \frac{1}{\sqrt{g}}(1-v)(1-w)\mathbf{a}_u,$$

$$\mathbf{e}_5 = \frac{1}{\sqrt{g}}(u)(w)\mathbf{a}_v, \qquad \mathbf{e}_6 = \frac{1}{\sqrt{g}}(1-u)(w)\mathbf{a}_v,$$

$$\mathbf{e}_7 = \frac{1}{\sqrt{g}}(u)(1-w)\mathbf{a}_v, \qquad \mathbf{e}_8 = \frac{1}{\sqrt{g}}(1-u)(1-w)\mathbf{a}_v,$$

$$\mathbf{e}_9 = \frac{1}{\sqrt{g}}(u)(v)\mathbf{a}_w, \qquad \mathbf{e}_{10} = \frac{1}{\sqrt{g}}(1-u)(v)\mathbf{a}_w,$$

$$\mathbf{e}_{11} = \frac{1}{\sqrt{g}}(u)(1-v)\mathbf{a}_w, \qquad \mathbf{e}_{12} = \frac{1}{\sqrt{g}}(1-u)(1-v)\mathbf{a}_w. \tag{6.190}$$

These covariant unitary vector expansions for the VIE formulation (as opposed to a contravariant form in finite element formulations; Antilla and Alexopoulos 1994) have the advantage of zero divergence inside the element. As in the case of the zeroth-order basis functions, this property satisfies the rational behavior of the electric fields. As opposed to a contravariant expansion (i.e., a covariant projection

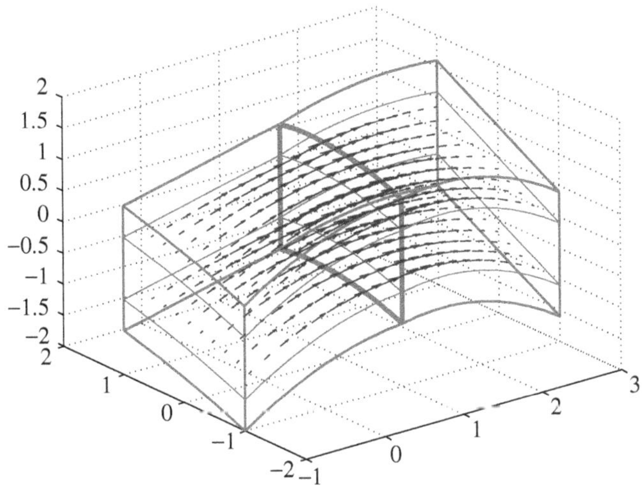


**FIGURE 6-53** Illustration of the first-order electric field basis functions.

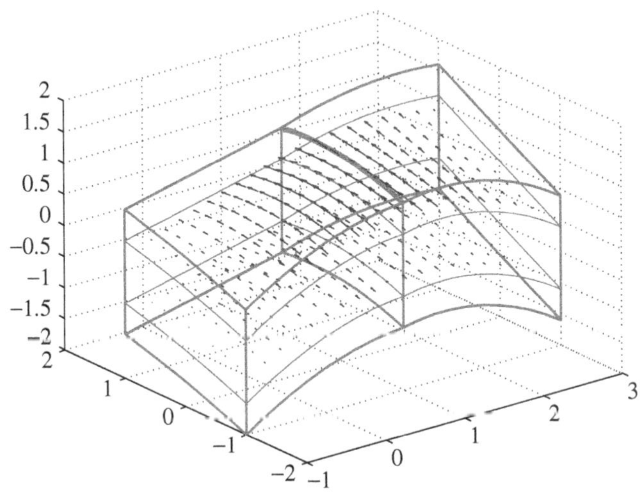


**FIGURE 6-54** Illustration of the second-order electric field basis functions.

form), this expansion does not have continuous tangential components across the common faces of the elements having a common edge. Figure 6-54 illustrates the behavior of the vector field inside a pair of curvilinear hexahedra (sharing a common edge) at two different isoparametric cuts.

## 6.4.6 Scattering by Dielectric Bodies

In this section, we examine the three previously outlined volumetric basis functions to compute the scattering from a dielectric shell. Figure 6-55 depicts the computed bistatic RCS curves for a dielectric shell of outer radius $0.2\lambda_0$ and thickness $0.02\lambda_0$. As seen, all three results agree well with the analytical solution. We note that although the same volume mesh is used in all three cases, the system sizes for different basis functions are inherently different. For this specific example, zeroth-order basis functions resulted in 648 unknowns, the first-order basis functions led to 864 unknowns, and the second-order basis functions resulted in 1, 082 unknowns. Correspondingly, the matrix fill times were different, with the zeroth order being the fastest and the second order being the slowest.

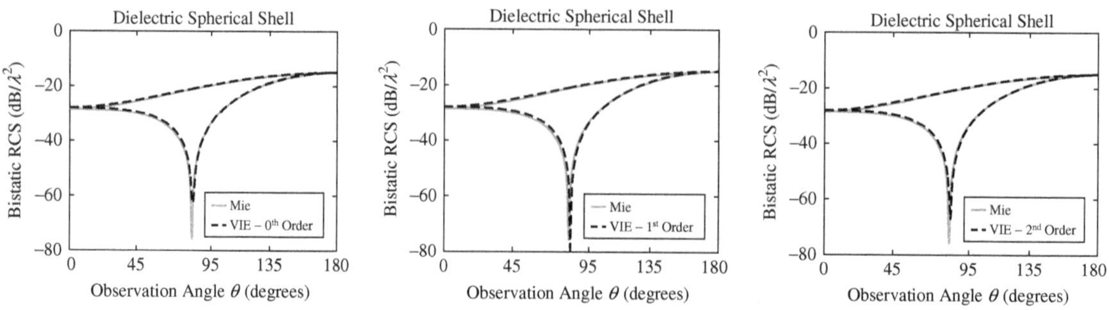


**FIGURE 6-55** ▪ Bistatic RCS of a $0.2\lambda_0$ radius, $0.02\lambda_0$ thick spherical shell with $\epsilon_r = 2.2$ using three different orders of basis functions (backscatter is at 0°).

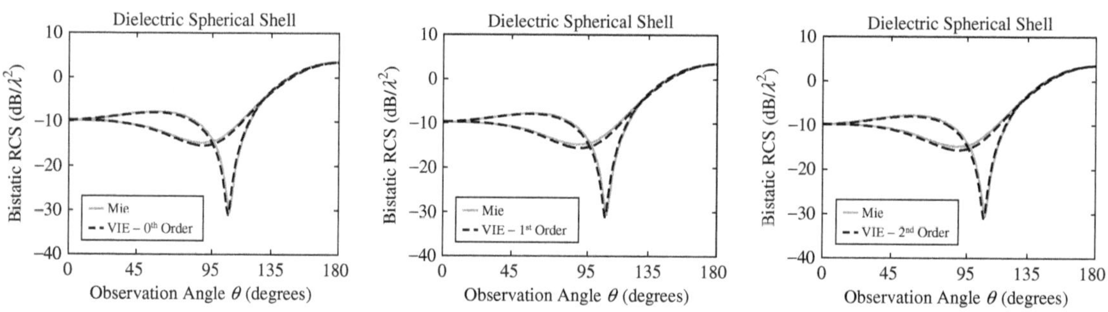


**FIGURE 6-56** ▪ Bistatic RCS of a $0.4\lambda_0$ radius, $0.04\lambda_0$ thick spherical shell with $\epsilon_r = 2.2$ using three different orders of basis functions (backscatter is at 0°).

For an electrically larger shell, the denser mesh with 384 hexahedra was used, resulting in 1152 unknowns for the zeroth-order basis functions. The bistatic RCS curves are given in Figure 6-56. As seen, all three methods are in very good agreement with the analytical solution.

Another geometry considered is a dielectric cube of side length $0.5\lambda_0$ and $\epsilon_r = 2.2$. The bistatic RCS of this dielectric cube using the three different orders of basis functions is plotted in Figure 6-57. As observed, all results are observed to be in good to fair agreement (no exact solution exists for the cube).

The examples demonstrate that accurate RCS results can be obtained regardless of the expansion order for the unknown electric fields. Based on this conclusion, the lowest-order basis function can be chosen for purely dielectric structures. However, when dealing with magnetic materials, second-order basis functions must be used due to the differentiability requirement on the unknown electric field intensity. This point will be discussed in detail in the following subsections.

### 6.4.7 VIE Solution for Magnetically Permeable Structures

Having derived the volume integral representation (6.179), we proceed with its numerical implementation and validation for composite structures where both dielectric and magnetic material parameters are involved. Specifically, this section describes the MoM procedure to cast (6.179) in the form $\left[Z_{ji}\right]\{x_i\} = \left\{b_j\right\}$.

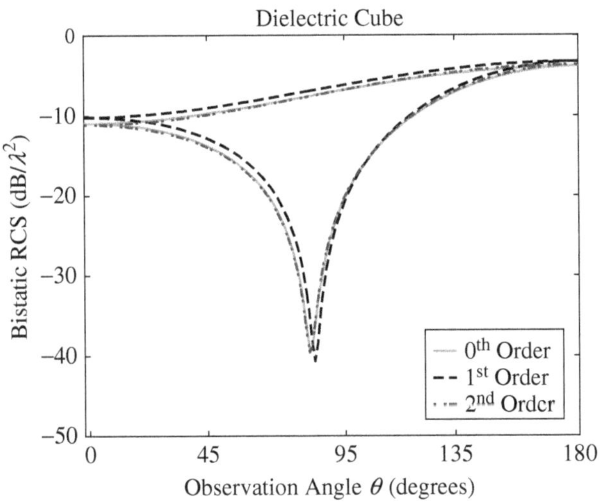


**FIGURE 6-57** Bistatic RCS for a dielectric cube ($0.5\lambda_0$ per side length) with $\epsilon_r = 2.2$ using three different orders of basis functions.

To discretize (6.179), we employ the same parametric hexahedral modeling and second-order electric field basis functions given in (6.190).

The discretized electric field within the inhomogeneous domain $V$ is constructed using the connectivity of the hexahedral mesh in the same manner as typical finite element ordering schemes (an unknown is assigned to each edge in the mesh). The hypersingularity in $\overline{\Gamma}(\mathbf{r}, \mathbf{r}')$ is relaxed through the use of the divergence theorem as discussed previously. The electric field basis functions are expressed in contravariant projection form, that is, in terms of covariant unitary vectors. In this form, it is easier to evaluate the divergence of the basis functions in parametric coordinates using

$$\nabla \cdot \mathbf{e} = \frac{1}{\sqrt{g}} \left\{ \frac{\partial}{\partial u} \left( [\mathbf{e} \cdot \mathbf{a}^u] \sqrt{g} \right) + \frac{\partial}{\partial v} \left( [\mathbf{e} \cdot \mathbf{a}^v] \sqrt{g} \right) + \frac{\partial}{\partial w} \left( [\mathbf{e} \cdot \mathbf{a}^w] \sqrt{g} \right) \right\} = 0. \tag{6.191}$$

However, the evaluation of $\nabla \times \mathbf{e}$ is rather cumbersome for a contravariant projection representation. The curl operation in parametric coordinates requires the covariant components of the basis functions. Specifically

$$\nabla \times \mathbf{e} = \frac{1}{\sqrt{g}} \left\{ \left[ \frac{\partial (\mathbf{e} \cdot \mathbf{a}_w)}{\partial v} - \frac{\partial (\mathbf{e} \cdot \mathbf{a}_v)}{\partial w} \right] \mathbf{a}_u + \left[ \frac{\partial (\mathbf{e} \cdot \mathbf{a}_u)}{\partial w} - \frac{\partial (\mathbf{e} \cdot \mathbf{a}_w)}{\partial u} \right] \mathbf{a}_v + \left[ \frac{\partial (\mathbf{e} \cdot \mathbf{a}_v)}{\partial u} - \frac{\partial (\mathbf{e} \cdot \mathbf{a}_u)}{\partial v} \right] \mathbf{a}_w \right\} \tag{6.192}$$

in which the covariant components must be evaluated using the metric of the parametric transformation that forms the curvilinear element. That is, we must employ the expression

$$\begin{Bmatrix} \mathbf{e} \cdot \mathbf{a}_u \\ \mathbf{e} \cdot \mathbf{a}_v \\ \mathbf{e} \cdot \mathbf{a}_w \end{Bmatrix} = \begin{bmatrix} (\mathbf{a}_u \cdot \mathbf{a}_u) & (\mathbf{a}_u \cdot \mathbf{a}_v) & (\mathbf{a}_u \cdot \mathbf{a}_w) \\ (\mathbf{a}_v \cdot \mathbf{a}_u) & (\mathbf{a}_v \cdot \mathbf{a}_v) & (\mathbf{a}_v \cdot \mathbf{a}_w) \\ (\mathbf{a}_w \cdot \mathbf{a}_u) & (\mathbf{a}_w \cdot \mathbf{a}_v) & (\mathbf{a}_w \cdot \mathbf{a}_w) \end{bmatrix} \begin{Bmatrix} \mathbf{e} \cdot \mathbf{a}^u \\ \mathbf{e} \cdot \mathbf{a}^v \\ \mathbf{e} \cdot \mathbf{a}^w \end{Bmatrix} \tag{6.193}$$

in (6.192). This leads to a rather cumbersome expression, requiring parametric derivatives of $\sqrt{g}$. As an example, the computation of $\frac{\partial(\mathbf{e} \cdot \mathbf{a}_w)}{\partial v}$ for the case when

$\mathbf{e} = \frac{1}{\sqrt{g}}(1-v)(1-w)\mathbf{a}_u$ gives

$$\begin{aligned}
\frac{\partial\,(\mathbf{e}\cdot\mathbf{a}_w)}{\partial v} &= \frac{\partial}{\partial v}[(\mathbf{a}_w\cdot\mathbf{a}_u)(\mathbf{e}\cdot\mathbf{a}^u)+(\mathbf{a}_w\cdot\mathbf{a}_v)(\mathbf{e}\cdot\mathbf{a}^v)+(\mathbf{a}_w\cdot\mathbf{a}_w)(\mathbf{e}\cdot\mathbf{a}^w)]\\
&= \frac{\partial}{\partial v}\left[\left(\frac{\partial\mathbf{r}}{\partial w}\cdot\frac{\partial\mathbf{r}}{\partial u}\right)\left(\frac{1}{\sqrt{g}}(1-v)(1-w)\right)\right]\\
&= \left[\frac{\partial}{\partial v}\left(\frac{\partial\mathbf{r}}{\partial w}\cdot\frac{\partial\mathbf{r}}{\partial u}\right)\right]\left(\frac{1}{\sqrt{g}}(1-v)(1-w)\right)\\
&\quad+\left(\frac{\partial\mathbf{r}}{\partial w}\cdot\frac{\partial\mathbf{r}}{\partial u}\right)\left[\frac{\partial}{\partial v}\frac{1}{\sqrt{g}}\right](1-v)(1-w)-\frac{1}{\sqrt{g}}\left(\frac{\partial\mathbf{r}}{\partial w}\cdot\frac{\partial\mathbf{r}}{\partial u}\right).
\end{aligned} \tag{6.194}$$

Nevertheless, once $\nabla\times\mathbf{e}$ is evaluated, the singular integrals can be evaluated using the same singularity annihilation method noted earlier (Sertel and Volakis, 2002). The specific matrix elements are of the form

$$\begin{aligned}
Z_{ji} &= \langle\mathbf{e}_j(\mathbf{r}),\mathbf{e}_i(\mathbf{r})\rangle - k_0^2\left\langle\mathbf{e}_j(\mathbf{r}),\int_v d\mathbf{r}'\overline{\Gamma}(\mathbf{r},\mathbf{r}')\cdot(\overline{\epsilon}_r-\overline{\mathbf{I}})\cdot\mathbf{e}_i(\mathbf{r}')\right\rangle\\
&\quad+\left\langle\mathbf{e}_j(\mathbf{r}),\int_v d\mathbf{r}'\nabla'\times\overline{\Gamma}(\mathbf{r},\mathbf{r}')\cdot(\overline{\mu}_r^{-1}-\overline{\mathbf{I}})\cdot\nabla'\times\mathbf{e}_i(\mathbf{r}')\right\rangle
\end{aligned} \tag{6.195}$$

and the right-hand side of $[Z]\{x_i\}=\{b_i\}$ is given by $b_j=\langle\mathbf{e}_j(\mathbf{r}),\mathbf{E}^i(\mathbf{r})\rangle$.

Having presented the implementation of the integral equation (6.179) for the volume fields $\mathbf{E}$, we proceed to validate this implementation. First, we consider calculations for purely magnetic materials. In this regard, we consider a magnetically permeable cube (plane wave incidence is normal to the cube's face). The specific cube has a side length of $0.5\lambda_0$ and $\mu_r = 2.2$. As seen in Figure 6-58, the VIE and finite

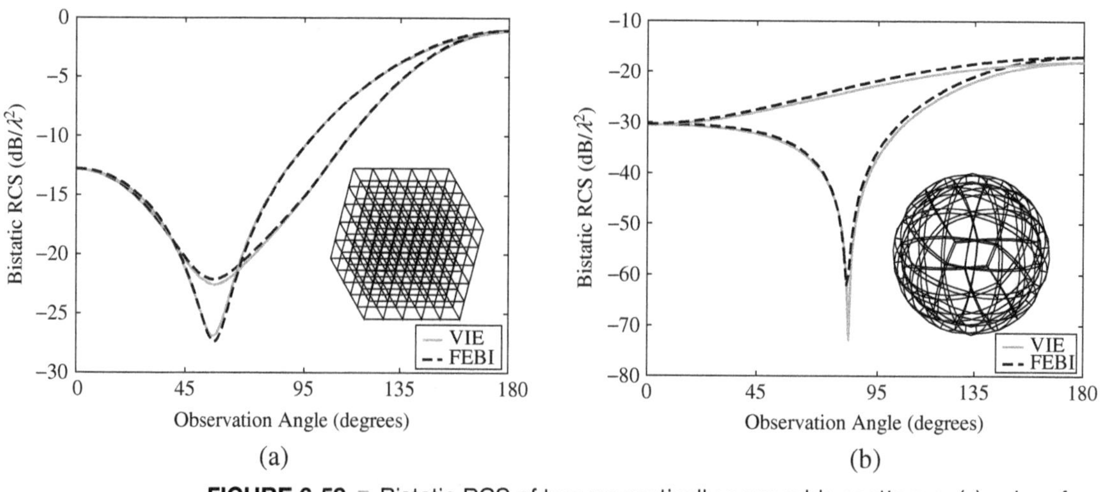


**FIGURE 6-58** ■ Bistatic RCS of two magnetically permeable scatterers: (a) cube of side length $0.5\lambda$ and $\mu_r = 2.2$ and (b) spherical shell of $\mu_r = 2.2$, $0.2\lambda$ outer radius, and $0.18\lambda$ inner radius.

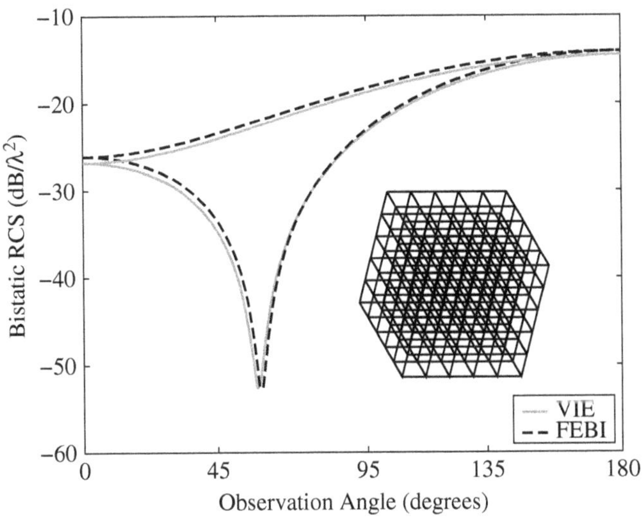


**FIGURE 6-59** Bistatic RCS of a homogeneous composite cube of side length $a = 0.2\lambda$, and $\epsilon_r = 1.5$, $\mu_r = 2.2$ (incidence is along the negative $z$-axis, and the plot is a cut in the $yz$-plane).

element-boundary integral (FE-BI) data are in good agreement and demonstrate the validity of (6.179). Also shown in Figure 6-58b is the bistatic RCS of a spherical shell having an outer radius of $0.2\lambda_0$, an inner radius of $0.18\lambda_0$, and a relative permeability of $\mu_r = 2.2$. Again, the agreement between the FE-BI and VIE curves is good.

To validate the VIE for composite structures, we now consider three different configurations. Our first example is a composite cube of side length $0.2\lambda_0$ (see Figure 6-59). The incident plane wave is propagating in the negative $z$-direction, and the bistatic RCS of the cube is plotted in Figure 6-59. For this special case, the covariant and contravariant components of the basis functions are scaled versions of each other, $\sqrt{g}$ is constant, and hence the curl operation is fairly straightforward to calculate. As seen, the agreement between the VIE and FE-BI solutions is very good for both polarizations.

As a second example, we again consider a thin spherical shell having outer radius of $0.2\lambda_0$, thickness $0.02\lambda_0$, and the same material parameters ($\epsilon_r = 1.5$, $\mu_r = 2.2$) as in the previous example. For modeling, the mesh is constructed entirely as a thin layer of distorted (nonrectangular) curvilinear elements, and both the VIE and FE-BI solution used the same mesh. The bistatic RCS is shown in Figure 6-60, and it is clear that the solutions are again in full agreement.

As our third example, we evaluate the RCS of a solid sphere of radius $0.15\lambda_0$. The permittivity and permeability of the sphere are again $\epsilon_r = 1.5$ and $\mu_r = 2.2$, respectively. We used the same volumetric mesh as shown in Figure 6-61 for both the VIE and FE-BI methods. Again, excellent agreement is seen between the VIE and FE-BI solutions.

We close this section by noting that, in contrast to the SIE, the VIE is a second kind integral equation, and its iterative solution converges quickly. As an example, for a system of 1082 VIE unknowns (second example), the CGS solver converged in five iterations for a relative error of $10^{-2}$. Also the overall solution was completed in only 0.7 seconds using a 1 GHz PIII processor. For the third example (880 VIE unknowns), the iterative CGS solver converged in 13 iterations and took 1.1 seconds.

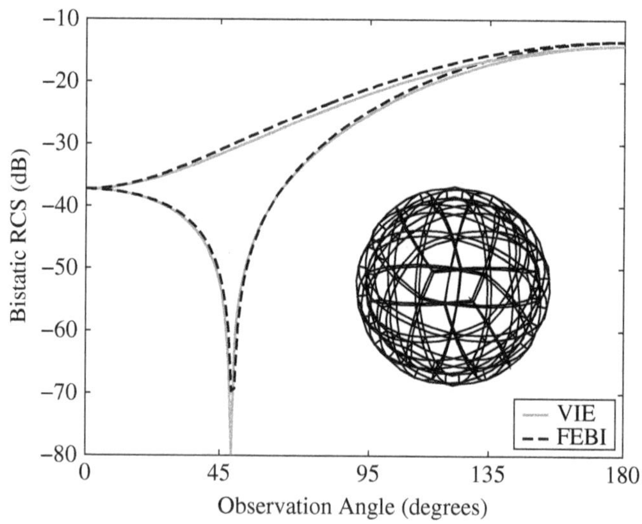


**FIGURE 6-60** Bistatic RCS of a composite spherical shell. The outer shell radius is $r_o = 0.2\lambda$, its thickness is $d = 0.02\lambda$, and its relative constitutive parameters are ($\epsilon_r = 1.5$, $\mu_r = 2.2$); backscatter is at 0°.

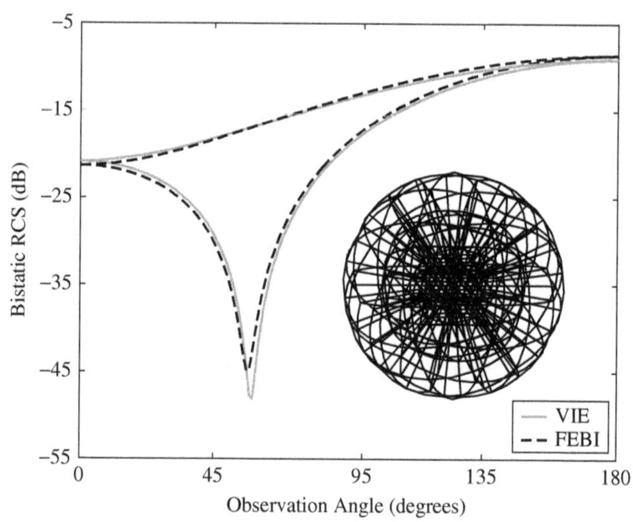


**FIGURE 6-61** Bistatic RCS (incidence at zero degrees) for a homogeneous composite sphere of radius $r = 0.15\lambda$. The sphere has the relative constitutive parameters ($\epsilon_r = 1.5$, $\mu_r = 2.2$); backscatter is at 0°.

### 6.4.7.1 Volume-Surface Integral Equations

When the problem geometry involves multiple material regions embedded in a high-contrast background, it may be more attractive to modify the VIE formulations previously given to reduce its high computational cost of discretizing the high-contrast background. An alternative is the VSIE described in detail in Volakis, Sertel, and Usner (2006). In particular, the VSIE formulation is very attractive for high-contrast materials that have slight variations in their permittivity or permeabilities.

Considering the geometry shown in Figure 6-62, the VSIE formulation relies on the following factorization of the material permittivity

$$\overline{\epsilon}_r(\mathbf{r}) = \epsilon_b\overline{\mathbf{I}} + \overline{\epsilon}_\delta(\mathbf{r}) = \epsilon_b[\overline{\mathbf{I}} + \epsilon_b^{-1}\overline{\epsilon}_\delta(\mathbf{r})] = \epsilon_b\overline{\epsilon}_\Delta(\mathbf{r}). \tag{6.196}$$

Similarly, the permeability $\overline{\mu}_r$ is written as

$$\overline{\mu}_r(\mathbf{r}) = \mu_b\overline{\mathbf{I}} + \overline{\mu}_\delta(\mathbf{r}) = \mu_b[\overline{\mathbf{I}} + \mu_b^{-1}\overline{\mu}_\delta(\mathbf{r})] = \mu_b\overline{\mu}_\Delta(\mathbf{r}). \tag{6.197}$$

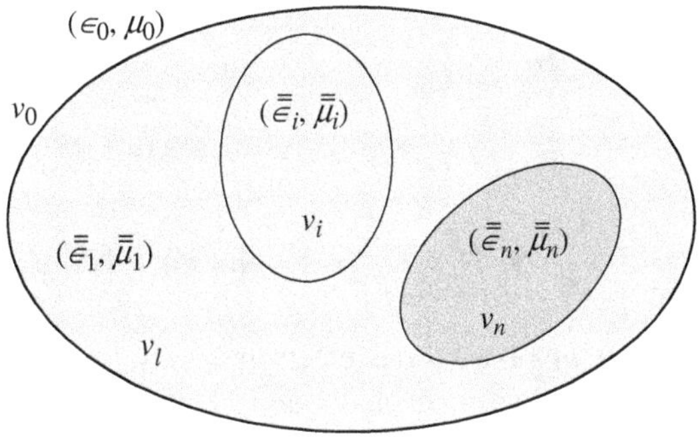

**FIGURE 6-62** Space partitioning for the generalized VSIE.

These peculiar factorizations, (6.196) and (6.197), allow for the variation in material parameters, denoted by $\overline{\epsilon}_\Delta(\mathbf{r})$ and $\overline{\mu}_\Delta(\mathbf{r})$, to be factored out and represented as a volume integral. More importantly, the highly oscillatory fields within the material need not be represented by a volume integral since they are modeled by the surface integrals with the corresponding Green's functions. Referring to Figure 6-62, the following VSIEs can be constructed for each homogeneous material domain $V_i$

$$\mathbf{E}_i(\mathbf{r}) + \oint_{S_i} \left[\nabla G_i \times (\hat{\mathbf{n}}_i \times \mathbf{E}_i) + jk_0\mu_b^i \overline{\mathbf{\Gamma}}_i \cdot (\hat{\mathbf{n}}_i \times \tilde{\mathbf{H}}_i)\right] dV$$
$$- k_i^2 \int_{V_i} \overline{\mathbf{\Gamma}}_i \cdot (\overline{\epsilon}^i_\Delta - \overline{\mathbf{I}})\mathbf{E}_i \, dS - jk_0\mu_b^i \int_{V_i} \nabla g_i \times \left[(\overline{\mu}^i_\Delta - \overline{\mathbf{I}})\tilde{\mathbf{H}}_i\right] dV = \mathbf{E}_i^{inc}(\mathbf{r}) \tag{6.198}$$

where $\overline{\mathbf{\Gamma}}_i$ and $G_i$ refer to the background Green's functions in each homogeneous domain $V_i$. For ease of notation, we also introduced $\tilde{\mathbf{H}}_i = Z_i\mathbf{H}_i$, where $Z_i = \sqrt{\mu_b^i/\epsilon_b^i}$ is the intrinsic impedance of the materials in the corresponding volumes. Also, $\mathbf{E}_i^{inc}(\mathbf{r})$ is the incident field in $V_i$ due to sources within $V_i$, and (6.198) applies to all subdomains $V_i$, $i = 1, 2, \ldots$ within the solution domain.

This VSIE can be recast into a succinct statement

$$\mathbf{E}_i(\mathbf{r}) - \mathbf{E}_i^s(\mathbf{r}) = \mathbf{E}_i^{inc}(\mathbf{r}) \tag{6.199}$$

where we use the surface equivalent currents

$$\tilde{\mathbf{J}}_i = \hat{\mathbf{n}}_i \times \tilde{\mathbf{H}}_i, \qquad \text{and} \qquad \mathbf{M}_i = -\hat{\mathbf{n}}_i \times \mathbf{E}_i \tag{6.200}$$

to express the integrals in (6.198) in the more familiar form

$$\mathbf{E}_i^s = \Psi_i\left(\overline{\epsilon}^i_\Delta, \mathbf{E}_i\right) + \Omega_i\left(\overline{\mu}^i_\Delta, \nabla \times \mathbf{E}_i\right) + \Phi_j^i(\mathbf{M}_i) - \mu_b^i\Theta_j^i(\tilde{\mathbf{J}}_i) \tag{6.201}$$

in terms of the usual operators

$$\Phi_j^i(\mathbf{Y}) = \int_{S_j} \nabla' G_{k_i}(\mathbf{r}, \mathbf{r}') \times \mathbf{Y}(\mathbf{r}') \, dr' \tag{6.202}$$

$$\Theta_j^i(\mathbf{Y}) = jk_0 \int_{S_j} G_{k_i}(\mathbf{r}, \mathbf{r}')\mathbf{Y}(\mathbf{r}') \, dr'$$
$$- \frac{j}{k_0}\nabla \int_{S_j} G_{k_i}(\mathbf{r}, \mathbf{r}')\nabla' \cdot \mathbf{Y}(\mathbf{r}') \, dr' \tag{6.203}$$

$$\Omega_i(\overline{\alpha}, \mathbf{X}) = \int_{V_i} \nabla' G_{k_i}(\mathbf{r}, \mathbf{r}') \times [\overline{\alpha}^{-1}(\mathbf{r}') - \overline{\mathbf{I}}]\mathbf{X}(\mathbf{r}')\, dr' \tag{6.204}$$

$$\Psi_i(\overline{\alpha}, \mathbf{X}) = k_i^2 \int_{V_i} G_{k_i}(\mathbf{r}, \mathbf{r}')[\overline{\alpha}(\mathbf{r}') - \overline{\mathbf{I}}]\mathbf{X}(\mathbf{r}')\, dr' - \nabla \int_{V_i} \nabla' G_{k_i}(\mathbf{r}, \mathbf{r}') \cdot [\overline{\alpha}(\mathbf{r}') - \overline{\mathbf{I}}]\mathbf{X}(\mathbf{r}')\, dr'. \tag{6.205}$$

Notice that (6.199) involves three electromagnetic quantities as unknowns, namely, the electric fields $\mathbf{E}_i$ within each domain $V_i$, and the surface currents $\tilde{\mathbf{J}}_i$, and $\mathbf{M}_i$ over the boundaries $S_i$ of each subdomain. Hence, three equations are needed to solve for these three unknowns, the first of which is given in (6.199) as a mixed volume-surface integral enforced over each subdomain $V_i$. The second equation results from the boundary condition on the electric fields across each boundary $S_i$, viz.

$$\frac{1}{2}\mathbf{t}_i \cdot \mathbf{E}_i(\mathbf{r}) - \mathbf{t}_i \cdot \mathbf{E}_i^s(\mathbf{r}) = \mathbf{t}_i \cdot \mathbf{E}_i^{inc}(\mathbf{r}) \tag{6.206}$$

where $\mathbf{t}_i$ is the tangential unit vector on $S_i$. We also note that the integral in (6.202) has now become a principal value integral excluding the observation point $\mathbf{r}$. The final equation is essentially the dual of (6.206), which is a statement of the boundary condition on tangential magnetic fields across $S_i$ given as

$$\frac{1}{2}\mathbf{t}_i \cdot \tilde{\mathbf{H}}_i(\mathbf{r}) - \mathbf{t}_i \cdot \tilde{\mathbf{H}}_i^s(\mathbf{r}) = \mathbf{t}_i \cdot \tilde{\mathbf{H}}_i^{inc}(\mathbf{r}) \tag{6.207}$$

where

$$\tilde{\mathbf{H}}_i^s(\mathbf{r}) = -jk_0\epsilon_b^i\left[\Psi_i\left(\left(\overline{\mu}_\Delta^i\right)^{-1}, \nabla \times \mathbf{E}_i\right) + \Omega_i\left(\left(\overline{\epsilon}_\Delta^i\right)^{-1}, \mathbf{E}_i\right)\right] - \Phi_j^i(\tilde{\mathbf{J}}_i) - \epsilon_b^i\Theta_j^i(\mathbf{M}_i). \tag{6.208}$$

A detailed MoM implementation based on (6.199), (6.206), and (6.207) can be found in Volakis, Sertel, and Usner (2006).

## 6.5 | SCATTERING EXAMPLES

To demonstrate the presented moment method techniques, in this section we consider several radar scattering examples. Surface meshes of the related test geometries (using curvilinear quadrilaterals) are given in Figure 6-63.

In each of these meshes, the density is such that the edge length is maintained around $\lambda_0/10$ ($\lambda_0$ being the free-space wavelength at the simulation frequency). The commercially available meshing package PATRAN (MSC Software, 2010) was used to generate these surface meshes. We note that the meshes in Figure 6-63a, 6-63c and 6-63e are of the paver type and Figure 6-64 compares such a mesh to an isoparametric one. As depicted, the isoparametric mesh in Figure 6-64a generates more uniform structured meshes. However, depending on the curvature of the surface being gridded, some of the element may be severely distorted leading to ill-conditioned matrices. In contrast, the paver mesh (such as the one shown in Figure 6-64b) is rid of extreme distortions within the mesh. Next, we provide numerical results using both kinds of meshes for the same geometry and demonstrate that each mesh provide comparable accuracy.

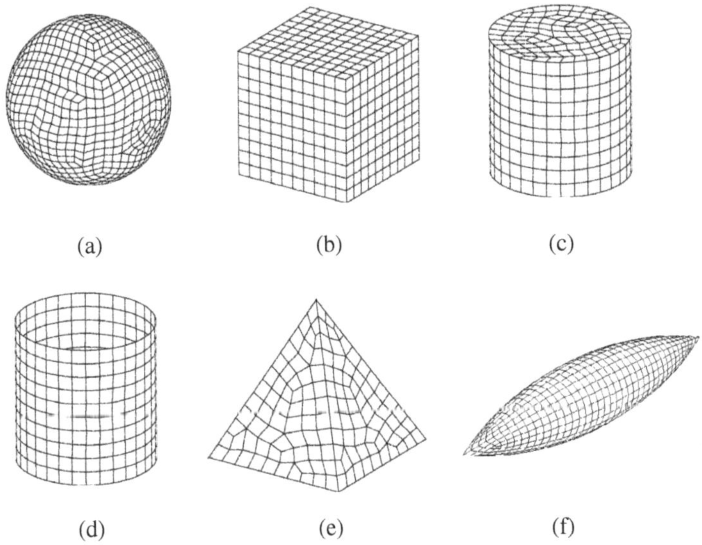


**FIGURE 6-63** Surface grids using quadrilateral elements for several geometries: (a) sphere, (b) cube, (c) cylinder, (d) open-ended cylinder, (e) pyramid, and (f) ogive.

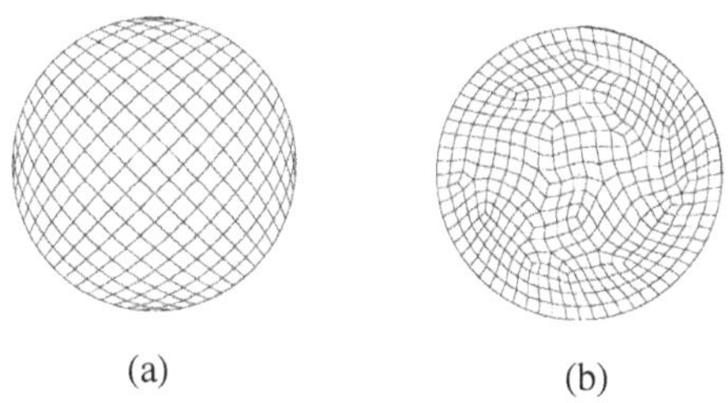


**FIGURE 6-64** Meshes (or grids) for a disk using: (a) isoparametric mesh and (b) paver mesh.

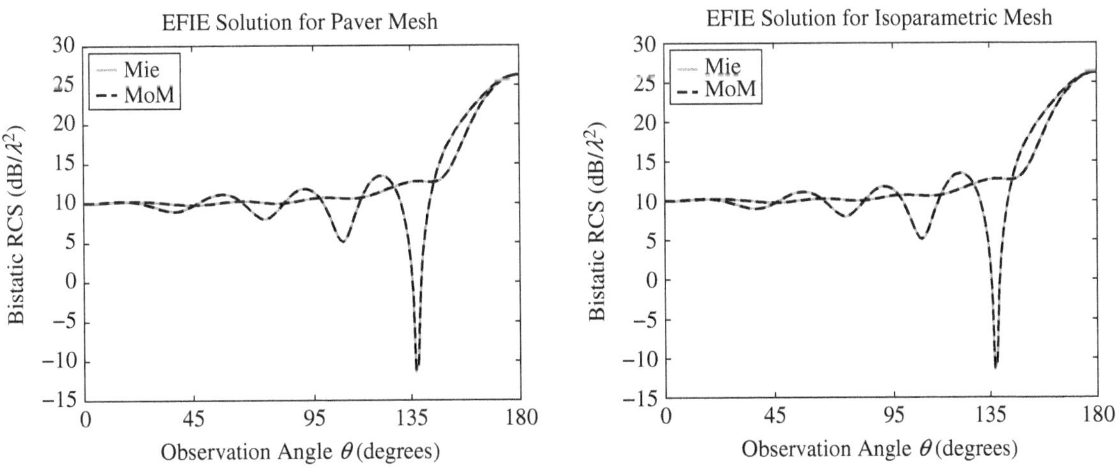


**FIGURE 6-65** Bistatic scattering results for the sphere using the EFIE.

The first test geometry being considered is a PEC sphere of radius 1 m, depicted in Figure 6-63a. This is a unique 3-D geometry that lends itself to an analytical solution (Mie series) for its radar scattering. Hence, a reference solution is available for evaluating the numerical solution accuracy.

Figure 6-65 compares the bistatic RCS for this 1 m radius sphere at 300 MHz using the paver and isoparametric meshes for both E- and H-polarizations. Clearly,

**TABLE 6-5** Percent root mean square (RMS) error in computing the RCS of a sphere of 1 m diameter using meshes having average edge length of 0.3λ, 0.2λ, 0.1λ (EFIE solution).

| | Paver | Paver | Paver | Iso | Iso | Iso |
|---|---|---|---|---|---|---|
| Discretization | 0.3 λ | 0.2 λ | 0.1 λ | 0.3 λ | 0.2 λ | 0.1 λ |
| Matrix size | 192 | 486 | 2118 | 192 | 423 | 1728 |
| *E*-polarization | 16.05 % | 4.01 % | 0.39 % | 16.58 % | 5.26 % | 0.56 % |
| *H*-polarization | 18.00 % | 4.73 % | 0.43 % | 18.31 % | 5.77 % | 0.65 % |

the MoM solution is in good agreement with the Mie series solution (Harrington, 1968). It is important to note that this level of accuracy is unattainable using lower order geometry modeling such as flat triangular elements as given in Rao, Wilton and Glisson (1982) and Sertel and Gürel (1997). A way to demonstrate the accuracy of the numerical solution is to increase the mesh density and observe possible changes to the solution as the geometry fidelity is improved. Table 6-5 summarizes numerical error versus average edge length (0.3λ, 0.2λ, 0.1λ). As seen, the error is rather large (16%–18%) for meshes with elements having average length of 0.3λ but reduces to less than 1% for elements with edge length of 0.1λ. This comparison also demonstrates the typical choice of 0.1λ edge length for typical discretizations.

For the EFIE solution leading to the results in Table 6-5, a CGS (Volakis, Chatterjee, and Kempel, 1998; Saad, 1996) iterative solver was employed. The CGS requires two matrix-vector products per iteration and provides the best convergence among other iterative solvers. More specifically, for the paver mesh (2,118 unknowns), the EFIE convergence was achieved in 82 iterations for the $E$-polarization (81 for $H$-polarization) to reach a prespecified residual error of $10^{-3}$.

To further speed up the convergence and avoid ill-conditioning of the EFIE (due to spurious solutions around internal resonances of the geometry), the CFIE must be adopted. As noted, the CFIE is simply a linear combination of the EFIE and MFIE, viz. CFIE $= \alpha$ EFIE $+ (1-\alpha)\frac{i}{k\eta}$ MFIE, where $\alpha$ is a constant of choice. Figure 6-66 gives the RCS results for the same setup as in Figure 6-65 with $\alpha = 0.5$. The corresponding RMS error for this CFIE solution are given in Table 6-6. By

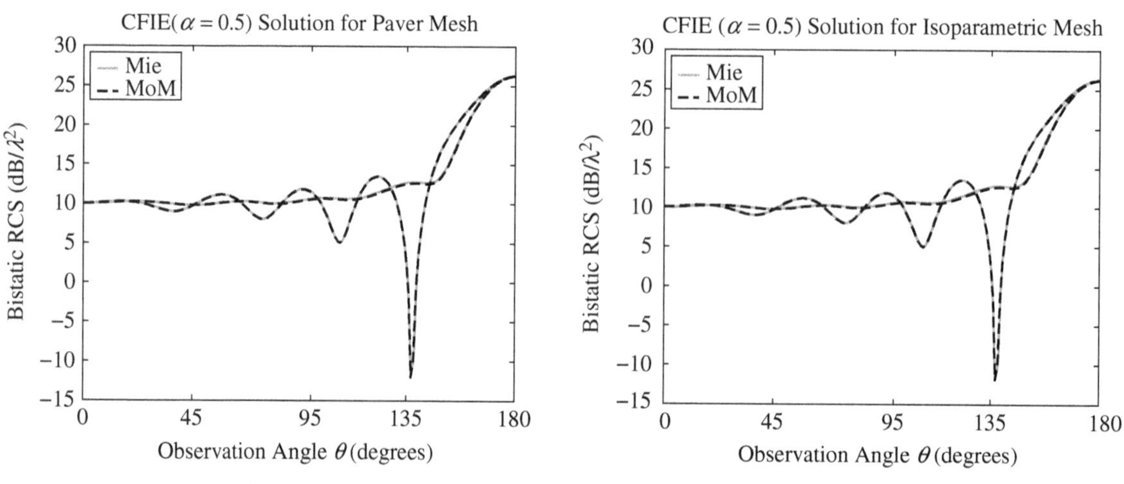


**FIGURE 6-66** Bistatic scattering from a 1m. radius sphere at 300 MHz obtained via the CFIE (with $\alpha = 0.5$). (a) Paver and (b) isoparametric meshes.

**TABLE 6-6** Percent RMS error for the computed RCS of a sphere for different mesh densities (CFIE solution).

| | Paver | Paver | Paver | Iso | Iso | Iso |
|---|---|---|---|---|---|---|
| Discretization | 0.3 λ | 0.2 λ | 0.1 λ | 0.3 λ | 0.2 λ | 0.1 λ |
| Matrix size | 192 | 486 | 2118 | 192 | 423 | 1728 |
| $\theta$-polarization | 16.04 % | 5.08 % | 0.82 % | 16.01 % | 5.70 % | 1.00 % |
| $\phi$-polarization | 16.26 % | 4.23 % | 0.67 % | 16.46 % | 5.05 % | 0.82 % |

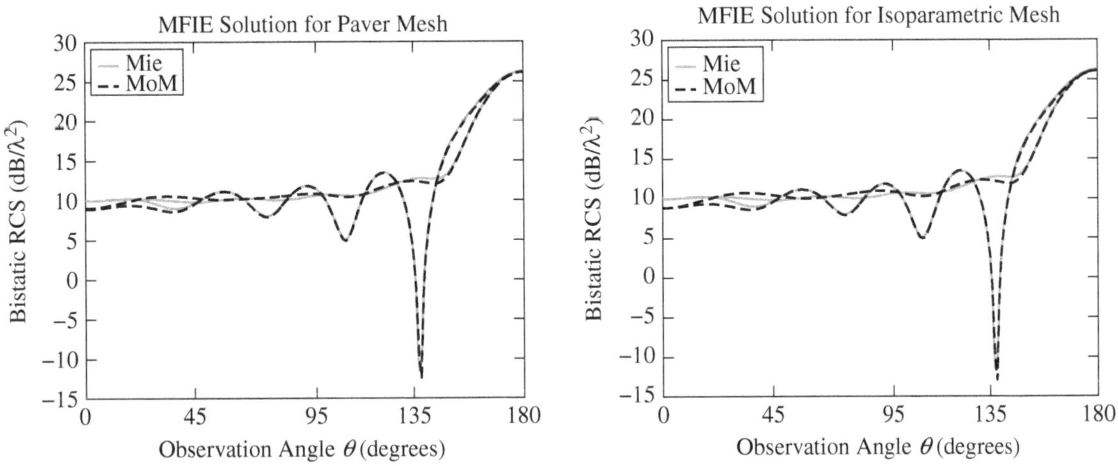


**FIGURE 6-67** Bistatic RCS of a sphere 1 m in diameter at 300 MHz obtained via the MFIE (with $\alpha = 0.5$): (a) paver and (b) isoparametric meshes.

comparison, the CFIE achieved convergence within 15 iterations (vs. 81 for the EFIE), (i.e., significantly less than the EFIE) to achieve an error of less than $10^{-3}$. Figure 6-67 gives the RCS results based on the MFIE. The MFIE is also prone to spurious solutions. But unlike the EFIE, the MFIE fails to provide correct far-field results at interior resonance frequencies. The curves in Figure 6-67 are provided for completeness only.

As a second test geometry, we consider the scattering by a cube where the edges are 1 meter long. As can be understood, there is no advantage in using curvilinear elements to grid the cube's surface. This geometry was chosen as a test case to represent targets with sharp edges. The bistatic RCS for the cube (both polarizations) using the EFIE and the CFIE are given in Figure 6-68. Since there is no analytical solution for this problem, solutions with different meshes are used to demonstrate convergence. The agreement for both integral equations is excellent, and the RMS error between the solutions using edge lengths of $0.1\lambda$ and $0.075\lambda$ are only 0.38% to 0.37% for the EFIE and 0.48% to 0.53% for the CFIE. The nominal $0.1\lambda$ sampling rate resulted in 1,200 unknowns, and convergence was achieved within 137 − 102 iterations for the EFIE and 13 iterations for the CFIE using the CGS method. The oversampled mesh (edge length $0.075\lambda$) generated 2,028 unknowns, and the EFIE converged in 116 iterations compared with 14 for the CFIE.

The PEC cylinder is the third test geometry considered. It is 1 meter long and 1 meter in diameter. Also, as before, all calculations are done at 300 MHz. Curvilinear elements were used to mesh the surface with the principle plane bistatic RCS shown

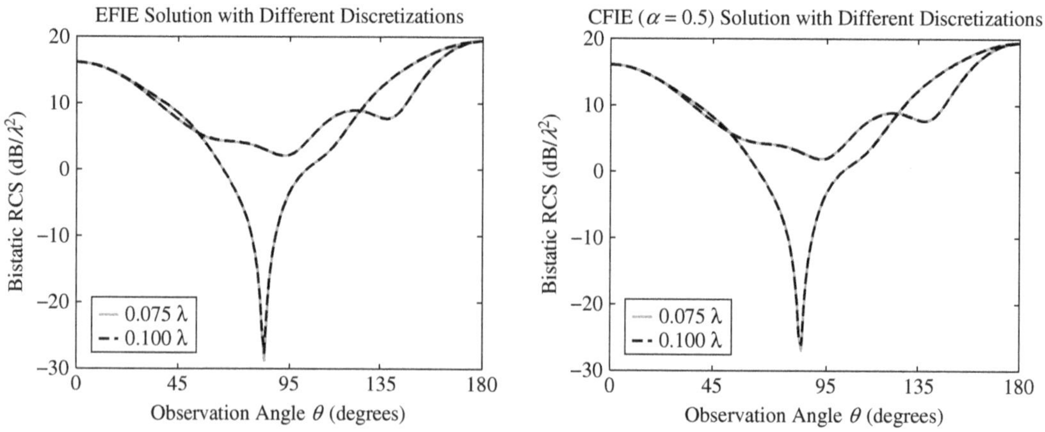


**FIGURE 6-68** ■ Bistatic RCS for a cube (1 m side at 300 MHz) based on (a) the EFIE and (b) the CFIE and solutions using different discterizations.

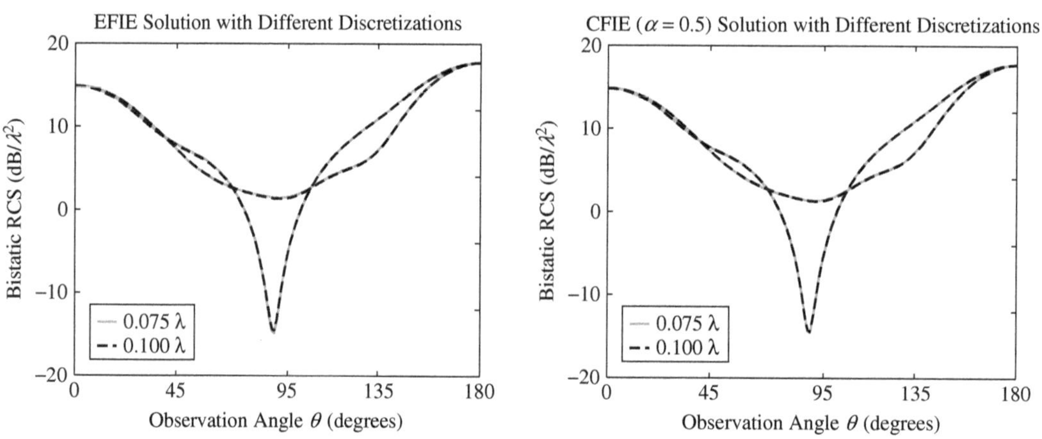


**FIGURE 6-69** ■ Bistatic RCS curves for a cylinder (1 m long and 1 m diameter) at 300 MHz using (a) the EFIE and (b) the CFIE.

in Figure 6-69. As done for the cube, the convergence of the RCS is examined and verified using higher-density meshing ($0.1\lambda$ and $0.075\lambda$ per edge length). We observe that the difference between the two paver mesh solutions are only 0.31% to 0.34% for the EFIE and 0.47% to 0.50% for CFIE. For the isoparametric meshes, the corresponding errors are 0.30% to 0.29% for the EFIE and 0.36% to 0.42% for the CFIE. We also note here that for the oversampled mesh (0.075 per edge length), the difference between the EFIE and CFIE solutions is about 0.70%.

As a fourth test geometry, we consider a PEC open-ended cylinder (cylinder without the top cap). For this case, as the geometry is open, only EFIE can be used. The associated bistatic RCS results are given in Figure 6-70, and the agreement between the two mesh densities is excellent, indicating convergence. We also note the effect on the RCS after the top cap of the cylinder is opened. Specifically, the backscattering ($\theta = 0$) is reduced by 3 dB after removing the top cover of the cylinder.

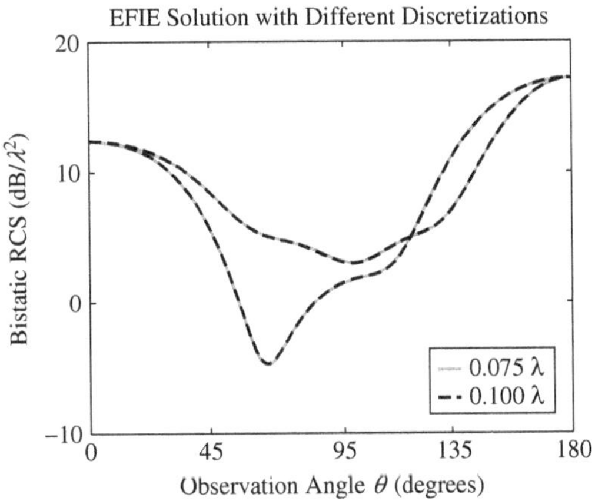


**FIGURE 6-70** Bistatic RCS for a cylinder with an open end (1 m long and 1 m wide) at 300 MHz using EFIE.

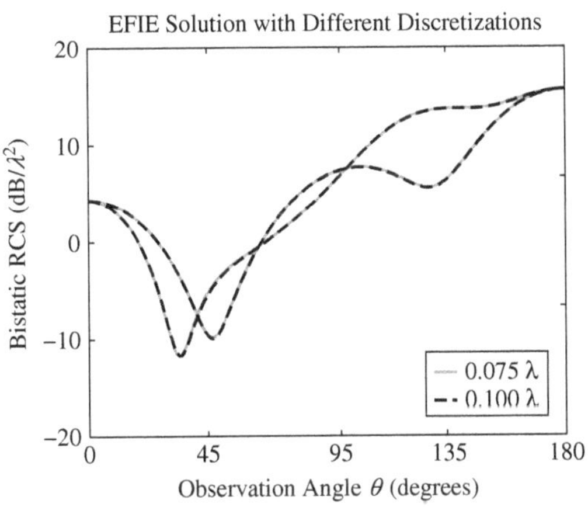


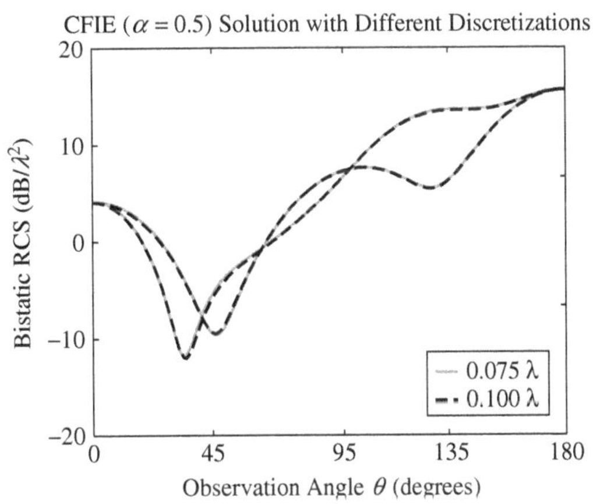


**FIGURE 6-71** Bistatic scattering results for the pyramid using the EFIE and the CFIE ($\alpha = 0.5$).

The fifth geometry considered is the pyramid (1 meter edge length) shown in Figure 6-63(e). Two different RCS curves for meshes with $0.1\lambda$ and $0.075\lambda$ per edge length are given in Figure 6-71 (300 MHz). We note that compared with the closed cylinder there is less than 10 dB backscatter return. The paver and isoparametric meshes resulted in 704 and 960 unknowns at $\lambda/10$ sampling and 1,222 and 1,766 unknowns at $\lambda/15$ sampling, respectively. For these meshes, the EFIE converged in 69 CGS iterations, and CFIE took only 15 CGS iterations for the 704 unknown problem. For the oversampled mesh (1,766 unknowns), the EFIE took 145 to 142 iterations compared with only 24 iterations for the CFIE.

Our sixth and final test geometry is a 1 meter long, 0.2 meters thick ogive geometry (Woo, Wang, Schuh, and Sanders, 1993). The ogive has a curved, elongated body and two sharp tips. The sharp tips present a modeling challenge, and consequently the geometry is well suited for validating electromagnetic computer codes. The computed bistatic RCS at 300 MHz is given in Figure 6-72, and we note that dynamic range in these curves is about 50 dB. In spite of this large dynamic range, the predicted results for both polarizations agree well for both mesh densities (with

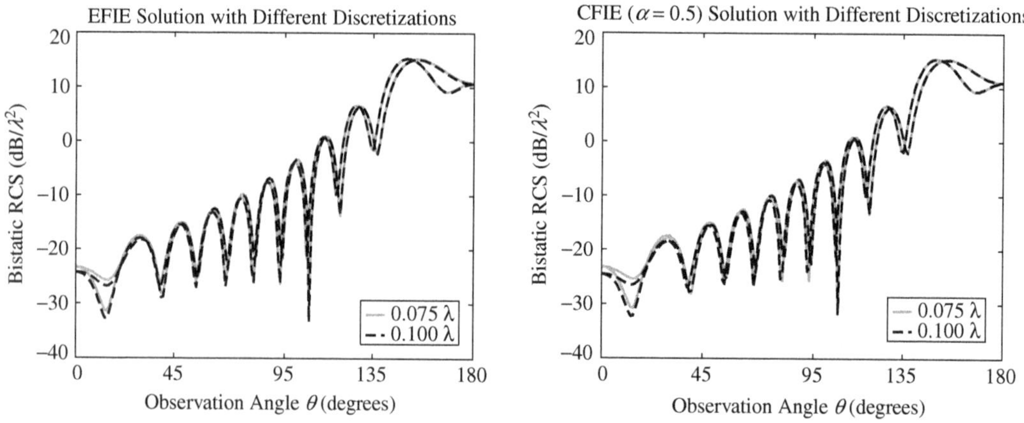


**FIGURE 6-72** ■ Bistatic RCS for an ogive 1 m long and 0.4 m thick. Curves are given for (a) the EFIE and (b) the CFIE.

**FIGURE 6-73** ■ Induced surface currents on test targets in Figure 6-63.

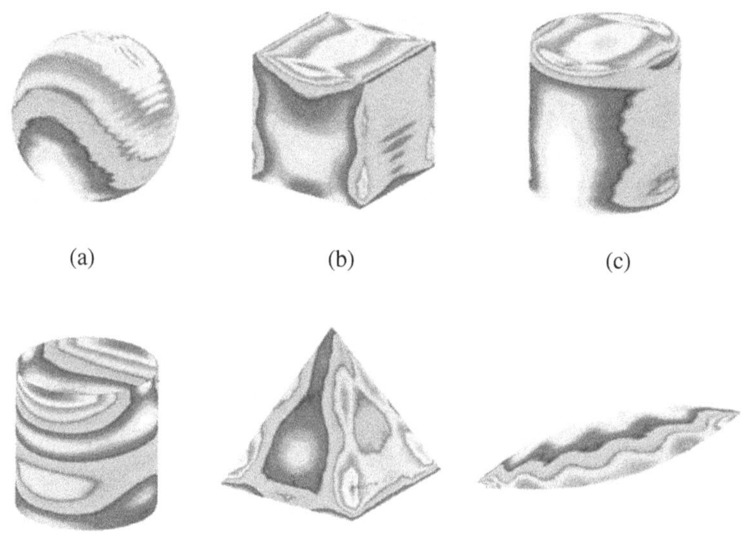


0.1λ and 0.075λ average edge lengths). We note that the EFIE convergence was achieved in 313 to 363 iterations compared with 30 to 31 iterations when the CFIE is employed (for a system of 8,872 unknowns).

We conclude this section by plotting induced surface current densities for all of the previously provided test geometries in Figure 6-73. These correspond to the incidences noted with the plotted RCS curves. Usually, induced surface currents are not observable in RCS measurements. However, computed distributions such as those given in Figure 6-73 can facilitate our understanding of scattering mechanisms.

For realistic structures with sharp edges and tips, highly curved surfaces, and large electrical sizes, the matrix system quickly increases to many unknowns. As an example, the EFIE for the ogive (8,872 unknowns) required 630 MB of computer memory and 45 minutes of solution time using a 1 GHz Pentium III processor.

To reduce this $O(N^2)$ memory requirement and large CPU requirements for MoM solutions, in the next chapter we outline the fast multipole and its multilevel version.

## 6.6 | STEP BY STEP MOMENT METHOD EXAMPLE

In this section, we present a detailed example to illustrate various aspects of the moment method. We consider the scattering by an infinitely thin PEC plate depicted in Fig. 6-74. As shown, the plate lies in the $x-y$ plane and is $3 \times 1$ meters in aperture. For excitation, an $x$-polarized plane wave, traveling in the $-z$ direction is assumed. The frequency of the incident plane wave is assumed to be 30MHz ($\lambda_0 = 10$ meters). Of interest is to find the induced surface currents on the plate. These contribute to the scattered field in accordance with (2.48). To compute the induced surface current **J** on the plate's surface, we proceed to solve the integral equation given in (3.42a) using the moment method (see Chapter 3).

The first step in the moment method implementation is to approximate the geometry of the scatterer (plate in this case) by a grid of finite elements. Here, we use a triangular mesh and the associated RWG basis functions as detailed in Section 6.1.3 to represent **J**. For illustrative purposes, we keep the mesh small, consisting of only 6 triangular elements. This mesh (see Fig. 6-75) is described

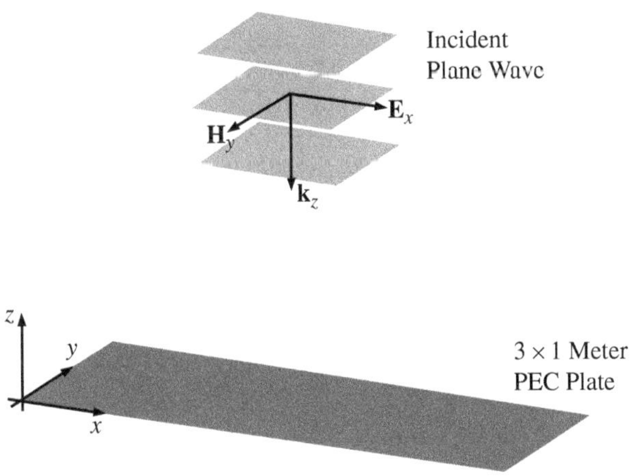


FIGURE 6-74 Geometry setup for computing the scattering by a $3 \times 1$ meter PEC plate.

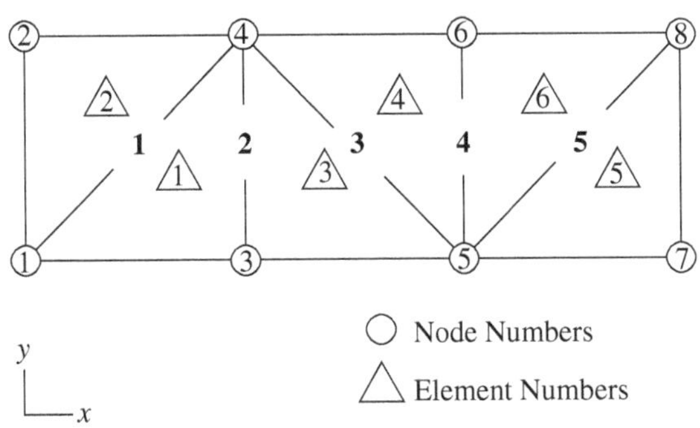


FIGURE 6-75 Triangular element mesh of the PEC plate illustrating the node, element, and basis function numbering.

by nodes and their interconnectivity. The node numbering scheme is given in a table as:

| Triangle # | Global Triangle Nodes $(n, p, q)$ |
|---|---|
| 1 | (4, 3, 1) |
| 2 | (1, 2, 4) |
| 3 | (3, 4, 5) |
| 4 | (6, 5, 4) |
| 5 | (8, 7, 5) |
| 6 | (5, 6, 8) |

Additionally, the coordinates of all the nodes in the mesh are also needed, as listed below:

| Node # | Node Coordinates $(x, y, z)$ |
|---|---|
| 1 | (0.0, 0.0, 0.0) |
| 2 | (0.0, 1.0, 0.0) |
| 3 | (1.0, 0.0, 0.0) |
| 4 | (1.0, 1.0, 0.0) |
| 5 | (2.0, 0.0, 0.0) |
| 6 | (2.0, 1.0, 0.0) |
| 7 | (3.0, 0.0, 0.0) |
| 8 | (3.0, 1.0, 0.0) |

These two lists provide a description of the geometry. Also, the shared edges internal to the mesh can be identified. The example mesh in Fig. 6-75 has 5 internal edges, numbered from 1 to 5. In addition, a current direction must also be assigned for each edge. Specifically, the basis function #2 defined on elements 1 and 3 defines a current that flows from element 1 to element 3. The vector field for the $2^{nd}$ and $3^{rd}$ RWG basis functions are depicted in Fig 6-76.

Having defined the mesh geometry, the next step is the computation of the impedance matrix entries, using (6.16). To do so, a suitable numerical quadrature can be used to numerically evaluate the double surface integrals. For example, we

**FIGURE 6-76** Vector plots of two basis functions defined on the triangular mesh of the PEC plate.

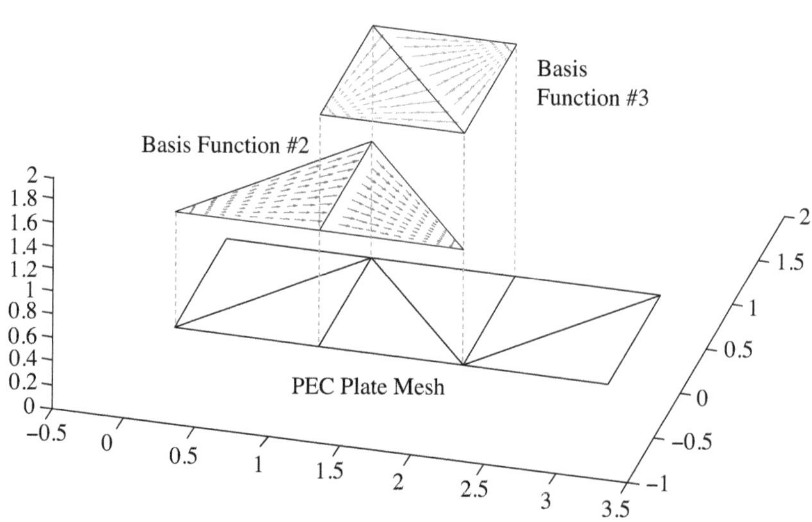

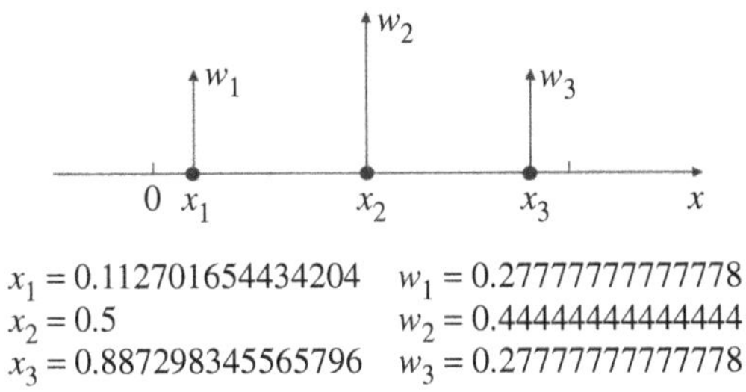


**FIGURE 6-77** 3-point Gaussian quadrature sample locations and weights.

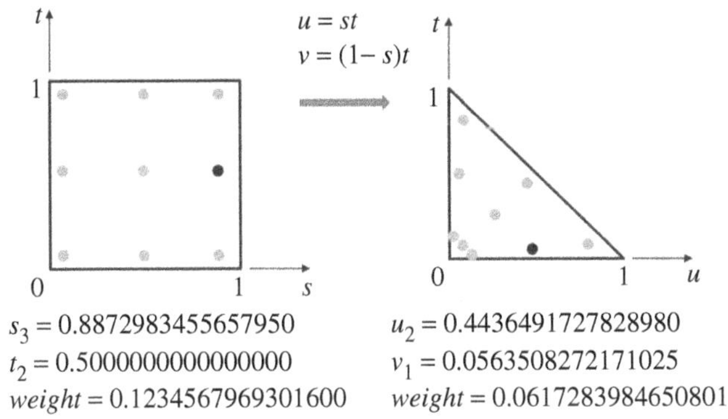


**FIGURE 6-78** Mapping of a 2-dimensional, 9-point Gaussian quadrature onto a triangular surface using a 3-point Gaussian quadrature.

may use a 9-point Gaussian quadrature for the outer surface integrals and a 16-point Gaussian quadrature for the inner integrals in (6.18). Alternatively, a 2-dimensional quadrature rule can be employed as a Cartesian product of 1-dimensional Gaussian quadratures (see Fig. 6-77).

Of course, the Cartesian product is typically employed over a rectangular integration area. However, the rectangular integration domain can be easily mapped to a triangular surface via the transformation shown in Fig. 6-78.

As a simple numerical computation of the MoM integrals, let us consider the evaluation of the excitation vector $b$ in (6.17) for a plane wave is incidence on the plate. An $x$-polarized incidence is assumed, propagating in the $-z$ direction, given by $\mathbf{E}^{inc} = \hat{x} E_0 e^{-ikz}$. The entries of the subject integral are evaluated using the 9-point Gaussian quadrature giving,

$$b_m = \frac{i}{k_0 Z_0} \int_{S_m^+ + S_m^-} \mathbf{j}_m(\mathbf{r}) \mathbf{E}^{inc}(\mathbf{r}) = \frac{i}{k_0 Z_0} \sum_{i=1}^{9} w_i \mathbf{j}_m(\mathbf{r}_i) \mathbf{E}^{inc}(\mathbf{r}_i). \tag{6.209}$$

Here, $\mathbf{r}_i$ are the quadrature locations and $w_i$ are the associated weights. Upon integrating, the resulting right hand side vector entries $\{b_m\}$ are:

| Basis Index | Excitation Vector Entry |
|---|---|
| 1 | (0.0, −6.666667) |
| 2 | (0.0, 13.333333) |
| 3 | (0.0, 6.666667) |
| 4 | (0.0, 13.333333) |
| 5 | (0.0, −6.666667) |

The impedance matrix can be evaluated in a similar way. For the EFIE formulation, the resulting impedance matrix is symmetric when Galerkin's testing is employed (i.e. RWG functions are used for both basis and testing functions). Thus, it is typically advantageous to store the matrix in a compressed form, rather than storing the whole matrix. Compressed storage is also critical when the MoM implementations are carried out via fast integral methods (see Chapter 7).

A common and straightforward method for storing sparse matrices is the compressed row storage. Here, the matrix entries are stored in a long, 1-dimensional array. For instance, for a symmetric matrix, one can store only the lower triangular half. Thus, instead of storing all $N^2$ entries, only $N^2/2$ elements are stored, reducing computer storage requirements by 1/2. Two auxiliary integer arrays are also needed for storing general sparse matrices. These are the *column-index* and the *row-pointer* arrays. The column-index array is of the same length as the compressed impedance matrix. The row-pointer array is of length $N$, and contains integer locations of the beginning of the corresponding row in the impedance matrix. As an example, the 5 basis functions in Fig. 6-75, result in an impedance matrix that has a total of 15 entries (only lower triangular part is stored due to symmetry). The row-pointer index for the lower triangular half is therefore $I_{pointer} = \{1, 2, 4, 7, 11, 16\}$. The corresponding column-index becomes $J_{index} = \{1, 1, 2, 1, 2, 3, 1, 2, 3, 4, 1, 2, 3, 4, 5\}$.

For the example above, the following numerical values of the impedance matrix entries are computed by the numerical quadrature:

| Column Index | Compressed $Z$ Matrix |
|---|---|
| 1 | (−10.13010, 0.091087580) |
| 1 | (−4.153840, −0.088780761) |
| 2 | (−11.65531, 0.2279114) |
| 1 | (−0.9522870, −0.0017802715) |
| 2 | (4.154089, 0.088780761) |
| 3 | (−10.13011, 0.091087103) |
| 1 | (−0.4132653, −0.087215662) |
| 2 | (1.904318, 0.1492724) |
| 3 | (4.153667, 0.088780403) |
| 4 | (−11.65531, 0.2279118) |
| 1 | (0.066926748, 0.072052717) |
| 2 | (−0.4132658, −0.087215781) |
| 3 | (−0.9522865, −0.0017806292) |
| 4 | (−4.153914, −0.088779569) |
| 5 | (−10.13010, 0.091087580) |

The solution vector obtained from the above impedance matrix and excitation vector can be obtained using standard matrix solvers (such as the LU decomposition or iterative solvers). The solution for the given $Z$ matrix and the excitation vector is

| Basis Index | Excitation Vector Entry |
|---|---|
| 1 | (−0.9178263, 3.6446535) |
| 2 | (1.3772673, −5.4526926) |
| 3 | (1.4508530, −5.7775457) |
| 4 | (1.3772242, −5.4525274) |
| 5 | (−0.9178200, 3.6446325) |

Having the solution of the MoM system, the resulting surface current on the plate can be constructed as a linear combination of the RWG basis functions and

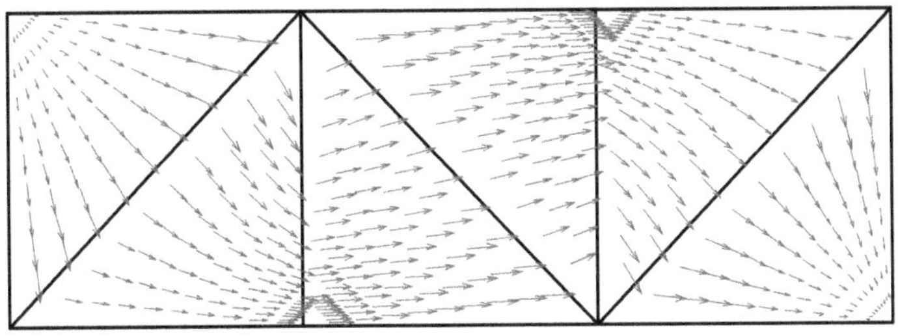

**FIGURE 6-79** ■ Quiver plot of the induced surface current magnitude on a PEC plate computed as a sum of the RWG basis functions.

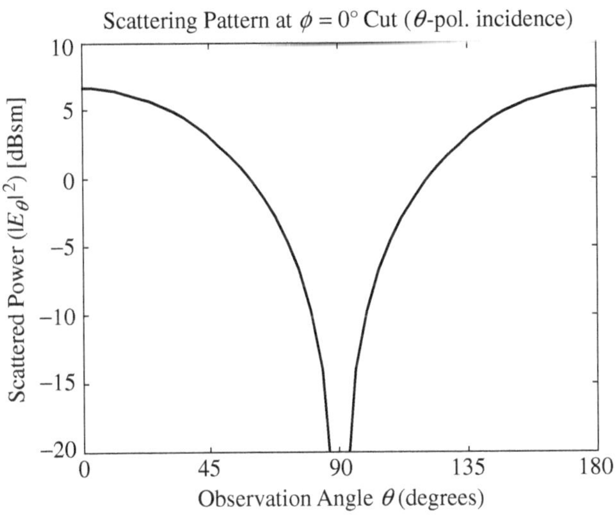


**FIGURE 6-80** ■ Scattered electric field plotted as a function of $\theta$ angle in the $x-z$ plane.

is depicted in Fig. 6-79. Using the currents, the radiation integral in (6.149) is next evaluated to compute the scattered far-zone fields given in Fig. 6-80 for the $x-z$ plane cut.

# BIBLIOGRAPHY

Abramowitz, M. and Stegun, I. A. (1972). *Handbook of Mathematical Functions.* National Bureau of Standards.

Adams, R. J. (2004). Physical and analytical properties of stabilized electric field integral equation. *IEEE Trans. Antennas Propagat.*, *52*(2), 362–372.

Andriulli, F., Cools, K., Bagci, H., Olyslager, F., Buffa, A., Christiansen, S., and Michielssen, E. (2008). A multiplicative Calderon preconditioner for the electric field integral equation. *IEEE Trans. on Antennas and Propagation*, *56*(8), 2398–2412.

Antilla, G. E. and Alexopoulos, N. G. (1994). Scattering from complex 3D geometries by a curvilinear hybrid finite element-integral equation approach. *J. Opt. Soc. Am. A.*, *11*(4), 1445–1457.

Arvas, E., Rahhal-Arabi, A., and Sadigh, A. (1991). Scattering from multiple conducting and dielectric bodies of arbitrary shape. *IEEE Antennas Propagation Society Magazine*, *33*, 29–36.

Barkeshli, K. and Volakis, J. L. (1990). On the implementation and accuracy of the conjugate gradient Fourier transform method. *IEEE Antennas Propagat. Mag.*, *32*(2), 20–26.

Bleszynski, E., Bleszynski, M., and Jaroszewicz, T. (1996). AIM: Adaptive integral method for solving large-scale electromagnetic scattering and radiation problems. *Radio Sci.*, *31*(5), 1225–1251.

Bluck, M. J. and Walker, S. P. (2007). High-order discrete Helmholtz decompositions for the electric field integral equation. *IEEE Trans. on Antennas and Propagation*, *55*(5), 1338–1347.

Buffa, A. and Christiansen, S. (2007). A dual finite element complex on the barycentric refinement. *Math. Comput.*, *76*, 1743–1769.

Carr, M., Bleszynski, M., and Volakis, J. L. (2004). A near-field preconditioner and its performance in conjunction with the BiCGstab(ell) solver. *IEEE Antennas and Propagation Magazine*, *46*(2), 23–30.

Chew, W. C. (1995). *Waves and Fields in Inhomogeneous Media*. Piscataway, NJ: IEEE Press.

Chew, W. C., Jin, J., Lu, C., Michielssen, E., and Song, J. M. (1997). Fast solution methods in electromagnetics. *IEEE Trans. Antennas Propagat.*, *45*(3), 533–543.

Christiansen, S. and Nedelec, J.-C. (2003). A preconditioner for the electric field integral equation based on Calderon formulas. *SIAM J. Numer. Anal.*, *40*, 1100–1135.

Coifman, R., Rokhlin, V., and Wandzura, S. (1993). The fast multipole method for the wave equation: A pedestrian prescription. *IEEE Antennas and Propagat. Mag.*, *35*(3), 7–12.

Duffy, M. G. (1982). Quadrature over a pyramid or cube of integrands with a singularity at a vertex. *SIAM J. Numer. Anal.*, *19*(6), 1260–1262.

Eibert, T. and Hansen, V. (1996). Calculation of unbounded field problems in free space by a 3-D FEM/BEM-hybrid approach. *Journal of Electromagnetic Waves and Applications*, *10*(1), 61–78.

Farin, G. (1997). *Curves and Surfaces for Computer Aided Geometric Design*. San Diego: Academic Press.

Graglia, R. D. (1987). Static and dynamic potential integrals for linearly varying source distributions in two- and three-dimensional problems. *IEEE Trans. on Antennas and Propagation*, *35*(6), 662–669.

Graglia, R. D. (1988). The use of parametric elements in the moment method solution of static and dynamic volume integral equations. *IEEE Trans. on Antennas and Propagation*, *36*(5), 636–646.

Graglia, R. D. (1993). On the numerical integration of the linear shape functions times the 3-D green's function or its gradient on a plane triangle. *IEEE Trans. on Antennas and Propagation*, *41*(10), 1448–1455.

Graglia, R. D., Wilton, D. R., and Peterson, A. F. (1997). Higher order interpolatory vector bases for computational electromagnetics. *IEEE Trans. Antennas Propagat.*, *45*(3), 329–342.

Graglia, R. D., Wilton, D. R., Peterson, A. F., and Gheorma, I.-L. (1998). Higher order interpolatory vector bases on prism elements. *IEEE Trans. Antennas Propagat.*, *46*(3), 442–450.

Harrington, R. F. (1968). *Field Computations by Moment Method*. New York: MacMillan Co.

Horng, T., McKinzie, W. E., and Alexopoulos, G. (1992). Full-wave spectral-domain analysis of compansation of microstrip discontinuities using triangular subdomain basis functions. *IEEE Trans. on Microwave Theory and Techniques*, *40*(12), 2137–2147.

Hsiao, G. C. and Kleinman, R. E. (1997). Mathematical foundations for error estimation in numerical solutions of integral equations in electromagnetics. *IEEE Trans. Antennas Propagat.*, *45*(3), 316–328.

Jin, J.-M. and Volakis, J. L. (1991). A finite element-boundary integral formulation for scattering by three-dimensional cavity-backed apertures. *IEEE Trans. Antennas and Propagat.*, *39*, 97–104.

Jin, J.-M., Volakis, J. L., and Liepa, V. V. (1989). A moment method solution of a volume-surface integral equation using isoparametric elements and point matching TE scattering. *IEEE Trans. Microwave Theory Tech.*, *37*(10), 1641–1645.

Jorgensen, E., Volakis, J. L., Meincke, P., and Breinbjerg, O. (2004). Higher order hierarchical Legendre basis functions for electromagnetic modeling. *IEEE Trans. on Antennas and Propagation*, 2985–2995.

Khayat, M. A. and Wilton, D. R. (2005). Numerical evaluation of singular and near-singular potential integrals. *IEEE Trans. Antennas Propagat.*, *53*(10), 3180–3190.

Lee, J.-F., Lee, R., and Burkholder, R. J. (2003). Loop star basis functions and a robust preconditioner for EFIE scattering problems. *IEEE Trans. on Antennas and Propagation*, *51*(8), 1855–1863.

Livesay, D. E. and Chen, K. (1974). Electromagnetic fields induced inside arbitrarily shaped biological bodies. *IEEE Trans. Microwave Theory Tech.*, *22*(12), 1273–1280.

Lu, C. C. (2001). Indoor radio wave propagation modeling using multilevel fast multipole algorithm. *Microwave Opt. Tech. Lett.*, *29*(3), 168–175.

Makarov, S. (2002). *Antenna and EM Modeling with Matlab*. New York: Wiley-Interscience.

MSC Software (2010). Patran: CAE modeling and pre/post processing. Available at http://www.mscsoftware.com/Contents/Products/CAE-Tools/Patran.aspx.

Notaros, B. M. and Popovic, B. D. (1996). General entire-domain method for analysis of dielectric scatterers. *IEE Proceedings - Microwaves, Antennas and Propagation*, *143*(6), 498–504.

Notaros, B. M., Popovic, B. D., Weem, J. P., Brown, R. A., and Popovic, Z. (2001). Efficient large-domain MoM solutions to electrically large practical EM problems. *IEEE Transactions on Microwave Theory and Techniques*, *49*(1), 151–159.

Özdemir, T. and Volakis, J. L. (1997). Triangular prisms for edge-based vector finite elements analysis of conformal antennas. *IEEE Trans. Antennas Propagat.*, *45*(5), 788–797.

Peterson, A. F. (1991). Analysis of heterogeneous electromagnetic scatterers: research progress of the past decade. *Proc. IEEE*, *79*(10), 1431–1441.

Popović, B. D. and Kolundžija, B. M. (1994). *Analysis of Metallic Antennas and Scatterers*, vol. 38 of *IEE Electromagnetic Wave Series*. Institution of Engineering and Technology (IET).

Putnam, J. M. and Medgyesi-Mitchang, L. N. (1991). Combined field formulation for inhomogeneous two- and three-dimensional bodies: The junction problem. *IEEE Trans. Antennas Propagat.*, *39*, 667–672.

Rao, S. and Wilton, D. (1991). Transient scattering by conducting surfaces of arbitrary shape. *IEEE Transactions on Antennas and Propagation*, *39*(1), 56–61.

Rao, S., D. Wilton, and Glisson, A. (1982). Electromagnetic scattering by surfaces of arbitrary shape. *IEEE Trans. on Antennas and Propagation*, *AP-30*(3), 409–418.

Richmond, J. H. (1965). Scattering by a dielectric cylinder of arbitrary cross section shape. *IEEE Trans. Antennas Propagat.*, *13*(3), 334–341.

Rogers, D. F. and Adams, J. A. (1990). *Mathematical Elements for Computer Graphics*. New York: McGraw Hill.

Saad, Y. (1996). *Iterative Methods for Sparse Linear Systems*. Boston: PWS Pub. Co.

Sertel, K. and Gürel, L. (1997). A comparison of surface modeling techniques. *IEEE AP-S Conference Digest*, *3*, 1834–1837.

Sertel, K. and Volakis, J. L. (2000). Incomplete LU preconditioner for FMM implementation. *Microwave Opt. Tech. Lett.*, *26*(4), 265–267.

Sertel, K. and Volakis, J. L. (2002). Method of moments solution of volume integral equations using parametric geometry. *Radio Sci.*, *37*(1), 1–7.

Song, J. M. and Chew, W. C. (1995). Moment method solutions using parametric geometry. *Journal of Electromagnetic Waves and Applications*, *9*(1/2), 71–83.

Stephanson, M. B. and Lee, J.-F. (2009). Preconditioned electric field integral equation using Calderon identities and dual loop/star basis functions. *IEEE Trans. on Antennas and Propagation*, *57*(4), 1274–1279.

Topsakal, E., Kindt, R., Sertel, K., and Volakis, J. (2001). Evaluation of the BICGSTAB(l) algorithm for the finite-element/boundary-integral method. *IEEE Antennas Propagation Society Magazine*, *43*, 124–131.

Volakis, J., Sertel, K., and Usner, B. (2006). *Frequency Domain Hybrid Finite Element Methods in Electromagnetics (Synthesis Lectures on Computational Electromagnetics).* Morgan and Claypool Publishers.

Volakis, J. L. (1992). Alternative field representations and integral equations for modeling inhomogeneous dielectrics. *IEEE Trans. Microwave Theory and Techn.*, 604–608.

Volakis, J. L., Chatterjee, A., and Kempel, L. (1998). *Finite Element Methods for Electromagnetics.* New York: IEEE Press.

Wandzura, S. (1992). Electric current basis functions for curved surfaces. *Electromagnetics*, *12*(1), 77–91.

Wilton, D. R., Rao, S. M., Glisson, A. W., Schaubert, D. H., Al-Bundak, O. M., and Butler, C. M. (1984). Potential integrals of uniform and linear source distributions on poygonal and polyhedral domains. *IEEE Trans. on Antennas and Propagation*, *AP-32*(3), 276–281.

Woo, A. C., Wang, H. T. G., Schuh, M. J., and Sanders, M. L. (1993). EM programmer's notebook-benchmark radar targets for the validation of computational electromagnetics programs. *IEEE Antennas and Propagation Magazine*, *35*(1), 84–89.

CHAPTER 7

# Fast Multipole Method and Its Multilevel Implementation

### Chapter Outline



As noted in the previous chapter, standard method of moments (MoM) implementations quickly reach the limits of computer time and memory resources. Even though better geometry modeling methods can enable solutions of the same problem using less computer resources, fast and low-memory solution methods must be employed in connection with iterative solvers to tackle electrically large real-life problems. The most time-consuming step in any iterative algorithm, such as the conjugate gradient (CG), biconjugate gradient (BiCG), quasi-minimal residual (QMR), and generalized minimal residual (GMRES) routines (see Saad, 1996; Topsakal, Kindt, Sertel, and Volakis, 2001, and the references therein), is the matrix-vector product. All fast methods have therefore focused on the rapid implementation of the $[Z]\{x\}$ matrix-vector product. Fast methods are often referred to as matrix compression algorithms, and $k$-space methods (Bojarski, 1971) were among the first such approaches to be employed with iterative solvers. Although $k$-space methods lead to $O(N \log N)$ memory and computational complexity, their application is restricted to systems/geometries that can be approximated with circulant matrices. Originally, this requirement could be fulfilled only using uniform discretizations of the integral equation. However, recently introduced fast integral methods such as the fast multipole method (FMM) (Coifman, Rokhlin, and Wandzura, 1993; Song, Lu, and Chew, 1997; Bindiganavale and Volakis, 1997; Lu and Jin, 1996; Chew, Jin, Lu, Michielssen, and Song, 1997; Sheng, Jin, Song, Lu, and Chew, 1998) and the adaptive integral method (AIM) (Eibert, Volakis, Wilton, and Jackson, 1999; Bleszynski, Bleszynski, and Jaroszewicz, 1996; Anastassiu, Smelyanskiy, Bindiganavale, and Volakis, 1998; Bindiganavale and Volakis, 1998) are rid of restrictions for uniform discretization of the original geometries. These methods were shown to deliver memory and central processing unit (CPU) reduction down to $O(N^{3/2})$ or better. Windowed fast multipole method (FMM) (Burkholder and Kwon, 1996), ray propagation fast multipole algorithm (Wagner and Chew, 1994), fast far-field approximation (Lu and Chew, 1995), and multilevel FMM (Song, Lu, and Chew, 1997) can reduce CPU time down to $O(N^{4/3})$ or even $O(N \log N)$ at the expense of further approximations. Early electromagnetic applications of FMM concentrated on pure integral equation approaches, but uses of FMM in the context of hybrid finite element-boundary integral (FE-BI) formulations have also been reported (Lu and Jin, 1996; Sheng, Jin, Song, Lu, and Chew, 1998).

AIM can be considered as the natural extension of the $k$–space methods and was introduced for arbitrary surface and volumetric scattering problems (Bleszynski, Bleszynski, and Jaroszewicz, 1996). It is certainly the method of choice for planar surface integral (or boundary integral) implementations. For that case, only two-dimensional fast Fourier transform (FFT) algorithms need be employed, and the method results in a low $O(N \log N)$ complexity. Thus, the speed up of AIM is considerably better than that of FMM or multilevel FMM (MLFMM) for planar boundaries (Song, Lu, and Chew, 1997), even for relatively small numbers of unknowns. However, for arbitrary three-dimensional geometries, AIM requires a three-dimensional grid and consequently a three-dimensional FFT implementation. As would be expected, most of the FFT grid points are not used in boundary integral discretizations since they lie away from the actual surface. This redundancy is a limiting factor for AIM. In contrast, for surface integral equations, FMM is more suitable and also equally applicable volumetric integral equations.

As mentioned in the previous chapter, the MoM matrix system requires $O(N^2)$ computer storage, where $N$ refers to the number of unknowns and the matrix dimension. Since typical discretizations require at least 10 samples (unknowns) per wavelength ($\lambda$), random access memory (RAM) requirements quickly become unmanageable for realistic geometries that range 100s of wavelengths in linear dimensions. Moreover, as noted, a direct lower-upper (LU) solution requires $O(N^3)$ operations, implying heavy CPU requirements, even for the largest of parallel machines. In contrast, iterative solution requires $O(N^2)$ flops per iteration, with the cost being mainly due to the matrix-vector multiplications in each iteration step. However, a closer examination of the matrix elements lead to faster and more efficient algorithms to evaluate the matrix-vector product indirectly. In this chapter, we outline the FMM and its multilevel version. We will particularly focus on curvilinear elements and conformal basis functions.

## 7.1 | FAST MULTIPOLE METHOD

When the MoM matrix system is solved iteratively, a search vector (approximating the solution vector) is generated at each iteration. As this search vector represents the basis function expansion, the product of that search vector with each matrix row is equivalent to computing the integral of the scattered, or radiated fields. In other words, each entry of the MoM matrix $Z_{ji} = \langle \mathbf{t}_j, \mathcal{L}(\mathbf{j}_i) \rangle$, where $\mathcal{L}$ is the integral operator, corresponds to the reaction between the field generated by the basis function $\mathbf{j}_i$ and the testing function $\mathbf{t}_j$.

The FMM relies on a mathematical manipulation of the free-space Green's function so that the reaction between the collective field of a group of basis functions and a group of testing functions can be evaluated at lower computational costs (by reusing

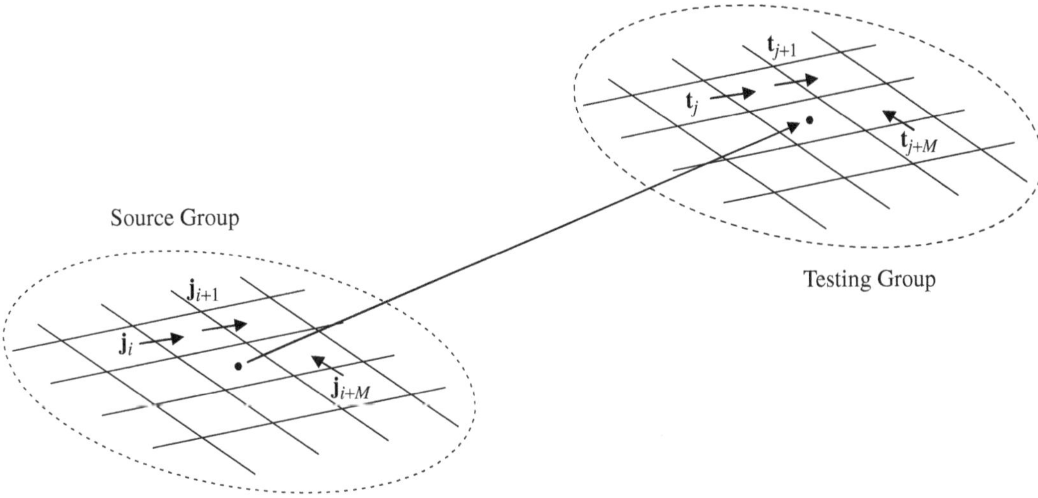


**FIGURE 7-1** Groups of basis and testing functions.

information). To show this, let us assume that basis functions $\mathbf{j}_i, \mathbf{j}_{i+1}, \ldots, \mathbf{j}_{i+M}$ are in the near vicinity of each other, as depicted in Figure 7-1. Similarly, we assume that the testing functions $\mathbf{t}_j, \mathbf{t}_{j+1}, \ldots, \mathbf{t}_{j+M}$ are in the near vicinity of each other as well. Furthermore, let us assume that these two groups are separated by a distance larger than the physical sizes of both groups, as already depicted in Figure 7-1.

By introducing a spherical multipole expansion to represent the fields radiated by the basis group over the testing group, it is possible to reduce the operation count necessary to evaluate the matrix-vector product. This spherical multipole expansion is based on Gegenbauer's addition theorem (Abramowitz and Stegun, 1972) given by

$$\frac{e^{-jk|\mathbf{r}+\mathbf{d}|}}{|\mathbf{r}+\mathbf{d}|} = -jk \sum_{l=0}^{\infty} (-1)^l (2l+1) j_l(kd) h_l^{(2)}(kr) P_l(\hat{d} \cdot \hat{r}). \tag{7.1}$$

In this, $j_l$ is the spherical Bessel function of order $l$ (not to be confused with $\mathbf{j}_i$ representing the current basis functions), $h_l^{(2)}$ is the spherical Hankel function of the second kind of order $l$, and $P_l$ is the Legendre polynomial of order $l$ (see Figure 7-2). This expansion is the backbone of the FMM algorithm, and its role in carrying out the matrix-vector products is discussed next.

Using (7.1) and the plane wave expansion of the product

$$4\pi(-j)^l j_l(kd) P_l(\hat{d} \cdot \hat{r}) = \int d^2\hat{k} e^{-j\mathbf{k}\cdot\mathbf{d}} P_l(\hat{k} \cdot \hat{r}) \tag{7.2}$$

it can be shown that

$$\frac{e^{-jk|\mathbf{r}+\mathbf{d}|}}{|\mathbf{r}+\mathbf{d}|} \approx \frac{-jk}{4\pi} \int d^2\hat{k} e^{-j\mathbf{k}\cdot\mathbf{d}} T_L(kr, \hat{k} \cdot \hat{r}) \tag{7.3}$$

where from (7.1)

$$T_L(kr, \hat{k} \cdot \hat{r}) = \sum_{l=0}^{L} (-j)^l (2l+1) h_l^{(2)}(kr) P_l(\hat{k} \cdot \hat{r}) \tag{7.4}$$

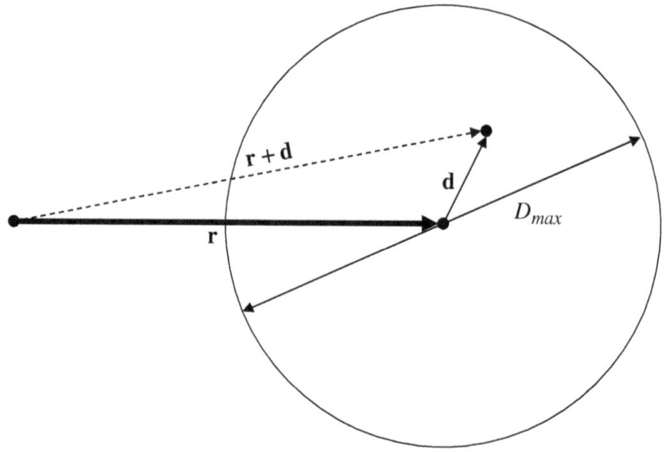


**FIGURE 7-2** Vector definitions for Gegenbauer's addition theorem.

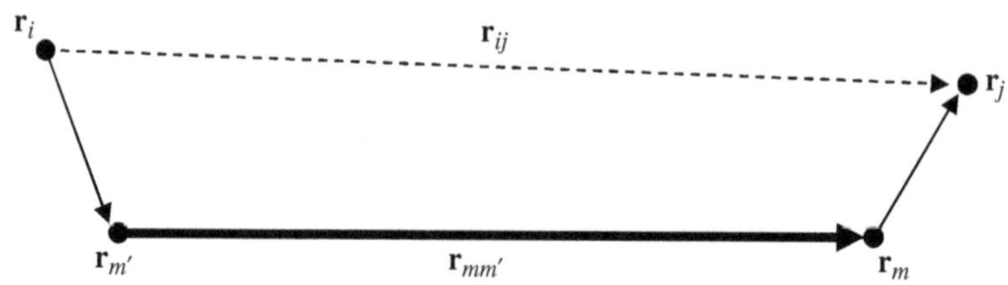


**FIGURE 7-3** Vector definitions for FMM expansion.

where the infinite upper limit of the summation was replaced by a finite integer $L$; that is, the infinite sum was truncated up to the first $(L + 1)$ multipole terms. This is a critical step in the approximation of the integral and implies that (7.4) is valid for $|\mathbf{r}| > |\mathbf{d}|$ in (7.1). However, (7.1) would require many terms for $|\mathbf{r}| \approx |\mathbf{d}|$. In practice $|\mathbf{d}|$ is chosen small to use fewer terms in the multipole sum for faster executions. Once the translation sum $T_L(kr, \hat{k} \cdot \hat{r})$ is computed for a given $\mathbf{r}$, (7.3) can be used to compute the field anywhere inside a sphere of radius $D_{max} < 2|\mathbf{r}|$ very quickly. Hence, the field generated by a source outside of the testing cluster, can be evaluated over all testing functions inside the testing cluster by employing (7.3) using the precomputed translation operator (Coifman, Rokhlin, and Wandzura, 1993). As the translation operators (7.4) can be precomputed and available a priori for many groups, there are significant CPU savings in computing the matrix elements.

For the more relevant vector construct, we refer to Figure 7-3. Using the identity (7.1), the scalar Green's function can be approximated as

$$\frac{e^{-jkr_{ji}}}{r_{ji}} \approx \frac{-jk}{4\pi} \int d^2\hat{k} e^{-j\mathbf{k}\cdot\mathbf{r}_{jm}} T_L(kr_{mm'}, \hat{k} \cdot \hat{r}_{mm'}) e^{j\mathbf{k}\cdot\mathbf{r}_{im'}}. \tag{7.5}$$

Using (7.5), it is straightforward to approximate the MoM matrix elements (for the electric field integral equation [EFIE])

$$Z_{ji} = \int_S d\mathbf{r}\mathbf{t}_j(\mathbf{r}) \cdot \int_S d\mathbf{r}' \overline{\mathbf{G}}(\mathbf{r}, \mathbf{r}') \cdot \mathbf{j}_i(\mathbf{r}') \tag{7.6}$$

using the translation operators as

$$Z_{ji} \approx \frac{-jk}{4\pi} \int d^2\hat{k} \mathbf{V}_{jm}(\hat{k}) \cdot T_L(kr_{mm'}, \hat{k} \cdot \hat{r}_{mm'}) \mathbf{V}^*_{im'}(\hat{k}) \tag{7.7}$$

where

$$\mathbf{V}_{im'}(\hat{k}) = \int_S d\mathbf{r}' e^{-j\mathbf{k}\cdot\mathbf{r}_{im'}}[\bar{\mathbf{I}} - \hat{k}\hat{k}] \cdot \mathbf{j}_i(\mathbf{r}_i)$$

$$\mathbf{V}_{jm}(\hat{k}) = \int_S d\mathbf{r} e^{-j\mathbf{k}\cdot\mathbf{r}_{jm}}[\bar{\mathbf{I}} - \hat{k}\hat{k}] \cdot \mathbf{t}_j(\mathbf{r}_j) \tag{7.8}$$

and

$$T_L(kr_{mm'}, \hat{k}\cdot\hat{r}_{mm'}) = \sum_{l=0}^{L}(-j)^l(2l+1)h_l^{(2)}(kr_{mm'})P_l(\hat{k}\cdot\hat{r}_{mm'}). \tag{7.9}$$

We remark that a key aspect of the revised $Z_{ji}$ expression is that it separates the primed and unprimed sections of the expression into $\mathbf{V}_{im'}$ and $\mathbf{V}_{jm}$. Because $T_L(kr_{mm'}, \hat{k}\cdot\hat{r}_{mm'})$ and $\mathbf{V}_{im'}$ can be computed a priori, for different distances $r_{mm'}$, (7.7) can be computed much faster for a group of basis functions clustered around $\mathbf{r}_{m'}$ and a group of testing functions clustered around $\mathbf{r}_m$ for $r_{mm'} > d$. For near-zone interactions (i.e., $r_{mm'} < d$), (7.6) must be evaluated using the techniques discussed in the previous chapter. The fast evaluation of $Z_{ji}$ using (7.7) is the essence of FMM.

The terms in (7.8) are known in FMM nomenclature as the signature functions. These are precomputed, and symmetry provides for further storage savings since (7.8) is computed only for the basis functions in Galerkin's testing. However, this is true only for the symmetric EFIE operator. When the magnetic field integral equation (MFIE) operator is considered, then

$$Z_{ji} = -\int_S d\mathbf{r}\mathbf{t}_j \cdot \left[\frac{1}{2}\mathbf{j}_i(\mathbf{r}) + \oint_S d\mathbf{r}'\hat{n}' \times \mathbf{j}_i(\mathbf{r}') \times \nabla g(\mathbf{r}, \mathbf{r}')\right] \tag{7.10}$$

and the signatures for the basis and testing functions take the forms

$$\mathbf{V}_{im'}(\hat{k}) = \int_S d\mathbf{r}' e^{-j\mathbf{k}\cdot\mathbf{r}_{im'}}\{\hat{k} \times [\hat{n}' \times \mathbf{b}_i(\mathbf{r}_i)]\}$$

$$\mathbf{V}_{jm}(\hat{k}) = \int_S d\mathbf{r} e^{-j\mathbf{k}\cdot\mathbf{r}_{jm}}[\bar{\mathbf{I}} - \hat{k}\hat{k}] \cdot \mathbf{t}_j(\mathbf{r}_j). \tag{7.11}$$

Here, the identity $[\bar{\mathbf{I}} - \hat{k}\hat{k}] \cdot [\hat{k} \times \mathbf{a}] = \hat{k} \times \mathbf{a}$ was used to reach the same expression for the EFIE in (7.8).

The translation operator, being dependent only on the prechosen vector $\mathbf{r}_{mm'}$ between the points $\mathbf{r}_m$ and $\mathbf{r}'_m$, is key to reducing the $O(N^2)$ complexity of the MoM matrix-vector product in the iterative solver. By grouping the basis functions into a prespecified number of clusters and by reusing the translation operator to compute the interactions of the basis and testing functions, the $O(N^2)$ complexity can be reduced down to $O(N^\gamma)$ where $\gamma \leq 1.5$. For the conventional FMM algorithm (Coifman, Rokhlin, and Wandzura, 1993) $\gamma = 1.5$, but $\gamma$ can be reduced further and perhaps close to $\gamma = 1$ for the multilevel version of FMM. As an example, Song, Lu, and Chew (1997) applied a nested grouping strategy along with a flat triangular discretization and Rao-Wilton-Glisson (RWG) (Rao, Wilton and Glisson, 1982) basis functions and introduced a multilevel FMM algorithm with $O(N \log N)$ computational complexity and $O(N)$ memory requirement. The lower complexity is due to forming multilevel groupings of the basis functions (by grouping smaller groups in larger groups) and reusing the translation operators computed between groups at each level (to be in the next section). As can be realized, there is no restriction on any type or form of basis and testing functions in applying FMM. Curvilinear surface elements and conformal basis functions as defined in Chapter 6 can be used equally well. As mentioned before, curved surface modeling has the advantage of reducing geometry modeling error in the solution procedure. This aspect also allows a more realistic (unbiased) evaluation of the error performance of the FMM algorithm (as geometry errors are minimized and therefore mostly due to the numerical solution).

A most important parameter in the FMM implementation is the number of terms kept in the evaluation of $T_L$ in (7.9). This parameter $L$ (Coifman, Rokhlin, and Wandzura, 1993), is semiempirically chosen as

$$L = kD_{max} + \alpha_L \ln(kD_{max} + \pi) \tag{7.12}$$

where $D_{max}$ is the maximum diameter of the clusters, and $\alpha_L$ is an accuracy control parameter. For $\alpha_L = 5$, the resulting value of $L$ is reported to provide single precision accuracy in evaluating $T_L$. Choosing $\alpha_L = 10$, double precision can be obtained in evaluating $T_L$ (Coifman, Rokhlin, and Wandzura, 1993). Typically, however, it has been reported (Song, Lu, and Chew, 1997) that setting $\alpha_L = 1$ gives acceptable accuracy for radar scattering computations. Nevertheless, for geometries where discretization leads to poorly conditioned matrices (e.g., geometries with sharp corners or fine detail), computation errors within FMM are amplified. On the other hand, keeping $\alpha_L$ small will result in faster execution time but may produce large error in the final solution. It is therefore a good practice to choose the level of the FMM approximation carefully so that acceptable accuracy is achieved.

In addition to carrying out an accurate but fast evaluation of $T_L$, the spectral integral in (7.3) must also be carefully evaluated. Typically, its numerical evaluation is carried out using $L$ points over the $\theta$ angular sector (Gaussian quadrature) and $2L$ points in the $\phi$ sector (trapezoidal rule). This choice of integration and associated weights ensures that spherical harmonics are exactly integrated up to order $2L$ (Coifman, Rokhlin, and Wandzura, 1993). Using such a numerical evaluation, the spectral integral in (7.7) can be rewritten as

$$Z_{ji} \approx \frac{-jk}{4\pi} \sum_{\hat{k}_p} w_{\hat{k}_p} \mathbf{V}_{jm}(\hat{k}_p) \cdot T_L(k_p r_{mm'}, \hat{k}_p \cdot \hat{r}_{mm'}) \mathbf{V}^*_{im'}(\hat{k}_p) \tag{7.13}$$

where $\hat{k}_p$ denotes the $2L^2$ spectral integration points, and $w_{\hat{k}_p}$ refers to the associated integration weights.

As already noted, the signature functions $\mathbf{V}_{im'}(\hat{k}_p)$ associated with all basis functions are also pre-computed and stored for all discrete $\hat{k}$ directions. These pre-computed signature functions map each basis function onto a signature in $k$-space, with corrections to the phase centers for each cluster. However, unlike the Green's function in real space, the translation operator in the $k$-space is diagonal. Thus, the collective signature function of a source or testing group can simply be constructed by adding the signatures of all basis functions forming that group.

The numerical implementation of the matrix-vector product using FMM consists of three successive steps as illustrated in Figure 7-4.

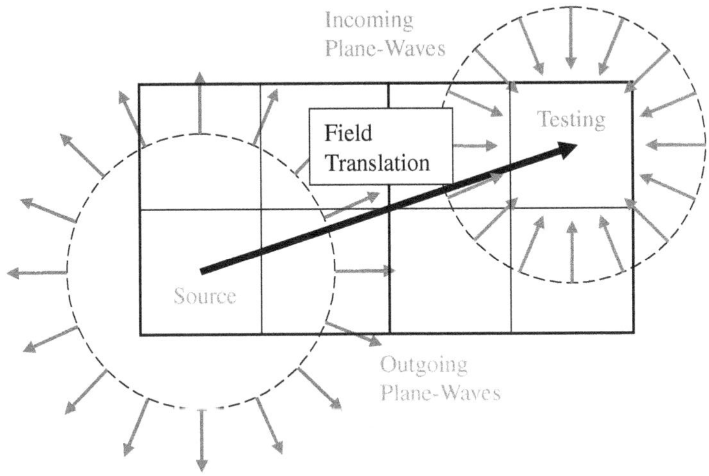


**FIGURE 7-4** Illustration of single-level interactions between source and testing clusters in FMM.

**Step 1.** The radiation signatures of all groups are formed by summing the signatures of all basis functions within each group

$$\mathbf{V}_{m'}(\hat{k}_p) = \sum_{\mathbf{j}_i \in G_{m'}} \mathbf{V}^*_{im'}(\hat{k}_p) \tag{7.14}$$

where $G_{m'}$ denotes the set of all basis functions in the source group $m'$ (see Figure 7-1). This is called the *aggregation* step.

**Step 2.** For the well-separated testing and basis groups, group signatures of all source groups are translated over to the testing group via the operation

$$\mathbf{V}_m(\hat{k}_p) = \sum_{m' \in F_{m'}} T_L(k_p r_{mm'}, \hat{k}_p \cdot \hat{r}_{mm'}) \mathbf{V}_{m'}(\hat{k}_p) \tag{7.15}$$

where $F_{m'}$ is the set of all source groups in the far zone of the $m$th testing group. Equation (7.15) basically transforms the outgoing plane wave signatures of the source groups onto an incoming plane wave signature on the testing group. This plane wave or spectral transformation is evocative of the Fourier transform process done for uniformly gridded testing and basis functions.

**Step 3.** The third and final sweep is that of *disaggregation*. This step transforms the field signatures over to the testing functions. As already noted, (7.7) is valid only for well-separated source and testing pairs. Hence, the interaction of basis and testing functions in proximity must be done using the original MoM procedure (i.e., using the near-field matrix generated via (7.6)). With this disaggregation step, the field over all testing functions inside the testing domain is computed via

$$\sum_i Z^{far}_{ji} x_i \approx \frac{-jk}{4\pi} \sum_{\hat{k}_p} w_{\hat{k}_p} \mathbf{V}_{jm}(\hat{k}_p) \cdot \mathbf{V}_m(\hat{k}_p). \tag{7.16}$$

Clearly, for smaller group sizes, fewer multipole terms are required. For example, for a maximum group size of $D_{max} = \lambda$, using (7.12) with $\alpha_L = 2$ would result in $L + 1 = 12$ terms ($L = kD_{max} + 2\ln(kD_{max} + \pi) = 10.77$). However, for

$D_{max} = \lambda/2$, we need only $L + 1 = 8$ terms. The corresponding spectral integral requires $2L^2 = 242$ samples for $D_{max} = \lambda$ and 98 samples for $D_{max} = \lambda/2$. Of course, to form smaller groups, we need to increase the number of groups, implying increased CPU resources for the translation operations. More specifically, if $N$ basis functions are grouped into $M$ groups, each group will consist of approximately $N/M$ basis functions. Hence, the near-field matrix will contain $O(N^2/M)$ nonzero entries representing the near-field interactions of the basis and testing functions. The remaining matrix-vector product is to be evaluated via the three-step FMM process. At the aggregation and disaggregation sweeps $N \times 2L^2$ operations are executed. On the other hand, the translation step requires $2L^2$ for each basis and testing group pair, resulting in a total number of $M^2 \times 2L^2$ operations. Thus, the FMM part of the matrix-vector product consists of $O(N \times 2L^2) + O(M^2 \times 2L^2)$ operations. In addition, the number of terms in the truncated translation operator, $L$, is roughly proportional to the maximum electrical diameter of the groups $kD_{max}$ (from (7.12)). Moreover, for surface geometries, the surface area of each group is proportional to $D_{max}^2$, as well as the number of basis functions in each group $N/M$, resulting in $L^2 \propto D_{max}^2 \propto N/M$. Hence, the complexities of the aggregation and disaggregation steps can be identified to be proportional to $O(N^2/M)$, and that of the translation step becomes $O(NM)$.

From this, the overall complexity of the FMM matrix-vector product is governed by three competing complexities, namely, that of aggregation/disaggregation $(O(N^2/M))$, translation $(O(NM))$, and near-field matrix-vector product $(O(N^2/M))$ stages. Thus, the overall complexity is $O(N^2/M + NM)$. To minimize the overall complexity, $M$ is set to $M = \sqrt{N}$, giving $O(N^{3/2})$ computational complexity for the FMM matrix-vector product. Also, only $O(N^2/M)$ nonzero matrix entries must be computed and stored instead of $O(N^2)$, implying concurrent memory savings. However, we still need to compute and store the signature functions and the translation operator. The overall storage requirement of the signature functions is also $O(2L^2N) = O(N^2/M)$, and that of the translation operator is $O(2L^2M^2) = O(NM)$. Summing these, the total storage complexity of the solution method still reduces down to $O(N^{3/2})$ with the choice $M = \sqrt{N}$.

The aforementioned CPU and memory estimates are asymptotic in the sense that they represent values approached only for very large $N$. The actual efficiency of the implementation depends on the choices of various parameters and the geometry of the radiator or scatterer. These choices control the constant in front of the asymptotic behavior in determining CPU requirements. However, choices of the FMM parameters for faster implementation inevitably lead to less accurate answers. Clearly, keeping $L$ as small as possible reduces the CPU time but not the order of $O(N^{3/2})$. That is, if the CPU time is given by $CN^{3/2}$ ($C = constant$) for the matrix-vector product evaluation, lowering $L$ implies smaller value for the constant $C$. Lower $C$ values imply that the CPU time crossover point between standard MoM and FMM implementations occurs for lower values of $N$.

In addition to the previously given parameter choices, the group size and geometrical location also impact the overall FMM accuracy and efficiency (speed). That is, if the element clustering is not uniform, the FMM implementation will be inefficient due to the presence of very large or very small clusters. This affects computation time as the number of terms used to accurately evaluate the translation sum will depend on the largest cluster. A good clustering example is shown in Figure 7-5

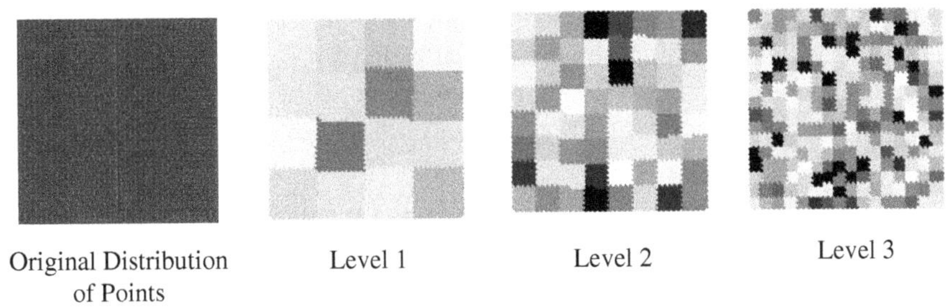


**FIGURE 7-5** Three clusterings of points lying on a square grid.

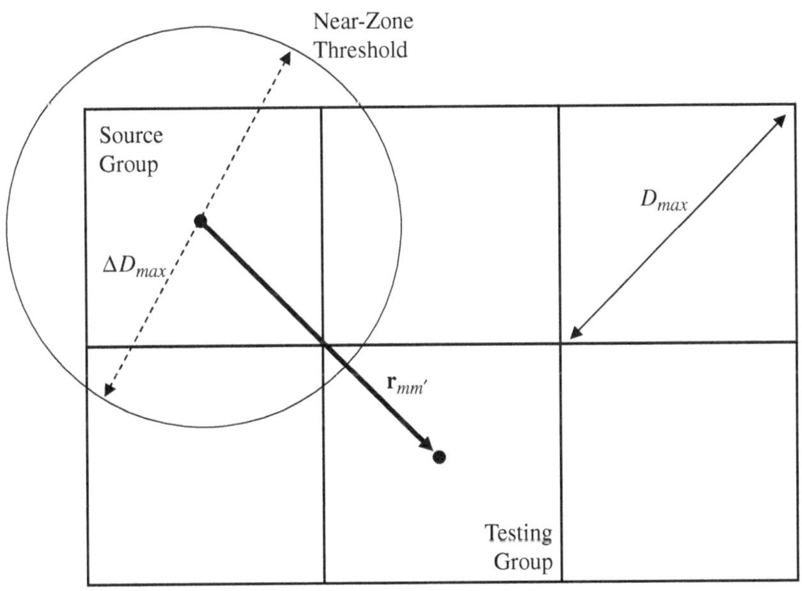


**FIGURE 7-6** Illustration of the near-zone threshold used in FMM implementations.

where all groups are approximately of the same size. As will be discussed in the multilevel FMM implementation, a structured grouping having translational symmetry, can provide further savings in storage since the translation operator will be identical due to translational symmetries. We close this section by noting additional parameter recommendations in implementing FMM. Specifically, if $r_{mm'}$ is the distance between clusters $m$ and $m'$, the FMM approximation is employed when

$$r_{mm'} > \Delta D_{max} \quad \text{and} \quad kr_{mm'} > L \tag{7.17}$$

where $D_{max}$ denotes the maximum diameter of the source and testing clusters, and $\Delta$ is a control parameter. Also, the condition $kr_{mm'} > L$ must be satisfied to suppress errors in evaluating the translation sum. Specifically, for small $kr_{mm'}$ values the Hankel functions in (7.4) tend to be very large, rendering the calculation of $T_L$ numerically unstable. For example, if $\Delta = 1$, then even neighboring clusters will be considered in the far zone (and FMM will be used for their interactions). But, for $\Delta = 1.5$, the matrix elements in the neighboring (touching) clusters will be evaluated without approximation (see Figure 7-6). Basically, higher $\Delta$ implies that more interactions will be carried out without approximation, increasing the size of the near-field matrix. Further, the concurrent requirement of $kr_{mm'} > L$ leads to more terms in the near-field matrix (when the same mesh is used at lower frequencies).

It is understandable that for many problems large sections of the MoM matrix may not carry the FMM approximation and afforded speed. The next section presents a multilevel nested strategy for the three governing FMM operations.

## 7.2 MULTILEVEL FAST MULTIPOLE METHOD

In this section, we first outline a multilevel grouping scheme leading to the MLFMM implementation of the surface integral equation. The tree structure of the grouping and the procedure necessary for the implementation of the MLFMM matrix-vector product are outlined here. An implicit assumption is that each basis function is associated with a point in space (e.g., the mid-point of the edge associated with the basis function) and that clustering is carried out in terms of points in space, rather than using actual basis functions themselves (see Figure 7-7).

Referring to Figure 7-8, the first step in forming the multilevel clustering is to enclose the given geometry into a cubic box. To do so, the maximum linear dimension $d_0^{max}$ of a given distribution of points (associated with the basis functions) in space

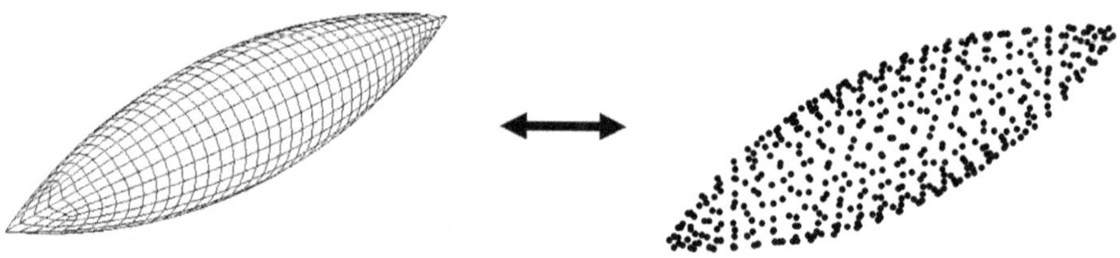

**FIGURE 7-7** ■ Association of basis functions to points on the geometry surface (midpoints of edges forming the surface mesh).

**FIGURE 7-8** ■ Zeroth-level cube enclosing the geometry.

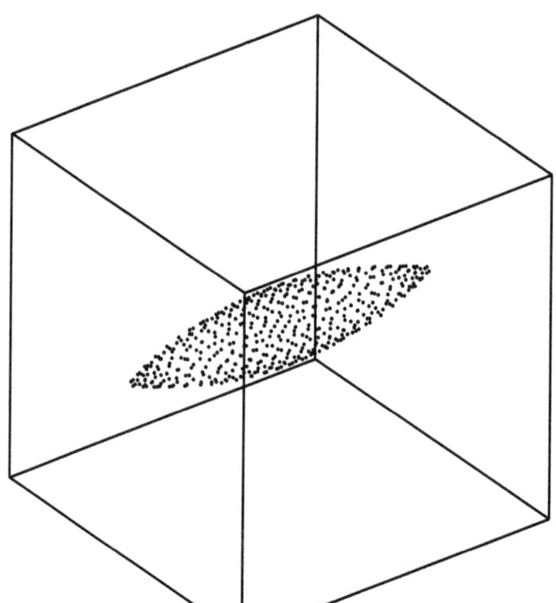

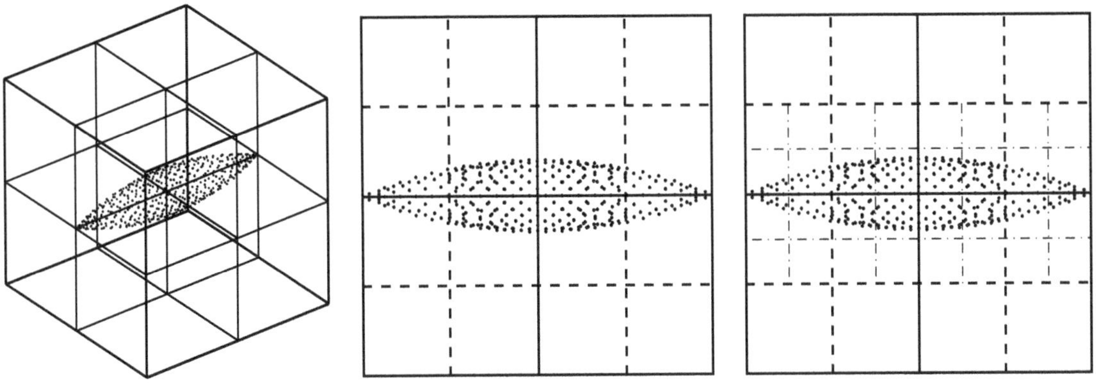

**FIGURE 7-9** First-, second-, and third-level grouping example.

is calculated. This is denoted as the 0th level cluster and encloses all the geometry of the problem as shown in Figure 7-8.

To form the next level (i.e., level 1), the level 0 cube is subdivided into eight smaller cubes (see Figure 7-9). These eight smaller cubes, or clusters, are called "children," and the original cluster is referred to as the parent. Likewise, each cluster in level 1 can be further subdivided in a similar manner to form level 2 clustering, and so on. Hence, at each level, the side length of the cubes are $d_l^{max} = d_0^{max}/2^l$, where $l$ denotes the level number. Thus, the maximum linear dimension of the clusters for the $l$th level is $D_l^{max} = \sqrt{3}d_l^{max}$. This subdivision is carried out until the $\mathcal{N}$th level is reached when $D_{\mathcal{N}}^{max} \leq \lambda/\pi$, $\lambda$ being the free-space wavelength at the frequency of operation.

As would be expected, in creating the clustering subdivisions, only non-empty cubes are kept. That is, if a cube at level $l$ is empty, it is simply discarded. This multilevel grouping (by forming eight children from each parent) is referred to as the *oct-tree* clustering process. At the finest level $\mathcal{N}$, each basis function is associated with a cluster and the parent–children relationship of all levels is stored in the oct-tree structure. It is understood that the electrical sizes of the clusters at each level are different; thus, the number of multipoles in the translation operators will be different for each level. Therefore, as already noted, this impacts CPU cost and memory savings. Later, we will address this issue using the two-level FMM implementation as reference.

## 7.3 MLFMM FORMULATION

Considering the two-level vector construct depicted in Figure 7-10, we introduce the decomposition

$$\mathbf{r}_{ji} = \mathbf{r}_j - \mathbf{r}_i = \mathbf{r}_{m'i}^{(2)} + \mathbf{r}_{m'}^{(1)(2)} + \mathbf{r}_{mm'} + \mathbf{r}_m^{(2)(1)} + \mathbf{r}_{jm}^{(2)} \tag{7.18}$$

This decomposition is critical to reevaluating the translation operator (7.4). To do so, we first proceed to rewrite the scalar Green's function as

$$\frac{e^{-jkr_{ji}}}{r_{ji}} = \frac{-jk}{4\pi} \int d^2\hat{k} e^{-j\mathbf{k}\cdot\left(\mathbf{r}_m^{(2)(1)}+\mathbf{r}_{jm}^{(2)}\right)} T_L(kr_{mm'}, \hat{k}\cdot\hat{r}_{mm'}) e^{j\mathbf{k}\cdot\left(\mathbf{r}_{im'}^{(2)}+\mathbf{r}_{m'}^{(2)(1)}\right)} \tag{7.19}$$

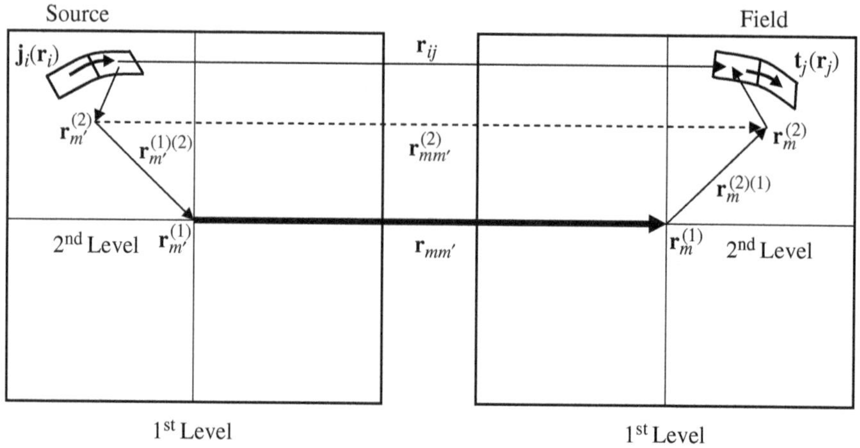


**FIGURE 7-10** Two-level construct for implementing the multilevel FMM.

This should be compared with (7.3) with

$$T_L(kr_{mm'}, \hat{k}\cdot\hat{r}_{mm'}) = \sum_{l=0}^{L}(-j)^l(2l+1)h_l^{(2)}(kr_{mm'})P_l(\hat{k}\cdot\hat{r}_{mm'}) \tag{7.20}$$

being the translation operator, and the superscripts (2) and (1) in $\mathbf{r}^{(1)(2)}$ denoting the grouping levels. Specifically, (1) refers to the coarser-level grouping, and (2) refers to the finer grouping. Hence, the dyadic Green's function can be written as

$$\overline{\mathbf{G}}(\mathbf{r}_j, \mathbf{r}_i) = \frac{-jk}{4\pi}\int d^2\hat{k}\,[\mathbf{I}-\hat{k}\hat{k}]\,e^{-j\mathbf{k}\cdot\mathbf{r}_m^{(2)(1)}}e^{-j\mathbf{k}\cdot\mathbf{r}_{jm}^{(2)}}T_L(kr_{mm'}, \hat{k}\cdot\hat{r}_{mm'}) \tag{7.21}$$
$$[\mathbf{I}-\hat{k}\hat{k}]\,e^{j\mathbf{k}\cdot\mathbf{r}_{im'}^{(2)}}e^{j\mathbf{k}\cdot\mathbf{r}_{m'}^{(2)(1)}}.$$

Upon discretizing the spectral integral (7.19) using numerical quadrature at level 1, we have

$$Z_{ji} = \frac{-jk}{4\pi}\sum_{\hat{k}^{(1)}} w_{\hat{k}^{(1)}}e^{-j\mathbf{k}\cdot\mathbf{r}_m^{(2)(1)}}\mathbf{V}_{jm}(\hat{k}^{(2)})T_L(k^{(1)}r_{mm'}, \hat{k}^{(1)}\cdot\hat{r}_{mm'})\mathbf{V}_{im'}^*(\hat{k}^{(2)})e^{j\mathbf{k}\cdot\mathbf{r}_{m'}^{(2)(1)}} \tag{7.22}$$

where, as in the standard FMM,

$$\mathbf{V}_{im'}(\hat{k}) = \int_S d\mathbf{r}_i e^{-j\mathbf{k}\cdot\mathbf{r}_{im'}^{(2)}}[\mathbf{I}-\hat{k}\hat{k}]\,\mathbf{j}_i(\mathbf{r}_i) \tag{7.23}$$

and

$$\mathbf{V}_{jm}(\hat{k}) = \int_S d\mathbf{r}_j e^{-j\mathbf{k}\cdot\mathbf{r}_{jm}^{(2)}}[\mathbf{I}-\hat{k}\hat{k}]\,\mathbf{t}_j(\mathbf{r}_j) \tag{7.24}$$

are the signature functions (often referred to as the Fourier transforms of the basis functions) for the EFIE.

We observe that signature functions $\mathbf{V}_{im'}(\hat{k}^{(2)})$ in (7.22) are the $k$-space discretizations at level 2. However, the evaluation of (7.22) requires their $k$-space values at level 1; that is, $\mathbf{V}_{im'}(\hat{k}^{(1)})$ (since the translation operator is generated for the sampling at level 1). This dilemma can be circumvented by a simple interpolation of $\mathbf{V}_{im'}(\hat{k}^{(2)})$ to compute $\mathbf{V}_{im'}(\hat{k}^{(1)})$.

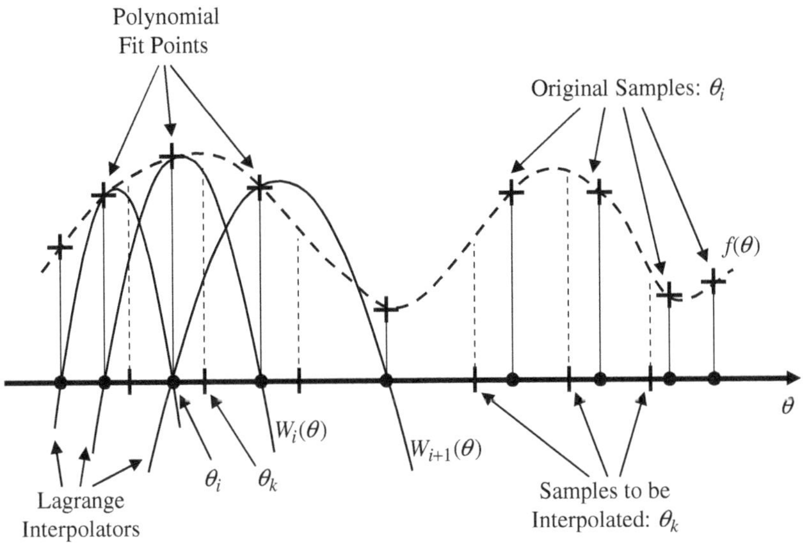


**FIGURE 7-11** Illustration of interpolation matrices in FMM.

Referring to Figure 7-11 for a smooth function $f(\theta)$, with given samples $f(\theta_i)$ at $\theta_i, i = 1, \ldots, L_{(2)}$, we can generate a different sampling at $\theta_k, k = 1, \ldots, L_{(1)}$ values through interpolation. For this, we start by generating an approximation to $f(\theta)$, viz.

$$\tilde{f}(\theta) = \sum_i W_i(\theta) f(\theta_i) \tag{7.25}$$

using the known values $f(\theta_i)$ and the interpolators $W_i(\theta)$ (see Figure 7-11). To calculate the new sampling of $f(\theta)$ at a new set of values $\theta_k$, (7.25) is simply evaluated at $\theta_k$, giving

$$\tilde{f}(\theta_k) = \sum_i W_i(\theta_k) f(\theta_i), k = 1, \ldots, L_{(2)}. \tag{7.26}$$

Of course, this can be cast in a matrix form as $\{\tilde{f}_k\} = [W_{ki}]\{f_i\}$, where $f_i = f(\theta_i)$ and $W_{ki} = W_i(\theta_k)$ are the matrix elements. For a general case, the interpolation matrix $\{W_{ki}\}$ is fully populated. However, for a computationally cheap interpolation, the matrix $[W_{ki}]$ should be as sparse as possible to keep the cost of interpolation as low as possible. One of the simplest and most popular interpolation techniques is polynomial interpolation, such as shown in Figure 7-11. For this case,the generated interpolation matrix is sparse since only a few interpolators around the new sample $\theta_k$ are included in the computation of $\tilde{f}(\theta_k)$.

Returning now to the signature functions, the same interpolation approach can be employed (on the unit sphere, i.e., on $k$-space). Specifically, we can write the level-1 signature functions in terms of the level-2 signature functions using the interpolatory expression

$$\mathbf{V}_{im'}\left(\hat{k}^{(1)}\right) = W\left\{\mathbf{V}_{im'}\left(\hat{k}^{(2)}\right)\right\} = \mathbf{W}_{(1),(2)}\mathbf{V}_{im'}\left(\hat{k}^{(2)}\right) \tag{7.27}$$

where $[\mathbf{W}_{(1),(2)}]$ refers to the interpolation coefficients between levels 1 and 2. Correspondingly, we can also write

$$\mathbf{V}_{im'}\left(\hat{k}^{(2)}\right) = \mathbf{W}^T_{(1),(2)}\mathbf{V}_{im'}\left(\hat{k}^{(1)}\right) \tag{7.28}$$

where the superscript $T$ implies matrix transposition. This interpolation strategy enables the generation of signatures at the parent clusters (level-1) using the collective signatures of the children clusters in level 2. However, the signatures of the "children" must be phase corrected through a simple phase-shifting operation as given in (7.22).

Based on this, the matrix-vector products for the two-level grouping are carried out as follows. Aggregations are first computed for the source clusters at the finer level 2 for each $\hat{k}^{(2)}$ direction. The $k$-space sampling of the aggregated source field signatures are then interpolated to compute the signatures at the required $k$-space samples for level 1 using (7.27). The phase centers of the aggregated field signatures are then shifted to the center of the parent source clusters (at the coarser level 1) by simply multiplying with $e^{j\mathbf{k}\cdot\mathbf{r}_{m'}^{(2)(1)}}$ as in (7.22). The same procedure is applied to all nonempty children corresponding to source clusters at level 1. After translating the signatures of each level-2 cluster (children) to level 1, the same operation is carried out in reverse order to compute the radiated fields and therefore the matrix-vector product. Namely, the phase center of the translated field is shifted to the center of the field cluster at the finer level on multiplying with $e^{-j\mathbf{k}\cdot\mathbf{r}_m^{(2)(1)}}$. This phase shifted field is then interpolated down to the $k$-space points at the finer level 2 using the transpose of the interpolation matrix. Once the incoming signatures of all level-2 clusters are computed, we multiply them with the signatures of the testing functions in the testing group (at level 2) to compute the actual MLFMM contribution to the matrix-vector product. Figure 7-12 depicts the steps of this two-level FMM.

As mentioned already, the complexity of the MLFMM relies on adopting a sparse interpolation technique to approximate the scattering or radiation at different granularity levels. In doing so, we note that as the signature functions are smooth on the unit sphere, a $4 \times 4$th order two-dimensional piecewise polynomial interpolation (that results in a sparse interpolation matrix) can be adopted. But as the problem size increases, more accurate interpolations may become necessary (Chien and Alpert, 1997; Darve, 2000; Yarvin and Rokhlin, 1998). Obviously, these are more costly than simple polynomial interpolation, implying slower execution speeds.

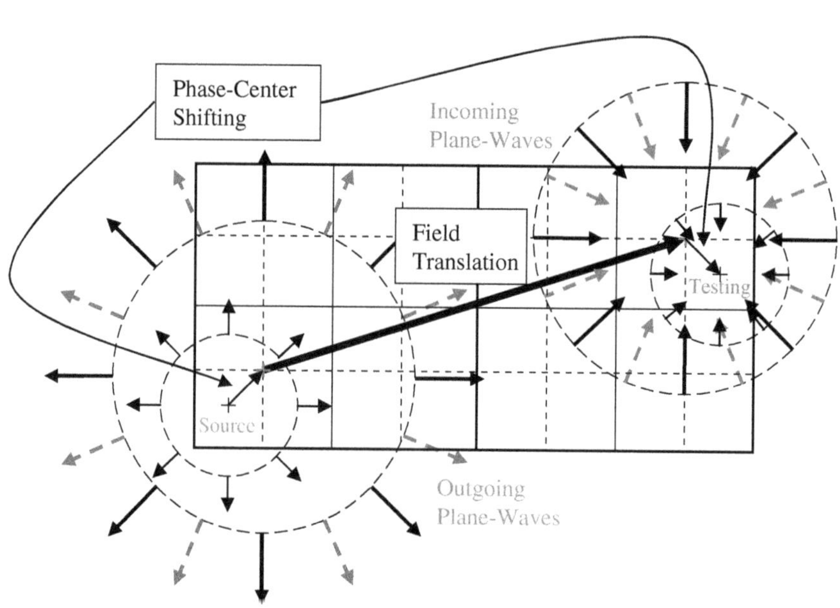


**FIGURE 7-12** Illustration of the interactions between source and testing groups in implementing a two-level FMM.

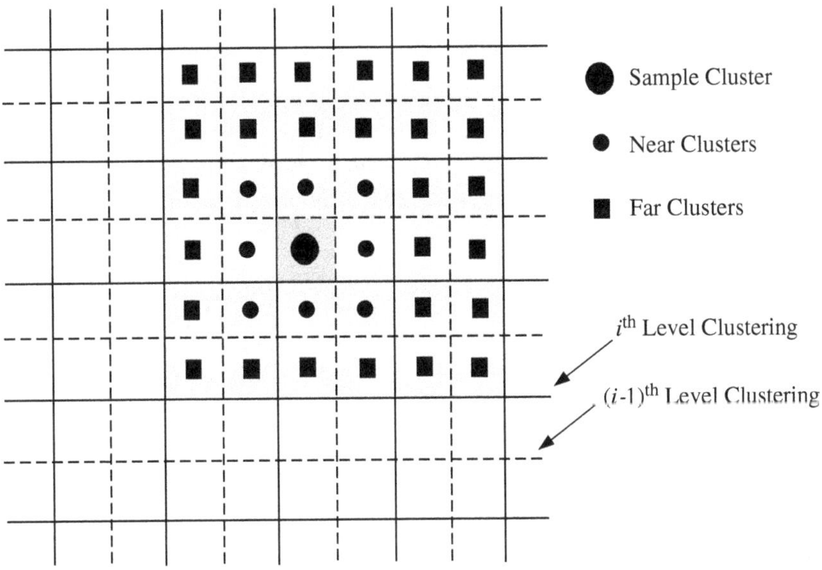


**FIGURE 7-13** Illustration of multilevel clustering for near- and far-zone interactions using a two-dimensional depiction.

As already noted, to speed up computations, interactions between source and testing functions in clusters $i$ and $j$, respectively, we use the translations level $(l+1)$ whenever the parents of clusters $i$ and $j$ are in the near zone of each other. In other words, if the parents of clusters $i$ and $j$ are far away from each other in level $(l)$, the interaction of those basis and testing functions are carried out using the translations in level $(l)$. The situation is exemplified in Figure 7-13, where the source cluster in level $(l+1)$ is denoted by a large circle. All interactions of basis functions in the source cluster and testing functions in clusters marked with a square are carried out using the signatures and translations on level $(l+1)$. In addition, the interactions of basis functions in the source cluster on level $(l+1)$ with all testing functions in the near-zone clusters (denoted by small circles) are carried out in the finer level $(l+2)$. If level $(l+1)$ is the finest level, then these interactions are carried out directly using the near-field matrix. Moreover, the interactions of basis functions in the source cluster with all testing functions in the remaining clusters in level $(l)$ are carried out using the signatures and translations in the coarser level $(l)$ for all clusters whose parents are in near zones. Thus, it is necessary that the multilevel grouping algorithm provide the neighborhood information of all clusters at all levels. The MLFMM rule for each level is to compute all interactions in that level between all far-zone clusters whose parents are in near zones.

Thus, we can summarize MLFMM steps as follows.

**Step 1.** Compute the collective signatures of all source functions at the finest level using the individual signatures of the basis functions.

**Step 2.** Calculate all coarser level cluster signatures through interpolation and shifting using the signatures of the children.

**Step 3.** For each level $(l)$, translate the signatures of source clusters onto far-zone testing clusters when their respective parents lie in near zones at level $(l-1)$.

**Step 4.** Starting with the coarsest level, shift and interpolate the translated incoming signatures onto the children at the all finer levels.

**Step 5.** At the finest level, compute the interactions using the signatures of the testing functions and the incoming plane wave signatures.

This multilevel tree structure leads to $O(N \log N)$ CPU and memory complexity. Also, due to the translational symmetry in the grouping, additional memory saving can be achieved in storing the translation operators. Significantly larger MoM problems can now be solved using MLFMM as demonstrated in the next section.

## 7.4 RADIATION AND SCATTERING EXAMPLES

In this section, we present several examples and complexity evaluations of the implemented MLFMM solvers for electromagnetic scattering and radiation problems. For the presented data, the MLFMM strategy is employed using curved biquadratic surface elements and conformal rooftop basis functions as described in the previous chapter.

Our first example is the radiation of a Hertzian dipole placed above a perfectly electrically conducting (PEC) sphere. As this problem has a closed-form solution, it can be used for validation and as a benchmark. Figure 7-14 displays the induced currents on the sphere's surface due to a vertical Hertzian (infinitesimal) dipole placed on the surface. Three different sphere sizes are shown ($4\lambda$, $8\lambda$, and $10\lambda$ diameter spheres). For all these solutions, curvilinear surface patches were used to model the sphere resulting in corresponding discretizations of 10 elements/$\lambda$, 5 elements/$\lambda$, and 3.5 elements/$\lambda$. Clearly, the latter two discretizations are low, but because of the better geometry fitting (owed to the curvilinear patches) the resulting radiation patterns are still accurate. The normalized (numerically computed vs. exact) radiation patterns are given in Figure 7-15. The exact analytical solution was computed using a Mie series solution given in (Harrington, 1968).

As we are interested to evaluate the speed up attained by the FMM and MLFMM implementations, here we consider the CPU timing in carrying out the sphere radiation problem. Table 7-1 shows the three-level FMM CPU timings and solver performance for the solutions in Figure 7-14. It is important to note that, although the same mesh was used to solve the same problem at larger frequencies, the execution time for the matrix-vector product increases as the number of multipoles $L$ also

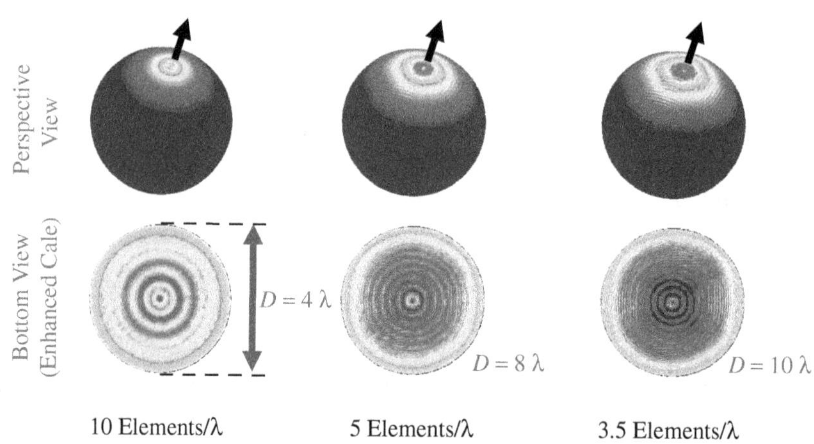


**FIGURE 7-14** Numerically computed (using curvilinear patch discretization) surface currents on a sphere excited by a Hertzian dipole.

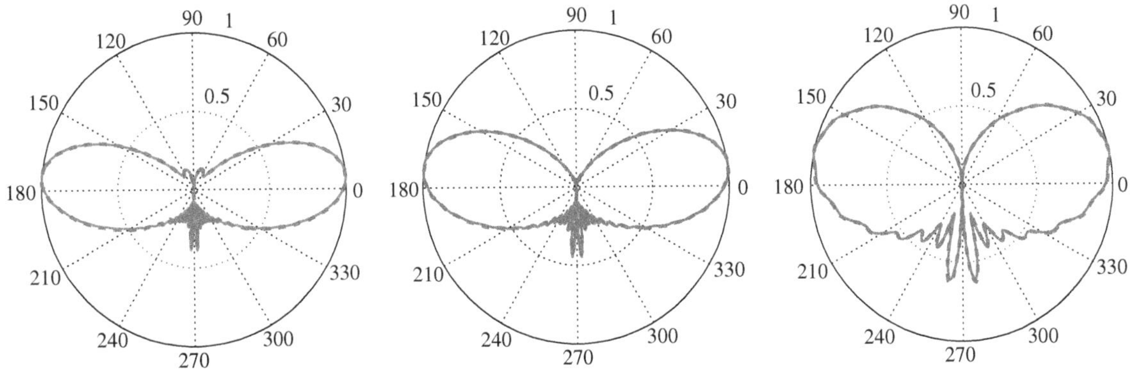


**FIGURE 7-15** ■ Radiation patterns (linear-scale) of a Hertzian dipole over the spheres given in Figure 7-14 (solid: Mie solution, dashed: MLFMM) using curvilinear quadrilaterals.

**TABLE 7-1** ■ Computational requirements for a fixed number of unknowns ($N = 11{,}164$) at increasing frequencies

| Mesh Size | Near-Field Matrix | Number of Multipoles $L$ at Each Level | Number of Iterations | Time Per Iteration | Solution Time |
|---|---|---|---|---|---|
| 10 per $\lambda$ | $5.0 \times 10^6$ | 5,7,10 | 28 | 3.9s. | 108.8s. |
| 5 per $\lambda$ | $2.5 \times 10^6$ | 7,11,18 | 19 | 6.9s. | 130.4s. |
| 3.5 per $\lambda$ | $2.0 \times 10^6$ | 8,14,22 | 21 | 10.1s. | 212.3s. |

**TABLE 7-2** ■ Computational requirements for a fixed $\lambda/10$ sampling in computing the radiation patterns in Figure 7-15.

| Mesh Size | Problem Size | Time Per Iteration $O(N \log N)$ |
|---|---|---|
| 10 per $\lambda$ | 10,000 | 3.9s. |
| 10 per $\lambda$ | 40,000 | 15.9s. |
| 10 per $\lambda$ | 90,000 | 38.4s. |

increases. That is, the FMM sampling requirements seem to be decoupled from the spatial sampling requirements. However, if the sampling rate is kept fixed at $\lambda/10$ for the three sphere sizes, the problem size for the larger sphere would have been considerably larger. Table 7-2 summarizes the CPU and solver performance for a constant sampling rate of $\lambda/10$. As seen, at the highest frequency (corresponding to a $10\lambda$ diameter sphere), a sampling rate of $\lambda/10$ results in 90,000 unknowns and requires 38.4 seconds of CPU time per matrix-vector product compared with 10.1 seconds using the lower sampling rate of 3.5 elements/$\lambda$. Hence, using curvilinear elements along with MLFMM gives a four-fold speed-up in the matrix-vector product computation (for this specific example).

As a second computational example we consider the scattering by a PEC plate $30\lambda$ long in size. Figure 7-16 depicts the computed bistatic radar cross section for

**FIGURE 7-16** Bistatic radar cross section of a flat plate using three-level FMM (comparison to physical optics is provided).

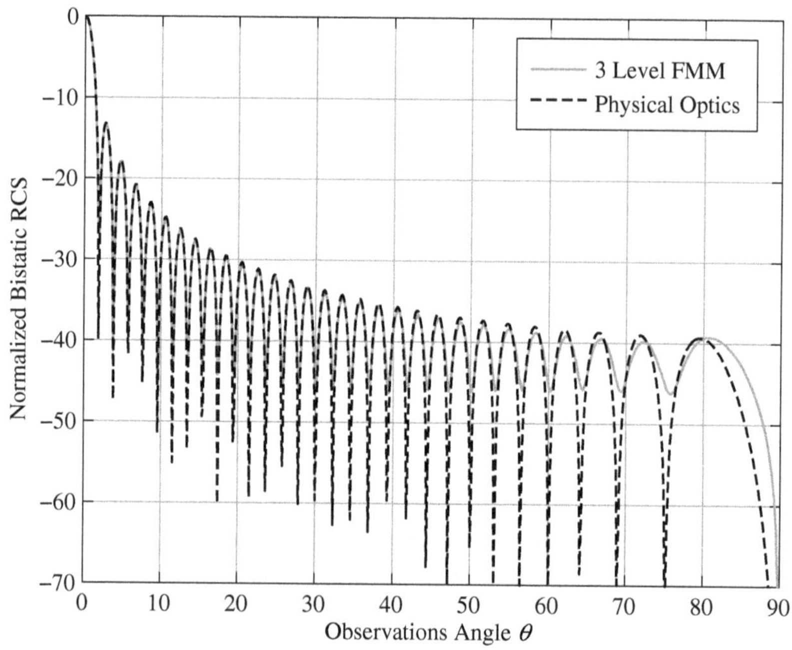


**FIGURE 7-17** Complexities of various levels of FMM in carrying out the solution in Figure 7-16 (e.g., FMM-1 stands for single-level FMM, FMM-2 stands for two-level FMM).

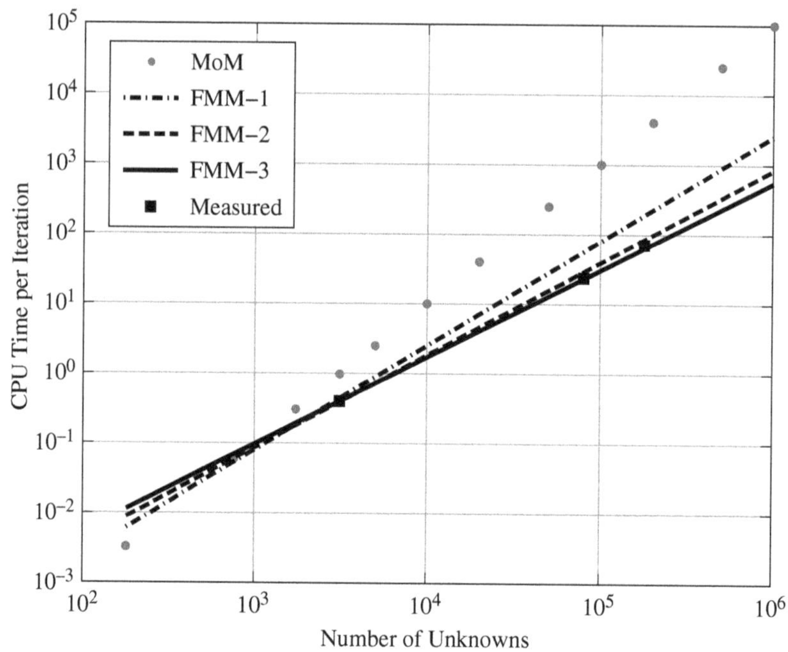


broadside plane wave incidence. As expected, the scattering pattern has 29 lobes plus the main lobe in the backscatter direction. Given that the plate is very large, the approximate physical optics (PO) solution is in good agreement with the three-level MLFMM solution. Figure 7-17 depicts the computational complexities of increasing levels of FMM for this plate scattering problem. The curves start with an $O(N^{3/2})$

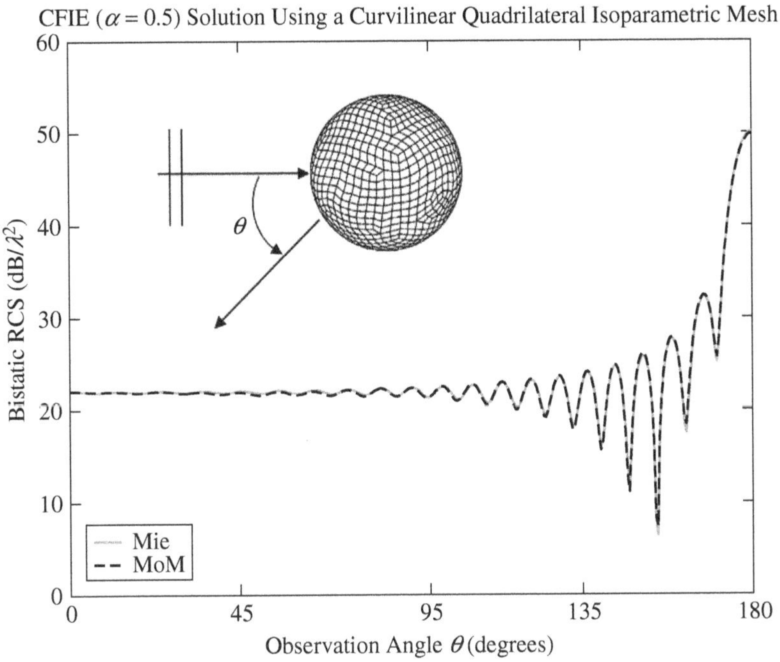


**FIGURE 7-18** Bistatic radar cross section for a PEC sphere, 8λ in diameter, using 7,500 unknowns to discretize with curvilinear quadrilateral elements.

complexity for the single-level FMM and tend toward $O(N \log N)$ complexity for the multilevel case.

Figure 7-18 depicts the bistatic scattering pattern for a sphere, our third example. The sphere was 8λ in diameter and the solution was carried out using five-level FMM with ($\alpha_L = 2$ in (7.12)) and curvilinear quadrilateral discretization. The resulting matrix system had 7,500 unknowns, and the number of multipoles for each level were {9, 11, 14, 20, 32}. The overall solution required 447 MB in memory and took 74 seconds in CPU time to fill the near-field matrix and 17 CGS iterations (972 seconds) to converge to a residual error less than $10^{-3}$. The bistatic radar cross section is plotted in Figure 7-18, and as seen it is accurate to within 2.9% of the analytical Mie series data.

The spatial sampling used for the data in Figure 7-18 was below the nominal λ/10 for this computation (afforded by the curvilinear elements). A sampling of λ/10 actually resulted in 28,812 unknowns (an increase by nearly a factor of 4). The MLFMM solution of this larger matrix system is given in Figure 7-19. Again, the computation was carried out using a five-level MLFMM with $L = \{7, 8, 12, 18, 30\}$. The corresponding required total memory was 508 MB and the near-field matrix was filled in 1034 seconds (on an Intel-PIII with 1 GHz clock speed). Specifically, the CGS solver took 20 iterations to converge, and the total solution was completed in 18.5 minutes (1110 seconds). As seen from Figure 7-19, the MLFMM solution is very accurate and within 0.4% of the analytical solution (Harrington, 1968). This example demonstrates that at a given frequency, the MLFMM solution takes about the same time regardless of the spatial sampling (primarily due to the $k$-space nature of the formulation). As will be mentioned later, this property is very useful in solving volume integral equations (Sertel and Volakis, 2002; Sancer, Sertel, Volakis, and Van Alstine, 2006) involving high-contrast material parameters (where the sampling

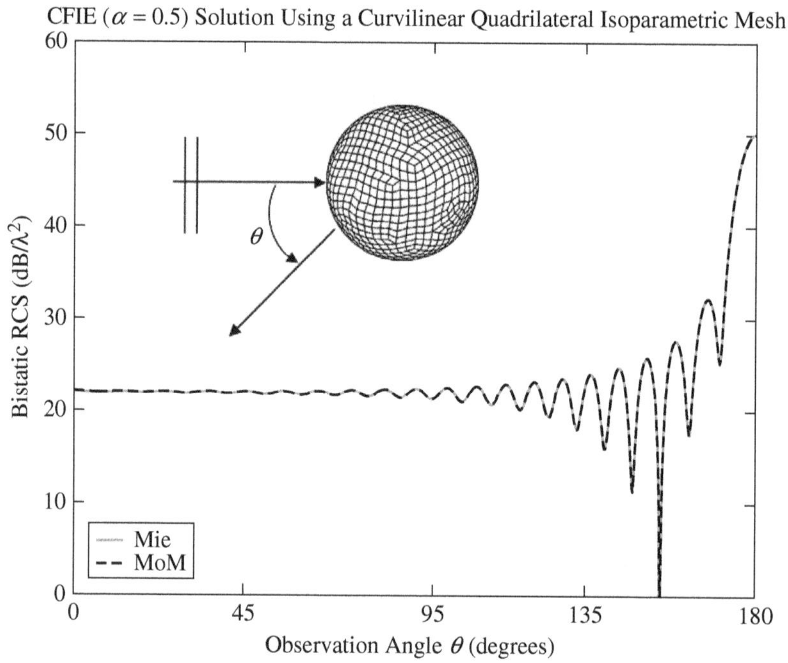


**FIGURE 7-19** Bistatic radar cross section of PEC sphere, 8λ in diameter, using 28, 812 unknowns and a five-level FMM for the MoM solution.

rate is determined by the wavelength inside the inhomogeneity). Of importance in volume integral equations is that the $k$-space sampling is based on the free-space Green's function. Thus, the number of samples is independent of the specific material parameters of the volume.

When the 28,812 unknown mesh is used to solve the sphere scattering problem of Figure 7-19 at twice the frequency (i.e., for a 16λ diameter sphere), the solver generates six MLFMM levels with $L = \{9, 11, 14, 20, 32, 55\}$. The corresponding solution required 1.1 GB of memory, and the near-field matrix was filled in 933 seconds. Also, the CGS solver converged in 18 iterations taking a total time of 1.15 hours. The computed radar cross section is shown in Figure 7-20 and is within 2.6% of the reference Mie series solution.

When we use a nominal discretization at $\lambda/10$ intervals, the resulting unknowns become 110,190. The corresponding computed bistatic radar cross section is now within 0.6% of the reference solution and is depicted in Figure 7-21. That is, the improvement in error is not as much, and this is due to employed curvilinear elements to model the sphere's surface. Were we to use planar patches to model the sphere, such a higher discretization would have been necessary. Also, were we to use a traditional MoM solution (Gaussian elimination or LU decomposition), 97 GB of memory would have been required. This is obviously well beyond the storage capacity of typical computing facilities.

Nevertheless, even with the MLFMM, when the storage exceeds several GB, we must still resort to using multiprocessor supercomputing resources. Porting of the MLFMM algorithm on distributed memory supercomputers requires a careful balanced distribution of the solver load among the individual processors and minimization of interprocessor communications. Several algorithms that achieve moderately efficient parallel implementations of MLFMM have recently been reported

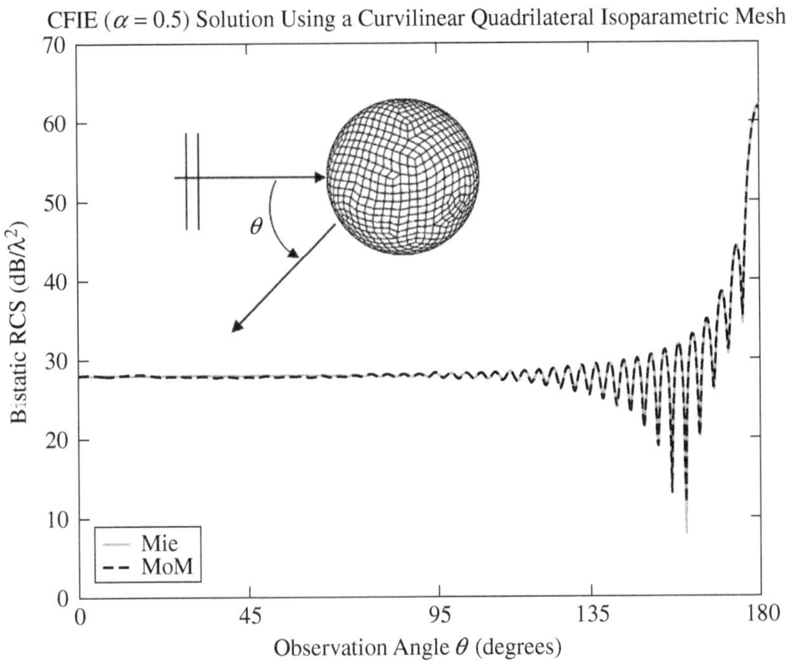


**FIGURE 7-20** Bistatic radar cross section of PEC sphere, 16λ in diameter, using 28, 812 unknowns and a six-level FMM for the MoM solution.

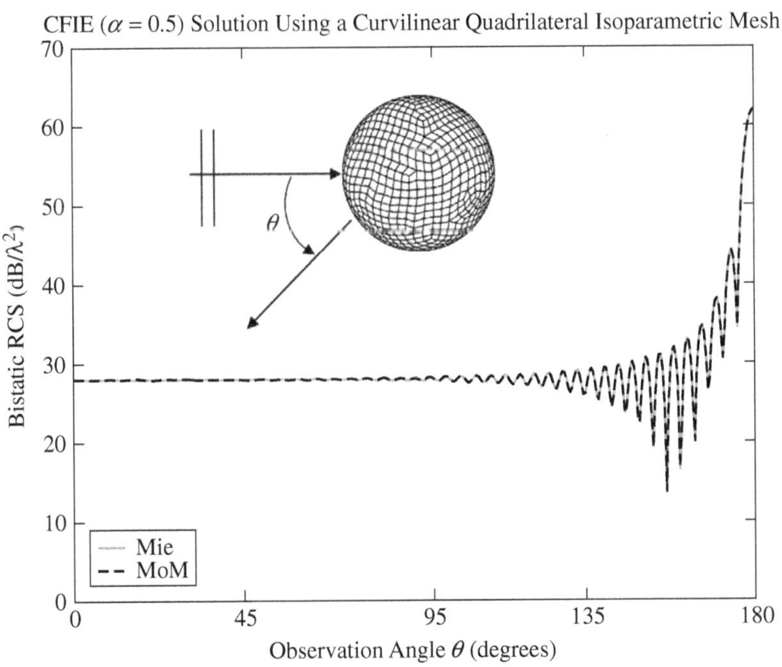


**FIGURE 7-21** Bistatic radar cross section of PEC sphere, 16λ in diameter, using 110, 190 unknowns and a six-level FMM for the MoM solution.

(Velamparambil and Chew, 2005; Ergul and Gurel, 2009). Alternatively, a single-level FMM, accelerated with fast Fourier transform (for the translations step) can be easily ported onto massively parallel supercomputing platforms (as well as clusters of commodity personal computers and even graphics processing units [GPUs]) (Waltz, Sertel, Carr, Usner, and Volakis, 2007) to tackle extremely large problems.

## 7.5 | MLFMM FOR VOLUME INTEGRAL EQUATIONS

In this section, we describe the implementation of the MLFMM for dielectric structures using zeroth-order basis functions outlined in Chapter 6. More specifically, the surface integral implementation will be combined with the volume integral equation (VIE) implementation of the integral equation given at the end of Chapter 6. Using this combined surface/volume integral solution, the goal is to provide practical solutions to material loaded scatterers and antenna radiation problems.

To implement the MLFMM for VIEs, let us consider a dielectric sphere of unit radius. In proceeding, another surface mesh is first formed using a meshing package, such as the MSC-PATRAN (MSC Software, 2010) to produce a grid consisting of, for example, quadrilateral elements or triangular elements. The size of the surface elements is, of course, determined by the wavelength of interest (the rule of thumb is 10 elements per linear wavelength is needed, but this number can be lowered using curvilinear elements, for a given accuracy, as already discussed). Starting with the surface mesh, a volumetric grid (using, e.g., cubical elements or hexahedra as mentioned in Chapter 6) is generated. Tables 7-3 and 7-4 provide a comparison of the number of unknowns for surface and volume discretizations as a function of gridding edge length $\Delta$ used. As seen, the unknowns for volume discretization grow as $1/\Delta^3$, whereas for surfaces they grow as $1/\Delta^2$. Of greater importance is that, for high-contrast dielectrics, $\Delta \approx \lambda/(10\sqrt{\epsilon_r})$ , implying a significantly smaller element size for an accurate solution. Therefore, for VIE formulations, the problem

**TABLE 7-3** ■ Surface mesh details for a sphere where $\Delta$ is the sampling interval.

| $\Delta$ | Quadrilaterals | Triangles |
|---|---|---|
| 0.250 | 150 | 300 |
| 0.200 | 216 | 432 |
| 0.150 | 384 | 768 |
| 0.100 | 864 | 1728 |
| 0.075 | 1536 | 3072 |
| 0.050 | 3750 | 7500 |
| 0.040 | 5766 | 11532 |

**TABLE 7-4** ■ Volumetric mesh details for a sphere where $\Delta$ is the sampling interval.

| $\Delta$ | Hexahedra | Tetrahedra |
|---|---|---|
| 0.250 | 125 | 625 |
| 0.200 | 216 | 1080 |
| 0.150 | 512 | 2560 |
| 0.100 | 1728 | 8640 |
| 0.075 | 4096 | 20480 |
| 0.050 | 15625 | 78125 |
| 0.040 | 29791 | 148955 |

size quickly grows to many thousands, even for a physically small geometry. To overcome such high computational needs, the MLFMM speed-up becomes even more important.

The core equations of the VIE solution were presented earlier in Chapter 6. The MLFMM algorithm is primarily the same as that for scattering integral equations (SIEs) with some minor adjustments. As before, far-zone elements for the matrix resulting from (6.187) are approximated using (Coifman, Rokhlin, and Wandzura, 1993)

$$Z_{ji} = \frac{-jk}{4\pi} \sum_{\hat{k}} w_{\hat{k}} \mathbf{V}_{jm}(\hat{k})\, T_L(kr_{mm'}, \hat{k} \cdot \hat{r}_{mm'})\, \mathbf{V}^{*}_{im'}(\hat{k}) \tag{7.29}$$

where in the VIE case

$$\mathbf{V}_{jm}(\hat{k}) = \int_V dv' e^{-j\mathbf{k}\cdot\mathbf{r}_{jm}} [\mathbf{I} - \hat{k}\hat{k}]\mathbf{e}_j(\mathbf{r}_{jm}) \tag{7.30}$$

and

$$\mathbf{V}_{im'}(\hat{k}) = \int_V dv' e^{-j\mathbf{k}\cdot\mathbf{r}_{im'}} [\mathbf{I} - \hat{k}\hat{k}]\mathbf{e}_i(\mathbf{r}_{im'}) \tag{7.31}$$

are the signature functions. That is, the only difference between the SIE-MLFMM signature functions in (7.23) and (7.24) and the VIE-MLFMM signature functions is the domains of the integration. By merely incorporating these signatures in the MLFMM algorithm as outlined for the SIE solution, the MLFMM can be directly adapted to VIEs.

Once the signature functions for each basis function are computed and stored, the rest of the MLFMM algorithm is identical to that of the surface formulation discussed in Section 7.3. However, the MLFMM parameters must be readjusted for optimal performance as the unknowns are distributed in a volume rather than on a surface. Consequently, the cost of the MLFMM in carrying out the matrix-vector products for the near-field section of the matrix will differ from that of the SIE-MLFMM implementation. Fine-tuning the implementation is also necessary as different clustering approaches may be needed depending on the machine used for more optimal CPU performance. For example, the MLFMM parameters (i.e., the number of multipoles and bandwidth of the near-field matrix) are predetermined using the size and neighborhood information of the clusters. Clustering information must also be considered in a parallel implementation of the MLFMM solver to balance the computational load across computer nodes. For example, for the oct-tree subdivision algorithm outlined previously (for surface formulations), assuming a locally flat geometry leads to $3 \times 3 = 9$ "near" clusters for each level, whereas the number of far clusters is $6 \times 6 - 3 \times 3 = 27$ (see Figure 7-13 for clusters). For volumetric discretizations, for each level, these numbers increase significantly. The near clusters are now $3 \times 3 \times 3 = 27$, and the far clusters are $6 \times 6 \times 6 - 3 \times 3 \times 3 = 189$. Hence, the computational burden of VIEs may be as much as three times higher for the near-field matrix-vector product and eight times higher for the far-field portion compared with SIE-MLFMM implementations.

In the following example, a CGS solver was used to carry out some numerical example solution of the VIE-MLFMM. We note that, as the VIE in (6.174) is a second

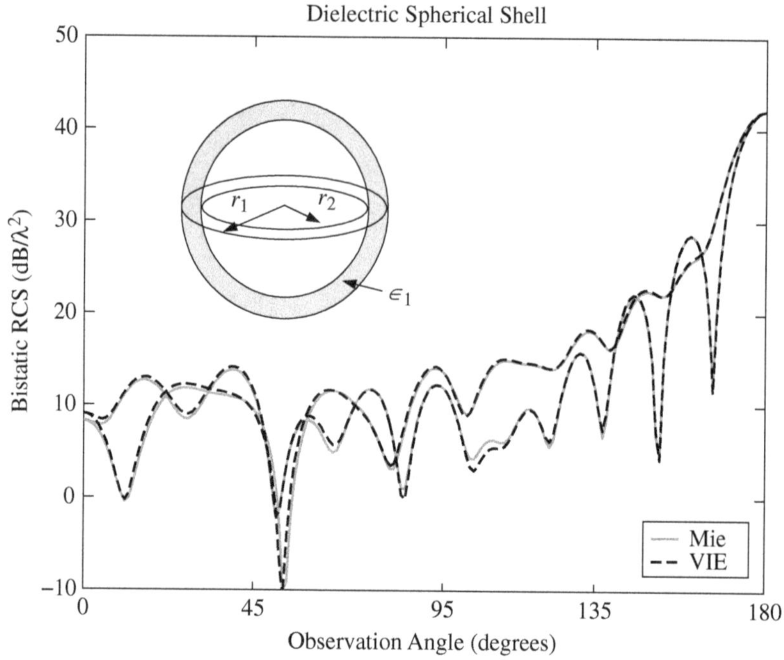


**FIGURE 7-22** Bistatic radar cross section for a $2\lambda$ radius dielectric spherical shell ($\epsilon_r = 2.75 - 0.3j$) computed using a VIE-MLFMM solver and compared with the Mie series solution.

kind integral equation, good convergence is expected and indeed observed. In all cases, the number of multipoles in the MLFMM implementation were determined using the empirical formula $L = kD_{max} + 2\log(kD_{max} + \pi)$ (Coifman, Rokhlin, and Wandzura, 1993).

The first example is that of scattering by a spherical dielectric shell of radius $2\lambda$ and thickness $0.2\lambda$ having $\epsilon_r = 2.75 - 0.3j$. Figure 7-22 shows the computed bistatic radar cross section compared with the analytic data obtained from a Mie series solution. The excellent agreement (within 2% root mean square [rms] error) between the computed and analytic curves (over a 40 dB dynamic range) can be attributed to the higher-order geometry modeling using curvilinear elements. This specific shell geometry was modeled using 57,624 curvilinear hexahedra and resulted in 172,872 unknowns with $\Delta \approx 0.07$. The MLFMM used six levels to converge to a residual of $10^{-2}$ within only 32 iterations. The number of multipoles used at each level were {4, 5, 7, 10, 16, 28} and the convergence rate of order $N/5000$ is impressive and point to the well-conditioned system generated by second-kind integral equations.

As a second example, we consider the scattering by a dielectric sphere enclosed by another dielectric shell. The sphere is of radius $0.9\lambda$ and has a dielectric constant of $\epsilon_r = 1.75 - 0.3j$, whereas the encapsulating shell is of thickness $0.1\lambda$ and has a dielectric constant of $\epsilon_r = 1.25 - 1.25j$. This geometry was modeled using 23,125 elements (69,375 unknowns), and the simulation was done using five MLFMM levels. The employed CGS solver converged in 231 iterations and the number of multipoles used were {4, 5, 7, 10, 16}. Again, the convergence rate is remarkably fast and accurate (within 2.5% rms error) using nominal sampling (see Figure 7-23).

Our third example is intended to demonstrate the accuracy of the VIE-MLFMM solver for high-contrast dielectric materials. The geometry is again a dielectric sphere

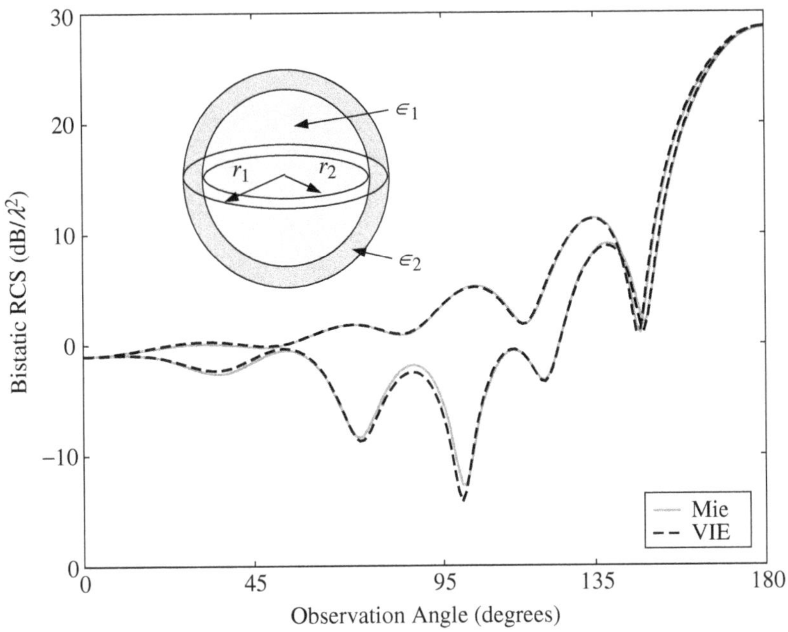


**FIGURE 7-23** Bistatic radar cross section for a dielectric coated dielectric sphere ($\epsilon_1 = 1.75 - 0.3j$, $\epsilon_2 = 1.25 - 1.25j$) computed using the VIE-MLFMM solver and compared with a Mie series solution.

coated now in two layers of high-contrast dielectrics. The permittivity of the sphere's core is $\epsilon_{r1} = 1.25$ up to a radius of $0.18\lambda_0$, $\lambda_0$ being the free-space wavelength. For the first layer coating, the thickness is $0.02\lambda_0$ and $\epsilon_{r2} = 15.0$ representing a contrast of 12 to 1 at the boundary between the layer and the sphere's core. The outermost layer also has a thickness of $0.02\lambda_0$ and a permittivity $\epsilon_{r3}$ (to be chosen $\epsilon_{r3} = 1$ or $\epsilon_{r3} = 2.2$). Considering the wavelength in the high permittivity shell ($\epsilon_{r2} = 15.0$), the volumetric mesh for this region is constructed by specifying the maximum edge length to be around $\lambda_0/(10\sqrt{\epsilon_{r2}}) \approx 0.0258\lambda_0$. Consequently, even though the electrical size of the geometry is only $0.4\lambda_0$, the resulting matrix system is on the order of tens of thousands of unknowns. By comparison, a PEC sphere of the same electrical size would require only a few hundred unknowns, and a full MoM solution can be done on a laptop or desktop personal computer. This example clearly demonstrates the need to use fast integral methods for volumetric scatterers or radiating structures. We consider two separate choices of $\epsilon_{r3}$ for the outermost layer. In one case, this layer was assumed to be a dummy air layer ($\epsilon_{r3} = 1.0$). The bistatic RCS for $\epsilon_{r3} = 1.0$ is given in Figure 7-24 and is seen to be in very good agreement with the analytical Mie series data. For the other case, the outermost layer was assumed to have $\epsilon_{r3} = 2.2$, and the bistatic radar cross section curve is again given in Figure 7-24. It is interesting to observe that a change in $\epsilon_{r3}$ from 1.0 to 2.2 ($0.02\lambda_0$ thick) leads to as much as 4 dB difference between the curves at their low scattering values.

Based on the previous examples, Figure 7-25 gives estimates of the CPU and memory requirements of the VIE-MLFMM as a function of unknowns. The actual CPU times refer to a 1 GHz Pentium III personal computer (today's personal computers have 3 GHz CPU speed). It is not surprising that compared with an FE–BI solver (Antilla and Alexopoulos, 1994), the VIE solver converges much faster and is also free of internal resonances. However, since the generation of the matrix

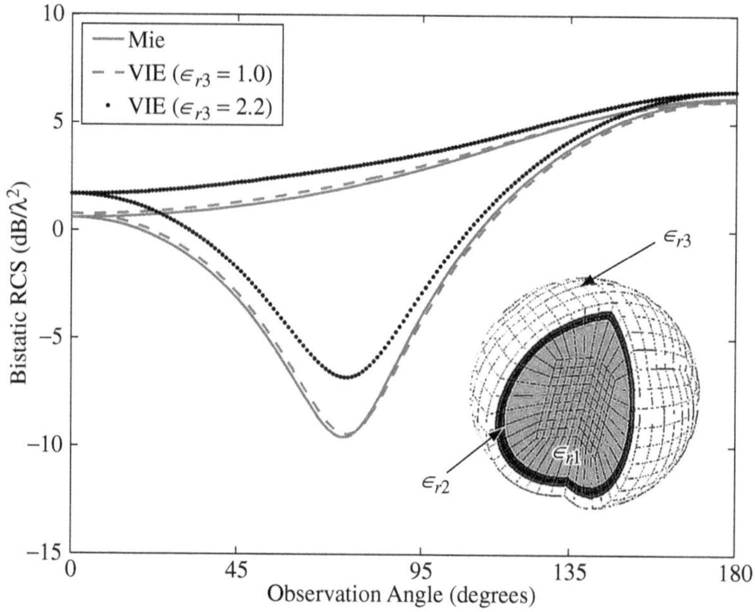


**FIGURE 7-24** Bistatic radar cross section for a two-layer dielectric coated sphere using the VIE-MLFMM solver ($\epsilon_{r1} = 1.25$, $\epsilon_{r2} = 15$, and $\epsilon_{r3}$ refers to the outermost layer).

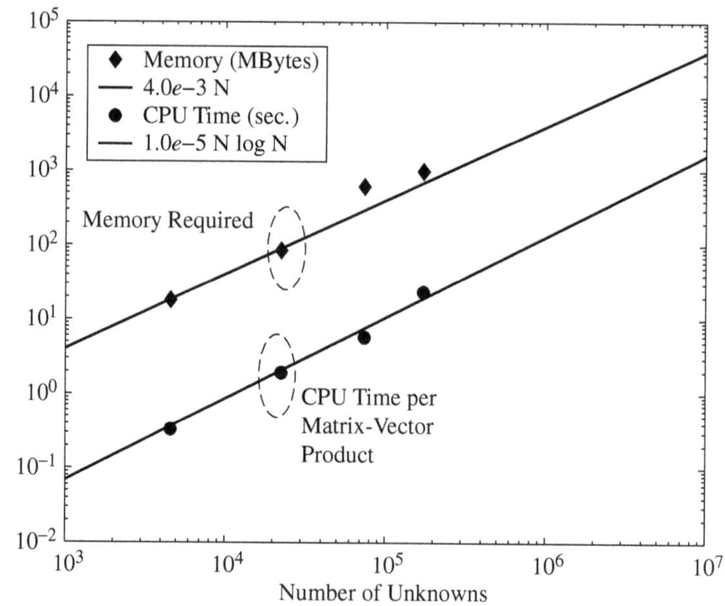


**FIGURE 7-25** Estimates of the CPU and memory requirements in using the VIE-MLFMM solver.

entries $Z_{ji}$ involves double volume integrals, more time is required to fill the system compared with the FE–BI approach. Also, the system matrix for the VIE formulation is completely dense, whereas the FE–BI formulation leads to partly sparse and partly dense matrix. It is also mentioned that the MLFMM becomes more efficient as the number of volume unknowns increases. Hence, the VIE is also attractive for high-contrast dielectrics where higher sampling and more unknowns are required.

# BIBLIOGRAPHY


Abramowitz, M. and Stegun, I. A. (1972). *Handbook of Mathematical Functions*. National Bureau of Standards.

Anastassiu, H. T., Smelyanskiy, M., Bindiganavale, S., and Volakis, J. L. (1998). Scattering from relatively flat surfaces using the adaptive integral method. *Radio Sci.*, *33*(1), 7–16.

Antilla, G. E. and Alexopoulos, N. G. (1994). Scattering from complex 3D geometries by a curvilinear hybrid finite element-integral equation approach. *J. Opt. Soc. Am. A.*, *11*(4), 1445–1457.

Bindiganavale, S. S. and Volakis, J. L. (1997). Comparison of three FMM techniques for solving hybrid FE/BI systems. *IEEE Antennas and Propagat. Mag.*, *39*(4), 47–60.

Bindiganavale, S. S. and Volakis, J. L. (1998). Scattering and radiation from planar structures containing small features using the adaptive integral method (AIM). *IEEE Trans. Antennas Propagat.*, *46*(12), 1867–1878.

Bleszynski, E., Bleszynski, M., and Jaroszewicz, T. (1996). AIM: Adaptive integral method for solving large-scale electromagnetic scattering and radiation problems. *Radio Sci.*, *31*(5), 1225–1251.

Bojarski, N. N. (1971). *k*-space formualtion of the electromagnetic scattering problems. *Tech. Report AFAL-TR-71-5*.

Burkholder, R. J. and Kwon, D. H. (1996). High frequency asymptotic acceleration of the fast multipole method. *Radio Sci.*, *31*(5), 1199–1206.

Chew, W. C., Jin, J., Lu, C., Michielssen, E., and Song, J. M. (1997). Fast solution methods in electromagnetics. *IEEE Trans. Antennas Propagat.*, *45*(3), 533–543.

Chien, R. J. and Alpert, B. K. (1997). A fast spherical filter with uniform resolution. *J. Comp. Phys.*, *136*, 580–584.

Coifman, R., Rokhlin, V., and Wandzura, S. (1993). The fast multipole method for the wave equation: A pedestrian prescription. *IEEE Antennas and Propagat. Mag.*, *35*(3), 7–12.

Darve, E. (2000). The fast multipole method: numerical implementation. *J. Comp. Phys.*, *160*(1), 195–240.

Eibert, T. F., Volakis, J. L., Wilton, D. R., and Jackson, D. R. (1999). Hybrid FE/BI modeling of 3D doubly periodic structures utilizing triangular prismatic elements and a MPIE formulation accelerated by the Ewald transformation. *IEEE Trans. Antennas Propagat.*, *47*(5), 843–850.

Ergul, O. and Gurel, L. (2009). A hierarchical partitioning strategy for an efficient parallelization of the multilevel fast multipole algorithm. *IEEE Trans. Antennas Propagat.*, *57*(6), 1740–1750.

Harrington, R. F. (1968). *Field Computations by Moment Method*. New York: MacMillan Co.

Lu, C. C. and Chew, W. C. (1995). Fast far-field approximation for calculating the RCS of large objects. *Microwave Opt. Tech. Lett.*, *8*(5), 238–241.

Lu, N. and Jin, J.-M. (1996). Application of fast multipole method to finite–element boundary–integral solution of scattering problems. *IEEE Trans. Antennas Propagat.*, *44*(6), 781–786.

MSC Software (2010). Patran: CAE modeling and pre/post processing. Available at http://www.mscsoftware.com/Contents/Products/CAE-Tools/Patran.aspx.

Rao, S., D. Wilton, and Glisson, A. (1982). Electromagnetic scattering by surfaces of arbitrary shape. *IEEE Trans. on Antennas and Propagation*, *AP-30*(3), 409–418.

Saad, Y. (1996). *Iterative Methods for Sparse Linear Systems*. Boston: PWS Pub. Co.

Sancer, M. I., Sertel, K., Volakis, J. L., and Van Alstine, P. (2006). On volume integral equations. *IEEE Trans. Antennas Propagat.*, *54*, 1488–1495.

Sertel, K. and Volakis, J. L. (2002). Method of moments solution of volume integral equations using parametric geometry. *Radio Sci.*, *37*(1), 1–7.

Sheng, X. Q., Jin, J.-M., Song, J. M., Lu, C. C., and Chew, W. C. (1998). On the formulation of hybrid finite-element and boundary-integral methods for 3-D scattering. *IEEE Trans. Antennas Propagat.*, *46*(3), 303–311.

Song, J. M., Lu, C. C., and Chew, W. C. (1997). Multilevel fast multipole algorithm for electromagnetic scattering by large complex objects. *IEEE Trans. Antennas Propagat.*, *45*(10), 1488–1493.

Topsakal, E., Kindt, R., Sertel, K., and Volakis, J. (2001). Evaluation of the BICGSTAB(l) algorithm for the finite-element/boundary-integral method. *IEEE Antennas Propagation Society Magazine*, *43*, 124–131.

Velamparambil, S. and Chew, W.-C. (2005). Analysis and performance of a distributed memory multilevel fast multipole algorithm. *IEEE Trans. Antennas Propagat.*, *53*(8), 2719–2727.

Wagner, R. L. and Chew, C. W. (1994). A ray-propagation fast multipole algorithm. *Microwave Opt. Tech. Lett.*, *7*(10), 435–438.

Waltz, C., Sertel, K., Carr, M., Usner, B., and Volakis, J. (2007). Massively parallel fast multipole method solutions of large electromagnetic scattering problems. *IEEE Trans. Antennas Propagat.*, *55*(6), 1810–1816.

Yarvin, N. and Rokhlin, V. (1998). A generalized one-dimensional fast multipole method with application to filtering of spherical harmonics. *J. Comp. Phys.*, *147*, 594–609.

# Appendix: Integral Equations for Microstrip Antennas

## Chapter Outline



Microstrip antennas are among the most popular due to their conformality and ease of fabrication. Their geometry is depicted in Figure A-1 and consists of a metallized surface residing on a grounded dielectric substrate. They were introduced in the 1950s by Deschamps (Deschamps, 1953) in the USA and by Gutton and Baissinot (Gutton, H. and Baissinot, G., 1955) in France shortly thereafter. However, their practical consideration did not occur till the early 1970s when new demands for compact, low profile and lightweight antennas emerged. This prompted extensive analytical and experimental investigations to accurately compute typical radiation parameters as input impedance, gain, pattern and feed configurations (Munson, R.E., 1974; Uzunoglu, Alexopoulos, and Fikioris, 1979; Carver, K. and Mink, J., 1981; Pozar, 1982; Pozar, D., 1992; Schaubert, D.H., 1992)

The procedure for a moment method analysis of microstrip antennas is the same as that described in Chapter 6 for scattering by plates. Referring to Figure A-2, the currents on the printed patch surface are represented by a set of basis functions (rooftop piece-wise sinusoidal), followed by Galerkin's testing to ensure the boundary condition of vanishing electric tangential fields on the perfectly conducting patch. However, due to the presence of the grounded substrate (as depicted in Figure A-1, a Sommerfeld type Green's function (Sommerfeld, 1941) as presented in Section 2.7 (see Chapter 2, Problems 14 and 15) must be employed. That is, instead of using the free space Green's function for the moment method analysis as in Chapter 6, the substrate Green's function must be employed. As a result, the resulting integrals for populating the impedance matrix are much more challenging to evaluate numerically.

Perhaps one of the earliest moment method analyses of patch antennas (on substrates), using surface basis to represent the patch currents was presented by Newman and Tulyathan (1981). For their analysis, they employed volumetric basis (see Chapter 5) to model the substrate under the patch. Although this approach avoids use of the substrate Green's function, it is rather time consuming. However, by the early 80s, use of the substrate Green's function for a moment method analysis

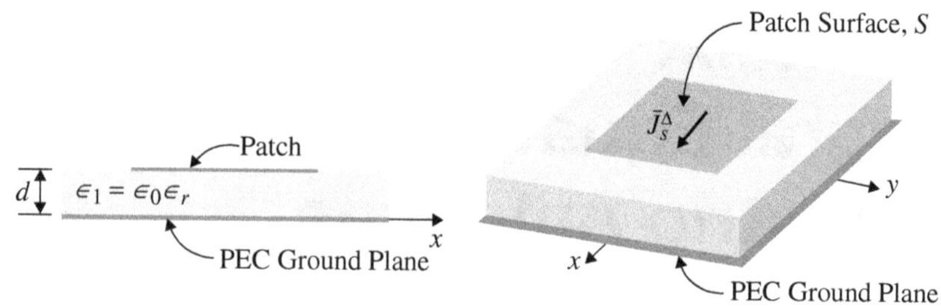


**FIGURE A-1** Geometry of the microstrip patch on a grounded dielectric slab (having a relative dielectric constant of $\epsilon_r$).

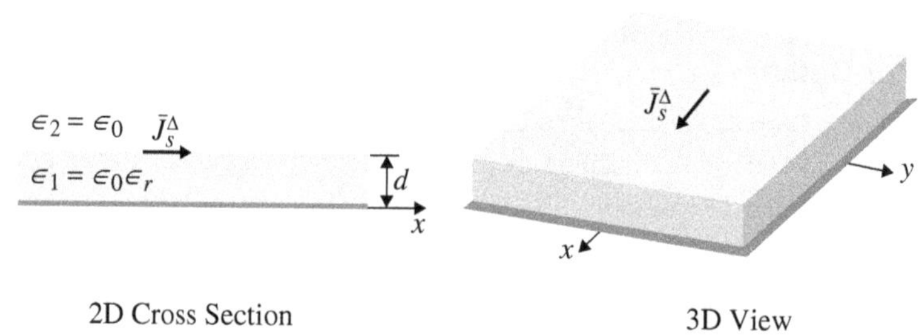


**FIGURE A-2** Delta source on a uniform substrate for computing the dyadic Green's function.

of patch antennas and arrays had become a popular approach. This is the preferred method and is discussed below.

Moment method formulations with substrate Green's functions can be divided in two approaches: (a) those employing Sommerfeld integrals to represent the Green's function (Rana, I.E. and Alexopoulos, N.G., 1981; Katehi and Alexopoulos, 1983; Alexopoulos, N.G. and Jackson, D.R., 1984; Mosig and Gandial, 1985), and (b) those employing spectral representations of the Green's function (Bailey, M.C. and Deshpande, M.D., 1982; Deshpande, M.D. and Bailey, M.C., 1982; Pozar, 1982; Itoh, T. and Menzel, W., 1981; Bailey, M.C. and Deshpande, M.D., 1984) as discussed in Section 2.7. As already noted in Chapter 2, spectral domain representations of the Green's function are easier to work with. Therefore, the analysis presented in this Appendix will adopt this representation and more specifically the approach presented in (Deshpande, M.D. and Bailey, M.C., 1982; Pozar, 1982; Jin, Volakis, and Liepa, 1989; Jin and Volakis, 1990). The moment method solution for the patch (as in Figure A-1) will be constructed using rooftop basis functions. However, other basis functions, such as those presented in Chapter 6 can be employed.

## A.1 | DYADIC GREEN'S FUNCTION FOR A GROUNDED SUBSTRATE

To construct an integral equation for patch currents, it is necessary to first evaluate the patch radiated fields in presence of the substrate. We can express this field using the representation

$$\mathbf{E}(x, y, z) = \iint_S \bar{\bar{\Gamma}}(x, y, z|x', y') \cdot \mathbf{J}_S(x', y')\, dx'dy'. \tag{A.1}$$

In this (see Chapter 2, Eqn. 2.98)

$$\overline{\mathbf{\Gamma}} = \hat{x}\hat{x}\Gamma_{xx} + \hat{y}\hat{x}\Gamma_{yx} + \hat{y}\hat{y}\Gamma_{yy} + \hat{x}\hat{y}\Gamma_{xy} + \hat{z}\hat{x}\Gamma_{zx} + \hat{z}\hat{y}\Gamma_{zy} \tag{A.2}$$

is the pertinent Green's function for the slab, and

$$\mathbf{J}_S = J_x(x, y)\hat{x} + J_y(x, y)\hat{y} \tag{A.3}$$

is the surface current of the patch residing on the substrate at $z = d$.

By definition, the Green's function is the radiated field of a delta source

$$\mathbf{J}_S^{\Delta} = \hat{x}\delta(x - x_0)\delta(y - y_0), \qquad z = d. \tag{A.4}$$

Therefore, we consider the radiation problem depicted in Figure A-2. From (A.4) we have

$$\mathbf{E}(x, y, z) = \hat{x}\Gamma_{xx} + \hat{y}\Gamma_{yx}. \tag{A.5}$$

Below, we enforce the appropriate boundary conditions to find $\Gamma_{xx}$ and $\Gamma_{yx}$. This is more easily done in the spectral domain. Therefore, as in (2.148), we introduce the Fourier pair

$$\tilde{f}(k_x, k_y) = F\{f(x, y)\} = \int_{-\infty}^{\infty}\int_{-\infty}^{\infty} f(x, y)e^{-j(k_x x + k_y y)}\,dx\,dy \tag{A.6}$$

$$f(x, y) = F^{-1}\{\tilde{f}(k_x, k_y)\} = \frac{1}{(2\pi)^2}\int_{-\infty}^{\infty}\int_{-\infty}^{\infty} \tilde{f}(k_x, k_y)e^{j(k_x x + k_y y)}\,dk_x\,dk_y \tag{A.7}$$

implying that (see (2.214))

$$\tilde{E}_x = \frac{jk_x}{\beta^2}\frac{\partial}{\partial z}\tilde{E}_z + \frac{\omega\mu_0 k_y}{\beta^2}\tilde{H}_z \tag{A.8}$$

$$\tilde{E}_y = \frac{jk_y}{\beta^2}\frac{\partial}{\partial z}\tilde{E}_z - \frac{\omega\mu_0 k_x}{\beta^2}\tilde{H}_z \tag{A.9}$$

$$\tilde{H}_x = \frac{jk_x}{\beta^2}\frac{\partial}{\partial z}\tilde{H}_z - \frac{\omega\epsilon_i k_y}{\beta^2}\tilde{E}_z \tag{A.10}$$

$$\tilde{H}_y = \frac{jk_y}{\beta^2}\frac{\partial}{\partial z}\tilde{H}_z + \frac{\omega\epsilon_i k_x}{\beta^2}\tilde{E}_z \tag{A.11}$$

with $\beta^2 = k_x^2 + k_y^2$, $k_i^2 = k_1^2 = \epsilon_r k_0^2 - \beta^2$ in the substrate, and $k_i^2 = k_2^2 = k_0^2 - \beta^2$ in air (above the substrate).

To compute the spectral fields $\tilde{E}_{x,y}$ and $\tilde{H}_{x,y}$ for the delta excitation in (A.4), we must enforce the boundary conditions

$$\tilde{E}_x = \tilde{E}_y = 0, \qquad z = 0 \tag{A.12}$$

$$\tilde{E}_x\,|_{z=d^+} - \tilde{E}_x\,|_{z=d^-} = 0, \qquad z = d \tag{A.13}$$

$$\tilde{E}_y\,|_{z=d^+} - \tilde{E}_y\,|_{z=d^-} = 0, \qquad z = d \tag{A.14}$$

$$\tilde{H}_x\,|_{z=d^+} - \tilde{H}_x\,|_{z=d^-} = 0, \qquad z = d \tag{A.15}$$

and

$$\tilde{H}_y\,|_{z=d^+} - \tilde{H}_y\,|_{z=d^-} = -F\{J_x^{\Delta}\} = e^{-jk_x x_0}e^{-jk_y y_0}. \tag{A.16}$$

Following Section 2.7, a set of field representations that can satisfy the above boundary conditions are

$$\tilde{E}_z = E_{z_0}^{+}(k_x, k_y)e^{-jk_2 z}, \qquad z > d \tag{A.17}$$

$$\tilde{H}_z = H_{z_0}^{+}(k_x, k_y)e^{-jk_2 z}, \qquad z > d \tag{A.18}$$

$$\tilde{E}_z = E_{z_1}^{-}(k_x, k_y)\cos k_1 z + E_{z_2}^{-}(k_x, k_y)\sin k_1 z, \qquad 0 < z < d \tag{A.19}$$

$$\tilde{H}_z = H_{z_1}^{-}(k_x, k_y)\sin k_1 z + H_{z_2}^{-}(k_x, k_y)\cos k_1 z, \qquad 0 < z < d. \tag{A.20}$$

Upon enforcing the boundary conditions (A.12)–(A.16), it follows (Pozar, 1987) that $E_{z_1}^{-} = H_{z_2}^{-} = 0$ and

$$E_{z_0}^{+} = \frac{k_x k_1 \sin k_1 d}{j\omega\epsilon_0 T_m} e^{-jk_x x_0} e^{-jk_y y_0} e^{jk_2 d} \tag{A.21}$$

$$H_{z_0}^{+} = \frac{k_y \sin k_1 d}{jT_e} e^{-jk_x x_0} e^{-jk_y y_0} e^{jk_2 d} \tag{A.22}$$

$$E_{z_1}^{-} = j\frac{k_2 e^{-jk_2 d}}{k_1 \sin k_1 d} E_{z_0}^{+} \tag{A.23}$$

$$H_{z_1}^{-} = j\frac{e^{-jk_2 d}}{\sin k_1 d} H_{z_0}^{+} \tag{A.24}$$

with

$$T_e = k_1 \cos k_1 d + jk_2 \sin k_1 d \tag{A.25}$$

$$T_m = \epsilon_r k_2 \cos k_1 d + jk_1 \sin k_1 d. \tag{A.26}$$

We note that the zeros of the function $T_{e,m}$ are associated with the surface waves supported by the substrate (Harrington, 1961).

On substituting the constants (A.21)–(A.24) into (A.17)–(A.20) and then into (A.8) and (A.9), it follows that

$$\tilde{E}_x = \tilde{\Gamma}_{xx} = \frac{-jZ_0}{k_0} \frac{\left(\varepsilon_r k_0^2 - k_x^2\right) k_2 \cos k_1 d + jk_1 \left(k_0^2 - k_x^2\right) \sin k_1 d}{T_e T_m}, \qquad z > 0$$
$$\cdot (\sin k_1 d) e^{-jk_x x_0} e^{-jk_y y_0} e^{-jk_2(z-d)} \tag{A.27}$$

$$\tilde{E}_y = \tilde{\Gamma}_{yx} = \frac{-jZ_0}{k_0} \frac{k_x k_y \left(k_2 \cos k_1 d + jk_1 \sin k_1 d\right)}{T_e T_m}, \qquad z > 0.$$
$$\cdot (\sin k_1 d) e^{-jk_x x_0} e^{-jk_y y_0} e^{-jk_2(z-d)} \tag{A.28}$$

Next, we take the inverse Fourier transform of these fields to obtain the spatial form of the Green's function given by

$$\Gamma_{xx} = \frac{-jZ_0}{4\pi^2 k_0} \int_{-\infty}^{\infty} \int_{-\infty}^{\infty} \frac{(\epsilon_r k_0^2 - k_x^2) k_2 \cos k_1 d + jk_1 (k_0^2 - k_x^2) \sin k_1 d}{T_e T_m}$$
$$\cdot \sin k_1 d e^{-jk_2(z-d)} e^{jk_x(x-x_0)} e^{jk_y(y-y_0)}\, dk_x\, dk_y, \quad z > d \tag{A.29}$$

$$\Gamma_{yx} = \frac{-jZ_0}{4\pi^2 k_0} \int_{-\infty}^{\infty} \int_{-\infty}^{\infty} \frac{k_x k_y \left(k_2 \cos k_1 d + jk_1 \sin k_1 d\right)}{T_e T_m}$$
$$\cdot (\sin k_1 d) e^{-jk_2(z-d)}\, e^{jk_x(x-x_0)}\, e^{jk_y(y-y_0)}\, dk_x\, dk_y, \quad z > d. \qquad \text{(A.30)}$$

The other two components of $\overline{\Gamma}$, namely $\Gamma_{yy}$ and $\Gamma_{xy}$, can be found by a simple interchange in coordinates. We have

$$\Gamma_{yy} = \frac{1}{(2\pi)^2} \int_{-\infty}^{\infty} \int_{-\infty}^{\infty} \tilde{\Gamma}_{yy}\, e^{jk_x x} e^{jk_y y}\, dk_x\, dk_y \qquad \text{(A.31)}$$

$$\Gamma_{xy} = \Gamma_{yx} \qquad \text{(A.32)}$$

with

$$\tilde{\Gamma}_{yy} = \frac{-jZ_0}{4\pi^2 k_0} \frac{\left(\varepsilon_r k_0^2 - k_y^2\right) k_2 \cos k_1 d + jk_1 \left(k_0^2 - k_y^2\right) \sin k_1 d}{T_e T_m}$$
$$\times \sin k_1 d e^{-jk_x x_0} e^{-jk_y y_0} e^{-jk_2(z-d)}. \qquad \text{(A.33)}$$

From (A.17) and (A.19), we can also find the Green's function components $\Gamma_{zx}$ and $\Gamma_{zy}$ in (A.2). Specifically, within the substrate ($0 < z < d$), we have

$$\tilde{\Gamma}_{zx} = \frac{-jZ_0}{k_0} \frac{jk_x k_2 \cos k_1 z}{T_m} e^{-jk_x x_0} e^{-jk_y y_0}, \quad 0 < z < d \qquad \text{(A.34)}$$

$$\tilde{\Gamma}_{zy} = \frac{-jZ_0}{k_0} \frac{jk_y k_2 \cos k_1 z}{T_m} e^{-jk_x x_0} e^{-jk_y y_0}, \quad 0 < z < d. \qquad \text{(A.35)}$$

Their spatial form is then given by

$$\Gamma_{zx}(x, y, z|x_0, y_0) = \frac{1}{(2\pi)^2} \int_{-\infty}^{\infty} \int_{-\infty}^{\infty} \tilde{\Gamma}_{zx} e^{jk_x x} e^{jk_y y} dk_x dk_y. \qquad \text{(A.36)}$$

By reciprocity, it also follows that

$$\Gamma_{xz} = \Gamma_{zx}, \quad \Gamma_{yz} = \Gamma_{zy}. \qquad \text{(A.37)}$$

## A.2 | MOMENT METHOD FORMULATION

Having an expression for the substrate Green's function, we now proceed to formulate the moment method procedure and solve for the currents on the patch (see Chapter 6). To do so, we refer to Figure A-1 and introduce the surface current density

$$\mathbf{J}_S(x, y) = \hat{x} J_x(x, y) + \hat{y} J_y(x, y) \qquad \text{(A.38)}$$

over the area occupied by the patch.

As usual, to compute the current $\mathbf{J}_S$ we proceed to enforce the boundary condition $\hat{z} \times \mathbf{E} = 0$ on the patch's surface $S_{patch}$. This gives

$$E_x = E_x^r + E_x^{source} = 0, \quad z = d$$
$$E_y = E_y^r + E_y^{source} = 0, \quad z = d \qquad \text{(A.39)}$$

where $\mathbf{E}^r$ denotes the radiated field (caused by $\mathbf{J}_S$) of the patch in presence of the substrate and $\mathbf{E}^{source}$ refers to the excitation field (see Chapter 4). Upon introducing the weighted residual method, the integral equations for constructing the moment method matrix is given by

$$\iint_{S_{patch}} \mathbf{E}^r \cdot \mathbf{J}_t dx\, dy = -\iint_{S_{patch}} \mathbf{E}^{source} \cdot \mathbf{J}_t dx\, dy \qquad \text{(A.40)}$$

where $\mathbf{J}_t$ refers to the moment method testing function.

To introduce the expansion and testing functions, we will assume a rectangular patch whose left edge is at $x = 0$ and right edge at $x = b$ as in Figure A-3. The patch is then discretized into small rectangular elements of size $\Delta x \times \Delta y$ as depicted in Figure A-3. For this solution, we chose roof-top basis functions to represent $\mathbf{J}_S$ (piecewise sinusoidal basis functions were chosen in (Pozar, 1982, 1987)).

Specifically, we introduce the basis representation

$$J_x(x, y) = \sum_{n=1}^{N} I_n^x J_n^x(x, y) \qquad \text{(A.41)}$$

where

$$J_n^x = \frac{\sin k_0 \left(\Delta x - |x - x_n|\right)}{\sin k_0 \Delta x} P_{2\Delta y}\left(y - y_n\right), \qquad \text{(A.42)}$$
$$x_n - \Delta x < x < x_n + \Delta x, \qquad y_n - \Delta y < y < y_n + \Delta y$$

are the basis functions, $I_n^x$ are the constant coefficients of the expansion, and $P_{2\Delta y}$ refers to the pulse functions of width $2\Delta y$. Similarly, for the $y$-directed currents

$$J_y\left(x, y\right) = \sum_{n=1}^{N} I_n^y J_n^y(x, y) \qquad \text{(A.43)}$$

where

$$J_n^y(x, y) = \frac{\sin k_0 \left(\Delta y - |y - y_n|\right)}{\sin k_0 \Delta y} P_{2\Delta x}\left(x - x_n\right) \qquad \text{(A.44)}$$
$$x_n - \Delta x < x < x_n + \Delta x, \qquad y_n - \Delta y < y < y_n + \Delta y$$

and $I_n^y$ are the corresponding constant coefficients of the expansion.

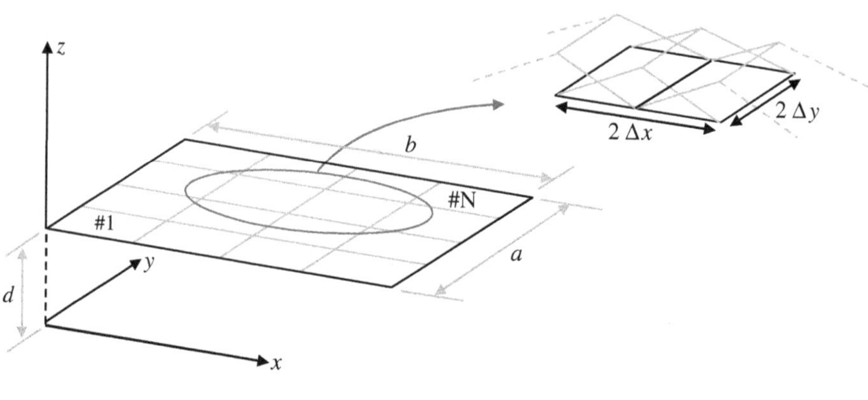


**FIGURE A-3** Rectangular patch discretization and roof-top basis functions.

The radiated field due to each expansion basis can be found using (A.1) giving

$$E_x^{r_n}(x, y, z) = \int_{x_n-\frac{\Delta x}{2}}^{x_n+\frac{\Delta x}{2}} \int_{y_n-\frac{\Delta y}{2}}^{y_n+\frac{\Delta y}{2}} \left[\Gamma_{xx} I_n^y J_n^x(x', y') + \Gamma_{xy} I_n^y J_n^y(x', y')\right] dx'\, dy' \quad (A.45)$$

$$E_y^{r_n}(x, y, z) = \int_{x_n-\frac{\Delta x}{2}}^{x_n+\frac{\Delta x}{2}} \int_{y_n-\frac{\Delta y}{2}}^{y_n+\frac{\Delta y}{2}} \left[\Gamma_{yy} I_n^y J_n^y(x', y') + \Gamma_{yx} I_n^y J_n^x(x', y')\right] dx'\, dy'. \quad (A.46)$$

For Galerkin's testing, we set

$$\mathbf{J}_t(x, y) = \mathbf{J}_m(x, y) = \begin{cases} \hat{x} J_m^x(x, y) \\ \text{or} \\ \hat{y} J_m^y(x, y) \end{cases} \qquad m = 1, 2, \ldots, N \quad (A.47)$$

and proceed to substitute (A.39) to get the integral equations

$$\sum_{n=1}^{N} \iint_{S_m} E_x^{r_n}(x, y, d)\, J_m^x(x, y)\, dx\, dy = -\iint_{S_m} E_x^{source}(x, y, d)\, J_m^x(x, y)\, dx\, dy$$

$$\sum_{n=1}^{N} \iint_{S_m} E_y^{r_n}(x, y, d)\, J_m^y(x, y)\, dx\, dy = -\iint_{S_m} E_y^{source}(x, y, d)\, J_m^y(x, y)\, dx\, dy. \quad (A.48)$$

In (A.48), $S_m$ denotes the area occupied by the weighting functions or test mode, $J_m(x,y)$. In particular, $S_m$ is defined by the area included in the region $x_m - \Delta x < x < x_m + \Delta x$, $y_m - \Delta y < y < y_m + \Delta y$.

The integral equations (A.48) lead to the system of equations

$$\sum_{n=1}^{N} \left(I_n^x Z_{nm}^{xx} + I_n^y Z_{nm}^{xy}\right) = -V_m^x, \qquad m = 1, 2, \ldots, N$$

$$\sum_{n=1}^{N} \left(I_n^x Z_{nm}^{yx} + I_n^y Z_{nm}^{yy}\right) = -V_m^y, \qquad m = 1, 2, \ldots, N. \quad (A.49)$$

In these

$$Z_{nm}^{xx} = \frac{1}{(2\pi)^2} \int_{-\infty}^{\infty} \int_{-\infty}^{\infty} \iint_{S_n} \iint_{S_m} \widetilde{\Gamma}'_{xx} e^{jk_x x} e^{jk_y y} \cdot \frac{\sin[k_0(\Delta x - |x' - x_n|)]}{\sin k_0 \Delta x} \cdot \frac{\sin[k_0(\Delta x - |x' - x_m|)]}{\sin k_0 \Delta x}\, dx\, dy\, dx'\, dy'\, dk_x\, dk_y \quad (A.50)$$

$$Z_{nm}^{xy} = Z_{nm}^{yx} = \int_{-\infty}^{\infty}\int_{-\infty}^{\infty}\iint_{S_n}\iint_{S_m} \widetilde{\Gamma}'_{xy} e^{jk_x x} e^{jk_y y} \cdot \frac{\sin[k_0(\Delta y - |y' - y_n|)]}{\sin k_0 \Delta y} \cdot \frac{\sin[k_0(\Delta x - |x - x_m|)]}{\sin k_0 \Delta x} \, dx\, dy\, dx'\, dy'\, dk_x\, dk_y \quad (A.51)$$

$$Z_{nm}^{yy} = \int_{-\infty}^{\infty}\int_{-\infty}^{\infty}\iint_{S_n}\iint_{S_m} \widetilde{\Gamma}'_{yy} e^{jk_x x} e^{jk_y y} \cdot \frac{\sin[k_0(\Delta y - |y' - y_n|)]}{\sin k_0 \Delta y} \cdot \frac{\sin[k_0(\Delta y - |y - y_m|)]}{\sin k_0 \Delta y} \, dx\, dy\, dx'\, dy'\, dk_x\, dk_y \quad (A.52)$$

with

$$\widetilde{\Gamma}'_{pq} = \widetilde{\Gamma}_{pq} e^{jk_x x_0} e^{jk_y y_0} \quad (A.53)$$

$p = \{x, y, z\}$, $q = \{x, y, z\}$ and $\widetilde{\Gamma}_{pq}$ as given in (A.27)–(A.37). Also, the excitation vectors are given by

$$V_m^x = \iint_{S_m} E_x^{source}(x, y, d) J_m^x(x, y)\, dx\, dy \quad (A.54)$$

$$V_m^y = \iint_{S_m} E_y^{source}(x, y, d)\, J_m^y(x, y)\, dx\, dy. \quad (A.55)$$

The computation of $V_m^{x,y}$ requires knowledge of the excitation field $E_{x,y}^{source}$ due to the source current $\mathbf{J}_{source}(x, y)$. This is discussed in Chapters 4 and 6. Typically, the excitation current is $z$-directed. In that case, it is instructive to invoke reciprocity and rewrite $V_m^{x,y}$ as

$$V_m^x = \iiint_{V_{source}} E_m^{zx}(x, y, z) J_z^{source}(x, y, z)\, dx\, dy\, dz \quad (A.56)$$

$$V_m^y = \iiint_{V_{source}} E_m^{zy}(x, y, z) J_z^{source}(x, y)\, dx\, dy\, dz \quad (A.57)$$

where $V_{source}$ denotes the volume of the source current $J_z^{source}$, $E_m^{zx}$ denotes the $z$ components of the field due to the mode $J_m^x$, and likewise $E_m^{zy}$ denotes the field due to the $J_m^y$ mode. From (A.34)–(A.37) they are given by

$$E_m^{zx} = \int_{-\infty}^{\infty}\int_{-\infty}^{\infty}\iint_{S_m} \widetilde{\Gamma}'_{zx} e^{jk_x x} e^{jk_y y} J_m^x(x', y')\, dx'\, dy', \qquad 0 < z < d \quad (A.58)$$

$$E_m^{zy} = \int_{-\infty}^{\infty}\int_{-\infty}^{\infty}\iint_{S_m} \widetilde{\Gamma}'_{zy} e^{jk_x x} e^{jk_y y} J_m^y(x', y')\, dx'\, dy', \qquad 0 < z < d \quad (A.59)$$

and as before $\widetilde{\Gamma}'_{zx} = \widetilde{\Gamma}_{zx} e^{jk_x x_0} e^{jk_y y_0}$, etc.

The integrations of $S_n$ and $S_m$ can be evaluated analytically. In particular, these integrations can be eliminated by introducing the Fourier transform of the modes given by

$$\mathcal{F}\left(J_n^x\right) = \iint_{S_n} J_n^x\,(x, y)\, e^{-jk_x x} e^{-jk_y y}\, dx\, dy \tag{A.60}$$

$$\mathcal{F}\left(J_n^y\right) = \iint_{S_n} J_n^y\,(x, y)\, e^{-jk_x x} e^{-jk_y y}\, dx\, dy. \tag{A.61}$$

Using these, the impedance matrix entries can be rewritten as

$$Z_{nm}^{xx} = \frac{1}{4\pi^2} \int_{-\infty}^{\infty}\int_{-\infty}^{\infty} \widetilde{\Gamma}'_{xx} \mathcal{F}\left(J_n^x\right) \mathcal{F}^*\left(J_m^x\right) dk_x\, dk_y \tag{A.62}$$

$$Z_{nm}^{xy} = \frac{1}{4\pi^2} \int_{-\infty}^{\infty}\int_{-\infty}^{\infty} \widetilde{\Gamma}'_{xy} \mathcal{F}\left(J_n^y\right) \mathcal{F}^*\left(J_m^x\right) dk_x\, dk_y \tag{A.63}$$

$$Z_{nm}^{yy} = \frac{1}{4\pi^2} \int_{-\infty}^{\infty}\int_{-\infty}^{\infty} \widetilde{\Gamma}'_{yy} \mathcal{F}\left(J_n^y\right) \mathcal{F}^*\left(J_m^y\right) dk_x\, dk_y \tag{A.64}$$

in which "$*$" indicates complex conjugation. Similarly, from (A.54) through (A.55),

$$V_m^x = \frac{1}{4\pi^2} \int_{-\infty}^{\infty}\int_{-\infty}^{\infty} \widetilde{\Gamma}'_{zx} e^{jk_x x_0} e^{jk_y y_0} \mathcal{F}\left(J_m^x\right) dk_x\, dk_y \tag{A.65}$$

$$V_m^y = \frac{1}{4\pi^2} \int_{-\infty}^{\infty}\int_{-\infty}^{\infty} \widetilde{\Gamma}'_{zy} e^{jk_x x_0} e^{jk_y y_0} \mathcal{F}\left(J_m^y\right) dk_x\, dk_y. \tag{A.66}$$

It remains to evaluate the integrals (A.60) through (A.64). These are all of the same form, and in particular their integrands have the same denominator. It is well known that this denominator becomes singular. This occurs at the zeros of $T_e$ and $T_m$ that correspond to the transverse electric (TE) and transverse magnetic (TM) surface wave poles and these poles complicate the numerical evaluation of the integrals (Rana, I.E. and Alexopoulos, N.G., 1981; Katehi and Alexopoulos, 1983; Alexopoulos, N.G. and Jackson, D.R., 1984) accommodate the evaluation, we introduce the polar transformation

$$k_x = \beta \cos\alpha, \qquad k_y = \beta \sin\alpha. \tag{A.67}$$

Thus, the integrals in (A.62) to (A.66) may be written in the form

$$\int_{-\infty}^{\infty}\int_{-\infty}^{\infty} (\cdot)\, dk_x\, dk_y = \int_0^{2\pi}\int_0^{\infty} (\cdot)\, d\beta\, d\alpha. \tag{A.68}$$

Unless $\epsilon_r = 1$, there will be at least one propagating TM mode and perhaps a few more depending on the thickness of the substrate and the value of $\epsilon_r$. For real $\epsilon_r$, the zeros of $T_e$ and $T_m$ occur at real values of $\beta = \beta_p$ and can be shown to satisfy

$$k_0 \le \beta_0 < \sqrt{\varepsilon_r} k_0.$$

If $\epsilon_r$ is complex, then $\beta_0$ moves off the real axis to $\beta_0 - j\gamma$. To avoid numerical difficulties in the evaluation of (A.68), a standard procedure is to isolate the section near the pole location as follows:

$$\int_0^{2\pi}\int_0^{\infty}(\cdot)\,d\beta\,d\alpha=\int_0^{2\pi}\left\{\int_0^{\beta_0-\delta}(\cdot)\,d\beta+\int_{\beta_0-\delta}^{\beta_0+\delta}(\cdot)\,d\beta+\int_{\beta_0+\delta}^{\infty}(\cdot)\,d\beta\right\}d\alpha \tag{A.69}$$

where

$$I_\delta=\int_{\beta_0-\delta}^{\beta_0+\delta}(\cdot)\,d\beta=\int_{\beta_0-\delta}^{\beta_0+\delta}\frac{f(\beta)}{g(\beta)}d\beta \tag{A.70}$$

can be evaluated analytically for small $\delta$. If $g(\beta)$ has a zero at $\beta=\beta_0-j\gamma$, it is instructive to introduce the Taylor series expansion

$$g(\beta)\cong(\beta-\beta_0+j\gamma)g'(\beta_0-j\gamma)\,. \tag{A.71}$$

Upon substituting this two term expansion to (A.70) and setting $f(\beta)\approx f(\beta_0-j\gamma)$ yields

$$I_\delta=\frac{jf(\beta_0-j\gamma)}{g'(\beta_0-j\gamma)}\tan^{-1}\left\{\frac{-2\delta\gamma}{\gamma^2-\delta^2}\right\}. \tag{A.72}$$

In closing, we note that several other integration techniques as discussed in Chapter 6 can be used. Also, different forms of the Green's functions (such as those based on complex images (Chow, Yang, Fang, and Howard, 1991; Shubair, R.M. and Chow, Y.L., 1993; Aksun, M.I., 1996; Kipp, R.A. and Chan, C.H., 1994)) lead to different integrals. These integrals may be easier to evaluate numerically or may require other numerical approaches.

## A.3 FAR-ZONE FIELD EVALUATION

To find the far-zone radiated fields, one can proceed by evaluating the Fourier integrals in (A.29) through (A.32) using the stationary-phase formula (Felsen and Marcuvitz, 1973; Senior and Volakis, 1995). Another much easier approach is to employ the reciprocity theorem (Harrington, 1961). Here we discuss both approaches.

Consider the integral

$$I(\Omega)=\int_{-\infty}^{\infty}\int_{-\infty}^{\infty}f(x,y)\,e^{j\Omega q(x,y)}dx\,dy,\qquad \Omega>0 \tag{A.73}$$

where $q(x,y)$ is a real function and has a simple, real stationary point $(x_s,y_s)$ defined by

$$\frac{\partial q}{\partial x}=0\text{ and }\frac{\partial q}{\partial y}=0\text{ at }(x_s,y_s)\,. \tag{A.74}$$

Provided $f$ is regular near $(x_s,y_s)$, then an asymptotic approximation of $I(\Omega)$ for large $\Omega$ is given by (see Section 2.7)

$$I(\Omega)\sim f(x_s,y_s)e^{j\Omega q(x_s,y_s)}\left(\frac{2\pi}{\Omega}\right)\frac{e^{j(\pi/4)\sigma}}{\left|\det\left(\partial^2 q/\partial x_s,\partial y_s\right)\right|^{1/2}} \tag{A.75}$$

where

$$\sigma = \mathrm{sgn} d_1 + \mathrm{sgn} d_2 \tag{A.76}$$

and $(d_1, d_2)$ are the eigenvalues of the matrix comprising the elements of $|\partial^2 q/\partial x_s, \partial y_s|$.

The stationary-phase formula in (A.75) can be applied for the evaluation of (A.29) through (A.33). To do so, we start with the expression

$$\overline{\Gamma} = \frac{1}{4\pi^2} \int_{-\infty}^{\infty} \int_{-\infty}^{\infty} \tilde{\overline{\Gamma}}(k_x, k_y)\, e^{j(k_2 d - k_x x' - k_y y')} \cdot e^{jr(k_x \sin\theta\cos\phi + k_y \sin\theta\sin\phi - k_2\cos\theta)}\, dk_x\, dk_y \tag{A.77}$$

where $(r, \theta, \phi)$ are the usual spherical coordinates describing the observation point and $\overline{\Gamma}$ is the dyadic matrix given in (2.98). Comparing (A.77) with (A.73), the following identifications are possible:

$$f = \frac{1}{4\pi^2} \tilde{\overline{\Gamma}}(k_x, k_y)\, e^{j(k_2 d - k_x x' - k_y y')}$$
$$\Omega = r$$
$$q = k_x \sin\theta\cos\phi + k_y \sin\theta\sin\phi - k_2\cos\theta.$$

The stationary point is found in accordance with (A.74) and is given by

$$k_{xs} = -k_0 \sin\theta\cos\phi,\ k_{ys} = -k_0\sin\theta\sin\phi \tag{A.78}$$

and together with

$$\left|\det\left(\partial^2 q/\partial k_{xs}, \partial k_{ys}\right)\right|^{1/2} = \frac{1}{k_0\cos\theta}, \qquad \sigma = 2 \tag{A.79}$$

we obtain

$$\overline{\Gamma} \approx \frac{jk_0\cos\theta}{2\pi r} e^{-jk_0 r} \tilde{\overline{\Gamma}}(k_{xs}, k_{ys}) e^{jk_0(d\cos\theta + x'\sin\theta\cos\phi + y'\sin\theta\sin\phi)} \tag{A.80}$$

for large $\mathbf{r}$. The transverse component of the far field radiated by a single patch supporting a current $\mathbf{J}_s\,(x, y)$ is then given by

$$\mathbf{E}_{ff} = \frac{jk_0\cos\theta}{2\pi r} e^{-jk_0 r} \tilde{\overline{\Gamma}}(k_{xs}, k_{ys})\, e^{jk_0 d\cos\theta} \cdot \iint_{S_0} \mathbf{J}_s(x', y') e^{jk_0(\sin\theta\cos\phi x' + \sin\theta\sin\phi y')}\, dx'\, dy'. \tag{A.81}$$

Note that $f$ is not regular at $\theta = \pi/2$, and therefore the previous asymptotic expression is not valid there.

We now consider the second approach. In accordance with the reciprocity theorem, if $\mathbf{E}_1$ is the field due to $\mathbf{J}_1$ and $\mathbf{E}_2$ is the field due to $\mathbf{J}_2$, then

$$\iiint \mathbf{E}_1 \cdot \mathbf{J}_2\, dV = \iiint \mathbf{E}_2 \cdot \mathbf{J}_1\, dV \tag{A.82}$$

holds. Assume now that a current element $Il$ is situated at point $(r, \theta, \phi)$ and oriented in the $\hat{\theta}$ direction. This current element, if away from the dielectric surface, produces a spherical wave of the form

$$\mathbf{E}^i = -j\frac{k_0 Z_0 Il}{4\pi r} e^{-jk_0(|\mathbf{r}-\mathbf{r}'|)}\hat{\theta} \tag{A.83}$$

where $\mathbf{r}'$ is a point near the origin. We observe that (A.83) represents a TM polarization plane wave when $\mathbf{r} \to 0$. Therefore, the plane wave reflection coefficient $R^{TM}$ (see Section 2.7) can be employed to find the reflected field near the origin. The total field at the dielectric surface near the origin is then given by

$$\mathbf{E}_1 = -j\frac{k_0 Z_0 Il}{4\pi r} e^{-jk_0 r}(1 - R^{TM})e^{jk_0(d\cos\theta + x'\sin\theta\cos\phi + y'\sin\theta\sin\phi)}\hat{\theta}. \tag{A.84}$$

We now let $\mathbf{e} = \mathbf{E}_2$ to denote the electric field at $(r, \theta, \phi)$ produced by the patch current $\mathbf{J}_s = \mathbf{J}_2$. In accordance with the reciprocity theorem, we then have

$$\iiint \mathbf{E}_2 \cdot \mathbf{J}_1 dv = \mathbf{e} \cdot Il\hat{\theta} = -j\frac{k_0 Z_0 Il}{4\pi r} e^{-jk_0 r}\left(1 - R^{TM}\right) e^{jk_0 d\cos\theta}$$

$$\cdot \iiint_{S_v} e^{jk_0(x'\sin\theta\cos\phi + y'\sin\theta\sin\phi)}\hat{\theta} \cdot \mathbf{J}_s(x', y')\, dx'\, dy' \tag{A.85}$$

which yields

$$e_\theta = -j\frac{k_0 Z_0}{4\pi r} e^{-jk_0 r}(1 - R^{TM})\, e^{jk_0 d\cos\theta}$$

$$\cdot \iint_{S_p} e^{jk_0(x'\sin\theta\cos\phi + y'\sin\theta\sin\phi)}\hat{\theta} \cdot \mathbf{J}_s(x', y')\, dx'\, dy'. \tag{A.86}$$

Similarly, by orienting the element current in the $\hat{\phi}$-direction, we find that

$$e_\phi = -j\frac{k_0 Z_0}{4\pi r} e^{-jk_0 r}(1 + R^{TE})\, e^{jk_0 d\cos\theta}$$

$$\cdot \iint_{S_p} e^{jk_0(x'\sin\theta\cos\phi + y'\sin\theta\sin\phi)}\hat{\phi} \cdot \mathbf{J}_s(x', y')\, dx'\, dy'. \tag{A.87}$$

It can be shown that the field given by (A.86) through (A.87) is the same as that in (A.81). They all exhibit a feature of geometrical optics and are, thus, accurate provided the observation point is not near the dielectric surface where the surface wave contribution may become appreciable. Therefore, in the region $0 \le \theta < \pi/2$, we have

$$\begin{aligned} e_\theta &= -j\frac{k_0 Z_0}{4\pi}\frac{e^{-jk_0 r}}{r}(1 - R^{TM})\, e^{jk_0 d\cos\theta}(Q_x \cos\theta\cos\phi + Q_y\cos\theta\sin\phi) \\ e_\phi &= -j\frac{k_0 Z_0}{4\pi}\frac{e^{-jk_0 r}}{r}(1 + R^{TE})\, e^{jk_0 d\cos\theta}(-Q_x\sin\phi + Q_y\cos\phi). \end{aligned} \tag{A.88}$$

In these, the components $Q_{x,y}$ are the integrals given by

$$Q_{x,y} = \iint_{S_p} J_{x,y} e^{jk_0(x\sin\theta\cos\phi + y\sin\theta\sin\phi)}\, dx\, dy \tag{A.89}$$

# BIBLIOGRAPHY

Aksun, M. I. (1996). A robust approach for the derivation of closed form green's functions. *IEEE Trans. Microwave Theory and Techn.*, *44*, 651–658.

Alexopoulos, N. G. and Jackson, D. R. (1984). Fundamental substrate (cover) effects on printed circuit antennas. *IEEE Trans. Antennas Propagat.*, *AP-32*(8), 807–816.

Bailey, M. C. and Deshpande, M. D. (1982). Integral equation formulation of microstrip antennas. *IEEE Trans. Antennas Propagat.*, *AP-30*(4).

Bailey, M. C. and Deshpande, M. D. (1984). Analysis of rectangular microstrip antennas. *NASA Technical Paper*, (2276).

Carver, K. and Mink, J. (1981). Microstrip antenna technology. *IEEE Trans. Antennas Propagat.*, 2–24.

Chow, Y. L., Yang, J. J., Fang, D. G., and Howard, G. E. (1991). A closed form spatial green's function for the thick microstrip substrate. *IEEE Trans. Microwave Theory Techn.*, *39*, 588–592.

Deschamps, G. (1953). Microstrip microwave antennas. *Presented at the 3rd USAF Symp. on Antennas*.

Deshpande, M. D. and Bailey, M. C. (1982). Input impedance of microstrip antennas. *IEEE Trans. Antennas Propagat.*, *AP-30*(4).

Felsen, L. B. and Marcuvitz, N. (1973). *Radiation and Scattering of Waves*. New York: Prentice Hall.

Gutton, H. and Baissinot, G. (1955). Flat aerial for ultra high frequencies. *French patent No. 703113*.

Harrington, R. F. (1961). *Time-Harmonic Electromagnetic Fields*. New York: McGraw-Hill.

Itoh, T. and Menzel, W. (1981). A full wave analysis method for open microstrip structures. *IEEE Trans. Antennas Propagat.*, *AP-29*(1), 63–68.

Jin, J. J. and Volakis, J. L. (1990). Electromagnetic scattering by a perfectly conducting patch array on a dielectric slab. *IEEE Trans. Antennas Propagat.*, *38*(4), 556–563.

Jin, J.-M., Volakis, J. L., and Liepa, V. V. (1989). A moment method solution of a volume-surface integral equation using isoparametric elements and point matching TE scattering. *IEEE Trans. Microwave Theory Tech.*, *37*(10), 1641–1645.

Katehi, P. B. and Alexopoulos, N. G. (1983). Real axis integration of sommerfeld type integrals with application to printed circuit antennas. *J. Math Physics*, *24*.

Kipp, R. A. and Chan, C. H. (1994). Complex image method for sources in bounded regions of multilayer structures. *IEEE Trans. Microwave Theory and Techn.*, *42*, 860–865.

Mosig, J. R. and Gandial, F. E. (1985). General integral equation formulation for microstrip antennas and scatterers. *IEEE Proceedings, Part H*.

Munson, R. E. (1974). Conformal microstrip antennas and microstrip phased arrays. *IEEE Trans. Antennas Propagat.*, *AP-22*(1), 74–77.

Newman, E. H. and Tulyathan, P. (1981). Analysis of microstrip antennas using moment method. *IEEE Trans. Antennas Propagat.*, *AP-29*(1), 47–53.

Pozar, D. (1982). Input impedance and mutual coupling of rectangular microstrip antennas. *IEEE Trans. Antennas Propagat.*, *30*.

Pozar, D. (1987). Radiation and scattering from a microstrip patch on a uniaxial substrate. *IEEE Trans. Antennas Propagat.*, *35*, 613–621.

Pozar, D. (1992). Microstrip antennas. *IEEE Proc.*, *80*(2), 79–91.

Rana, I. E. and Alexopoulos, N. G. (1981). Current and input impedance of printed dipoles. *IEEE Trans. Antennas Propagat.*, *AP-29*(1), 99–105.

Schaubert, D. H. (1992). Microstrip antennas. *Electromagnetics*, *12*, 381–401.

Senior, T. B. A. and Volakis, J. L. (1995). *Approximate Boundary Conditions in Electromagnetics*. IET Press.

Shubair, R. M. and Chow, Y. L. (1993). Efficient computation of the periodic green's function in layered dielectric media. *IEEE Trans. Microwave Theory and Techn.*, 498–502.
Sommerfeld, A. (1941). *Partial Differential Equations*. Vol. VI: New York: Academic.
Uzunoglu, N. K., Alexopoulos, N. G., and Fikioris, J. G. (1979). Radiation properties of microstrip dipoles. *IEEE Trans. Antennas Propagat.*, *27*, 853–858.

# Index

Page numbers followed by *f* and *t* indicate figures and tables, respectively.

## N



## O



## P

## U



## V

www.ingramcontent.com/pod-product-compliance
Lightning Source LLC
LaVergne TN
LVHW062110200926
848484LV00007B/1635